*Compiled by*

**Thomas J. Glover**

**First Edition**

Sequoia Publishing, Inc.
Morrison, Colorado, U.S.A.

This POCKET REF belongs to:

NAME: DON WELSH

HOME ADDRESS:

HOME PHONE:

WORK PHONE: 425 7991

BUSINESS ADDRESS:

COORS BREWING CO

In case of accident or serious illness, please notify:
Name:
Phone Number:


6th Printing - June 1992



Library of Congress
Catalog Card Number: 89 - 90848

ISBN 0-9622359-0-3

# Preface

Sequoia Publishing, Inc. has made a serious effort to provide accurate information in this book. However, the probability exists that there are errors and misprints and that variations in data values may also occur depending on field conditions. Information included in this book should only be considered as a general guide and Sequoia Publishing, Inc. does not represent the information as being exact.

The publishers would appreciate being notified of any errors, omissions, or misprints which may occur in this book. Your suggestions for future editions would also be greatly appreciated.

The information in this manual was collected from numerous sources and if not properly acknowledged, Sequoia Publishing, Inc. would like to express its appreciation for those contributions. See page 2 for specific trade name, trade mark, and credit information.

Many thanks to my wonderful wife Georgia and my two children, Trish and Carrie, for all their patience, understanding and help in making this book possible.

Sequoia Publishing, Inc.
Department 101
P.O. Box 620820
Littleton, Colorado 80162-0820
(303) 972-4167

# PERSONAL INFORMATION

| Name ⇒ | | | | | |
|---|---|---|---|---|---|
| Birthday | | | | | |
| Anniversary | | | | | |
| Hat | | | | | |
| Suit/Dress | | | | | |
| Neck | | | | | |
| Shoulder to Waist | | | | | |
| Chest | | | | | |
| Shirt Size | | | | | |
| Sleeve Length | | | | | |
| Gloves | | | | | |
| Waist | | | | | |
| Hips | | | | | |
| Inseam | | | | | |
| Sock or Hose | | | | | |
| Shoe | | | | | |
| Ring Size | | | | | |
| | | | | | |
| | | | | | |
| | | | | | |
| | | | | | |

# TABLE OF CONTENTS

# REFERENCES, TRADE NAMES & TRADE MARKS

Index463The following are Trade Names and Trade Marks included in Pocket Ref. If we missed your Trade Name or Trade Mark, we apologize, please let us know and we will insert it in the next printing.

AWS – American Welding Society
ANSI – American National Standards Institute
ASCII – American Standard Code for Information Interchange
Brown & Sharp
Cedarapids – Iowa Manufacturing Co.
Commodore 64 – Commodore Computers
Diablo 630 – Xerox Corporation
Epson, FX–80 – Epson America Inc
Hayes – Hayes Microcomputer Products, Inc.
HP, HP-IB, Hewlett-Packard, Laserjet – Hewlett–Packard Company
IBM , AT, XT, PC, PS/2, PC Convertible, PC Jr. - International Business Machines Corporation
ISO – International Standards Organization
Macintosh, Apple IIc, Apple – Apple Computer, Inc.
Metropolitan Life Insurance Company
Microsoft and MS–DOS – Microsoft Corporation
NCHS – National Center for Health Statistics
NEC, Pinwriter – NEC Corporation
NEMA – National Electrical Manufacturers Association
Pioneer – Portec Pioneer Division
ROMEX –
SAE – Society of Automotive Engineers

**Some of the references used in writing Pocket Ref include the following (They are all excellent references and should be added to any good reference library):**

Arco's New Complete Woodworking Handbook – J. T. Adams, Arco
Builders Vest Pocket Reference – W.J. Hornumg, Prentice–Hall,Inc
Cedarapids Reference Book – Iowa Manufacturing Company
Dana's Manual of Minerology – E.S. Dana, John Wiley & Sons
Electronic Engineers Master Catalog – Hearst Business Communications Inc.
Field Geologists Manual – Australian Institute of Mining and Met.
Grainger Catalog– W. W. Grainger, Inc
Handbook of Chemistry & Physics – The Chemical Rubber Co.
Handbook of Physical Calculations – Jan J. Tuma, McGraw–Hill
Machinery's Handbook – E.O. & F.D. Jones, Industrial Press Inc
Machinists' & Draftsmen's Handbook – A.M. Wagener & H.R. Arthur
Mechanical Engineers' Handbook – McGraw–Hill Book Co., Inc.
National Electrical Code – National Fire Protection Association
Pioneer Facts and Figures – Portec, Pioneer Division
Scientific Tables – Ciba–Geigy Ltd, New York
Standard Math Tables – The Chemical Rubber Co.
Technical Reference Handbook – E. P. Rasis, American Tech. Pub.
The World Almanac 1989 – Pharos Books
Water Well Handbook – K.E. Anderson, Missouri Water Well Assn.

NOTE: There are many more references, most of which are referenced on specific pages in Pocket Ref. If we have omitted a reference, we apologize, please let us know and we will include it in the next printing of Pocket Ref.

# POCKET REF

# AIR and GASES



(See WATER for Tank Volumes & Pollution, p. 377)

# COMPOSITION OF AIR

| Component of Air | Symbol | Content – %Volume |
|---|---|---|
| Nitrogen | $N_2$ | 78.084 percent |
| Oxygen | $O_2$ | 20.947 percent |
| Argon | Ar | 0.934 percent |
| Carbon dioxide | $CO_2$ | 0.033 percent |
| Neon | Ne | 18.2 parts/million |
| Helium | He | 5.2 parts/million |
| Krypton | Kr | 1.1 parts/million |
| Sulfur dioxide | $SO_2$ | 1.0 parts/million |
| Methane | $CH_4$ | 2.0 parts/million |
| Hydrogen | $H_2$ | 0.5 parts/million |
| Nitrous oxide | $N_2O$ | 0.5 parts/million |
| Hydrogen | $H_2$ | 0.5 parts/million |
| Xenon | Xe | 0.09 parts/million |
| Ozone | $O_3$ | 0.0 to 0.07 parts/million |
| Ozone – Winter | $O_3$ | 0.0 to 0.02 parts/million |
| Nitrogen dioxide | $NO_2$ | 0.02 parts/million |
| Iodine | $I_2$ | 0.01 parts/million |
| Carbon monoxide | CO | 0.0 to trace |
| Ammonia | $NH_3$ | 0.0 to trace |

(Nitrogen, Oxygen, Argon, Carbon dioxide together: > 99.998%)

The above table is an average for clean, dry air at sea level.
1 part/million = 0.0001 percent.

# PHYSICAL PROPERTIES OF AIR

Density of dry air at Standard Temperature and Pressure:
1.2929 kilograms/cu meter = 0.0807 pounds/cu foot

Universal Gas Constant (R): 0.0821 liter–atmosphere/ °K /mole

Standard Temperature & Pressure (STP):
Standard Temperature = 0°C = 32°F = 273.15°K
Standard Pressure = 760 mm Hg =
14.70 pounds/sq inch = 2116.22 pounds/sq foot =
29.92 inch Hg = $1.01325 \times 10^5$ N/m$^2$

Speed of sound in dry air at STP:
331.4 meters/sec = 1089 ft/sec = 742.5 miles/hr

ICAO Sea Level Air Standard Values:
Atmospheric pressure = 760 mm Hg = 14.7 lbs/sq inch
Temperature = 15°C = 288.15°K = 59°F

## WEIGHTS OF GASES

| Gas | Weight (gms/liter) | Weight (lb/cu ft) |
|---|---|---|
| Air | 1.2928 | 0.08071 |
| Air @ 59°F | 1.2256 | 0.07651 |
| Argon | 1.7840 | 0.111368 |
| Carbon Dioxide | 1.9770 | 0.123416 |
| Carbon Monoxide | 1.2500 | 0.078033 |
| Helium | 0.1785 | 0.011143 |
| Hydrogen | 0.0899 | 0.005612 |
| Neon | 0.9002 | 0.056196 |
| Nitrogen | 1.2506 | 0.075261 |
| Oxygen | 1.4290 | 0.089207 |

All weights listed above assume a dry gas at standard temperature and pressure (0°C and 760 mm Hg).

## STANDARD ATMOSPHERE

The unit "1 Standard Atmosphere" is defined as the pressure equivalent to that exerted by a 76mm column of mercury at 0°C (32°F), at sea level, and at standard gravity (32.174 ft/sec$^2$). Atmospheric pressure is simply the weight of a column of air per area unit as measured from the top of the atmosphere to the reference point being measured. Atmospheric pressure decreases as altitude increases. Equivalents to 1 atmosphere are as follows:

76 centimeters of mercury
29.921 inches of mercury
10.3326 meters of water
406.788 inches of water
33.899 feet of water
14.696 pounds per square inch
2,116.2 pounds per square foot

# GENERAL GAS LAWS & FORMULAS

## Perfect Gas Law:

$$PV = nRT$$

P = Pressure in atmospheres
V = Volume in liters  n = Number of moles
R = Gas constant (0.0821 liter-atmospheres/°K/mole)
T = Temperature in degrees K

| | |
|---|---|
| If constant pressure | V1/V2 = T1/T2 |
| If constant temperature | P1/P2 = V2/V1 |
| If constant volume | P1/P2 = T1/T2 |

## Boyle's Law

If temperature is kept constant, the volume of a given mass of gas is inversely proportional to the pressure which is exerted upon it.

$$\frac{\text{Initial Pressure}}{\text{Initial Volume}} = \frac{\text{Pressure Change}}{\text{Volume Change}}$$

## Charles' Law

If the pressure is constant, the volume of a given mass of gas is directly proportional to the absolute temperature.

$$\frac{\text{Initial Volume}}{\text{Initial Temperature }^{\circ}\text{K}} = \frac{\text{Volume Change}}{\text{Final Temperature }^{\circ}\text{K}}$$

## Dalton's Law of Partial Pressures

The pressure which is exerted on the walls of a vessel is the sum of the pressures which each gas would exert if it were present alone.

## Graham's Law of Diffusion

Relative rates of diffusion of two gases are inversely proportional to the square roots of their densities.

## Avogadro's Law

Equal volumes of gases, measured under the same conditions of temperature and pressure, contain equal numbers of molecules.

## Air Velocity in a Pipe:

$$V = \sqrt{\frac{25{,}000\ DP}{L}}$$

V = Air velocity in feet per second
D = Pipe inside diameter in inches
L = Length of pipe in feet
P = Pressure loss due to air friction in ounces/square inch
Approximate values of P are as follows:

| Velocity Ft/Sec | Pipe Diameter in inches, 10 feet long 1 | 2 | 4 | 6 | 10 |
|---|---|---|---|---|---|
| 1 | 0.004 | 0.0002 | 0.0001 | 0.00006 | 0.00004 |
| 2 | 0.0016 | 0.0008 | 0.0004 | 0.0002 | 0.00016 |
| 5 | 0.0100 | 0.005 | 0.0025 | 0.0016 | 0.001 |
| 10 | 0.04 | 0.02 | 0.01 | 0.0066 | 0.004 |
| 15 | 0.09 | 0.45 | 0.0225 | 0.015 | 0.0089 |
| 20 | 0.16 | 0.08 | 0.04 | 0.026 | 0.016 |
| 25 | 0.25 | 0.125 | 0.0625 | 0.041 | 0.025 |
| 30 | 0.36 | 0.18 | 0.09 | 0.06 | 0.036 |

(Formula from B.F. Sturtevant Co)

## Air Volume Discharged from Pipe:

$$CFM = 60VA$$

CFM = Air volume in cubic feet per minute
V = Air velocity in feet per second as determined in the equation at the top of this page.
A = Cross section area of pipe in square feet.

## Theoretical Horsepower to Compress Air:

$$HP = CFM \times PSI \times 0.0007575$$

HP = Compressor Horsepower
CFM = Air flow in cubic feet per minute
PSI = Air pressure in pounds per square inch
(assumes Atmospheric Pressure = 14.7 psi, Temp = 60°F)

# DENSITY OF MOIST AIR

| mm Hg | Air Temperature (Dew Point = 10°C) 0°C | 10°C | 20°C | 40°C | 60°C |
|---|---|---|---|---|---|
| 1000 | 1.695 | 1.635 | 1.579 | 1.479 | 1.390 |
| 975 | 1.653 | 1.594 | 1.540 | 1.441 | 1.355 |
| 950 | 1.610 | 1.553 | 1.500 | 1.404 | 1.320 |
| 925 | 1.568 | 1.512 | 1.461 | 1.367 | 1.285 |
| 900 | 1.525 | 1.471 | 1.421 | 1.330 | 1.250 |
| 875 | 1.482 | 1.430 | 1.381 | 1.293 | 1.215 |
| 850 | 1.440 | 1.389 | 1.342 | 1.256 | 1.181 |
| 825 | 1.397 | 1.348 | 1.302 | 1.219 | 1.146 |
| 800 | 1.355 | 1.307 | 1.262 | 1.182 | 1.111 |
| 775 | 1.312 | 1.266 | 1.223 | 1.145 | 1.076 |
| 760 | 1.287 | 1.241 | 1.199 | 1.122 | 1.055 |
| 750 | 1.270 | 1.225 | 1.183 | 1.108 | 1.041 |
| 725 | 1.227 | 1.184 | 1.144 | 1.071 | 1.006 |
| 700 | 1.185 | 1.143 | 1.104 | 1.033 | 0.971 |
| 675 | 1.142 | 1.102 | 1.064 | 0.996 | 0.937 |
| 650 | 1.100 | 1.061 | 1.025 | 0.959 | 0.902 |
| 625 | 1.057 | 1.020 | 0.985 | 0.922 | 0.867 |
| 600 | 1.015 | 0.979 | 0.945 | 0.885 | 0.832 |
| 575 | 0.972 | 0.938 | 0.906 | 0.848 | 0.797 |
| 550 | 0.930 | 0.897 | 0.866 | 0.811 | 0.762 |
| 525 | 0.887 | 0.856 | 0.827 | 0.774 | 0.727 |
| 500 | 0.845 | 0.815 | 0.787 | 0.737 | 0.692 |
| 475 | 0.802 | 0.774 | 0.747 | 0.700 | 0.658 |
| 450 | 0.760 | 0.733 | 0.708 | 0.663 | 0.623 |
| 425 | 0.717 | 0.692 | 0.668 | 0.625 | 0.588 |
| 400 | 0.674 | 0.651 | 0.628 | 0.588 | 0.553 |
| 375 | 0.632 | 0.610 | 0.589 | 0.551 | 0.518 |
| 350 | 0.589 | 0.569 | 0.549 | 0.514 | 0.483 |
| 325 | 0.547 | 0.528 | 0.510 | 0.477 | 0.448 |
| 300 | 0.504 | 0.487 | 0.470 | 0.440 | 0.414 |
| 275 | 0.462 | 0.446 | 0.430 | 0.403 | 0.379 |
| 250 | 0.419 | 0.405 | 0.391 | 0.366 | 0.344 |
| 225 | 0.377 | 0.363 | 0.351 | 0.329 | 0.309 |
| 200 | 0.334 | 0.322 | 0.311 | 0.292 | 0.274 |
| 175 | 0.292 | 0.281 | 0.272 | 0.254 | 0.239 |
| 150 | 0.249 | 0.240 | 0.232 | 0.217 | 0.204 |
| 125 | 0.207 | 0.199 | 0.193 | 0.180 | 0.169 |
| 100 | 0.164 | 0.158 | 0.153 | 0.143 | 0.135 |
| 75 | 0.122 | 0.117 | 0.113 | 0.106 | 0.100 |

Moist air density (gms/liter) = 1.2929 x (273.13/T) x ((P–Vp)/760)

T = Absolute air temperature (°Kelvin)

P = Barometric pressure (mm of mercury)

Vp = Vapor pressure of water (see table in WATER Chapter)

# ELEVATION vs AIR & WATER

| Elevation Meters | Elevation Feet | US Std Atmosphere Temp °F | US Std Atmosphere Pressure lbs/sq in | Boiling Point H20(°F) | Speed Sound m/sec |
|---|---|---|---|---|---|
| −1000 | −3280 | 70.7 | 16.52 | 217.9 | 344.1 |
| −500 | −1640 | 64.9 | 15.59 | 215.1 | 342.2 |
| 0 | 0 | 59.0 | 14.70 | 212.0 | 340.3 |
| 250 | 820 | 56.1 | 14.26 | 210.6 | 339.3 |
| 500 | 1640 | 53.2 | 13.85 | 208.9 | 338.4 |
| 750 | 2461 | 50.2 | 13.44 | 207.5 | 337.4 |
| 1000 | 3281 | 47.3 | 13.03 | 206.1 | 336.4 |
| 1250 | 4101 | 44.4 | 12.64 | 204.4 | 335.5 |
| 1500 | 4921 | 41.5 | 12.26 | 203.0 | 334.5 |
| 1750 | 5742 | 38.5 | 11.89 | 201.6 | 333.5 |
| 2000 | 6562 | 35.6 | 11.53 | 199.9 | 332.5 |
| 2500 | 8202 | 29.8 | 10.83 | 197.1 | 330.6 |
| 3000 | 9843 | 23.9 | 10.17 | 194.0 | 328.6 |
| 3500 | 11483 | 18.1 | 9.54 | 190.4 | 326.6 |
| 4000 | 13123 | 12.2 | 8.94 | 188.1 | 324.6 |
| 4500 | 14764 | 6.4 | 8.38 | 184.8 | 322.6 |
| 5000 | 16404 | 0.5 | 7.84 | 181.9 | 320.5 |
| 5500 | 18045 | −5.3 | 7.33 | 178.9 | 318.5 |
| 6000 | 19685 | −11.1 | 6.85 | 175.8 | 316.5 |
| 6500 | 21325 | −17.0 | 6.39 | 172.8 | 314.4 |
| 7000 | 22966 | −22.8 | 5.96 | 169.7 | 312.3 |
| 7500 | 24606 | −28.6 | 5.56 | 166.6 | 310.2 |
| 8000 | 26247 | −34.5 | 5.17 | 163.6 | 308.1 |
| 8500 | 27887 | −40.3 | 4.81 | 160.5 | 305.9 |
| 9000 | 29528 | −46.2 | 4.47 | 157.5 | 303.8 |
| 9500 | 31168 | −52.0 | 4.15 | 154.4 | 301.7 |
| 10000 | 32808 | −57.8 | 3.84 | 151.3 | 299.5 |
| 11000 | 36089 | −69.5 | 3.29 | 145.0 | 295.2 |
| 12000 | 39370 | −69.5 | 2.81 | 138.9 | 295.1 |
| 13000 | 42651 | −69.5 | 2.41 | 133.0 | 295.1 |
| 14000 | 45932 | −69.5 | 2.06 | 127.0 | 295.1 |
| 15000 | 49212 | −69.5 | 1.74 | 121.3 | 295.1 |
| 16000 | 52493 | −69.5 | 1.50 | 115.7 | 295.1 |
| 17000 | 55774 | −69.5 | 1.28 | 110.3 | 295.1 |
| 18000 | 59055 | −69.5 | 1.10 | 104.7 | 295.1 |
| 19000 | 62336 | −69.5 | .94 | 99.7 | 295.1 |
| 20000 | 65617 | −69.5 | .80 | 94.5 | 295.1 |
| 25000 | 82021 | −60.9 | .37 | 70.5 | 298.4 |
| 30000 | 98425 | −52.0 | .17 | 49.3 | 301.7 |
| 32000 | 104986 | −48.5 | .13 | 41.5 | 303.0 |

1 mm Hg @ 0°C = 0.019336 lbs/sq in = 1 Torr

# AIR TOOL REQUIREMENTS

| Tool Category | Tool CFM | Tool Category | Tool CFM |
|---|---|---|---|
| Air Filter Cleaner | 3 | Hydr Lift 8000 lb | 6* |
| Air Hammer | 4 | Hydr Floor Jack | 6* |
| Air Hoist 1000# | 5 | Impact Wrenches: | |
| Air Motor 1 hp | 6–10 | 1/4 inch drive | 3 |
| Air Motor 2 hp | 12–15 | 3/8 inch drive | 2–5 |
| Air Motor 3 hp | 18–20 | 1/2 inch drive | 4–8 |
| Bead Breaker | 12* | 3/4 inch drive | 7–9 |
| Bench Rammer | 5 | 1 inch drive | 10 |
| Body Polisher | 2 | 1 1/4 inch drive | 14 |
| Body Orbital Sander | 5 | Nutsetter – 3/8 inch | 3–6 |
| Brake Tester | 4 | Nutsetter – 3/4 inch | 5–8 |
| Burr Tool – small | 4 | Pneu. Garage Door | 3* |
| Burr Tool – large | 5–6 | Radiator Tester | 1 |
| Carbon Remover | 3 | Rammers – small | 3 |
| Chain Saw | 7–22 | Rammers – medium | 9 |
| Circle Saw – 8 inch | 12 | Rammers – large | 10 |
| Circle Saw – 12 inch | 17 | Sander – 5 in Pad | 8–10 |
| Compression Riviter | 1 | Sander – 7 in Pad | 15 |
| Die Grinder | 4–6 | Sander – 9 in Pad | 17–20 |
| Drill 1/16–3/8 inch | 4–6 | Screwdriver #2–#6 | 1–3 |
| Drill 3/8–5/8 inch | 7–8 | Screwdriver #6–up | 3–6 |
| Dust blow gun | 3 | Spray Cleaner | 5 |
| File/Saw Machine | 3–5 | Spray Paint Guns: | |
| Floor Rammer | 7 | Standard | 5 |
| Grease Gun | 3* | Production | 9 |
| Grinder – 2 in Horz | 5–10 | Touch–up | 4 |
| Grinder – 4 in Horz | 15 | Undercoat | 19 |
| Grinder – 6 in Horz | 15–17 | Tamper Backfill | 6–15 |
| Grinder – 8 in Horz | 20 | Tapper – 3/8 inch | 3–5 |
| Grinder – 5 in Vert | 8–10 | Tire Changer | 1* |
| Grinder – 7 in Vert | 14–15 | Tire inflation | 2* |
| Grinder – 9 in Vert | 17–20 | Tire Rim Stripper | 6* |
| Hammer – Chip | 8 | Tire Spreader | 1* |
| Hammer – Fender | 9 | Transmission Flusher | 3 |
| Hammer – Rivet | 8–15 | | |
| Hammer – Scale | 3–4 | | |
| Hammer – Tire | 12 | | |
| Hoist – Cyl type | 2 | | |

NOTE: Most tools listed above are rated at 90 to 100 pounds/sq inch, however, those items with a "*" next to the cfm rating require 125 to 160 pounds/sq inch.

Always check the manufacturers recommendations for both air pressure (psi – pounds per sq inch) and air flow (cfm – cubic feet per minute) requirements. The ratings listed above are only averages based on a 25% load factor (running 25% of the time).

# CFM vs PSI FOR NOZZLES

| Gage PSI | CFM Free Air Flow @ Nozzle Diameter (inch) 1/64 | 1/32 | 3/64 | 1/16 |
|---|---|---|---|---|
| 1 | 0.03 | 0.11 | 0.2 | 0.4 |
| 5 | 0.06 | 0.24 | 0.5 | 1.0 |
| 10 | 0.08 | 0.34 | 0.8 | 1.4 |
| 15 | 0.10 | 0.42 | 0.9 | 1.6 |
| 20 | 0.12 | 0.48 | 1.1 | 1.9 |
| 25 | 0.13 | 0.54 | 1.2 | 2.2 |
| 30 | 0.16 | 0.63 | 1.4 | 2.5 |
| 40 | 0.19 | 0.77 | 1.7 | 3.1 |
| 50 | 0.22 | 0.91 | 2.0 | 3.6 |
| 60 | 0.26 | 1.05 | 2.3 | 4.2 |
| 70 | 0.29 | 1.19 | 2.7 | 4.8 |
| 80 | 0.33 | 1.33 | 3.0 | 5.3 |
| 90 | 0.36 | 1.47 | 3.3 | 5.9 |
| 100 | 0.40 | 1.61 | 3.7 | 6.4 |
| 110 | 0.43 | 1.76 | 3.9 | 7.0 |
| 120 | 0.47 | 1.90 | 4.3 | 7.6 |
| 130 | 0.50 | 2.04 | 4.6 | 8.1 |
| 140 | 0.54 | 2.17 | 4.9 | 8.7 |
| 150 | 0.57 | 2.33 | 5.2 | 9.2 |
| 175 | 0.66 | 2.65 | 5.9 | 10.6 |
| 200 | 0.76 | 3.07 | 6.9 | 12.2 |

| PSI | 3/32 | 1/8 | 3/16 | 1/4 |
|---|---|---|---|---|
| 1 | 1.0 | 1.7 | 3.9 | 6.8 |
| 5 | 2.2 | 3.9 | 8.7 | 15.4 |
| 10 | 3.1 | 5.4 | 12.3 | 21.8 |
| 15 | 3.7 | 6.6 | 15.0 | 26.7 |
| 20 | 4.2 | 7.7 | 17.1 | 30.8 |
| 25 | 4.7 | 8.6 | 19.4 | 34.5 |
| 30 | 5.6 | 10.0 | 22.5 | 40.0 |
| 40 | 6.8 | 12.3 | 27.5 | 49.1 |
| 50 | 8.2 | 14.5 | 32.8 | 58.2 |
| 60 | 9.4 | 16.8 | 37.5 | 67.0 |
| 70 | 10.7 | 19.0 | 43.0 | 76.0 |
| 80 | 11.9 | 21.2 | 47.5 | 85.0 |
| 90 | 13.1 | 23.5 | 52.5 | 94.0 |
| 100 | 14.5 | 25.8 | 58.3 | 103.0 |
| 110 | 15.7 | 28.0 | 63.0 | 112.0 |
| 120 | 17.0 | 30.2 | 68.0 | 121.0 |
| 130 | 18.2 | 32.4 | 73.0 | 130.0 |
| 140 | 19.5 | 34.5 | 78.0 | 138.0 |
| 150 | 20.7 | 36.7 | 83.0 | 147.0 |
| 175 | 23.8 | 42.1 | 95.0 | 169.0 |
| 200 | 27.5 | 48.7 | 110.0 | 195.0 |

PSI = pounds/square inch; CFM = cubic feet/minute

# AIR HOSE FRICTION

| Hose Size (inch) | CFM thru 50 ft Hose | Gage Pressure – Pounds/sq inch 50 | 70 | 90 | 110 |
|---|---|---|---|---|---|
| | | PSI Loss Over 50 foot Hose Length | | | |
| 1/2 | 20 | 1.8 | 1.0 | 0.8 | 0.6 |
| | 30 | 5.0 | 3.4 | 2.4 | 2.0 |
| | 40 | 10.1 | 7.0 | 5.4 | 4.3 |
| | 50 | 18.1 | 12.4 | 9.5 | 7.6 |
| | 60 | + | 20.0 | 14.8 | 12.0 |
| | 70 | + | 28.4 | 22.0 | 17.6 |
| | 80 | + | + | 30.5 | 24.6 |
| | 90 | + | + | 41.0 | 33.3 |
| | 100 | + | + | + | 44.5 |
| | 110 | + | + | + | + |
| 3/4 | 20 | 0.4 | 0.2 | 0.2 | 0.1 |
| | 30 | 0.8 | 0.5 | 0.4 | 0.3 |
| | 40 | 1.5 | 0.9 | 0.7 | 0.5 |
| | 50 | 2.4 | 1.5 | 1.1 | 0.9 |
| | 60 | 3.5 | 2.3 | 1.6 | 1.3 |
| | 70 | 4.4 | 3.2 | 2.3 | 1.8 |
| | 80 | 6.5 | 4.2 | 3.1 | 2.4 |
| | 90 | 8.5 | 5.5 | 4.0 | 3.1 |
| | 100 | 11.4 | 7.0 | 5.0 | 3.9 |
| | 110 | 14.2 | 8.8 | 6.2 | 4.9 |
| | 120 | + | 11.0 | 7.5 | 5.9 |
| | 130 | + | + | 9.0 | 7.1 |
| 1 | 20 | 0.1 | 0.0 | 0.0 | 0.0 |
| | 30 | 0.2 | 0.1 | 0.1 | 0.1 |
| | 40 | 0.3 | 0.2 | 0.2 | 0.2 |
| | 50 | 0.5 | 0.4 | 0.3 | 0.2 |
| | 60 | 0.8 | 0.5 | 0.4 | 0.3 |
| | 70 | 1.1 | 0.7 | 0.6 | 0.4 |
| | 80 | 1.5 | 1.0 | 0.7 | 0.6 |
| | 90 | 2.0 | 1.3 | 0.9 | 0.7 |
| | 100 | 2.6 | 1.6 | 1.2 | 0.9 |
| | 110 | 3.5 | 2.0 | 1.4 | 1.1 |
| | 120 | 4.8 | 2.5 | 1.7 | 1.3 |
| | 130 | 7.0 | 3.1 | 2.0 | 1.5 |

PSI = Pressure in pounds/square inch
CFM = Air flow in cubic feet/minute

"+" means pressure loss is too great and therefore, the combination of Hose Size, CFM, and Gage Pressure is not recommended. Gage Pressure is the indicated air pressure, in pounds/square inch, at the source (ie, the air compressor receiver tank).

# AIR LINE RECOMMENDED SIZES

| Air Flow CFM | Length of Air Line in Feet 50 | 100 | 200 | 300 |
|---|---|---|---|---|
| | Recommended Air Line Size in Inches | | | |
| 1–5 | 1/2 | 1/2 | 1/2 | 1/2 |
| 6–10 | 1/2 | 3/4 | 3/4 | 3/4 |
| 11–15 | 3/4 | 3/4 | 3/4 | 3/4 |
| 16–20 | 3/4 | 3/4 | 3/4 | 3/4 |
| 21–25 | 3/4 | 3/4 | 1 | 1 |
| 26–30 | 3/4 | 3/4 | 1 | 1 |
| 31–35 | 3/4 | 1 | 1 | 1 |
| 36–40 | 1 | 1 | 1 | 1 |
| 41–59 | 1 | 1 | 1 | 1 |
| 60–79 | 1 | 1 | 1–1/4 | 1–1/4 |
| 80–100 | 1–1/4 | 1–1/4 | 1–1/2 | 1–1/2 |

# AIR RECEIVER CAPACITIES

| Tank Size (inches) | Tank Size (gallons) | Gauge Pressure on Tank (PSI) 0 | 100 | 150 | 200 |
|---|---|---|---|---|---|
| | | Cubic Feet Tank Capacity | | | |
| 12 x 24 | 10 | 1.3 | 11 | 15 | 19 |
| 14 x 36 | 20 | 2.7 | 21 | 30 | 39 |
| 16 x 36 | 30 | 4.0 | 31 | 45 | 59 |
| 20 x 48 | 60 | 8.0 | 62 | 90 | 117 |
| 20 x 63 | 80 | 10.7 | 83 | 120 | 156 |
| 24 x 68 | 120 | 16.0 | 125 | 180 | 234 |
| 30 x 84 | 240 | 32.0 | 250 | 360 | 467 |

If your tank is not listed in the above table, use the following formula to calculate the Tank Size (gallons) and then estimate that Cubic Feet Tank Capacity at a given pressure from the table above.

$$\text{Tank Gallons} = \frac{\text{Tank Height} \times (\text{Tank Radius})^2}{73.53}$$

Height and Radius are in inches.

# AIR POLLUTION SAFE LIMITS [1]

| Pollutant | Safe Exposure 8hr/day, 5days/week |
|---|---|
| Asbestos | 5 million parts/cubic foot |
| Benzene | 80 mg/m$^3$ |
| Bromine | 0.7 mg/m$^3$ |
| Cadmium | 0.2 mg/m$^3$ |
| Carbon dioxide | 9,000 mg/m$^3$ |
| Carbon disulfide | 60 mg/m$^3$ |
| Carbon monoxide | 55 mg/m$^3$ |
| Carbon tetrachloride | 65 mg/m$^3$ |
| Chlorine | 1.5 mg/m$^3$ |
| Chloroform | 240 mg/m$^3$ |
| Chromic acid | 0.1 mg/m$^3$ |
| Cresol | 22 mg/m$^3$ |
| Freon–12 | 4,950 mg/m$^3$ |
| Freon–21 | 4,200 mg/m$^3$ |
| Ethyl alcohol | 1,900 mg/m$^3$ |
| Ether | 1,200 mg/m$^3$ |
| Fluorine | 0.02 mg/m$^3$ |
| Formaldehyde | 6 mg/m$^3$ |
| Gasoline | 1,800 to 2,000 mg/m$^3$ |
| Hydrochloric acid | 7 mg/m$^3$ |
| Hydrogen cyanide | 11 mg/m$^3$ |
| Iodine | 1 mg/m$^3$ |
| Iron oxide | 10 mg/m$^3$ |
| Isopropyl alcohol | 980 mg/m$^3$ |
| Lead | 0.2 mg/m$^3$ |
| Manganese | 5 mg/m$^3$ |
| Mercury | 0.01 mg/m$^3$ |
| Methyl alcohol | 260 mg/m$^3$ |
| Naptha (coal tar) | 400 mg/m$^3$ |
| Naptha (petroleum) | 2,000 mg/m$^3$ |
| Nitric oxide | 30 mg/m$^3$ |
| Nitrogen dioxide | 9 mg/m$^3$ |
| Propane | 1,800 mg/m$^3$ |
| Selenium | 0.2 mg/m$^3$ |
| Sulfur dioxide | 13 mg/m$^3$ |
| Sulfuric acid | 1 mg/m$^3$ |
| Tellurium | 0.1 mg/m$^3$ |
| Tetraethyl lead | 0.075 mg/m$^3$ |
| Toluene | 750 mg/m$^3$ |
| Turpentine | 560 mg/m$^3$ |
| Vinyl chloride | 1,300 mg/m$^3$ |
| Zinc oxide | 5 mg/m$^3$ |

(1) *American Industrial Hygiene Association*

See also WATER Chapter for pollution, page 386

# POCKET REF

## Automotive



(See ELECTRICAL for more wiring tables, p. 107)
(See HARDWARE for Bolt Torques, page 249)
(See also GLUE for paint types, page 235)

# ANTIFREEZE TABLE

| Cooling System Capacity Quarts | Temperature Rating °F at Quarts of Antifreeze Required | | | |
|---|---|---|---|---|
| | 3 | 4 | 5 | 6 |
| 8 | -7 | -34 | -68 | |
| 9 | 0 | -21 | -50 | -84 |
| 10 | 4 | -12 | -34 | -62 |
| 11 | 8 | -6 | -23 | -47 |
| 12 | 10 | 0 | -15 | -34 |
| 13 | 13 | 3 | -9 | -25 |
| 14 | 15 | 6 | -5 | -17 |
| 15 | 16 | 8 | 0 | -12 |
| 16 | 17 | 10 | 2 | -7 |
| 17 | 18 | 12 | 5 | -4 |
| 18 | 19 | 14 | 7 | 0 |
| 19 | 20 | 15 | 9 | 2 |
| 20 | | 16 | 10 | 4 |

| Cooling System Capacity Quarts | Temperature Rating °F at Quarts of Antifreeze Required | | | |
|---|---|---|---|---|
| | 7 | 8 | 9 | 10 |
| 8 | | | | |
| 9 | | | | |
| 10 | -84 | | | |
| 11 | -69 | -84 | | |
| 12 | -58 | -74 | | |
| 13 | -45 | -66 | -84 | |
| 14 | -34 | -53 | -74 | -84 |
| 15 | -26 | -43 | -62 | -76 |
| 16 | -19 | -34 | -53 | -68 |
| 17 | -14 | -27 | -43 | -59 |
| 18 | -10 | -21 | -34 | -51 |
| 19 | -7 | -16 | -28 | -42 |
| 20 | -3 | -12 | -22 | -34 |

NOTE: Never use more that 70% antifreeze in a cooling system or the antifreeze and boiling properties of the mixture become unfavorable. Commercial automotive antifreeze is an ethylene glycol based solution that, when mixed with radiator water, lowers the temperature at which the radiator water will freeze and also increases the temperature at which the water will boil. A 50% solution of antifreeze and water will increase the boiling point to 265° F and a 70% solution will increase it to 276° F. Ethylene glycol is actually the chemical "1,2 Ethanediol" and has a chemical formula of $HOCH_2CH_2OH$. Antifreeze is poisonous and if swallowed, give two glasses of water, induce vomiting and call a physician. Portions of the above information are based on *Prestone Anti Freeze by Union Carbide Corp.*

# SPARK PLUG TORQUE

| Spark Plug Thread Size | Aluminum Head | | Iron Head | |
|---|---|---|---|---|
| | Ft Lbs | Kg M | Ft Lbs | Kg M |
| 10mm (Gasket) | 8-11 | 1.1-1.5 | 8-12 | 1.1-1.7 |
| 12 mm (Gasket) | 10-18 | 1.4-2.5 | 10-18 | 1.4-2.5 |
| 14 mm (Gasket) | 18-22 | 2.5-3.0 | 26-30 | 3.6-4.1 |
| 14 mm (Taper seat) | 7-15 | 1.0-2.1 | 7-15 | 1.0-2.1 |
| 18 mm (Gasket) | 28-34 | 3.9-4.7 | 32-38 | 4.4-5.2 |
| 18 mm (Taper seat) | 15-20 | 2.1-2.8 | 15-20 | 2.1-2.8 |
| 7/8-18 inch (Gasket) | 31-39 | 4.3-5.4 | 31-39 | 4.3-5.4 |

NOTE: If the engine manufacturers' torque specification is available, it should always take precedence over the values in the above table. Even with the above torque ranges, exercise care when tightening the spark plugs, since condition of the head threads, length of the threads and temperature all have an effect on the maximum torque. If a torque wrench is not available, simply tighten spark plug down to finger tight and then wrench tighten an additional 1/4 turn with gasket type plugs or 1/16 turn with taper seat plugs. It is recommended that a small amount of antiseize compound be used on threads in aluminum heads and a small amount of light weight oil be used in cast iron heads.

For additional information on torque ratings of various bolt specifications, see the HARDWARE Chapter, page 249.

# LEAD–ACID BATTERY SPECIFIC GRAVITY & CHARGE

| Acids Specific Gravity | Charge Level |
|---|---|
| 1.30 to 1.32 | Overcharged |
| 1.26 to 1.28 | 100% |
| 1.24 to 1.26 | 75% |
| 1.20 to 1.22 | 50% |
| 1.15 to 1.17 | 25% |
| 1.13 to 1.15 | Very low capacity |
| 1.11 to 1.12 | Discharged |

Battery Efficiency Changes with Temperature

| | |
|---|---|
| 80°F = 100% Charge | 10°F = 50% Charge |
| 50°F = 82% Charge | 0°F = 40% Charge |
| 30°F = 64% Charge | −10°F = 33% Charge |
| 20°F = 58% Charge | −20°F = 18% Charge |

# OIL VISCOSITY vs TEMPERATURE

| Engine Oil SAE Viscosity | -20 | 0 | 20 | 40 | 60 | 80 | 100 |
|---|---|---|---|---|---|---|---|
| | | | Outside Temperature °F | | | | |
| 20W-20 | no | no | yes | yes | yes | yes | yes |
| 20W-40 | no | no | yes | yes | yes | yes | yes |
| 20W-50 | no | no | yes | yes | yes | yes | yes |
| 10W-30 | no | yes | yes | yes | yes | yes | yes |
| 10W-40 | no | yes | yes | yes | yes | yes | yes |
| 10W | no | yes | yes | yes | yes | no | no |
| 5W-30 | yes | yes | yes | yes | yes | no | no |
| 5W-20 | yes | yes | no | no | no | no | no |

| Gear Oil SAE Viscosity | -20 | 0 | 20 | 40 | 60 | 80 | 100 |
|---|---|---|---|---|---|---|---|
| | | | Outside Temperature °F | | | | |
| 75W | yes | yes | yes | yes | no | no | no |
| 80W | yes | yes | yes | yes | yes | yes | no |
| 80W-90 | yes | yes | yes | yes | yes | yes | yes |
| 85W | no | no | yes | yes | yes | yes | no |
| 90 | no | no | no | no | yes | yes | yes |
| 140 | no | no | no | no | yes | yes | yes |

Note: Values listed above are for average conditions and may vary depending on the type of equipment being used. Manufacturers specifications should always take precedence over the above tables.

# AUTO HEADLIGHT WARNING

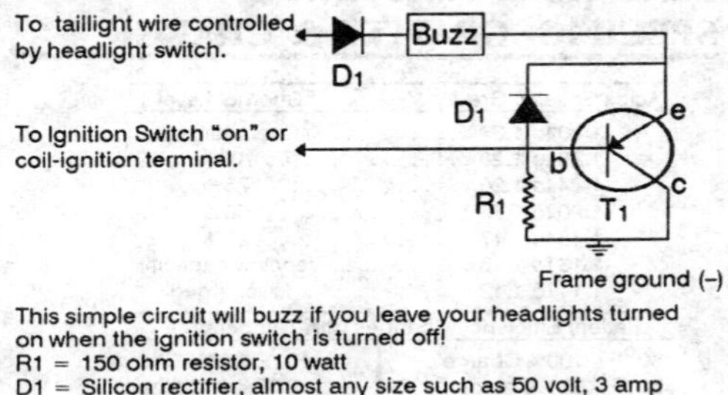


This simple circuit will buzz if you leave your headlights turned on when the ignition switch is turned off!
R1 = 150 ohm resistor, 10 watt
D1 = Silicon rectifier, almost any size such as 50 volt, 3 amp
Buzz = Small 12 volt buzzer
T1 = PNP 35 watt silicon switching transistor
"To taillight" is the wire that has 12 volts when the lights are ON
"To Ignition" is the wire that has 12 volts when the ignition is ON.

# AUTOMOTIVE AIR CONDITIONING

**WARNING:** Air conditioning systems can be dangerous. Make sure you know what you are doing ! Read directions for your gauges, freon, and the system you are recharging FIRST.

## General Recharging Steps:

1. If the system is completely empty, you will probably have to use a vacuum pump to evacuate the lines down to 28–29.5 in. of vacuum (reduce those numbers by 1 in of vacuum for every 1000 feet of altitude where you live). Vacuum pumps are expensive, you may want to have this task done at a garage.

2. Attach a pressure gauge to the low pressure side of the system and run the air conditioner for 15 minutes to stabilize the system. (Leave the high pressure side alone unless absolutely necessary to check the compressor). Use the following table to establish the desired interior cooling value (column 1); normally you should use about 35 psi for a temp of 38°F:

**Temperature / Gauge Pressure Chart for R–12 Freon**

| Evaporator Temp °F (Inside Car) | Low Pressure Reading in PSI (Evaporator side) | High Pressure Reading in PSI (Head side) | Condenser Temp °F |
|---|---|---|---|
| 16 | 18 | 132 | 72 |
| 18 | 20 | 137 | 74 |
| 20 | 21 | 144 | 76 |
| 22 | 22 | 152 | 78 |
| 24 | 24 | 160 | 80 |
| 26 | 25 | 165 | 82 |
| 28 | 27 | 170 | 84 |
| 30 | 28 | 175 | 86 |
| 32 | 30 | 180 | 88 |
| 34 | 32 | 185 | 90 |
| 36 | 33 | 189 | 92 |
| 38 | 35 | 193 | 94 |
| 40 | 37 | 200 | 96 |
| 42 | 39 | 210 | 98 |
| 44 | 41 | 220 | 100 |
| 46 | 43 | 228 | 102 |
| 48 | 45 | 236 | 104 |

3. If one or both of the low pressure or high pressure readings are HIGHER than shown in the table above for the desired interior temperature, **DO NOT ADD REFRIGERANT!** If it is not cooling properly at this point, you should seek professional help from a qualified service representative. Overcharging the system can be dangerous not only to yourself but also to the system. If outside air temp is less than condenser temp, no air conditioning!

*Automotive Air Conditioning by Boyce Dwiggins. 1983. Delmar Publishers. is an excellent reference for auto air conditioning work.*

# AUTOMOTIVE ELECTRIC WIRING

| Wire Gauge AWG | Maximum Wire Length (feet) for Car Wiring (1) | | | | | |
|---|---|---|---|---|---|---|
| | Current Load in Amps @ 12 Volts DC (2) | | | | | |
| | 1 | 2 | 4 | 6 | 8 | 10 |
| 20 | 106 | 53 | 26 | 17 | 13 | nr |
| 18 | 150 | 75 | 37 | 25 | 18 | 15 |
| 16 | 224 | 112 | 56 | 37 | 28 | 22 |
| 14 | 362 | 181 | 90 | 60 | 45 | 36 |
| 12 | 572 | 286 | 143 | 95 | 71 | 57 |
| 10 | 908 | 454 | 227 | 151 | 113 | 90 |
| 8 | 1452 | 726 | 363 | 241 | 181 | 145 |
| 6 | 2342 | 1171 | 585 | 390 | 292 | 234 |
| 4 | 3702 | 1851 | 925 | 616 | 462 | 370 |
| 2 | 6060 | 3030 | 1515 | 1009 | 757 | 606 |
| 1 | 7692 | 3846 | 1923 | 1280 | 961 | 769 |
| 0 | 9708 | 4854 | 2427 | 1616 | 1213 | 970 |

| Wire Gauge AWG | Maximum Wire Length (feet) for Car Wiring (1) | | | | | |
|---|---|---|---|---|---|---|
| | Current Load in Amps @ 12 Volts DC (2) | | | | | |
| | 12 | 15 | 20 | 50 | 100 | 200 |
| 20 | nr | nr | nr | nr | nr | nr |
| 18 | 12 | nr | nr | nr | nr | nr |
| 16 | 18 | 14 | nr | nr | nr | nr |
| 14 | 30 | 24 | 18 | nr | nr | nr |
| 12 | 47 | 38 | 28 | nr | nr | nr |
| 10 | 75 | 60 | 45 | nr | nr | nr |
| 8 | 120 | 96 | 72 | 29 | nr | nr |
| 6 | 194 | 155 | 117 | 46 | 23 | nr |
| 4 | 307 | 246 | 185 | 74 | 37 | nr |
| 2 | 503 | 403 | 303 | 121 | 60 | 30 |
| 1 | 638 | 511 | 384 | 153 | 76 | 38 |
| 0 | 805 | 645 | 485 | 194 | 97 | 48 |
| 000 | 1296 | 1039 | 781 | 312 | 156 | 78 |

(1) Maximum recommended wire lengths are based on a 1/2 volt maximum voltage drop over the length of the wire. If you want to determine lengths based on a different drop, simply multiply the table value by the appropriate factor (for example, if you want the values for a 1 volt drop, multiply the table value by 2).

(2) If you want the lengths for 6 volt or 24 volt systems, multiply the listed amperage by 0.5 for 6 volt or 2 for 24 volt and then select the wire length from the table.

"nr" means wire size is not recommended at selected current. To be safe, always pick one wire size larger than you need for the specified wire length at the required current level.

# TIRE SIZE vs LOAD RATING

| Tire Size (Bias, Bias-Belted, Radial) | Max Load lbs @ Cold Inflation 20 psi | 32 psi |
|---|---|---|
| ***Passenger Car Tires:*** | | |
| 145R- 12 inch | 600 | 780 |
| 6.00- 12 inch | 605 | 845 |
| 155R- 12 inch | 665 | 865 |
| 145/80R 12 inch | 617 | 783 |
| 155/80R 12 inch | 694 | 871 |
| 145R- 13 inch | 660 | 825 |
| 145R- 13 inch | 665 | 860 |
| 155R- 13 inch | 730 | 950 |
| 155/80R- 13 inch | 740 | 925 |
| 165/70R- 13 inch | 750 | 880 |
| 6.00-, 165R-, 165/75R- 13 inch | 770 | 1010 |
| 175/70R- 13 inch | 845 | 980 |
| 165/80R- 13 inch | 816 | 1025 |
| 195/60R- 13 inch | 825 | 1035 |
| 215/50R- 13 inch | 840 | 1050 |
| 175/75R- 13 inch | 850 | 1060 |
| A78-, A70-, AR78-, AR70-, 13 inch | 810 | 1060 |
| 205/60R- 13 inch | 835 | 1085 |
| 185/70R- 13 inch | 870 | 1090 |
| 195/65R- 13 inch | 880 | 1115 |
| 175/80R- 13 inch | 905 | 1135 |
| B13, 175R-, 205/60R- 13 inch | 890 | 1150 |
| 6.50-,B78-,B70-,175R-,BR78-,BR70-,BR60-13 | 890 | 1150 |
| 185/75R- 13 inch | 925 | 1170 |
| 195/70R- 13 inch | 948 | 1190 |
| 195/70R- 13 inch | 1045 | 1210 |
| 235/50R- 13 inch | 970 | 1220 |
| C-, C78-, C70-, CR78-, CR70-, 185/70R- 13 inch | 950 | 1230 |
| 185/80R- 13 inch | 990 | 1250 |
| 7.00-, 185R- 13 inch | 980 | 1270 |
| 205/70R- 13 inch | 1040 | 1300 |
| D70-, DR78-, DR70- 13 inch | 1010 | 1320 |
| 195R- 13 inch | 1060 | 1370 |
| E70-, ER78-, ER70-, ER60- 13 inch | 1070 | 1400 |
| 155R- 14 inch | 780 | 1010 |
| A70-, AR70- 14 inch | 810 | 1060 |
| 6.45-, 165R- 14 inch | 860 | 1120 |
| B78-, B70-, BR78-, BR70- 14 inch | 890 | 1150 |
| 6.95- 14 inch | 950 | 1230 |
| C78-, C70-, CR78-, CR70- 14 inch | 950 | 1230 |
| D78-, D70-, DR78-, DR70-, 195/70- 14 inch | 1010 | 1320 |
| 7.35-, 185R- 14 inch | 1040 | 1360 |
| E78-, E70-, ER78-, ER70-, 205/70- 14 inch | 1070 | 1400 |
| 7.75-, 195R- 14 inch | 1150 | 1500 |
| F78-,F70-,F60-,FR78-,FR70-,FR60-,215/70-14 | 1160 | 1500 |
| 8.25-, 205R- 14 inch | 1250 | 1620 |
| G78-,G70-,G60-,GR78-,GR70-,GR60- 14 inch | 1250 | 1620 |
| 8.55-, 215R- 14 inch | 1360 | 1770 |
| H78-, H70-, H60-, HR78-, HR70- 14 inch | 1360 | 1770 |
| 8.85-, 225R- 14 inch | 1430 | 1860 |
| 6.85-, C78-, C70-, 175R-, CR78-, CR70- 15 inch | 950 | 1230 |
| D78-, D70- 15 inch | 1010 | 1320 |
| 7.35-, 185R- 15 inch | 1070 | 1390 |
| E78-,E70-,E60-,ER78-,ER70-,ER60-,205/70-15 | 1070 | 1400 |
| 7.75-, 195R- 15 inch | 1150 | 1490 |
| F78-,F70-,F60-,FR78-,FR70-,FR60-,215/70-15 | 1160 | 1500 |
| 205R- 15 inch | 1240 | 1610 |
| G78-,G70-,G60-,GR78-,GR70-,GR60-,225/70-15 | 1250 | 1620 |
| 8.25- 15 inch | 1250 | 1620 |
| 215R- 15 inch | 1340 | 1740 |

# TIRE SIZE vs LOAD RATING

| Tire Size (Bias, Bias-Belted, Radial) | Max Load lbs @ Cold Inflation 20 psi | 32 psi |
|---|---|---|
| H78-,H70-,H60-,HR78-,HR70-,HR60- 15 inch | 1360 | 1770 |
| 8.55- 15 inch | 1360 | 1770 |
| 8.85-, 225R- 15 inch | 1430 | 1860 |
| J78-, J70-, J60-, JR78-, JR70- 15 inch | 1430 | 1860 |
| 9.00- 15 inch | 1460 | 1900 |
| K70-, KR70- 15 inch | 1460 | 1900 |
| 9.15-, 235R- 15 inch | 1510 | 1970 |
| L78-, L70-, L60-, LR78-, LR70-, LR60- 15 inch | 1520 | 1970 |
| 205/55R- 16 inch | 890 | 1150 |
| 225/50R- 16 inch | 1000 | 1300 |
| 245/50R- 16 inch | 1200 | 1510 |
| 255/50R- 16 inch | 1280 | 1610 |

| Light Truck Tires: (single tire) Tire Size (Bias, Bias-Belted, Radial) | Max Load lbs @ Cold Inflation 35 psi *Radial* / 30 psi *Bias* | 60 psi / 75 psi |
|---|---|---|
| E-, ER78-14LT | 1140 | 1620 |
| G-, GR78-14LT | 1260 | — |
| G-, GR78-15LT | 1310 | 1870 |
| H-, HR78-15LT | 1440 | 2060 |
| L-, LR78-15LT | 1600 | 2290 |
| F-, FR78-16LT | 1270 | 1820 |
| H-, HR78-16LT | 1510 | 2150 |
| L-, LR78-16LT | 1670 | 2380 |
| 6.70-15LT | 1210 | 2060 |
| 7.00-15LT | 1350 | 2320 |
| 6.50-16LT | 1270 | 2160 |
| 7.50-16LT | 1620 | 2780 |
| 8.00-16.5LT | 1360 | 2330 |
| 8.75-16.5LT | 1570 | 2680 |
| 9.50-16.5LT | 1860 | 3170 |
| 10.00-16.5LT | 1840 | 3135 |
| 12.00-16.5LT | 2370 | 4045 |
| LT195/75-14 | 1115 | 1625 |
| LT195/75-15 | 1165 | — |
| LT215/75-15 | 1345 | 1960 |
| LT235/75-15 | 1530 | 2230 |
| LT255/75-16 | 1920 | 2800 |

| | 30 psi *radial* / 20 psi *belted* | 50 psi / 45 psi |
|---|---|---|
| 27 x 8.50-14LT | 940 | 1515 |
| 30 x 9.50-15LT | 1240 | 1990 |
| 31 x 10.50-15LT | 1400 | 2250 |
| 31 x 11.50-15LT | 1455 | 2340 |
| 32 x 11.50-15LT | 1575 | 2530 |
| 33 x 12.50-15LT | 1755 | — |

NOTE: Always check the current manufacturers specifications on a tire to verify the above approximations. If you want detailed information on tires, sizes, recommendations for different cars, etc, use a book entitled *"1986 Tire Guide" (get one for the current year), published by Bennett Garfield Publishers, 1101-6 S. Rogers Circle, Boca Raton, FL, 33487, (407) 997-9229.* Cost of the book is only $5.00 and contains an abundance of information.

# TIRE MANUFACTURER CODES

| Manufacturer | Codes |
|---|---|
| Alliance Tire & Rubber (Israel) | CD |
| Armstrong Rubber | CE, CF, CH, CV |
| Avon Rubber (England) | AT |
| B.F. Goodrich | BA to BP |
| Bridgestone Tire | EH to EP |
| Carlisle Tire & Rubber | UU |
| CEAT (Italy) | HT, HU, HV |
| Continental Gummiwerke | CM to CU |
| Copper Tire & Rubber | UP, UT |
| Dayton Tire & Rubber | HX and HY |
| Denman Rubber Mfg | DY |
| Dunlop Tire & Rubber | DA to DU |
| Firestone Tire & Rubber | VA to VY, WA to WJ |
| Gates Rubber | BW, BX, BY |
| General Tire | AA to AH |
| Goodyear Tire & Rubber | MA to MY, NA to NY, PA to PF |
| Hung Ah Tire (Korea) | EF |
| Inoue Rubber (Japan) | CJ |
| IRI International Rubber | BV |
| Kelly–Springfield Tire | PH to PY, TA to TY, UA to UN |
| Kleber-Colombes | EV to EY |
| Kyowa Rubber Industry (Japan) | UV |
| Lee Tire & Rubber | JA to JY, KA to KY, LA to LF |
| Madras Rubber (India) | WT |
| Mansfield-Denman (Canada) | LV, WL |
| McCreary Tire & Rubber | CY |
| Metzeler A.G. | EA, EB, EC |
| Michelin Tire | FF to FY, HA to HP |
| Mitsuboshi Belting | LX, LY |
| Mohawk Rubber | CA, CB, CC |
| Nitto Tire (Japan) | EE |
| Okamoto Riken Gomu (Japan) | ED |
| Olympic Tire & Rubber (Aust.) | WM, WN |
| Phoenix Gummiwerke A.G. | AX, AY |
| Pirelli Tire | XA to XP |
| Samson Tire & Rubber | AW |
| Semperit Gummiwerke A.G. | BT, BU |
| Sieberling Tire and Rubber | AV |
| Sumitomo Rubber Industries | ET, EU |
| Toyo Rubber Industry | CW, CX |
| Trelleborg Rubber (Sweden) | LW |
| Uniroyal, Inc | AJ to AP, LH to LU, AU |
| Veith-Pirelli A.G. | XT |
| Vredestine (Netherlands) | DV, DW, DX |
| Yokohama Rubber | FA to FE |

**NOTE:** The number–letter tire code is located on the sidewall, near the rim and the letters "DOT", the code is "XXYY MMM999". The "XX" is the Manufacturer, the "YY" is the tire size (another letter type code), "MMM" is an optional tire type code, and "999" is the Date of Manufacture. In the above coding system, the letters "G", "I", "O", "Q", "R", "S", "Z", and the number "0" are never used. If you need more specific information on Manufacturer Codes, you can obtain a detailed list *(PART 574–TIRE CODE)* of the codes from the *Department of Transportation (DOT)*.

# TIRE SIZE CODES

## Conventional Coding:

### 9.50R–15C

9.50 = Tire section width
R = Radial ply construction (no letter if Bias)
15 = Rim diameter in inches
C = Load Range

### LR60–15B–HR

L = Load Range
R = Radial ply construction (no letter if Bias)
60 = Aspect ratio, 60 means the section height is 60% as great as the width
15 = Rim Diameter in inches
B = Load Range
HR = Speed rating: HR = 112 mph, SR = 130 mph, VR = 165 mph maximum allowable safe speed

## Metric Coding:

### P215/75R–15

P = P is passenger car, LT is light truck.
215 = Tire section width in millimeters
75 = Aspect ratio (see description above)
R = Radial ply construction (no letter if Bias)
15 = Rim diameter in inches

# LOAD RANGE vs PLY RATING

| Load Range | Ply Rating | Load Range | Ply Rating |
|---|---|---|---|
| A | 2 | G | 14 |
| B | 4 | H | 16 |
| C | 6 | J | 18 |
| D | 8 | L | 20 |
| E | 10 | M | 22 |
| F | 12 | N | 24 |

# AUTOMOTIVE FORMULAS

**Engine Displacement** = Stroke x Bore$^2$ x 0.7854 x Cylinders

Engine Displacement is in cubic inches if Bore and Stroke are in inches or cubic centimeters (cc's) if Bore and Stroke are in centimeters.

Cylinders is the number of engine cylinders (4, 6, 8, etc)

**CFM Engine Carburetor Air Flow:**

$$\text{4 Stroke Engine CFM} = \frac{\text{CID} \times \text{RPM} \times \text{VE}}{3456}$$

CFM is Cubic Feet per Minute air flow through the carburetor.
CID is the engine displacement in cubic inches.
RPM is the engine Revolutions Per Minute.
VE is the engine Volumetric Efficiency, use 1 for 100% efficient.
(For a **2 Stroke Engine**, divide by 1728, not 3456)

# POCKET REF

# Carpentry and Construction



(See also HARDWARE on page 249)
(See also PLUMBING on page 317)
(See also GLUE on page 235)

# SOFTWOOD LUMBER SIZES

| Nominal Size (Inches) | Actual Size Dry (Inches) | Actual Size Green (Inches) |
|---|---|---|
| **THICKNESS:** | | |
| 1 | 3/4 | 25/32 |
| 1–1/4 | 1 | 1–1/32 |
| 1–1/2 | 1–1/4 | 1–9/32 |
| 2 | 1–1/2 | 1–9/16 |
| 2–1/2 | 2 | 2–1/16 |
| 3 | 2–1/2 | 2–9/16 |
| 3–1/2 | 3 | 3–1/16 |
| 4 | 3–1/2 | 3–9/16 |
| 4–1/2 | 4 | 4–1/16 |
| 6 | 5–1/2 | 5–9/16 |
| 8 | 7–1/2 | 7–9/16 |
| **FACE WIDTH:** | | |
| 2 | 1–1/2 | 1–9/16 |
| 3 | 2–1/2 | 2–9/16 |
| 4 | 3–1/2 | 3–9/16 |
| 5 | 4–1/2 | 4–5/8 |
| 6 | 5–1/2 | 5–5/8 |
| 7 | 6–1/2 | 6–5/8 |
| 8 | 7–1/4 | 7–1/2 |
| 9 | 8–1/4 | 8–1/2 |
| 10 | 9–1/4 | 9–1/2 |
| 11 | 10–1/4 | 10–1/2 |
| 12 | 11–1/4 | 11–1/2 |
| 14 | 13–1/4 | 13–1/2 |
| 16 | 15–1/4 | 15–1/2 |

Dry lumber is defined as lumber with less than 19 percent moisture and green is greater than 19 percent. All sizes listed above, both nominal and actual, conform to standards set by the *American Softwood Lumber Standards.*

Lumber is sold by a "feet board measure" or "board foot" rating. 1 board foot = 144 square inches (12 inch x 12 inch x 1 inch).

Board feet = thickness (in) x face width (in) x length (in)/144
or = thickness (in) x face width (in) x length (ft)/12

The following are quick approximations for calculating board feet:
- for a 1 x 4, divide linear length (feet) by 3
- for a 1 x 6, divide linear length (feet) by 2
- for a 1 x 8, multiply linear length (feet) by 0.66
- for a 1 x 12, linear length (feet) = board feet
- for a 2 x 4, multiply linear length (feet) by 0.66
- for a 2 x 6, linear length (feet) = board feet
- for a 2 x 8, multiply linear length (feet) by 1.33
- for a 2 x 12, multiply linear length (feet) by 2

# SOFTWOOD LUMBER GRADING

Softwood grading is based on the appearance, strength and stiffness of lumber. Grading systems are established by a variety of associations in different parts of the country but they all must follow the US Department of Commerce American Lumber Standards. The grading system is quite long and very detailed *If you want more detailed information on softwood grading, obtain the book "Western Lumber Grading Rules 88" by the Western Wood Products Association, 522 S.W. Fifth, Portland, Oregon, 97204, (503)224-3930. The cost is only $7.00 and it is an excellent pocket reference.*

Softwood lumber comes from "conifer" trees, which means they have needle shaped leaves that stay green all year. Hardwoods come from "deciduous" trees, which means they have broad leaves and loose their leaves in the cold months. A list of tree types and their characteristics is given later in this chapter.

The first broad softwood classification is as follows:

**Rough Lumber** – Sawn, trimmed, and edged, but the faces are rough and show saw marks.

**Surfaced Lumber** (dressed) – Rough lumber that has been smoothed by a surfacing machine. Subcategories are based on the number of sides and edges that have been smoothed:

- **S1S** – Sawn 1 Side
- **S1E** – Sawn 1 Edge
- **S2S** – Sawn 2 Sides
- **S2E** – Sawn 2 Edges
- **S1S1E** – Sawn 1 Side and 1 Edge
- **S1S2E** – Sawn 1 Side and 2 Edges
- **S2S1E** – Sawn 2 Sides and 1 Edge
- **S4S** – Sawn 4 Sides
- **S/S** – Saw Sized (resawn)

**Worked Lumber** – Surfaced lumber that has been matched, patterned, shiplapped or any combination thereof.

Another broad softwood classification (which is not a subcategory of the first classification above) is as follows:

**Shop and Factory Lumber** – This is millwork lumber used for applications such as molding, door jambs, and window frames.

**Yard Lumber** – Lumber used for house framing, concrete forms, and sheathing. It is also known as structural lumber.

# SOFTWOOD LUMBER GRADING

Yard or Structural softwood lumber is further subdivided into the following categories, based on size:

**Boards** – Lumber must be no more than 1 inch thick and 4 to 12 inches wide.

**Planks** – Lumber must be over 1 inch thick and more than 6 inches wide.

**Timbers** – Lumber width and thickness must both be greater than 5 inches.

The most common softwood grading system places lumber into three main categories. Once again, bear in mind that some of these categories are very detailed and long; for example, the specific description of "#2 Common Board" is almost 2 pages long and covers details such as degree of cupping, twist, wane, knots, and raising of grain. The following descriptions cover the primary system only, see a grading manual for more detail:

1. **Select and Finish Materials** – These are "Appearance" grades and are used primarily for interior and exterior trim work, moldings, cabinets, and interior walls. Select grades are based on the best face and Finish grades are based on the best face and 2 edges.
   Select – B & BTR – 1 & 2 Clear
   C Select
   D Select
   Superior Finish VG, FG or MG
   Prime Finish VG, FG, MG
   E Finish
2. **Boards** – Five grades referred to as "Commons" (1 Common through 5 Common) are used for general building, crafts, form lumber, flooring, sheathing, etc. "Alternate Board Grades" include the following (in order from best to worst):
   Select Merchantable
   Construction
   Standard
   Utility
   Economy

   The final category of Boards is the "Stress Related Boards". These are special use products for light trusses, rafters, and box beams for factory built and mobile homes.

3. **Dimension Lumber** – This category is limited to surfaced softwood lumber that is 2 to 4 inches thick and is to be used as framing components. Category breakdowns are as follows:

   **Light Framing** – General framing and stud walls. Up to 4 inch wide. Grades are as follows:
   Construction (34% Bending Strength Ratio)
   Standard (19% Bending Strength Ratio)

   **Structural Light Framing** – This is suitable for higher stress applications such as roof trusses and concrete forms. Up to 4 inch wide. Grades are as follows:
   Select Structural (67% Bending Strength Ratio)
   No. 1 (55% Bending Strength Ratio)
   No. 2 (45% Bending Strength Ratio)
   No. 3 (26% Bending Strength Ratio)

   **Studs** – Load bearing and stud walls of 2 x 4 and 2 x 6 construction. Lengths are less than 10 feet. Up to 4 inch wide and 5 inch and over. Stud grade is a 26% Bending Strength Ratio.

   **Structural Joists and Planks** – Roof rafters, ceiling and floor joists. 5 inch and wider. Grades are as follows:
   Select Structural (65% Bending Strength Ratio)
   No. 1 (55% Bending Strength Ratio)
   No. 2 (45% Bending Strength Ratio)
   No. 3 (26% Bending Strength Ratio)

   **Timber** – Heavy beam support and floor and ceiling supports.

   **Appearance Framing** – High bend strength ratio, over 2 inch wide, and good appearance for special applications. Appearance Framing grade has a 55% Bending Strength Ratio.

If you are confused by the softwood grading scheme, don't feel bad, you're not alone! Grading is not an exact science since it deals with both visual and strength analysis. A maximum of 5% variation below grade is allowable between graders. Note that the above grading is only a small portion of the actual code, there are literally hundreds of different grades.

# HARDWOOD LUMBER SIZE & GRADE

Hardwood comes from "deciduous" trees, which have broad leaves and loose their leaves in the cold months. Oak and walnut constitute 50% of all hardwood production. Other common hardwoods include Basswood, Beech, Birch, Butternut, Chestnut, Cherry, Elm, Gum, Hickory, Maple, Mahogany, and Yellow Poplar. See the section later in this chapter that describes wood types and their general characteristics.

## HARDWOOD SIZES

| Nominal Size (Fraction In) | Rough Size (Inches) | Surface 2 Sides Actual Size Dry (Inches) |
|---|---|---|
| 4/4 | 1 | 13/16 |
| 5/4 | 1–1/4 | 1–1/16 |
| 6/4 | 1–1/2 | 1–5/16 |
| 7/4 | 1–3/4 | 1–1/2 |
| 8/4 | 2 | 1–3/4 |
| 10/4 | 2–1/2 | 2–1/4 |
| 12/4 | 3 | 2–3/4 |
| 14/4 | 3–1/2 | 3–1/4 |
| 16/4 | 4 | 3–3/4 |

## HARDWOOD GRADES

Grading is simpler than that used for Softwood and appearance is the prime consideration. Grades are based on the appearance of the poorest side, assuming that the board will be cut into pieces that are 2 to 7 feet long, each of which will have one clear face. There are numerous other requirements for grades of each of the various tree species, but the general grades of hardwood as determined by the National Hardwood Lumber Association are as follows (Listed in order from best to worst):

**First and Second (FAS)** – The best grade. Normally required for a natural or stained finish. A FAS board must be at least 6 inches wide, 8 to 16 feet long, and 83.3% clear on the worst face.

**Select – No. 1 Common** – Minimum 3 inches wide, 4 to 16 feet long, 66.66% clear wood.

**Select – No. 2 Common**

**Select – No. 3 Common**

*If you want detailed information on the grading of hardwood, obtain a copy of the Hardwood Rule Book, National Hardwood Association, P.O. Box 34518, Memphis, Tennessee, 38134. Cost of the book is $4.00. It is an excellent source book.*

# WOOD MOISTURE CONTENT

Moisture content in wood affects both the size and strength of lumber. In general, the physical properties of wood can be improved by seasoning or drying. Although dependent on the tree species type, the strength of wood decreases as the moisture content goes up. *The following table is from Circular 108 of the U.S. Forest Service.*

**MOISTURE vs COMPRESSIVE STRENGTH**

| % Moisture | Relative Maximum crushing strength compared to wood containing 2% moisture (compression parallel to the grain) | | |
|---|---|---|---|
| | Red Spruce | Longleaf Pine | Douglas Fir |
| 2 | 1.000 | 1.000 | 1.000 |
| 4 | 0.926 | 0.894 | 0.929 |
| 6 | 0.841 (c) | 0.790 | 0.850 |
| 8 | 0.756 | 0.702 | 0.774 |
| 10 | 0.681 | 0.623 | 0.714 |
| 12 | 0.617 | 0.552 | 0.643 |
| 14 | 0.554 (b) | 0.488 | 0.589 |
| 16 | 0.505 | 0.431 | 0.535 |
| 18 | 0.463 | 0.377 | 0.494 |
| 20 | 0.426 | 0.328 (a) | 0.458 |
| 22 | 0.394 | 0.278 | 0.428 |
| 24 | 0.362 | | 0.398 (a) |
| 26 | 0.335 | | |
| 28 | 0.314 | | |
| 30 | 0.292 | | |
| 32 | 0.271 | | |
| 34 | 0.255 | | |

(a) Green wood
(b) Air dried
(c) Kiln dried

The above table clearly indicates that high moisture content in wood significantly decreases the woods strength. As an example, Longleaf Pine has half the strength (0.552) with 16% moisture as is does with 2% moisture.

*Additional information can be obtained from U.S. Department of Agriculture Bulletin 282 and Technical Bulletin 479.*

# PLYWOOD & PANEL GRADING

Plywood is generally graded in terms of the quality of the veneer on both the front and back sides of the panel or by a "use type" name. Plywood is also grouped by the tree species type.

**The American Plywood Association (APA) Veneer Grades**

**N**..... Smooth surface "natural finish" veneer. Select, all heartwood or all sapwood. Free of open defects. Allows not more than 6 repairs, wood only, per 4 x 8 panel, made parallel to grain and well matched for grain and color.

**A**..... Smooth, paintable. Not more than 18 neatly made repairs, boat, sled, or router type, and parallel to grain, permitted. May be used for natural finish in less demanding applications.

**B**..... Solid surface. Shims, circular repair plugs and tight knots to 1 inch across grain permitted. Some minor splits permitted.

**C**..... **Plugged** Improved C veneer with splits limited to 1/8–inch width and knotholes and borer holes limited to 1/4 x 1/2 inch. Admits some broken grain. Synthetic repairs permitted.

**C**..... Tight knots to 1–1/2 inch. Knotholes to 1 inch across grain and some to 1–1/2 inch if total width of knots and knotholes is within specified limits. Synthetic or wood repairs. Discoloration and sanding defects that do not impair strength permitted. Limited splits allowed. Stitching permitted.

**D**..... Knots and knotholes to 2–1/2 inch width across grain and 1/2 inch larger within specified limits. Limited splits are permitted. Stitching permitted. Limited to Exposure 1 or Interior panels.

As an example, "C–D" grade panel would have one side conforming to the "C" grade and the other side conforming to the "D" grade. You must also specify the "Exposure Durability" (defined on the next page) to completely define the grade, e.g., EXTERIOR C–D.

NOTE: "CDX" is a very common grade of panel, but it does not have an "EXTERIOR" rating, it has an "EXPOSURE 1" rating.

*A full description of the plywood and panel code can be obtained from the American Plywood Association P.O. Box 11700. Tacoma, Washington, 98411. (206) 565–6600.*

# PLYWOOD & PANEL GRADING

## EXPOSURE DURABILITY

**EXTERIOR:** Fully waterproof bond and designed for applications subject to permanent exposure to weather or moisture.

**EXPOSURE 1:** Fully waterproof bond but not for permanent exposure to weather or moisture.

**EXPOSURE 2:** Interior type with intermediate glue under PS 1. Intended for protected construction applications where slight moisture exposure can be expected.

**INTERIOR:** Interior applications only.

## GROUP CLASSIFICATION OF SPECIES

| Group 1 | Group 2 | Group 3 | Group 4 | Group 5 |
|---|---|---|---|---|
| Apitong | Cedar–Port | Alder–Red | Aspen | Basswood |
| Beech–Amer. | Cedar-Oxford | Birch–Paper | Bigtooth | Poplar |
| Birch–Sweet | Cypress | Cedar-Alaska | Quaking | Balsam |
| Birch–Yellow | Douglas Fir 2 | Fir-Subalpine | Cativo | |
| Fir 1-Douglas | Fir | Hemlock-East | Cedar | |
| Kapur | Balsam | Maple-Bigleaf | Incense | |
| Keruing | Calif. Red | Pine | West Red | |
| Larch–West. | Grand | Jack | Cottonwood | |
| Maple-Sugar | Noble | Lodgepole | Eastern | |
| Pine | Pacific-Silver | Ponderosa | Black | |
| Caribbean | White | Spruce | West Poplar | |
| Ocote | Hemlock | Redwood | Pine | |
| Pine South. | Lauan | Spruce | East White | |
| Loblolly | Almon | Engelmann | Sugar | |
| Longleaf | Bagtikan | White | | |
| Shortleaf | Mayapis | | | |
| Slash | Red | | | |
| Tanoak-White | Tangile | | | |
| | Maple–Black | | | |
| | Mengkulang | | | |
| | Meranti–Red | | | |
| | Mersawa | | | |
| | Pine–Pond, Red, Virginia, Western White | | | |
| | Spruce–Black, Red, and Sitka | | | |
| | Sweetgum | | | |
| | Tamarack | | | |
| | Yellow Poplar | | | |

Group numbers are used to define the strength and stiffness of the panel, Group 1 being the strongest, Group 5 the weakest.

# WOOD CHARACTERISTICS

| Wood Name | 1988 Cost/ Brd Ft [1] | Density Lbs per Cubic Ft | Hard | Split Resist | Grain |
|---|---|---|---|---|---|
| Alder | $1.95 | 25–30 | Med | Good | Low |
| Ash | $2.80 | 40–45 | Hard | Good | Mod open |
| Aspen | $1.25 | 25 | Soft | Good | Mild fine |
| Balsa | $2.50 | 8 | V Soft | Good | Open |
| Basswood | $1.70 | 25–28 | Soft | Good | Low, fine |
| Beech | $1.90 | 45 | Hard | V Good | Mod, fine |
| Birch | $2.50 | 40–45 | Hard | V Good | Mod, fine |
| Butternut | | 27 | Med | Good | Mod |
| Cedar, East | | 29 | M Hard | Poor | Fine, knots |
| Cedar, West | $0.85/1.80 | 25 | Med | Poor | Fine |
| Cherry | $3.20 | 35 | M Hard | V Good | Mod, fine |
| Chestnut | | 30 | M Hard | Good | Mod, coarse |
| Cottonwood | | 25 | Med | Good | Low, fine |
| Cypress | $2.30 | 35 | M Hard | Poor | Wide, fine |
| Ebony | $24.00 | 50–65 | V Hard | V Good | V Low, fine |
| Elm, American | | 35 | M Hard | V Good | Mod, v fine |
| Elm, Rock | | 44 | Hard | V Good | Mod, v fine |
| Fir, Douglas | $0.30/1.35 | 35 | Med | Fair | Wide |
| Fir, White | $0.65 | 25 | Med | Fair | Wide |
| Gum, Black | | 36 | M Hard | V Good | Mod |
| Gum, Blue | | 50 | Hard | V Good | Mod, open |
| Gum, Red | $3.15 | 35 | M Hard | V Good | Mod |
| Hackberry | | 38 | M Hard | Poor | Coarse |
| Hickory | $1.70 | 40–55 | Hard | Good | Mod, pores |
| Holly | | 40 | M Hard | V Good | None, fine |
| Ligum Vitae | $6.00 lb | 80 | V Hard | V Good | Mod, v fine |
| Madrone | | 45 | Hard | Good | Mod, v fine |
| Magnolia | $1.70 | 35 | M Hard | Good | Fine |
| Mahogany | | | | | |
| African | | 30 | M Hard | Good | Open, figure |
| Cuban | | 40 | Hard | Good | Open, figure |
| Honduras | $4.50 | 35 | M Hard | Good | Open, figure |
| Phillippine | | | Not a Mahogany, see Phillippine. | | |
| Maple (hard) | $1.85 | 35–44 | M Hard | Good | Mod, fine |
| Myrtle | | 40 | Hard | Good | Mod, fine |
| Oak | $3.75 | | | | |
| Amer. Red | $3.10 | 45 | Hard | Good | Coarse, pores |
| Amer. White | $2.95 | 47 | Hard | Good | Coarse, pores |
| English Brown | | 45 | Hard | Good | Coarse, pores |
| Pecan | $1.70 | 47 | Hard | Good | Fine, pores |
| Persimmon | | 55 | Hard | V Good | V fine |
| Philippine: | | | | | |
| Red Luan | $2.60 | 36 | M Hard | Good | Mod, coarse |
| Tanguile | | 39 | M Hard | V Good | Mod, coarse |

# WOOD CHARACTERISTICS

| Wood Name | 1988 Cost/ Brd Ft [1] | Density Lbs per Cubic Ft | Hard | Split Resist | Grain |
|---|---|---|---|---|---|
| Pine, White: | 0.40/2.35 | | | | |
| Northern | | 25 | Soft | Poor | V coarse |
| Western | | 27 | Soft | Poor | Mod, fine |
| Poplar, Yellow | $1.65 | 30 | M Hard | Good | Mod, v fine |
| Redwood | $0.95/1.95 | 28 | Med | Poor | Fine |
| Rosewood: | | | | | |
| Bolivian | $5.90 | 50 | Hard | Good | Swirls,pores |
| East Indian | $11.00 | 55 | Hard | Good | Mod |
| Satinwood | $6.20 | 67 | V Hard | Good | Mod, fine |
| Spruce | $4.60 | 28 | Med | Poor | Mod, fine |
| Sycamore | $1.20 | 35 | M Hard | High | Mod, fine |
| Teak (Burma) | $8.00 | 45 | Hard | High | Mod to High |
| Walnut: | | | | | |
| Amer Black | $4.20 | 38 | Hard | Good | Mod, fine |
| Claro | | 30 | M Hard | Good | Mod, open |
| European | | 35 | M Hard | Good | Mod, open |
| Willow | $2.00 | 26 | Soft | Good | Mod, fine |
| Zebrawood | $6.20 | 48 | Hard | Good | High, fine |

**Hardness** is a relative term between the different species. "V Soft" is an abbreviation for Very Soft, "V Hard" is Very Hard, and "M Hard" is moderately hard.

**Split Resist** refers to the susceptibility the lumber has to splitting. The scale ranges from "V Good" (Very Good) to "Good" to "Fair" to "Poor".

**Grain** defines the general appearance of the wood grain. "Mod" is moderate, "High" is very pronounced grain, "pores" is large open pores, "fine" is fine grained, "V fine" is very fine grained, and "coarse" is coarse grained.

(1) The cost per board foot column will sometimes contain two values instead of one. The first number represents the cost of relatively low grade lumber such as "#3 Common" and the second number represents the cost of the higher grades such as "Select 1" or "Select 2". Hardwood costs are for First & Seconds (FAS) and are sawn 2 sides (S2S). Thicker hardwood boards are usually more expensive, e.g. 4/4 (1 inch thick) Red Oak is $3.09/board foot whereas 8/4 (2 inch thick) is $4.63/board foot. Price increases of 10% to 40% are not uncommon for double the thickness.

*An excellent book on woods is "Beautiful Woods", by the Frank Paxton Lumber Co. 4837 Jackson St. Denver, CO 80216, (303)399–6810, cost $5.00 (complete with color photos!). Also, "Know Your Woods" by Albert Constantine, 1959, ISBN 0–684–14115–9.*

# INSULATION VALUE OF MATERIALS

| Insulation Material | Thickness (inches) | k | C | R Value |
|---|---|---|---|---|
| Ground surface | | | 2.00 | 0.50 |
| Concrete | 1 | 12.00 | | 0.08 |
| Plaster | 1 | 8.00 | | 0.12 |
| Face Brick | 1 | 9.00 | | 0.11 |
| Brick – Low density | 1 | 5.00 | | 0.20 |
| Hollow Concrete Block | 8 | | 0.90 | 1.11 |
| Hollow Tile | 4 | | 1.00 | 1.00 |
| Stucco | 1 | 5.00 | | 0.20 |
| Metal Lath & Plaster | 3/4 | | 7.70 | 0.13 |
| Rock cork | 1 | 0.328 | | 3.05 |
| Celotex | 1 | 0.330 | | 3.03 |
| Cork Board | 1 | 0.30 | | 3.33 |
| Gypsum Board | 1/2 | | 2.20 | 0.45 |
| Plywood | 1/2 | | 1.60 | 0.62 |
| Most softwoods | 1 | 0.80 | | 1.25 |
| Most hardwoods | 1 | 1.10 | | 0.91 |
| Sawdust | 1 | 0.410 | | 2.44 |
| Redwood | 1 | 0.570 | | 1.75 |
| Asphalt Shingles | | | 2.27 | 0.44 |
| Built–up Roofing | 3/8 | | 3.00 | 0.33 |
| Wood Shingles | | | 1.06 | 0.94 |
| Structural Insulation Bd | 1/2 | | 0.76 | 1.32 |
| Glass wool | 1 | 0.266 | | 3.76 |
| Mineral Wool Bat | 3–4 | | 0.09 | 11.00 |
| Mineral Wool Bat | 5–6 | | 0.05 | 19.00 |
| Mineral Wool Bat | 6–7 | | 0.05 | 22.00 |
| Mineral Wool Bat | 8–9 | | 0.03 | 30.00 |
| Sheep's wool | 1 | 0.338 | | 2.96 |
| Balsam wool | 1 | 0.27 | | 3.70 |
| Polystyrene | 1 | | 0.28 | 3.57 |
| Air Space, nonreflective | 3/4 | | 1.00 | 1.01 |
| Air Space, reflective | 3/4 | | | 3.48 |

"k" is heat conductivity over a thickness of 1 inch and "C" is heat conductance over the specified thickness. "R Value" is the most common number used to compare the insulating properties of various material and is typically marked on the wrapper or container of the insulator. The "R Value" is effectively the materials resistance to heat–flow and is based on the "k" and "C" values. "R Values" based on "k" assume a thickness of 1 inch and "R Values" based on "C" are based on the thickness indicated above. *Values listed above are from the National Bureau of Standards and from the ASHRAE 1977 Fundamentals Handbook.* Refer to those references for more detailed information.

# MAXIMUM FLOOR JOIST SPANS

**S. Pine–Douglas Fir**, Max Load **40** lbs/sq ft, uniformly distributed

| Lumber Size (Inch) | | Inches | Max Feet Between Supports | |
|---|---|---|---|---|
| Nominal | Actual | On Center | Unplastered | Plastered |
| 2 x 6 | 1–5/8 x 5–5/8 | 12 | 12.00 | 10.00 |
| | | 16 | 10.50 | 9.08 |
| 3 x 6 | 2–5/8 x 5–5/8 | 12 | 15.00 | 11.66 |
| | | 16 | 13.08 | 10.66 |
| 2 x 8 | 1–5/8 x 7–1/2 | 12 | 15.92 | 13.25 |
| | | 16 | 13.92 | 12.08 |
| 3 x 8 | 2–5/8 x 7–1/2 | 12 | 19.66 | 15.33 |
| | | 16 | 17.33 | 14.00 |
| 2 x 10 | 1–5/8 x 9–1/2 | 12 | 19.92 | 16.66 |
| | | 16 | 17.33 | 15.25 |
| 3 x 10 | 2–5/8 x 9–1/2 | 12 | 24.58 | 19.25 |
| | | 16 | 21.66 | 17.66 |
| 2 x 12 | 1–5/8 x 11–1/2 | 12 | 23.92 | 20.08 |
| | | 16 | 20.92 | 18.42 |
| 3 x 12 | 2–5/8 x 11–1/2 | 12 | 29.33 | 23.08 |
| | | 16 | 25.92 | 21.25 |
| 2 x 14 | 1–5/8 x 13–1/2 | 12 | 27.66 | 23.42 |
| | | 16 | 24.33 | 21.42 |
| 3 x 14 | 2–5/8 x 13–1/2 | 12 | – | 26.92 |
| | | 16 | – | 24.83 |

**S. Pine–Douglas Fir**, Max Load **75** lbs/sq ft, uniformly distributed

| Lumber Size (Inch) | | Inches | Max Feet Between Supports | |
|---|---|---|---|---|
| Nominal | Actual | On Center | Unplastered | Plastered |
| 2 x 8 | 1–5/8 x 7–1/2 | 12 | 12.00 | 11.33 |
| | | 16 | 10.50 | 10.33 |
| 3 x 8 | 2–5/8 x 7–1/2 | 12 | 15.17 | 13.17 |
| | | 16 | 12.17 | 12.08 |
| 2 x 10 | 1–5/8 x 9–1/2 | 12 | 15.25 | 14.25 |
| | | 16 | 13.25 | 13.00 |
| 4 x 8 | 3–5/8 x 7–1/2 | 12 | 17.58 | 14.58 |
| | | 16 | 15.42 | 13.33 |
| 3 x 10 | 2–5/8 x 9–1/2 | 12 | 19.00 | 16.66 |
| | | 16 | 16.66 | 15.17 |
| 2 x 12 | 1–5/8 x 11–1/2 | 12 | 18.33 | 17.17 |
| | | 16 | 16.00 | 12.75 |
| 4 x 10 | 3–5/8 x 9–1/2 | 12 | 22.08 | 18.33 |
| | | 16 | 18.42 | 16.83 |
| 3 x 12 | 2–5/8 x 11–1/2 | 12 | 22.83 | 20.08 |
| | | 16 | 20.08 | 18.33 |
| 3 x 14 | 2–5/8 x 13–1/2 | 12 | 26.58 | 23.42 |
| | | 16 | 23.33 | 21.50 |

*Data from the National Lumber Manufacturers Association.*

# MAXIMUM FLOOR JOIST SPANS

**Western Hemlock**, Max Load **40** lbs/sq ft, uniformly distributed

| Lumber Size (Inch) Nominal | Actual | Inches On Center | Max Feet Between Supports Unplastered | Plastered |
|---|---|---|---|---|
| 2 x 6 | 1–5/8 x 5–5/8 | 12 | 11.50 | 9.50 |
| | | 16 | 10.00 | 8.66 |
| 3 x 6 | 2–5/8 x 5–5/8 | 12 | 14.33 | 11.17 |
| | | 16 | 12.50 | 10.17 |
| 2 x 8 | 1–5/8 x 7–1/2 | 12 | 15.25 | 12.66 |
| | | 16 | 13.33 | 11.58 |
| 3 x 8 | 2–5/8 x 7–1/2 | 12 | 18.83 | 14.66 |
| | | 16 | 17.58 | 13.42 |
| 2 x 10 | 1–5/8 x 9–1/2 | 12 | 19.08 | 16.00 |
| | | 16 | 16.66 | 14.58 |
| 3 x 10 | 2–5/8 x 9–1/2 | 12 | 23.50 | 18.42 |
| | | 16 | 20.75 | 16.92 |
| 2 x 12 | 1–5/8 x 11–1/2 | 12 | 22.92 | 19.25 |
| | | 16 | 20.08 | 17.58 |
| 3 x 12 | 2–5/8 x 11–1/2 | 12 | 28.08 | 22.08 |
| | | 16 | 24.83 | 20.33 |
| 2 x 14 | 1–5/8 x 13–1/2 | 12 | 26.500 | 22.50 |
| | | 16 | 23.33 | 20.50 |
| 3 x 14 | 2–5/8 x 13–1/2 | 12 | – | 29.75 |
| | | 16 | – | 23.75 |

**Western Hemlock**, Max Load **75** lbs/sq ft, uniformly distributed

| Lumber Size (Inch) Nominal | Actual | Inches On Center | Max Feet Between Supports Unplastered | Plastered |
|---|---|---|---|---|
| 2 x 8 | 1–5/8 x 7–1/2 | 12 | 11.50 | 10.83 |
| | | 16 | 10.00 | 9.83 |
| 3 x 8 | 2–5/8 x 7–1/2 | 12 | 14.50 | 12.66 |
| | | 16 | 12.66 | 11.58 |
| 2 x 10 | 1–5/8 x 9–1/2 | 12 | 14.58 | 13.66 |
| | | 16 | 12.66 | 12.42 |
| 4 x 8 | 3–5/8 x 7–1/2 | 12 | 16.92 | 14.00 |
| | | 16 | 14.83 | 12.75 |
| 3 x 10 | 2–5/8 x 9–1/2 | 12 | 18.25 | 15.92 |
| | | 16 | 15.92 | 14.58 |
| 2 x 12 | 1–5/8 x 11–1/2 | 12 | 17.50 | 16.42 |
| | | 16 | 15.17 | 15.08 |
| 4 x 10 | 3–5/8 x 9–1/2 | 12 | 21.17 | 17.58 |
| | | 16 | 18.58 | 16.17 |
| 3 x 12 | 2–5/8 x 11–1/2 | 12 | 21.92 | 19.17 |
| | | 16 | 19.25 | 17.58 |
| 3 x 14 | 2–5/8 x 13–1/2 | 12 | 25.25 | 22.25 |
| | | 16 | 22.33 | 20.50 |

*Data from the National Lumber Manufacturers Association.*

# MAXIMUM FLOOR JOIST SPANS

**Spruce**, Max Load **40** lbs/sq ft, uniformly distributed

| Lumber Size (Inch) Nominal | Actual | Inches On Center | Max Feet Between Supports Unplastered | Plastered |
|---|---|---|---|---|
| 2 x 6 | 1–5/8 x 5–5/8 | 12 | 10.92 | 9.08 |
| | | 16 | 9.50 | 8.25 |
| 3 x 6 | 2–5/8 x 5–5/8 | 12 | 13.66 | 10.50 |
| | | 16 | 12.00 | 9.66 |
| 2 x 8 | 1–5/8 x 7–1/2 | 12 | 14.50 | 12.00 |
| | | 16 | 12.66 | 11.00 |
| 3 x 8 | 2–5/8 x 7–1/2 | 12 | 17.92 | 13.92 |
| | | 16 | 15.75 | 12.75 |
| 2 x 10 | 1–5/8 x 9–1/2 | 12 | 18.25 | 15.17 |
| | | 16 | 15.92 | 13.83 |
| 3 x 10 | 2–5/8 x 9–1/2 | 12 | 22.42 | 17.50 |
| | | 16 | 19.75 | 16.08 |
| 2 x 12 | 1–5/8 x 11–1/2 | 12 | 21.83 | 18.25 |
| | | 16 | 19.25 | 16.75 |
| 3 x 12 | 2–5/8 x 11–1/2 | 12 | 25.42 | 20.92 |
| | | 16 | 22.42 | 19.33 |
| 2 x 14 | 1–5/8 x 13–1/2 | 12 | 25.25 | 21.17 |
| | | 16 | 22.25 | 19.50 |
| 3 x 14 | 2–5/8 x 13–1/2 | 12 | 30.00 | 24.42 |
| | | 16 | 27.50 | 22.50 |

**Spruce**, Max Load **75** lbs/sq ft, uniformly distributed

| Lumber Size (Inch) Nominal | Actual | Inches On Center | Max Feet Between Supports Unplastered | Plastered |
|---|---|---|---|---|
| 2 x 8 | 1–5/8 x 7–1/2 | 12 | 11.00 | 10.25 |
| | | 16 | 9.58 | 9.42 |
| 3 x 8 | 2–5/8 x 7–1/2 | 12 | 13.83 | 12.08 |
| | | 16 | 12.42 | 11.00 |
| 2 x 10 | 1–5/8 x 9–1/2 | 12 | 13.92 | 12.92 |
| | | 16 | 12.08 | 11.83 |
| 4 x 8 | 3–5/8 x 7–1/2 | 12 | 16.17 | 13.25 |
| | | 16 | 14.08 | 12.17 |
| 3 x 10 | 2–5/8 x 9–1/2 | 12 | 17.42 | 15.17 |
| | | 16 | 15.25 | 13.75 |
| 2 x 12 | 1–5/8 x 11–1/2 | 12 | 19.75 | 15.66 |
| | | 16 | 14.58 | 14.25 |
| 4 x 10 | 3–5/8 x 9–1/2 | 12 | 20.08 | 16.66 |
| | | 16 | 17.75 | 15.25 |
| 3 x 12 | 2–5/8 x 11–1/2 | 12 | 20.66 | 18.17 |
| | | 16 | 18.33 | 16.66 |
| 3 x 14 | 2–5/8 x 13–1/2 | 12 | 24.25 | 21.25 |
| | | 16 | 21.33 | 19.50 |

*Data from the National Lumber Manufacturers Association.*

# STRENGTH OF WOOD BEAMS [1]

| Wood | Stress in Pounds per Square Inch (PSI) | | | | | |
|---|---|---|---|---|---|---|
| | Bending Horizontal Shear | | Compression Perpendicular to Grain | | Compression Parallel to Grain | |
| | Dry | Wet | Dry | Wet | Dry | Wet |
| Cedar, red | 900 | 800 | 200 | 150 | 700 | 700 |
| Cedar, white | 750 | 650 | 175 | 140 | 550 | 500 |
| Chestnut | 950 | 850 | 300 | 200 | 800 | 700 |
| Cypress | 1300 | 1100 | 350 | 250 | 1100 | 1100 |
| Fir, Balsum | 900 | 750 | 150 | 125 | 700 | 600 |
| Fir, Douglas #1 | 1600 | 1400 | 350 | 250 | 1200 | 1100 |
| Fir, Douglas #2 | 1300 | 1100 | 300 | 225 | 1000 | 900 |
| Gum, red | 1100 | 900 | 300 | 200 | 800 | 750 |
| Hemlock, Eastern | 1000 | 900 | 300 | 225 | 700 | 700 |
| Hemlock, Western | 1300 | 1100 | 300 | 225 | 900 | 900 |
| Hickory | 1900 | 1500 | 600 | 400 | 1500 | 1200 |
| Maple, silver | 1000 | 900 | 250 | 250 | 800 | 700 |
| Maple, sugar | 1500 | 1300 | 500 | 375 | 1200 | 1100 |
| Oak, Red | 1400 | 1200 | 500 | 375 | 1000 | 900 |
| Pine, Eastern | 900 | 800 | 250 | 150 | 750 | 750 |
| Pine, Norway | 1100 | 1000 | 250 | 175 | 800 | 800 |
| Pine, Southern | 1450 | 1250 | 325 | 235 | 1100 | 1000 |
| Pine, Western | 900 | 800 | 250 | 150 | 750 | 750 |
| Redwood | 1200 | 1000 | 250 | 150 | 1000 | 900 |
| Spruce, red | 1100 | 900 | 250 | 150 | 800 | 750 |

(1) *U.S. Government test*

Note from the above table, that all strength ratings of wood decrease dramatically when the wood is wet!

The above table should only be used as a general guide and that the values shown will vary greatly depending on the actual field conditions, moisture content (see the Wood Moisture Content page in this chapter), etc. For specific information on the wood you are using, contact the supplier of you wood and ask for items such as the exact grading specifications, moisture content, and stress specifications (if available).

*Values for the above table were obtained from the Machinists and Draftsmen's Handbook by A.M. Wagener and H.R. Arthur. 1946. D. Van Nostrand Company. New York. page 423.*

# WOOD GLUING CHARACTERISTICS

Wood gluing is a very common practice today, but there are a large number of glue types from which to choose and each of the different types of wood have different gluing properties. See the chapter on GLUE, page 235, for specific information on each of the common glue types.

The following 4 groups define the relative difficulty with which various woods can be glued:

- **Easy: Works with many different types of glues and under many gluing conditions.** Aspen, Western Red Cedar, Chestnut, Cottonwood, Cypress, White Fir, Larch, Redwood, Spruce, Willow, Yellow Poplar.
- **Moderate: More restricted gluing conditions than the Easy category. Different types of glue work fine.** Red Alder, Basswood, Butternut, Eastern Red Cedar, Douglas Fir, American and Rock Elm, Hackberry, Western Hemlock, Magnolia, Mahogany, Pine, Sweet Gum.
- **Difficult: Well controlled gluing conditions are required but still works with many different glue types.** White Ash, Alaskan Cedar, Cherry, Dogwood, Silver Maple (soft), Red and White Oak, Pecan, Sycamore, Black and Water Typelo, Black Walnut.
- **Very Difficult: Requires special glues and very close control of gluing conditions.** American Beech, Sweet and Yellow Birch, Hickory, Sugar Maple (hard), Osage–orange, Persimmon.

"Gluing conditions" is a function of proper sanding, letting surfaces to be glued become tacky before joining, using clamps to hold glue positions, and drying in a warm, dry area. Heat lamps will sometimes aid in the drying process

In addition to the above, the following generalities are also true:

- Hardwoods are more difficult to glue than softwoods.
- Heartwood is more difficult to glue than sapwood.
- Heavy woods are more difficult to glue than lightweight woods.

# CONCRETE

Concrete is a mixture of aggregate (typically sand and gravel), Portland cement, and water. Characteristics of each of these components are as follows:

**Aggregate**: A mixture of sand and gravel ranging in size from dust to 2–1/2 inches. Rounded fragments are generally better and do not use fragments larger than 1/4 the thickness of the concrete unit you are pouring (e.g. for a 4 inch slab, don't use greater than 1 inch gravel). The larger the gravel the more cost effective the concrete and there will be less problems from shrinkage.

**Portland Cement:** Cement comes in 1 cubic foot bags that weigh 94 lbs. It can also be purchased in bulk trailer loads. There are 5 basic types of cement:

Type I: The most common type sold by building suppliers.
Type II: A "sulfate resistant" variety used in bridges & pilings.
Type III: Quick hardening, used for rush jobs and winter use.
Type IV: Slow hardening, low heat for large structures.
Type V: Very high "sulfate resistance". (near water)

**Water:** Use clean, impurity free water, not muddy water.

**Air:** A fourth component of some concrete is millions of tiny air bubbles entrained in the mixture. This component helps the concrete withstand the effects of freezing and thawing and also makes the concrete lighter. Machine mixing is a must.

The strength of concrete increases when:

1. The amount of cement in the mixture increases.
2. The amount of water relative to cement decreases.
3. The density of the concrete is higher.
4. The aggregate is coarser.

The most common problems encountered in making concrete are adding too much water or sand, and poor mixing.

Other factors affecting the quality of the finished product include mixing and curing. Thorough mixing of the concrete is absolutely necessary in order to produce the strongest, most durable pour. Curing of concrete is necessary in order for the material to harden properly. The concrete must be kept moist for a period of 7 days and the temperature must not drop below 50°F. Although after 28 days, there is normally very little increase in the strength of concrete, most concrete does not completely cure for years.

## Typical Concrete Mixtures by Volume

| Cement:Sand:Gravel Ratio | Application |
|---|---|
| 1:3:6 | Normal static loads, no rebar; not exposed |
| 1:2.5:5 | Normal foundations & walls; exposed |
| 1:2.5:4 | Basement walls |
| 1:2.5:3.5 | Waterproof basement walls |
| 1:2.5:3 | Floors (light duty), driveways |
| 1:2.25:3 | Steps, driveways, sidewalks |
| 1:2:4 | Lintels |
| 1:2:4 | Reinforced roads, buildings, walls; exposed |
| 1:2:3.5 | Retaining walls, drive ways |
| 1:2:3 | Swimming pools, fence posts |
| 1:1.75:4 | Floors (light duty) |
| 1:1.5:3 | Watertight, reinforced tanks & columns |
| 1:1:2 | High strength columns, girders, floors |
| 1:1:1.5 | Fence posts |

When mixing concrete, mix the sand and cement first until a uniform color is obtained then mix in the aggregate. Adding the correct amount of water is a difficult task. In the above table, the portion for water is about 1/2 but this will vary depending on whether the sand is dry, damp, or wet (the 1/2 ratio component is equal to about 6 gallons of water per sack of cement). Simply remember that you only want to add enough water to make the concrete mixture workable and that the less water in relation to cement, the stronger the final concrete will be.

The strength of concrete can also be increased by compacting or working the mixture into place. This is accomplished by walking in the wet mixture or tamping or vibrating. If vibrators are used be careful that you do not cause segregation of the aggregate.

## Recommended Thickness of Slabs

| Thickness (inches) | Application |
|---|---|
| 4 | Home basement floors, farm building floors |
| 4 to 5 | Home garage floors, porches |
| 5 to 6 | Sidewalks, barn and granary floors, small shed floors |
| 6 to 8 | Driveways |

## Calculating Cubic Volumes

Concrete and mortar are normally sold and used on a cubic volume basis (either cubic feet or cubic yards). Use the following to calculate the amount of concrete you need for a slab:

Cubic feet of Concrete = Slab thickness in feet x Slab width in feet x Slab length in feet

1 cubic yard = 27 cubic feet
1 cubic foot = 1,728 cubic inches

In using the above equations, note that the volume of the final concrete mixture is approximately 2/3's the volume of the original cement–aggregate mixture. This occurs because the sand and cement fill in the void spaces between the gravel fragments.

## Standard Steel Reinforcing Bar (re–bar)

| Bar # Number | Diameter (inches) | Pounds per foot |
|---|---|---|
| 2b | 1/4 | 0.17 |
| 3 | 3/8 | 0.38 |
| 4 | 1/2 | 0.67 |
| 5 | 5/8 | 1.04 |
| 6 | 3/4 | 1.50 |
| 7 | 7/8 | 2.04 |
| 8 | 1 | 2.67 |
| 9 | 1–1/8 | 3.40 |
| 10 | 1–1/4 | 4.30 |
| 11 | 1.41 | 5.31 |

## Coloring Concrete and Mortar

| Color | Color material | lbs / sack Cement |
|---|---|---|
| Black | Black oxide or mineral Black | 1 to 12 |
| Blue | Ultramarine Blue | 5 to 9 |
| Brown–Red | Red iron oxide | 5 to 9 |
| Bright Red | Mineral Turkey Red | 5 to 9 |
| Purple–Red | Indian red | 5 to 9 |
| Brown | Metallic Brown Oxide | 5 to 9 |
| Buff to yellow | Yellow ocher or yellow oxide | 2 to 9 |
| Green | Chromium oxide or ultramarine | 5 to 9 |

# MORTAR

Mortar is composed of basically the same material as concrete, except that its composition has been altered to increase the ease of workability and decrease the setting time of the mixture. As with concrete, the strength of mortar is a function of the proportions of its ingredients.

Mortar is a mixture of Portland cement, hydrated lime, sand (well graded and in a size range of 1/8 inch to 100 mesh) and water. Masonry cement, which already contains the hydrated lime, can be used instead of Portland cement.

The strength of mortar increases when:

1. The amount of cement in the mixture increases.
2. The amount of water relative to cement decreases. Unfortunately, there is no rule for the amount of water since it is a function of workability. Just use as little as as possible.
3. The amount of hydrated lime decreases.
4. The amount of Portland cement relative to masonry cement increases.
5. Brick with low water absorption is used. If brick absorbs water readily, the bricks must be wetted before mortaring.
6. Clean sand is used. Organic matter and salts in the sand will drastically decrease the mortar strength. A higher percentage of coarse to fine sand increases strength.
7. Special epoxies are available that can be mixed with the mortar. These increase both the strength and bonding power of the mortar and in many cases will create mortar that is stronger that the brick. Note that this adds to the cost of the mortar.

**The workability of mortar increases when:**

1. The amount of hydrated lime in the mixture increases.
2. The amount of water increases.

As with concrete, thorough mixing of the mortar is imperative. A power mixer is best, but small quantities can be mixed by hand. Once the mortar has been mixed, it will begin to cure and stiffen. If the mortar begins to stiffen within 2 to 2.5 hours of mixing (above 80°F outside air temperature) you can add a small amount of water to increase workability. After the 2.5 hour time limit, the mortar should be thrown away and a new batch mixed. If the outside air temperatures are below 80°F, the 2.5 hour time limit can be increased to approximately 3.5 hours.

# MORTAR

Type numbers are used to define the various mortar mixes. The following are the four common types with their volume proportions of Portland cement (see also concrete section for different types of Portland), masonry cement (Type II unless otherwise stated), lime, and sand:

Type M: General use for foundations, walls, sidewalks and other situations in contact with the ground or below grade. 28 day compression strength 4900–5400 psi, depending on amount of water used.
Portland Mix: 1 Portland; 1/4 hydrated lime; 3 sand
Masonry Mix: 1 Portland; 1 masonry, 6 sand

**Type S:** General use for high resistance to sideways or lateral stress. 28 day compression strength 2100–2800 psi, depending on amount of water used.
Portland Mix: 1 Portland; 1/2 hydrated lime; 4.5 sand.
Masonry Mix: 1/2 Portland; 1 masonry; 4.5 sand

**Type N:** General use above grade for severe exposure walls. 28 day compression strength 800–1200 psi, depending on amount of water used.
Portland Mix: 1 Portland; 1 hydrated lime, 6 sand
Masonry Mix: 1 masonry; 3 sand

**Type O:** Low strength load bearing walls where excessive moisture and freezing are not present. Compression strength must be below 100 psi.
Portland Mix: 1 Portland; 2 hydrated lime; 9 sand
Masonry Mix: 1 masonry (Types I or II); 3 sand

If you need a small amount of general use mortar, the following will make about 1 cubic foot: 16 lbs Portland cement, 8.5 lbs hydrated lime, 100 lbs dry sand, and 2 to 3 gallons of water.

The amount of mortar required for a job varies tremendously, but the following average quantities may be helpful:

**Mortar required for Common Brick (8 in x 3–3/4 in x 2–1/4 in) assuming 20 bricks per cubic foot:**

| Joint Thickness Inches | Cu ft Mortar/ 1000 brick | Cu ft mortar/ Cu ft brick |
|---|---|---|
| 1/4 | 9 | 0.2 |
| 3/8 | 14 | 0.3 |
| 1/2 | 20 | 0.4 |

With practice, you can lay 90–120 common bricks/hour.

# POCKET REF

# Chemistry & Physics



(See Oxide Conversions in GEOLOGY, page 212)
(See also WEIGHTS OF MATERIALS, page 389)

# ELEMENT TABLES

| Element Name | Symbol | Atomic Number | Atomic Weight | Valence | Redox Potential |
|---|---|---|---|---|---|
| Actinium | Ac | 89 | (227) | 3 | |
| Aluminum | Al | 13 | 26.9815 | 3 | $Al^{3+} = +1.66$ |
| Americium | Am | 95 | (243) | 3,4,5,6 | $Am^{3+} = +2.32$ |
| Antimony | Sb | 51 | 121.75 | 3,5 | |
| Argon | Ar | 18 | 39.948 | 0 | |
| Arsenic | As | 33 | 74.9216 | 3,5 | |
| Astatine | At | 85 | (210) | 1,3,5,7 | |
| Barium | Ba | 56 | 137.34 | 2 | $Ba^{2+} = +2.90$ |
| Berkelium | Bk | 97 | (247) | 3,4 | |
| Beryllium | Be | 4 | 9.0122 | 2 | $Be^{2+} = +1.85$ |
| Bismuth | Bi | 83 | 208.980 | 3,5 | |
| Boron | B | 5 | 10.811 | 3 | |
| Bromine | Br | 35 | 79.904 | 1,3,5,7 | $2Br^-, Br_2 = -1.066$ |
| Cadmium | Cd | 48 | 112.40 | 2 | $Cd^{2+} = +0.40$ |
| Calcium | Ca | 20 | 40.08 | 2 | $Ca^{2+} = +2.87$ |
| Californium | Cf | 98 | (252) | | |
| Carbon | C | 6 | 12.01115 | 2,4 | |
| Cassiopeium | Cp | see Lutetium | | | |
| Cerium | Ce | 58 | 140.12 | 3,4 | $Ce^{3+} = +2.48$ |
| Cesium | Cs | 55 | 132.905 | 1 | $Cs = +2.92$ |
| Chlorine | Cl | 17 | 35.453 | 1,3,5,7 | $2Cl^-, Cl_2 = -1.36$ |
| Chromium | Cr | 24 | 51.996 | 2,3,6 | $Cr^{3+} = +0.74$ |
| Cobalt | Co | 27 | 58.9332 | 2,3 | $Co^{2+} = +0.28$ |
| Columbium | Cb | see Niobium | | | |
| Copper | Cu | 29 | 63.546 | 1,2 | $Cu^{2+} = -0.34$ |
| Curium | Cm | 96 | (247) | 3 | |
| Dysprosium | Dy | 66 | 162.50 | 3 | |
| Einsteinium | Es | 99 | (254) | | |
| Emanation | Em | see Radon | | | |
| Erbium | Er | 68 | 167.26 | 3 | |
| Europium | Eu | 63 | 151.96 | 2,3 | |
| Fermium | Fm | 100 | (257) | | |
| Fluorine | F | 9 | 18.9984 | 1 | $2F^- F_2 = -2.85$ |
| Francium | Fr | 87 | (223) | 1 | |
| Gadolinium | Gd | 64 | 157.25 | 3 | $Gd^{3+} = +2.4$ |
| Gallium | Ga | 31 | 69.72 | 2,3 | $Ga^{3+} = +0.53$ |
| Germanium | Ge | 32 | 72.59 | 4 | |
| Glucinium | Gl | see Beryllium | | | |
| Gold | Au | 79 | 196.967 | 1,3 | $Au^{3+} = -1.50$ |
| Hafnium | Hf | 72 | 178.49 | 4 | $Hf^{4+} = +1.70$ |

| Element Name | Symbol | Atomic Number | Atomic Weight | Valence | Redox Potential |
|---|---|---|---|---|---|
| Hahnium | Ha | 105 | 262 | | |
| Helium | He | 2 | 4.0026 | 0 | |
| Holmium | Ho | 67 | 164.930 | 3 | |
| Hydrogen | H | 1 | 1.00797 | 1 | $H^+ = +2.10$ |
| Illinium | Il | see Promethium | | | |
| Indium | In | 49 | 114.82 | 3 | $In^{3+} = +0.34$ |
| Iodine | I | 53 | 126.9044 | 1,3,5,7 | $2I^-, I2 = -0.54$ |
| Iridium | Ir | 77 | 192.2 | 3,4 | |
| Iron | Fe | 26 | 55.847 | 2,3 | $Fe^{2+} = +0.44$ |
| Krypton | Kr | 36 | 83.80 | 0 | |
| Lanthanum | La | 57 | 138.91 | 3 | $La^{3+} = +2.52$ |
| Lawrencium | Lw/Lr | 103 | (256) | | |
| Lead | Pb | 82 | 207.19 | 2,4 | $Pb^{2+} = +0.13$ |
| Lithium | Li | 3 | 6.939 | 1 | $Li^+ = +3.04$ |
| Lutetium | Lu | 71 | 174.97 | 3 | $Lu^{3+} = +2.25$ |
| Magnesium | Mg | 12 | 24.305 | 2 | $Mg^{2+} = +2.37$ |
| Manganese | Mn | 25 | 54.9380 | 2,3,4,6,7 | $Mn^{2+} = +1.18$ |
| Mendelevium | Md | 101 | (257) | | |
| Mercury | Hg | 80 | 200.59 | 1,2 | $2Hg, 2+ = -0.79$ |
| Molybdenum | Mo | 42 | 95.94 | 3,4,6 | $Mo^{3+} = +0.20$ |
| Neodymium | Nd | 60 | 144.24 | 3 | $Nd^{3+} = +2.44$ |
| Neon | Ne | 10 | 20.179 | 0 | |
| Neptunium | Np | 93 | (237) | 4,5,6 | $Np^{3+} = +1.86$ |
| Nickel | Ni | 28 | 58.71 | 2,3 | $Ni^{2+} = +0.25$ |
| Niobium | Nb | 41 | 92.906 | 3,5 | $Nb^{3+} = +1.1$ |
| Niton | Nt | see Radon | | | |
| Nitrogen | N | 7 | 14.0067 | 3,5 | |
| Nobelium | No | 102 | (255) | | |
| Osmium | Os | 76 | 190.2 | 2,3,4,8 | |
| Oxygen | O | 8 | 15.9994 | 2 | |
| Palladium | Pd | 46 | 106.4 | 2,4,6 | $Pd^{2+} = -0.99$ |
| Phosphorus | P | 15 | 30.9738 | 3,5 | |
| Platinum | Pt | 78 | 195.09 | 2,4 | $Pt^{2+} = -1.2$ |
| Plutonium | Pu | 94 | (244) | 3,4,5,6 | $Pu^{3+} = +2.07$ |
| Polonium | Po | 84 | (210) | | |
| Potassium | K | 19 | 39.102 | 1 | $K^+ = +2.925$ |
| Praseodymium | Pr | 59 | 140.907 | 3 | |
| Promethium | Pm | 61 | (147) | 3 | |
| Protactinium | Pa | 91 | (231) | | |
| Radium | Ra | 88 | (226) | 2 | $Ra^{2+} = +2.92$ |
| Radon | Rn | 86 | (222) | 0 | |

# ELEMENT TABLES

| Element Name | Symbol | Atomic Number | Atomic Weight | Valence | Redox Potential |
|---|---|---|---|---|---|
| Rhenium | Re | 75 | 186.2 | | |
| Rhodium | Rh | 45 | 102.905 | 3 | $Rh^{3+}$ = −0.8 |
| Rubidium | Rb | 37 | 85.47 | 1 | $Rb^{+}$ = +2.92 |
| Ruthenium | Ru | 44 | 101.07 | 3,4,6,8 | |
| Rutherfordium | Rf | 104 | 261 | | |
| Samarium | Sm | 62 | 150.35 | 2,3 | $Sm^{3+}$ = +2.41 |
| Scandium | Sc | 21 | 44.956 | 3 | $Sc^{3+}$ = +2.08 |
| Selenium | Se | 34 | 78.96 | 2,4,6 | $Se^{2-}$-Se = +0.78 |
| Silicon | Si | 14 | 28.086 | 4 | |
| Silver | Ag | 47 | 107.868 | 1 | $Ag^{+}$ = +0.799 |
| Sodium | Na | 11 | 22.9898 | 1 | $Na^{+}$ = +2.71 |
| Strontium | Sr | 38 | 87.62 | 2 | $Sr^{2+}$ = +2.89 |
| Sulphur | S | 16 | 32.064 | 2,4,6 | $S^{2-}$,S = +0.92 |
| Tantalum | Ta | 73 | 180.948 | 5 | |
| Technetium | Tc | 43 | (99) | 6,7 | |
| Tellurium | Te | 52 | 127.60 | 2,4,6 | $Te^{2+}$ = +0.51 |
| Terbium | Tb | 65 | 158.924 | 3 | |
| Thallium | Tl | 81 | 204.37 | 1,3 | $Tl^{+}$ = +0.34 |
| Thorium | Th | 90 | 232.038 | 4 | $Th^{4+}$ = +1.9 |
| Thulium | Tm | 69 | 168.934 | 3 | |
| Tin | Sn | 50 | 118.69 | 2,4 | $Sn^{2+}$ = +0.14 |
| Titanium | Ti | 22 | 47.90 | 3,4 | $Ti^{2+}$ = +1.63 |
| Tungsten | W | 74 | 183.85 | 6 | |
| Uranium | U | 92 | 238.03 | 4,6 | $U^{3+}$ = +1.80 |
| Vanadium | V | 23 | 50.942 | 3,5 | $V^{2+}$ = +1.18 |
| Wolfram | W | see Tungsten | | | |
| Xenon | Xe | 54 | 131.30 | 0 | |
| Ytterbium | Yb | 70 | 173.04 | 2,3 | |
| Yttrium | Y | 39 | 88.905 | 3 | |
| Zinc | Zn | 30 | 65.37 | 2 | Zn = +0.76 |
| Zirconium | Zr | 40 | 91.22 | 4 | $Zr^{4+}$ = +1.53 |

Table of atomic weights based on carbon isotope $C^{12}$. Numbers in parenthesis represent the mass (Atomic Weight) of the most stable isotope. "Redox Potential" lists the end product (or beginning and end product if two are given) and electrode potential.

# ELEMENT PROPERTIES

| Element Name | State | Atomic Number | Density gm/cc | Melting Point °C | Boiling Point °C |
|---|---|---|---|---|---|
| Actinium | | 89 | 10.07 | 1050 | 3200 |
| Aluminum | | 13 | 2.6989 | 660 | 2467 |
| Americium | | 95 | 11.7 | 850? | |
| Antimony | | 51 | 6.69 | 630 | 1380 |
| Argon | gas | 18 | 1.78 | −189.2 | −185.7 |
| Argon | liquid | 18 | 1.402(−185°) | −189.2 | −185.7 |
| Argon | solid | 18 | 1.65(−223°) | −189.2 | −185.7 |
| Arsenic | xtal | 33 | 5.73 | 817 | |
| Arsenic | yellow | 33 | 1.97 | 817 | |
| Astatine | | 85 | | | |
| Barium | | 56 | 3.5 | 725 | 1140 |
| Berkelium | | 97 | | | |
| Beryllium | | 4 | 1.848 | 1278 | 2970 |
| Bismuth | | 83 | 9.747 | 271 | 1560 |
| Boron | | 5 | 2.34 | 2300 | 2550 |
| Bromine | gas | 35 | 7.59 | −7.2 | 58.78 |
| Bromine | liquid | 35 | 3.12 | −7.2 | 58.78 |
| Cadmium | | 48 | 8.65 | 321 | 765 |
| Calcium | | 20 | 1.55 | 842 | 1487 |
| Californium | | 98 | | | |
| Carbon | amorph. | 6 | 1.8-2.1 | 3550 | 4827 |
| Carbon | graphite | 6 | 1.9-2.3 | 3550 | 4827 |
| Carbon | diamond | 6 | 3.15-3.53 | 3550 | 4827 |
| Cerium | | 58 | 6.66 | 795 | 3468 |
| Cesium | | 55 | 1.87 | 28.5 | 690 |
| Chlorine | gas | 17 | 3.214 | −100.98 | −34.6 |
| Chlorine | liquid | 17 | 1.56(−33°) | −100.98 | −34.6 |
| Chromium | | 24 | 7.18 | 1890 | 2482 |
| Cobalt | | 27 | 8.9 | 1492 | 2900 |
| Copper | | 29 | 8.96 | 1083 | 2595 |
| Curium | | 96 | 7? | | |
| Dysprosium | | 66 | 8.536 | 1407 | 2600 |
| Einsteinium | | 99 | | | |
| Erbium | | 68 | 9.051 | 1497 | 2900 |
| Europium | | 63 | 5.259 | 826 | 1439 |
| Fermium | | 100 | | | |
| Fluorine | gas | 9 | 1.696 | −219.6 | −188.1 |
| Fluorine | liquid | 9 | 1.108(−188°) | −219.6 | −188.1 |
| Francium | | 87 | | | |
| Gadolinium | | 64 | 7.89 | 1312 | 3000 |
| Gallium | | 31 | 5.9 | 29.8 | 2403 |
| Germanium | | 32 | 5.32(25°) | 937 | 2830 |
| Gold | | 79 | 19.32 | 1063.0 | 2966 |
| Hafnium | | 72 | 13.29 | 2150 | 5400 |

# ELEMENT PROPERTIES

| Element Name | State | Atomic Number | Density gm/cc | Melting Point °C | Boiling Point °C |
|---|---|---|---|---|---|
| Helium | gas | 2 | 0.177 | −272.2 | −268.6 |
| Holmium | | 67 | 8.803 | 1461 | 2600 |
| Hydrogen | gas | 1 | 0.08988 | −259.14 | −252.5 |
| Hydrogen | liquid | 1 | 0.070(−252°) | −259.14 | −252.5 |
| Indium | | 49 | 7.31 | 156.6 | 2000 |
| Iodine | gas | 53 | 11.27 | 113.5 | 184.3 |
| Iodine | | 53 | 4.93 | 113.5 | 184.3 |
| Iridium | | 77 | 22.42 | 2443 | 4527 |
| Iron | | 26 | 7.87 | 1535 | 3000 |
| Krypton | gas | 36 | 3.733 | −156.6 | −152.3 |
| Lanthanum | | 57 | 5.98 | 920 | 3469 |
| Lawrencium | | 103 | | | |
| Lead | | 82 | 11.35 | 327 | 1744 |
| Lithium | | 3 | 0.534 | 179 | 1317 |
| Lithium | liquid | 3 | 0.507(200°) | 179 | 1317 |
| Lutetium | | 71 | 9.842 | 1652 | 3327 |
| Magnesium | | 12 | 1.738 | 651 | 1107 |
| Manganese | | 25 | 7.21 | 1244 | 2097 |
| Mendelevium | | 101 | | | |
| Mercury | | 80 | 13.54 | −38.87 | 356.58 |
| Molybdenum | | 42 | 10.22 | 2610 | 5560 |
| Neodymium | | 60 | 7.004 | 1024 | 3027 |
| Neon | gas | 10 | 0.8999 | −248.7 | −245.9 |
| Neon | liquid | 10 | 1.207(−245°) | −248.7 | −245.9 |
| Neptunium | | 93 | 18-20 | 640 | |
| Nickel | | 28 | 8.9(25°) | 1453 | 2732 |
| Niobium | | 41 | 8.57 | 2468 | 4927 |
| Nitrogen | gas | 7 | 1.2506 | −209.86 | −195.8 |
| Nitrogen | liquid | 7 | 0.808(−195°) | −209.86 | −195.8 |
| Nitrogen | solid | 7 | 1.026(−252°) | −209.86 | −195.8 |
| Nobelium | | 102 | | | |
| Osmium | | 76 | 22.57 | 3000 | 5000 |
| Oxygen | gas | 8 | 1.429 | −218.4 | −182.9 |
| Oxygen | liquid | 8 | 1.14(−182°) | −218.4 | −182.9 |
| Palladium | | 46 | 12.02 | 1552 | 2927 |
| Phosphorus | yellow | 15 | 1.82 | 44.1 | 280 |
| Phosphorus | red | 15 | 2.20 | 44.1 | 280 |
| Phosphorus | black | 15 | 2.25-2.69 | 44.1 | 280 |
| Platinum | | 78 | 21.45 | 1769 | 3827 |
| Plutonium | | 94 | 19.84(25°) | 639.5 | 3235 |
| Polonium | | 84 | 9.32 | 254 | 962 |
| Potassium | | 19 | 0.862 | 63.6 | 754 |
| Praseodymium | | 59 | 6.78 | 935 | 3127 |
| Promethium | | 61 | | 1035 | 2730 |

# ELEMENT PROPERTIES

| Element Name | State | Atomic Number | Density gm/cc | Melting Point °C | Boiling Point °C |
|---|---|---|---|---|---|
| Protactinium | | 91 | 15.37 | 1230? | |
| Radium | | 88 | 5? | 700 | 1737 |
| Radon | gas | 86 | 9.73 | −71 | −61.8 |
| Radon | liquid | 86 | 4.4(−62°) | −71 | −61.8 |
| Radon | solid | 86 | 4 | −71 | −61.8 |
| Rhenium | | 75 | 21.02 | 3180 | 5627 |
| Rhodium | | 45 | 12.41 | 1960 | 3727 |
| Rubidium | liquid | 37 | 1.475(39°) | 38.89 | 688 |
| Rubidium | solid | 37 | 1.532 | 38.89 | 688 |
| Ruthenium | | 44 | 12.41 | 2250 | 3900 |
| Samarium | | 62 | 7.54 | 1072 | 1900 |
| Scandium | | 21 | 2.992 | 1539 | 2727 |
| Selenium | | 34 | 4.79(25°) | 217 | 685 |
| Silicon | | 14 | 2.33 | 1410 | 2355 |
| Silver | | 47 | 10.5 | 960 | 2112 |
| Sodium | | 11 | 0.971 | 97.8 | 892 |
| Strontium | | 38 | 2.54 | 769 | 1384 |
| Sulphur | | 16 | 2.07 | 112.8 | 444.6 |
| Tantalum | | 73 | 16.6 | 2996 | 5425 |
| Technetium | | 43 | 11.5 | 2200 | |
| Tellurium | | 52 | 6.24 | 449 | 989 |
| Terbium | | 65 | 8.272 | 1356 | 2800 |
| Thallium | | 81 | 11.85 | 303 | 1457 |
| Thorium | | 90 | 11.66 | 1700 | 4000 |
| Thulium | | 69 | 9.332 | 1545 | 1727 |
| Tin | | 50 | 7.31 | 231.9 | 2270 |
| Titanium | | 22 | 4.54 | 1675 | 3260 |
| Tungsten | | 74 | 19.3 | 3380 | 5927 |
| Uranium | | 92 | 18.95 | 1132 | 3818 |
| Vanadium | | 23 | 6.11(18°) | 1890 | 3000 |
| Xenon | gas | 54 | 5.88 | −112 | −107 |
| Xenon | liquid | 54 | 3.52 | −112 | −107 |
| Ytterbium | | 70 | 6.977 | 824 | 1427 |
| Yttrium | | 39 | 4.45 | 1495 | 2927 |
| Zinc | | 30 | 7.13 | 419.5 | 907 |
| Zirconium | | 40 | 6.53 | 1852 | 3578 |

The elements above are in a "solid" state unless otherwise noted. Density of gases are measured at 760mm Hg and 0°C and other elements are measured at 20°C (unless otherwise indicated).

| IA | IIA | IIIB | IVB | VB | VIB | VIIB | VIII | VIII | VIII | IB | IIB | IIIB | IVA | VA | VIA | VIIA | VIIIA |
|---|---|---|---|---|---|---|---|---|---|---|---|---|---|---|---|---|---|
| 1 H | | | | | | | | | | | | | | | | | 2 He |
| 3 Li | 4 Be | | | | | | | | | | | 5 B | 6 C | 7 N | 8 O | 9 F | 10 Ne |
| 11 Na | 12 Mg | | | | | | | | | | | 13 Al | 14 Si | 15 P | 16 S | 17 Cl | 18 Ar |
| 19 K | 20 Ca | 21 Sc | 22 Ti | 23 V | 24 Cr | 25 Mn | 26 Fe | 27 Co | 28 Ni | 29 Cu | 30 Zn | 31 Ga | 32 Ge | 33 As | 34 Se | 35 Br | 36 Kr |
| 37 Rb | 38 Sr | 39 Y | 40 Zr | 41 Nb | 42 Mo | 43 Tc | 44 Ru | 45 Rh | 46 Pd | 47 Ag | 48 Cd | 49 In | 50 Sn | 51 Sb | 52 Te | 53 I | 54 Xe |
| 55 Cs | 56 Ba | 57 La | 72 Hf | 73 Ta | 74 W | 75 Re | 76 Os | 77 Ir | 78 Pt | 79 Au | 80 Hg | 81 Tl | 82 Pb | 83 Bi | 84 Po | 85 At | 86 Rn |
| 87 Fr | 88 Ra | 89 Ac | 104 Rf | 105 Ha | 106 | | | | | | | | | | | | |
| | | | | 58 Ce | 59 Pr | 60 Nd | 61 Pm | 62 Sm | 63 Eu | 64 Gd | 65 Tb | 66 Dy | 67 Ho | 68 Er | 69 Tm | 70 Yb | 71 Lu |
| | | | | 90 Th | 91 Pa | 92 U | 93 Np | 94 Pu | 95 Am | 96 Cm | 97 Bk | 98 Cf | 99 Es | 100 Fm | 101 Md | 102 No | 103 Lw |

# pH OF COMMON ACIDS

| Acids (pH < 7) | Molarity | pH |
|---|---|---|
| Acetic | N | 2.4 |
| Acetic | 0.1N | 2.9 |
| Acetic | 0.01N | 3.4 |
| Alum | 0.1N | 3.2 |
| Arsenious | Saturated | 5.0 |
| Benzoic | 0.1N | 3.0 |
| Boric | 0.1N | 5.3 |
| Carbonic | Saturated | 3.8 |
| Citric | 0.1N | 2.1 |
| Formic | 0.1N | 2.3 |
| Hydrochloric | N | 0.1 |
| Hydrochloric | 0.1N | 1.1 |
| Hydrochloric | 0.01N | 2.0 |
| Hydrocyanic | 0.1N | 5.1 |
| Hydrogen Sulfide | 0.1N | 4.1 |
| Lactic | 0.1N | 2.4 |
| Malic | 0.1N | 2.2 |
| Orthophosphoric | 0.1N | 1.5 |
| Oxalic | 0.1N | 1.3 |
| Succinic | 0.1N | 2.7 |
| Salicylic | Saturated | 2.4 |
| Sulfuric | N | 0.3 |
| Sulfuric | 0.1N | 1.2 |
| Sulfuric | 0.01N | 2.1 |
| Sulfurous | 0.1N | 1.5 |
| Tartaric | 0.1N | 2.0 |
| Trichloracetic | 0.1N | 1.2 |

# pH OF COMMON BASES

| Bases (pH > 7) | Molarity | pH |
|---|---|---|
| Ammonia | N | 11.6 |
| Ammonia | 0.1N | 11.1 |
| Ammonia | 0.01N | 10.6 |
| Barbital Sodium | 0.1N | 9.4 |
| Borax | 0.01N | 9.2 |
| Calcium Carbonate | Saturated | 9.4 |
| Calcium Hydroxide | Saturated | 12.4 |
| Ferrous Hydroxide | Saturated | 9.5 |
| Lime | Saturated | 12.4 |
| Magnesia | Saturated | 10.5 |
| Potassium Acetate | 0.1N | 9.7 |
| Potassium Bicarbonate | 0.1N | 8.2 |
| Potassium Carbonate | 0.1 | 11.5 |
| Potassium Cyanide | 0.1N | 11.0 |
| Potassium Hydroxide | N | 14.0 |
| Potassium Hydroxide | 0.1N | 13.0 |
| Potassium Hydroxide | 0.01N | 12.0 |
| Sodium Acetate | 0.1N | 8.9 |
| Sodium Benzoate | 0.1N | 8.0 |
| Sodium Bicarbonate | 0.1N | 8.4 |
| Sodium Carbonate | 0.1N | 11.6 |
| Sodium Hydroxide | N | 14.0 |
| Sodium Hydroxide | 0.1N | 13.0 |
| Sodium Hydroxide | 0.01N | 12.0 |
| Sodium Metasilicate | 0.1N | 12.6 |
| Sodium Sesquicarbonate | 0.1N | 10.1 |
| Trisodium Phosphate | 0.1N | 12.0 |

## pH INDICATORS

| Indicator | Acid Color | pH | Base Color |
|---|---|---|---|
| Cresol red #1 | red | 0.2-1.8 | yellow |
| Cresol purple #1 | red | 1.2-2.8 | yellow |
| Thymol blue | red | 1.2-2.8 | yellow |
| Metanil yellow | red | 1.2-2.3 | yellow |
| Tropaeolin LL | red | 1.4-3.2 | yellow |
| 2,6 Dinitrophenol | no color | 1.7-4.4 | yellow |
| Benzyl orange | red | 1.9-3.3 | yellow |
| 2,4 Dinitrophenol | no color | 2.0-4.7 | yellow |
| Benzo Yellow | red | 2.4-4.0 | yellow |
| p-Dimethylaminoazobenzene | red | 2.9-4.0 | yellow |
| Bromophenol blue | red | 3.0-4.6 | violet |
| Congo red | blue | 3.0-5.0 | red |
| Bromochlorophenol blue | yellow | 3.0-4.6 | purple |
| Methyl orange | red | 3.1-4.4 | yellow |
| Bromocresol green | yellow | 3.8-5.4 | blue |
| 2,5 Dinitrophenol | no color | 4.0-5.8 | yellow |
| Methyl red | red | 4.4-6.0 | yellow |
| Azolitmin (litmus) | red | 4.4-6.6 | blue |
| Propyl red | red | 4.6-6.6 | yellow |
| p-Nitrophenol | no color | 4.7-7.9 | yellow |
| Bromocresol purple | yellow | 4.8-6.8 | purple |
| Bromophenol red | yellow | 4.8-6.8 | purple |
| Chlorophenol red | yellow | 5.0-6.9 | purple |
| Bromothymol blue | yellow | 6.0-7.6 | blue |
| m-Nitrophenol | no color | 6.6-8.6 | yellow |
| Neutral red | red | 6.8-8.0 | yellow |
| Phenol red | yellow | 6.8-8.4 | red |
| Rosolic acid | brown | 6.9-8.0 | red |
| Cresol red #2 | yellow | 7.2-8.8 | purple |
| a-Naphtholphthalein | brown | 7.3-8.7 | green |
| Orange I | yellow | 7.6-8.9 | rose |
| m-Cresol purple #2 | yellow | 7.6-9.2 | purple |
| Thymol blue #2 | yellow | 8.0-9.6 | blue |
| o-Cresolphthalein | no color | 8.2-9.8 | red |
| Phenolphthalein | no color | 8.3-10.0 | red |
| Phthalein red | yellow | 8.6-10.2 | red |
| Thymolphthalein | no color | 9.3-10.5 | blue |
| Tolyl red | red | 10.0-11.6 | yellow |
| b-Naphthol violet | yellow | 10.0-12.0 | violet |
| Alizarin yellow R | yellow | 10.0-12.1 | brown |
| Alizarin yellow GG | yellow | 10.0-12.0 | orange |
| Nitramine | no color | 10.8-13.0 | brown |
| Parazo orange | yellow | 11.0-12.6 | orange |
| Poirrier blue | blue | 11.0-13.0 | red |
| Tropaeolin O | yellow | 11.1-12.7 | orange |
| Acyl blue | red | 12.0-13.6 | blue |

pH values are approximate values and have been rounded off to the nearest tenth. Values assume a temperature of 25°C (77°F).

# RADIOISOTOPE HALF LIVES

| Isotope | Half Life |
|---|---|
| $Ag^{110}$ | 252 days |
| $Am^{241}$ | 432 years |
| $Am^{243}$ | 7380 years |
| $Au^{198}$ | 2.69 days |
| $Ba^{137}$ | 11.2 years |
| $Ba^{140}$ | 12.79 days |
| $Be^{7}$ | 53.28 days |
| $Bi^{210}$ | 3.5 million years |
| $Br^{82}$ | 35.3 hours |
| $C^{14}$ | 5730 years |
| $Ca^{45}$ | 163 days |
| $Cd^{113}$ | 14.6 years |
| $Ce^{144}$ | 284.4 days |
| $Cl^{36}$ | 300,000 years |
| $Cm^{242}$ | 162.8 days |
| $Cm^{244}$ | 18.11 years |
| $Co^{60}$ | 5.27 years |
| $Cr^{51}$ | 27.71 days |
| $Cs^{134}$ | 2.06 years |
| $Cs^{137}$ | 30.17 years |
| $Cu^{64}$ | 12.71 hours |
| $Fe^{55}$ | 2.7 years |
| $Fe^{59}$ | 44.6 days |
| $Ga^{72}$ | 14.1 hrs |
| $Ge^{68}$ | 287 days |
| $H^{3}$ | 12.33 years |
| $Hg^{203}$ | 46.6 days |
| $I^{131}$ | 8.04 days |
| $In^{115}$ | 95.7 years |
| $Ir^{192}$ | 74.2 days |
| $K^{40}$ | 0.012 years |
| $K^{42}$ | 12.36 hrs |
| $Kr^{85}$ | 10.72 years |
| $La^{140}$ | 40.23 hours |
| $Mn^{54}$ | 312.5 days |
| $Mo^{99}$ | 66.02 hours |
| $Na^{22}$ | 2.6 years |
| $Na^{24}$ | 15.02 hours |
| $Ni^{63}$ | 100 years |
| $Np^{237}$ | 2.14 million yr |
| $P^{32}$ | 14.28 days |
| $Pb^{210}$ | 22.3 years |
| $Pd^{106}$ | 27.3 years |
| $Pd^{109}$ | 13.43 hours |
| $Pm^{147}$ | 2.62 years |
| $Po^{210}$ | 138.38 days |
| $Pr^{144}$ | 17.28 minute |
| $Pu^{238}$ | 87.74 years |
| $Rb^{86}$ | 18.65 days |
| $Rb^{87}$ | 27.83 years |
| $Rh^{103}$ | 100 years |
| $Rh^{106}$ | 29.9 seconds |
| $Ru^{106}$ | 368 days |
| $S^{35}$ | 87.2 days |
| $Sb^{125}$ | 2.73 years |
| $Sc^{46}$ | 83.8 days |
| $Se^{75}$ | 120 days |
| $Si^{32}$ | 280 years |
| $Sn^{113}$ | 115 days |
| $Sr^{89}$ | 50.52 days |
| $Sr^{90}$ | 29 years |
| $Ta^{182}$ | 115 days |
| $Tc^{99}$ | 213,000 years |
| $Te^{123}$ | 0.89 years |
| $Tl^{201}$ | 73 hours |
| $Tm^{170}$ | 129 days |
| $U^{238}$ | 4.5 billion years |
| (See more U on page 60) | |
| $V^{50}$ | 0.25 years |
| $Y^{90}$ | 64 hours |
| $Zn^{65}$ | 27.9 years |
| $Zr^{95}$ | 64 days |

# ELEMENTARY PARTICLES

| Particle Name | Lifespan Seconds | Mass MeV | Charge | Spin |
|---|---|---|---|---|
| **Baryons** | | | | |
| **Proton** | Stable | 938.3 | +1 | 1/2 |
| Antiproton | Stable | 938.3 | −1 | 1/2 |
| **Neutron** | $9.18 \times 10^2$ | 939.6 | 0 | 1/2 |
| Antineutron | $9.19 \times 10^2$ | 939.6 | 0 | 1/2 |
| Lambda hyperon | $3 \times 10^{-10}$ | 1115.6 | 0 | 1/2 |
| Lambda antihyperon | $3 \times 10^{-10}$ | 1115.6 | 0 | 1/2 |
| Positive sigma | 8 x 10–9 | 1189.4 | +1 | 1/2 |
| Neutral sigma | $10^{-11}$ | 1192.5 | 0 | 1/2 |
| Negative sigma | $1.5 \times 10^{-10}$ | 1195.3 | −1 | 1/2 |
| Neutral xi | $3 \times 10^{-10}$ | 1315 | 0 | 1/2 |
| Negative xi | $1.7 \times 10^{-19}$ | 1321 | −1 | 1/2 |
| Omega | $1.3 \times 10^{-10}$ | 1672 | −1 | 1/2 |
| **Gluons** | | | | |
| Red to blue | Stable | 0 | 0 | 1 |
| Red to green | Stable | 0 | 0 | 1 |
| Green to red | Stable | 0 | 0 | 1 |
| Green to blue | Stable | 0 | 0 | 1 |
| Blue to red | Stable | 0 | 0 | 1 |
| Blue to green | Stable | 0 | 0 | 1 |
| Neutral (1) | Stable | 0 | 0 | 1 |
| Neutral (2) | Stable | 0 | 0 | 1 |
| **Leptons** | | | | |
| Neutrino | Stable | 0 | 0 | 1/2 |
| Antineutrino | Stable | 0 | 0 | 1/2 |
| Neutrino–muon | Stable | 0 | 0 | 1/2 |
| Antineutrino–muon | Stable | 0 | 0 | 1/2 |
| Neutrino–tau | Stable | 0 | 0 | 1/2 |
| Antineutrino–tau | Stable | 0 | 0 | 1/2 |
| **Electron** | Stable | 0.511 | −1 | 1/2 |
| Positron | Stable | 0.511 | +1 | 1/2 |
| Muon | $2.2 \times 10^{-6}$ | 105.7 | −1 | 1/2 |
| Antimuon | $2.2 \times 10^{-6}$ | 105.7 | +1 | 1/2 |
| Tau | | 1750 | +1 | 1/2 |
| Antitau | | 1750 | −1 | 1/2 |

# ELEMENTARY PARTICLES

| Particle Name | Lifespan Seconds | Mass MeV | Charge | Spin |
|---|---|---|---|---|
| **Mesons** | | | | |
| Positive pi | $2.6 \times 10^{-8}$ | 139.6 | +1 | 0 |
| Negative pi | $2.6 \times 10^{-8}$ | 189.6 | −1 | 0 |
| Neutral pi | $8 \times 10^{-17}$ | 135 | 0 | 0 |
| Positive K | $1.2 \times 10^{-8}$ | 493.7 | +1 | 0 |
| K–zero–short | $9 \times 10^{-11}$ | 497.7 | 0 | 0 |
| K–zero–long | $5.2 \times 10^{-8}$ | 497.7 | 0 | 0 |
| Negative K | $1.2 \times 10^{-8}$ | 493.7 | −1 | 0 |
| Rho | | 773 | 0 | 1 |
| Omega | | 783 | 0 | 1 |
| Eta | | 2980 | 0 | 0 |
| Psi | | 3098 | | 1 |
| B meson | | 1228 | | 1 |
| D meson | | 1286 | | 1 |
| **Quarks** | | | | |
| Up | Stable | 1.0 | +2/3 | 1/2 |
| Down | Stable | 3.0 | −1/3 | 1/2 |
| Strange | Stable | 102.2 | −1/3 | 1/2 |
| Charm | | 1530 | +2/3 | 1/2 |
| Top | | 4600 | +2/3 | 1/2 |
| Bottom | $< 5 \times 10^{-12}$ | 4700 | −1/3 | 1/2 |
| **Vector Bosons** | | | | |
| Photon | Stable | 0 | 0 | 1 |
| Positive W | $10^{-20}$ | 81000 | +1 | 1 |
| Negative W | $10^{-20}$ | 81000 | −1 | 1 |
| Neutral Z | $10^{-20}$ | 93800 | 0 | 1 |
| Higgs particle | | $10^{6}$ | 0 | 0 |
| Graviton | Stable | 0 | 0 | 0 |

Note that there are many more subatomic particles than those listed in this table, but the above list contains most of the "common" particles. For detailed information see references such as the *Handbook of Chemistry and Physics by The Chemical Rubber Company*.

# URANIUM–238 DECAY SERIES

| Rnuclide | Element | Half–Life | Energy (MeV) |
|---|---|---|---|
| $^{238}U$ | Uranium | 4.5 billion yrs | 4.1-4.2 Alpha |
| $^{234}Th$ | Thorium | 24 days | 0.06-0.2 Beta |
| $^{234}Pa$ | Protactinium | 2.1 minutes | 2.3 Beta |
| $^{234}U$ | Uranium | 250,000 years | 4.7-4.8 Alpha |
| $^{230}Th$ | Thorium | 80,000 years | 4.6-4.7 Alpha |
| $^{226}Ra$ | Radium | 1,600 years | 4.6-4.8 Alpha |
| $^{222}Rn$ | Radon | 3.82 days | 5.5 Alpha |
| $^{218}Po$ | Polonium | 3.05 minutes | 6.0 Alpha |
| $^{214}Pb$ | Lead | 26.8 minutes | 0.7-1.0 Beta |
| $^{214}Bi$ | Bismuth | 19.7 minutes | 0.4-3.3 Beta |
| $^{214}Po$ | Polonium | 16 milliseconds | 7.7 Alpha |
| $^{210}Pb$ | Lead | 22 years | 1 Beta |
| $^{210}Bi$ | Bismuth | 5 days | 1.2 Beta |
| $^{210}Po$ | Polonium | 138 days | 5.3 Alpha |
| $^{206}Pb$ | Lead | Stable | |

# ALUMINUM TYPES

| Class | Description |
|---|---|
| F | As fabricated, temper from shaping process only. |
| O | Annealed, recrystallized, softest temper for wrought. |
| H1 | Strain hardened only. |
| H2 | Strain hardened and then partially annealed. |
| H3 | Strain hardened and then stabilized. |
| W | Solution heat–treated. |
| T2 | Annealed, cast only, improved ductility & stability. |
| T3 | Solution heat–treated and then cold worked, improved strength. |
| T4 | Solution heat–treated and naturally aged to a substantially stable condition. Not cold worked. |
| T5 | Artificially aged only. Rapid cool process like casting. |
| T6 | Solution heat–treated and then artificially aged. Not cold worked after heat treated. Common class. |
| T7 | Solution heat–treated and then stabilized, good growth control and residual stress. |
| T8 | Solution heat–treated, cold worked, and artificially aged. Cold worked to improve strength. |
| T9 | Solution heat–treated, artificially aged, and cold worked to improve strength. |
| T10 | Artificially aged and then cold worked. Rapid cooling after heat treatment the cold worked for strength. |
| TX51 | Stress–relieved by stretching. |
| TX52 | Stress–relieved by compressing. |
| TX53 | Stress–relieved by thermal treatment. |
| T42 | Wrought only, properties of T4. |
| T62 | Wrought only, properties of T6. |

# POCKET REF

# Computers

# COMPUTER ASCII CODES

The following ASCII (American Standard Code for Information Interchange) tables are used by most of the microcomputer industry. The codes occur in two sets: the "low–bit" set, from Dec 0 to Dec 127, and the "high–bit" set, from Dec 128 to Dec 255. The "low–bit" set is standard for almost all microcomputers but the "high–bit" set varies between the different computer brands. For instance, in the case of Apple computers and Epson printers, the "high–bit" set repeats the "low–bit" set except that the alphanumeric characters are italic. In the case of IBM and many other MSDOS systems, the "high–bit" set is composed of foreign language and box drawing characters and mathematic symbols.

| Hex | Dec | Description | Abbr | Character | Control |
|---|---|---|---|---|---|
| 00 | 0 | Null | Null | ● | Control @ |
| 01 | 1 | Start Heading | SOH | ☺ | Control A |
| 02 | 2 | Start of Text | STX | ☻ | Control B |
| 03 | 3 | End of Text | ETX | ♥ | Control C |
| 04 | 4 | End Transmit | EOT | ♦ | Control D |
| 05 | 5 | Enquiry | ENQ | ♣ | Control E |
| 06 | 6 | Acknowledge | ACK | ♠ | Control F |
| 07 | 7 | Beep | BEL | · | Control G |
| 08 | 8 | Back space | BS | ■ | Control H |
| 09 | 9 | Horizontal Tab | HT | O | Control I |
| 0A | 10 | Line Feed | LF | ◙ | Control J |
| 0B | 11 | Vertical Tab | VT | ♂ | Control K |
| 0C | 12 | Form Feed | FF | ♀ | Control L |
| 0D | 13 | Carriage Ret. | CR | ♪ | Control M |
| 0E | 14 | Shift Out | SO | ♫ | Control N |
| 0F | 15 | Shift In | SI | ☼ | Control O |
| 10 | 16 | Device Link Esc | DLE | ► | Control P |
| 11 | 17 | Dev Cont 1 X-ON | DC1 | ◄ | Control Q |
| 12 | 18 | Dev Control 2 | DC2 | ↕ | Control R |
| 13 | 19 | Dev Cont 3 X-OFF | DC3 | ‼ | Control S |
| 14 | 20 | Dev Control 4 | DC4 | ¶ | Control T |
| 15 | 21 | Negative Ack | NAK | § | Control U |
| 16 | 22 | Synchronous Idle | SYN | ▬ | Control V |
| 17 | 23 | End Trans Block | ETB | ↨ | Control W |
| 18 | 24 | Cancel | CAN | ↑ | Control X |
| 19 | 25 | End Medium | EM | ↓ | Control Y |
| 1A | 26 | Substitute | SUB | → | Control Z |
| 1B | 27 | Escape | ESC | ← | Control [ |

# COMPUTER ASCII CODES

| Hex | Dec | Description | Abbr | Character | Control |
|---|---|---|---|---|---|
| 1C | 28 | Cursor Right | FS | ∟ | Control / |
| 1D | 29 | Cursor Left | GS | ↔ | Control ] |
| 1E | 30 | Cursor Up | RS | ▲ | Control ^ |
| 1F | 31 | Cursor Down | US | ▼ | Control – |

| Hex | Dec | Character | Description |
|---|---|---|---|
| 20 | 32 | | Space (SP) |
| 21 | 33 | ! | Exclamation Point |
| 22 | 34 | " | Double Quote |
| 23 | 35 | # | Number sign |
| 24 | 36 | $ | Dollar sign |
| 25 | 37 | % | Percent |
| 26 | 38 | & | Ampersand |
| 27 | 39 | ' | Apostrophe |
| 28 | 40 | ( | Left parenthesis |
| 29 | 41 | ) | Right parenthesis |
| 2A | 42 | * | Asterisk |
| 2B | 43 | + | Plus sign |
| 2C | 44 | , | Comma |
| 2D | 45 | – | Minus sign |
| 2E | 46 | . | Period |
| 2F | 47 | / | Right or Front slash |
| 30 | 48 | 0 | Zero |
| 31 | 49 | 1 | One |
| 32 | 50 | 2 | Two |
| 33 | 51 | 3 | Three |
| 34 | 52 | 4 | Four |
| 35 | 53 | 5 | Five |
| 36 | 54 | 6 | Six |
| 37 | 55 | 7 | Seven |
| 38 | 56 | 8 | Eight |
| 39 | 57 | 9 | Nine |
| 3A | 58 | : | Colon |
| 3B | 59 | ; | Semicolon |
| 3C | 60 | < | Less than |
| 3D | 61 | > | Greater than |
| 3E | 62 | = | Equal sign |
| 3F | 63 | ? | Question mark |
| 40 | 64 | @ | "at" symbol |

## COMPUTER ASCII CODES

| Hex | Dec | Character | Description |
|---|---|---|---|
| 41 | 65 | A | |
| 42 | 66 | B | |
| 43 | 67 | C | |
| 44 | 68 | D | |
| 45 | 69 | E | |
| 46 | 70 | F | |
| 47 | 71 | G | |
| 48 | 72 | H | |
| 49 | 73 | I | |
| 4A | 74 | J | |
| 4B | 75 | K | |
| 4C | 76 | L | |
| 4D | 77 | M | |
| 4E | 78 | N | |
| 4F | 79 | O | |
| 50 | 80 | P | |
| 51 | 81 | Q | |
| 52 | 82 | R | |
| 53 | 83 | S | |
| 54 | 84 | T | |
| 55 | 85 | U | |
| 56 | 86 | V | |
| 57 | 87 | W | |
| 58 | 88 | X | |
| 59 | 89 | Y | |
| 5A | 90 | Z | |
| 5B | 91 | [ | Right bracket |
| 5C | 92 | \ | Left or Back Slash |
| 5D | 93 | ] | Left bracket |
| 5E | 94 | ^ | Caret |
| 5F | 95 | _ | Underline |
| 60 | 96 | ‘ | Accent |
| 61 | 97 | a | |
| 62 | 98 | b | |
| 63 | 99 | c | |
| 64 | 100 | d | |
| 65 | 101 | e | |
| 66 | 102 | f | |
| 67 | 103 | g | |

| Hex | Dec | Standard Character | Description |
|---|---|---|---|
| 68 | 104 | h | |
| 69 | 105 | i | |
| 6A | 106 | j | |
| 6B | 107 | k | |
| 6C | 108 | l | |
| 6D | 109 | m | |
| 6E | 110 | n | |
| 6F | 111 | o | |
| 70 | 112 | p | |
| 71 | 113 | q | |
| 72 | 114 | r | |
| 73 | 115 | s | |
| 74 | 116 | t | |
| 75 | 117 | u | |
| 76 | 118 | v | |
| 77 | 119 | w | |
| 78 | 120 | x | |
| 79 | 121 | y | |
| 7A | 122 | z | |
| 7B | 123 | { | Left bracket |
| 7C | 124 | \| | Vertical line |
| 7D | 125 | } | Right bracket |
| 7E | 126 | ~ | Tilde |
| 7F | 127 | DEL | Delete |

| Hex | Dec | Standard Character | IBM Set | Standard Description |
|---|---|---|---|---|
| 80 | 128 | Null | Ç | Null |
| 81 | 129 | SOH | ü | Start Heading |
| 82 | 130 | STX | é | Start of Text |
| 83 | 131 | ETX | â | End of Text |
| 84 | 132 | EOT | ä | End Transmit |
| 85 | 133 | ENQ | à | Enquiry |
| 86 | 134 | ACK | å | Acknowledge |
| 87 | 135 | BEL | ç | Beep |
| 88 | 136 | BS | ê | Back Space |
| 89 | 137 | HT | ë | Horiz Tab |
| 8A | 138 | LF | è | Line Feed |

# COMPUTER ASCII CODES

| Hex | Dec | Standard Character | IBM Set | Standard Description |
|---|---|---|---|---|
| 8B | 139 | VT | ï | Vertical Tab |
| 8C | 140 | FF | î | Form Feed |
| 8D | 141 | CR | ì | Carriage Return |
| 8E | 142 | SO | Ä | Shift Out |
| 8F | 143 | SI | Å | Shift In |
| 90 | 144 | DLE | É | Device Link Esc |
| 91 | 145 | DC1 | æ | Device Cont 1 X–ON |
| 92 | 146 | DC2 | Æ | Device Control 2 |
| 93 | 147 | DC3 | ô | Device Cont 3 X–OFF |
| 94 | 148 | DC4 | ö | Device Control 4 |
| 95 | 149 | NAK | ò | Negative Ack |
| 96 | 150 | SYN | û | Synchronous Idle |
| 97 | 151 | ETB | ù | End Transmit Block |
| 98 | 152 | CAN | ÿ | Cancel |
| 99 | 153 | EM | Ö | End Medium |
| 9A | 154 | SUB | Ü | Substitute |
| 9B | 155 | ESC | ¢ | Escape |
| 9C | 156 | FS | £ | Cursor Right |
| 9D | 157 | GS | ¥ | Cursor Left |
| 9E | 158 | RS | Pt | Cursor Up |
| 9F | 159 | US | ƒ | Cursor Down |
| A0 | 160 | Space | á | Space |
| A1 | 161 | ! | í | Italic Exclamation point |
| A2 | 162 | " | ó | Italic Double quote |
| A3 | 163 | # | ú | Italic Number sign |
| A4 | 164 | $ | ñ | Italic Dollar sign |
| A5 | 165 | % | Ñ | Italic Percent |
| A6 | 166 | & | ª | Italic Ampersand |
| A7 | 167 | ' | º | Italic Apostrophe |
| A8 | 168 | ( | ¿ | Italic Left parenthesis |
| A9 | 169 | ) | ⌐ | Italic Right parenthesis |
| AA | 170 | * | ¬ | Italic asterisk |
| AB | 171 | + | ½ | Italic plus sign |
| AC | 172 | , | ¼ | Italic comma |
| AD | 173 | – | ¡ | Italic minus sign |
| AE | 174 | . | « | Italic period |
| AF | 175 | / | » | Italic right slash |
| B0 | 176 | 0 | ░ | Italic Zero |
| B1 | 177 | 1 | ▒ | Italic One |

| Hex | Dec | Standard Character | IBM Set | Standard Description |
|---|---|---|---|---|
| B2 | 178 | *2* | ▓ | Italic Two |
| B3 | 179 | *3* | │ | Italic Three |
| B4 | 180 | *4* | ┤ | Italic Four |
| B5 | 181 | *5* | ╡ | Italic Five |
| B6 | 182 | *6* | ╢ | Italic Six |
| B7 | 183 | *7* | ╖ | Italic Seven |
| B8 | 184 | *8* | ╕ | Italic Eight |
| B9 | 185 | *9* | ╣ | Italic Nine |
| BA | 186 | *:* | ║ | Italic colon |
| BB | 187 | *;* | ╗ | Italic semicolon |
| BC | 188 | *<* | ╝ | Italic less than |
| BD | 189 | *=* | ╜ | Italic equal |
| BE | 190 | *>* | ╛ | Italic greater than |
| BF | 191 | *?* | ┐ | Italic question mark |
| C0 | 192 | *@* | └ | Italic "at" symbol |
| C1 | 193 | *A* | ┴ | Italic A |
| C2 | 194 | *B* | ┬ | Italic B |
| C3 | 195 | *C* | ├ | Italic C |
| C4 | 196 | *D* | ─ | Italic D |
| C5 | 197 | *E* | ┼ | Italic E |
| C6 | 198 | *F* | ╞ | Italic F |
| C7 | 199 | *G* | ╟ | Italic G |
| C8 | 200 | *H* | ╚ | Italic H |
| C9 | 201 | *I* | ╔ | Italic I |
| CA | 202 | *J* | ╩ | Italic J |
| CB | 203 | *K* | ╦ | Italic K |
| CC | 204 | *L* | ╠ | Italic L |
| CD | 205 | *M* | ═ | Italic M |
| CE | 206 | *N* | ╬ | Italic N |
| CF | 207 | *O* | ╧ | Italic O |
| D0 | 208 | *P* | ╨ | Italic P |
| D1 | 209 | *Q* | ╤ | Italic Q |
| D2 | 210 | *R* | ╥ | Italic R |
| D3 | 211 | *S* | ╙ | Italic S |
| D4 | 212 | *T* | ╘ | Italic T |
| D5 | 213 | *U* | ╒ | Italic U |
| D6 | 214 | *V* | ╓ | Italic V |
| D7 | 215 | *W* | ╫ | Italic W |
| D8 | 216 | *X* | ╪ | Italic X |

# COMPUTER ASCII CODES

| Hex | Dec | Standard Character | IBM Set | Description |
|---|---|---|---|---|
| D9 | 217 | *Y* | ┘ | Italic Y |
| DA | 218 | *Z* | ┌ | Italic Z |
| DB | 219 | *[* | █ | Italic left bracket |
| DC | 220 | ** | ▄ | Italic left or back slash |
| DD | 221 | *]* | ▌ | Italic right bracket |
| DE | 222 | *^* | ▐ | Italic caret |
| DF | 223 | | ▀ | Italic underline |
| E0 | 224 | *`* | α | Italic accent / alpha |
| E1 | 225 | *a* | β | Italic a / beta |
| E2 | 226 | *b* | Γ | Italic b / gamma |
| E3 | 227 | *c* | π | Italic c / pi |
| E4 | 228 | *d* | Σ | Italic d / sigma |
| E5 | 229 | *e* | σ | Italic e / sigma |
| E6 | 230 | *f* | μ | Italic f / mu |
| E7 | 231 | *g* | γ | Italic g / gamma |
| E8 | 232 | *h* | Φ | Italic h / phi |
| E9 | 233 | *i* | θ | Italic i / theta |
| EA | 234 | *j* | Ω | Italic j / omega |
| EB | 235 | *k* | δ | Italic k / delta |
| EC | 236 | *l* | ∞ | Italic l / infinity |
| ED | 237 | *m* | Ø | Italic m / slashed zero |
| EE | 238 | *n* | ∈ | Italic n |
| EF | 239 | *o* | ∩ | Italic o |
| F0 | 240 | *p* | ≡ | Italic p |
| F1 | 241 | *q* | ± | Italic q |
| F2 | 242 | *r* | ≥ | Italic r |
| F3 | 243 | *s* | ≤ | Italic s |
| F4 | 244 | *t* | ⌠ | Italic t |
| F5 | 245 | *u* | ⌡ | Italic u |
| F6 | 246 | *v* | ÷ | Italic v |
| F7 | 247 | *w* | ≈ | Italic w |
| F8 | 248 | *x* | ° | Italic x |
| F9 | 249 | *y* | • | Italic y |
| FA | 250 | *z* | • | Italic z |
| FB | 251 | *{* | √ | Italic left bracket |
| FC | 252 | *\|* | *n* | Italic vertical line |
| FD | 253 | *}* | 2 | Italic right bracket |
| FE | 254 | *~* | | Italic tilde |
| FF | 255 | Blank | Blank | Blank |

## GAME CONTROL CABLE - IBM PC

| Joystick Pin Number | Signal Description | Function | Signal Direction At Joystk |
|---|---|---|---|
| 1 | +5 Volts | Supply voltage | Input |
| 2 | Button 1 | Push Button 1 | Output |
| 3 | Position 0 | X Coordinate | Output |
| 4 | Ground | Ground | |
| 5 | Ground | Ground | |
| 6 | Position 1 | Y Coordinate | Output |
| 7 | Button 2 | Push Button 2 | Output |
| 8 | +5 Volts | Supply voltage | Input |
| 9 | +5 Volts | Supply voltage | Input |
| 10 | Button 3 | Push Button 3 | Output |
| 11 | Position 2 | X Coordinate | Output |
| 12 | Ground | Ground | |
| 13 | Position 3 | Y Coordinate | Output |
| 14 | Button 4 | Push Button 4 | Output |
| 15 | +5 Volts | Supply voltage | Input |

## GAME CONTROL CABLE - APPLE IIe

| Joystick Pin Number | Signal Description | Function | Signal Direction At Joystk |
|---|---|---|---|
| 1 | PB1 | Push Button 1 | Output |
| 2 | +5 Volt | Supply voltage | Input |
| 3 | Ground | Ground | |
| 4 | N.C. | No Connection | |
| 5 | PDL0 | Paddle 0 | Output |
| 6 | N.C. | No Connection | |
| 7 | PB0 | Push Button 0 | Output |
| 8 | PDL1 | Paddle 1 | Output |
| 9 | N.C. | No Connection | |

## GAME CONTROL - COMMODORE 64

| Joystick Pin Number | Signal Description | Function | Signal Direction At Joystk |
|---|---|---|---|
| 1 | JOYA0 or BO | Control A0 or B0 | Output |
| 2 | JOY A1 or B1 | Control A1 or B1 | Output |
| 3 | JOY A2 or B2 | Control A2 or B2 | Output |
| 4 | JOY A3 or B3 | Control A3 or B3 | Output |
| 5 | Pot AY or BY | Pot A or B Y axis | Output |
| 6 | Button A/LP or B | Button A or B or Light Pen | Output |
| 7 | +5 Volt, 50 ma | Supply Voltage | Input |
| 8 | Ground | Ground | |
| 9 | Pot AX or BX | Pot A or B X axis | Output |

# PARALLEL PRINTER INTERFACE

| Printer Pin Number | Signal Description | Function | Signal Direction At Printer |
|---|---|---|---|
| 1 | STROBE | Reads in the data | Input |
| 2 | DATA Bit 0 | Data line | Input |
| 3 | DATA Bit 1 | Data line | Input |
| 4 | DATA Bit 2 | Data line | Input |
| 5 | DATA Bit 3 | Data line | Input |
| 6 | DATA Bit 4 | Data line | Input |
| 7 | DATA Bit 5 | Data line | Input |
| 8 | DATA Bit 6 | Data line | Input |
| 9 | DATA Bit 7 | Data line | Input |
| 10 | ACKNLG | Acknowledge receipt of data | Output |
| 11 | Busy | Printer is busy | Output |
| 12 | Paper Empty | Printer out of paper | Output |
| 13 | SLCT | Online mode indicator | Output |
| 14 | Auto Feed XT | | Input |
| 15 | Not Used | Not Used | |
| 16 | Signal ground | Signal ground | |
| 17 | Frame ground | Frame ground | |
| 18 | +5 volts | +5 volts | |
| 19-30 | Ground | Return signals of pins 1–12, twisted pairs. | |
| 31 | Input Prime or INIT | Resets printer, clears buffer & initializes | Input |
| 32 | Fault or Error | Indicates offline mode | Output |
| 33 | Signal ground | External ground | |
| 34 | Not Used | Not Used | |
| 35 | +5 Volts | +5 Volts (3.3 K-ohm) | |
| 36 | SLCT IN | TTL high level | Input |

The above pinout is at the printer plug, computer side pinouts are on the next page. The "Parallel" or "Centronics" configuration for printer data transmission has become the de facto standard in the personal computer industry. This configuration was developed by a printer manufacturer (Centronics) as an alternative to serial data transmission. High data transfer rates are the main advantage of parallel and are attained by simultaneous transmission of all bits of a binary "word" (normally an ASCII code). Disadvantages of the parallel transfer are the requirement for 8 separate data lines and computer to printer cable lengths of less than 12 feet.

# PARALLEL PINOUTS @ COMPUTER

## IBM/ MS DOS DB25 Systems

| Computer Pin Number | Signal Description | Function | Signal Direction At Computer |
|---|---|---|---|
| 1 | STROBE | Reads in the data | Output |
| 2 | DATA Bit 0 | Data line | Output |
| 3 | DATA Bit 1 | Data line | Output |
| 4 | DATA Bit 2 | Data line | Output |
| 5 | DATA Bit 3 | Data line | Output |
| 6 | DATA Bit 4 | Data line | Output |
| 7 | DATA Bit 5 | Data line | Output |
| 8 | DATA Bit 6 | Data line | Output |
| 9 | DATA Bit 7 | Data line | Output |
| 10 | ACKNLG | Acknowledge receipt of data | Input |
| 11 | Busy | Printer is busy | Input |
| 12 | Paper Empty | Printer out of paper | Input |
| 13 | SLCT | Online mode indicator | Input |
| 14 | Auto Feed XT | | Output |
| 15 | Fault or Error | Indicates offline mode | Input |
| 16 | Input Prime or INIT | Resets printer, clears buffer & initializes | Output |
| 17 | SLCT IN | TTL high level | Output |
| 18-25 | Ground | Return signals of pins 1–12, twisted pairs. | |

# IBM VIDEO CARD PINOUTS

| Pin Number | Description |
|---|---|
| **Monochrome Display Adapter** | |
| 1 & 2 | Ground |
| 3, 4, & 5 | Not Used |
| 6 | + Intensity |
| 7 | + Video |
| 8 | + Horizontal Drive |
| 9 | – Vertical Drive |
| **Color/Graphics Display Adapter** | |
| 1 & 2 | Ground |
| 3 | Red |
| 4 | Green |
| 5 | Blue |
| 6 | + Intensity |
| 7 | Reserved |
| 8 | + Horizontal Drive |
| 9 | – Vertical Drive |

## Standard DB25 Pin Connector

| Serial Pin Number | Signal Description | Function | Signal Direction At Device |
|---|---|---|---|
| 1 | FG | Frame ground | |
| 2 | TD | Transmit Data | Output |
| 3 | RD | Receive Data | Input |
| 4 | RTS | Request to Send | Output |
| 5 | CTS | Clear to Send | Input |
| 6 | DSR | Data Set Ready | Input |
| 7 | SG | Signal Ground | |
| 8 | DCD | Data Carrier Detect | Input |
| 9 | +V | + DC test voltage | Input |
| 10 | – V | – DC test voltage | Input |
| 11 | QM | Equalizer Mode | Input |
| 12 | (S)DCD | 2nd Data Carrier Detect | Input |
| 13 | (S)CTS | 2nd Clear to Send | Input |
| 14 | (S)TD | 2nd Transmitted Data | Output |
| 15 | TC | Transmitter Clock | Input |
| 16 | (S)RD | 2nd Received Data | Input |
| 17 | RC | Receiver Clock | Input |
| 18 | Not used | Not used | |
| 19 | (S)RTS | 2nd Request to Send | Output |
| 20 | DTR | Data Terminal Ready | Output |
| 21 | SQ | Signal Quality Detect | Input |
| 22 | RI | Ring Indicator | Input |
| 23 | | Data Rate Selector | Output |
| 24 | (TC) | External Transmitter Clk | Output |
| 25 | Not used | Not used | |

## IBM Standard DB9 Pin Connector

| Serial Pin Number | Signal Description | Function | Signal Direction At Device |
|---|---|---|---|
| 1 | DCD | Data Carrier Detect | Input |
| 2 | RD | Receive Data | Input |
| 3 | SD | Transmit Data | Output |
| 4 | DTR | Data Terminal Ready | Output |
| 5 | SG | Signal Ground | |
| 6 | DSR | Data Set Ready | Input |
| 7 | RTS | Request to Send | Output |
| 8 | CTS | Clear to Send | Input |
| 9 | RI | Ring Indicator | Input |

# SERIAL I/O INTERFACES (RS232C)

## Macintosh DB9 Pin Connector

| Serial Pin Number | Signal Description | Function | Signal Direction At Device |
|---|---|---|---|
| 1 | GND | Ground | |
| 2 | | No connection | |
| 3 | SG | Signal Ground | |
| 4 | | No connection | |
| 5 | TD | Transmit Data | Output |
| 6 | DTR | Data Terminal Ready | Output |
| 7 | DSR | Data Set Ready | Input |
| 8 | | No connection | |
| 9 | RD | Receive Data | Input |

## Apple IIc Round Connector

| Serial Pin Number | Signal Description | Function | Signal Direction At Device |
|---|---|---|---|
| 1 | DTR | Data Terminal Ready | Output |
| 2 | TD | Transmit Data | Output |
| 3 | GND | Ground | |
| 4 | RD | Receive Data | Input |
| 5 | DSR | Data Set Ready | Input |

## Commodore 64 Round Connector

| Serial Pin Number | Signal Description | Function |
|---|---|---|
| 1 | SRQIN | Serial SRQIN |
| 2 | GND | Signal Ground |
| 3 | | Serial Attn I/O |
| 4 | | Serial Clock I/O |
| 5 | | Serial Data I/O |
| 6 | | Reset |

# IBM LIGHT PEN INTERFACE

| Pin # | Description |
|---|---|
| 1 | – Light Pen Input |
| 2 | No connection |
| 3 | – Light Pen Switch |
| 4 | Chassis Ground |
| 5 | +5 Volts |
| 6 | +12 Volts |

# NOTES ON SERIAL INTERFACING

Printers and asynchronous modems are relatively unsophisticated pieces of electronic equipment. Although all 25 pins of the **Standard DB25** serial connector are listed 2 pages back, only a few of the pins are needed for normal applications. The following list gives the necessary pins for each of the indicated applications.

1. "Dumb Terminals" – 1,2,3, & 7
2. Printers and asynchronous modems – 1,2,3,4,5,6,7,8, & 20
3. "Smart" and synchronous modems – 1,2,3,4,5,6,7,8,13,14, 15,17,20,22, & 24

Cable requirements also differ, depending on the particular hardware being used. The asynchronous modems normally use the 9 pin or 25 pin cables and are wired 1 to 1 (ie, pin 1 on one end of the cable goes to pin 1 on the other end of the cable.) Serial printers, however, have several wires switched in order to accommodate "handshaking" between computer and printer. The rewired junction is called a "Modem Eliminator". In the case of Standard DB25 the following are typical rewires:

**Standard IBM PC**

| DB25 @ Computer | DB25 @ Printer |
|---|---|
| 1 | 1 |
| 3 | 2 |
| 2 | 3 |
| 8 | 4 |
| 4 | 8 |
| 5 & 6 | 20 |
| 20 | 5 & 6 |
| 7 | 7 |

**Second Standard PC**

| DB25 @ Computer | DB25 @ Printer |
|---|---|
| 1 | 1 |
| 3 | 2 |
| 2 | 3 |
| 20 | 5, 6 & 8 |
| 7 | 7 |
| 5, 6 & 8 | 20 |

**PC to Terminal**

| | |
|---|---|
| 1 | 1 |
| 2 | 3 |
| 3 | 2 |
| 4 | 5 |
| 5 | 4 |
| 6 & 8 | 20 |
| 20 | 6 & 8 |
| 7 | 7 |

**Std Hewlett-Packard**

| | |
|---|---|
| 1 | 1 |
| 2 | 3 |
| 3 | 2 |
| 4 & 5 | 8 |
| 8 | 4 & 5 |
| 6 | 20 |
| 7 & 22 | 7 & 22 |
| 17 | 15 |
| 11 | 12 |
| 12 | 11 |
| 15 & 24 | 17 |
| 20 | 6 |

# HAYES COMPATIBLE MODEM SWITCH SETTINGS

| Switch | Position | Function |
|---|---|---|
| 1 | Off | Supports DTR line, RS232C |
| | On | Ignores DTR line, RS232C |
| 2 | Off | Word Result Codes |
| | On | Digital Result Codes: Ø = OK, 1 = Connect, 2 = Ring, 3 = No Carrier, 4 = Error, 5 = Connect 1200 (Extended Code Set) |
| 3 | Off | Display no result codes (Q1 Command) |
| | On | Display result codes (Q0 Command) |
| 4 | Off | Echo characters in command state (E1) |
| | On | No Echo unless 1/2 duplex (EØ) |
| 5 | Off | Auto Answer on 1st Ring |
| | On | Do not answer a call |
| 6 | Off | Read Status of Carrier Detect, RS232C |
| | On | Set Carrier Detect line TRUE at all times. |
| 7 | Off | Single Phone Line Jack, R11 |
| | On | Multiline Phone Line Jack, RJ12 or RJ13 |
| 8 | Off | Disable Smartmodem 1200 Command Recognition (Dumb Mode) |
| | On | Enable Smartmodem 1200 Command Recognition (Smart Mode) |

# HAYES COMPATIBLE MODEM COMMAND SETTINGS

| Command | Function |
|---|---|
| A | Immediate answer on ring |
| A/ | Repeat last command line (Replaces AT) |
| Cn | n = Ø is Transmitter off, n = 1 is on, (1 = default) |
| Bn | n = Ø is CCITT answer tone, n = 1 is US/Canada Tone |
| Dn | Dial telephone number<br>n = Ø to 9 for phone numbers<br>n = T is Touch Tone Dial, P is Pulse Dial<br>n = R is Originate Only, n = , is Pause<br>n = ! is xfer call to following extension<br>n = " is dial letters that follow<br>n = @ is Dial, Wait for answer, & continue<br>n = ; is Return to command mode after dialing |
| En | n = Ø is no character echo in command state<br>n = 1 is echo all characters in command state |
| Fn | n = Ø is Half Duplex: n = 1 if Full Duplex |
| Hn | n = Ø is On Hook (Hang Up), n = 1 is Off Hook<br>n = 2 is Special Off Hook |
| In | n = Ø is Display product code, n = 1 show Check Sum<br>n = 2 is show RAM test, n = 3 is show call time length<br>n = 4 is show current modem settings |
| Kn | n = Ø at AT13 show last call length, n = 1 show time |

(1) Courtesy of *Hayes Microcomputer Products, Inc. Norcross, GA*

# HAYES COMPATIBLE MODEM COMMAND SETTINGS

| Command | Function |
|---|---|
| Mn | n=Ø is Speaker always off, n=2 is always on<br>n=1 is Speaker on until carrier detected (default)<br>n=3 is Speaker on after dial through CONNECT |
| O | Return to on–line state |
| P | Pulse Dial |
| Qn | n=Ø is send Result Codes, n=1 is do not send code |
| R | Reverse mode (Originate Only) |
| SØ=n | n=Ø to 255 rings before answer (see switch 5) |
| S1=n | Counts rings from Ø to 255 |
| S2=n | Set escape code character, n=Ø to 127, 43 default |
| S3=n | Set carriage return character, n=Ø to 127, 13 default |
| S4=n | Set line feed character, n=Ø to 127, 10 default |
| S5=n | Set backspace character, n=Ø to 127, 8 default |
| S6=n | Wait time for dial tone, n=2 to 255 seconds |
| S7=n | Wait time for carrier, n=2 to 255 seconds |
| S8=n | Set duration of "," pause character, n=Ø to 255 sec. |
| S9=n | Carrier detect response time, n=1 to 255 1/10 secs. |
| S10=n | Delay time carrier loss to hang-up, n=1 to 255 1/10 s. |
| S11=n | Duration & space of Touch Tones, n=50 to 255 ms. |
| S12=n | Escape code guard time, n=50 to 255 1/50 seconds |
| S13=n | UART Status Register Bit Mapped (reserved) |
| S14=n | Option Register, Product code returned by AT1Ø |
| S15=n | Flag Register (reserved) |
| S16=n | Self test mode. n=Ø is data mode (default), n=1 is Analog Loopback, n=2 is dial test, n=4 is Test Pattern, n=5 is Analog Loopback and Test Pattern. |
| Sn ? | Send contents of Register n (Ø to 16) to Computer |
| T | Touch Tone Dial |
| Vn | n=Ø is send result codes as digits, n=1 is words |
| Xn | n=Ø send basic result codes Ø to 4<br>n=1 to 6 send extended result codes Ø to 12 |
| Z | Software reset and reset to default values |
| &Cn | n=Ø is carrier detect always active, n=1 active |
| &Dn | n=Ø is DTR always ignored, =1 DTR causes return to command, =2 DTR disconnects, =3 disconnect/reset |
| &F | Get Factory Configuration |
| &Gn | n=Ø Disable Guard Tone, =1 is 550hz, =2 is 1800hz |
| &Ln | n=Ø or 1 Speaker Volume Low, =2 medium, =3 high |
| &Pn | n=Ø Pulse Make/Break Ratio USA 39% / 61%<br>n=1 Pulse Make/Break Ratio UK 33% / 67% |
| &Rn | n=Ø is CTS always active, n=1 CTS always active |
| &Sn | n=Ø is DSR always active, n=1 DSR active at connect |
| &Tn | Test Commands: n=Ø end test, =1 local analog loopback, =3 local digital loopback, =4 enable Rmt digital loopback, =5 disable digital loopback, =6 request Rmt digital loop, =7 request Rmt dig loop & enter self test, =8 local analog loop & self test |
| &W | Write Configuration to Memory |
| &Zn | Store Phone Number |

# GPIB I/O INTERFACE (IEEE-488)

The HPIB/GPIB/IEEE–488 standard is a very powerful interface developed originally by Hewlett–Packard ( HP–IB ). The interface has been adopted by a variety of groups, such as IEEE, and is known by names such as HP–IB, GPIB, IEEE–488 and IEC Standard 625–1 (outside the US). Worldwide use of this standard has come about due to its ease of use, handshaking protocol, and precisely defined function.

Information management is handled by three device types, Talkers, Listeners, and Controllers. Talkers send information, Listeners receive data, and Controllers manage the interactions. Up to 15 devices can be interconnected, but are usually located within 20 feet of the computer. Additional extenders can be used to access more than 15 devices.

**GPIB 24 Line Bus**

| Pin Number | Signal Description | Function |
|---|---|---|
| 1 | DATA I/O 1 | Data line I/O bus |
| 2 | DATA I/O 2 | Data line I/O bus |
| 3 | DATA I/O 3 | Data line I/O bus |
| 4 | DATA I/O 4 | Data line I/O bus |
| 5 | EIO | End or Identify |
| 6 | DAV | Data valid |
| 7 | NRFD | Not Ready For Data |
| 8 | NDAC | Data Not Accepted |
| 9 | SRQ | Service Request |
| 10 | IFC | Interface Clear |
| 11 | ATN | Attention |
| 12 | Shield | or wire ground |
| 13 | DATA I/O 5 | Data line I/O bus |
| 14 | DATA I/O 6 | Data line I/O bus |
| 15 | DATA I/O 7 | Data line I/O bus |
| 16 | DATA I/O 8 | Data line I/O bus |
| 17 | REN | Remote Enable |
| 18 | Ground | Ground |
| 19 | Ground | Ground |
| 20 | Ground | Ground |
| 21 | Ground | Ground |
| 22 | Ground | Ground |
| 23 | Ground | Ground |
| 24 | Logic Ground | Logic Ground |

Devices can be set up in star, linear or other combinations and are easily set up using male/female stackable connectors.

# IBM PC MEMORY MAP

| Address Range | Size | Description |
|---|---|---|
| ØØØØØ-ØØ3FF | 1K | Interrupt Vectors |
| ØØ4ØØ-7FFFF | 512K | Bios, DOS, 512K RAM Expansion |
| 8ØØØØ-9FFFF | 128K | 128K RAM Expansion (Top of 640K) |
| AØØØØ-AFFFF | 64K | EGA Video Buffer |
| BØØØØ-B7FFF | 32K | Monochrome & other screen buffers |
| B8ØØØ-B8FFF | 32K | CGA and EGA Buffers |
| AT LIM Expanded Memory 64K page is between 768K and 896K | | |
| CØØØØ-C3FFF | 16K | EGA Video Bios |
| C4ØØØ-C7FFF | 16K | ROM Expansion Area |
| XT LIM Expanded Memory 64K page is between 800K and 960K | | |
| C8ØØØ-CCFFF | 20K | XT Hard Disk Controller Bios |
| CDØØØ-CFFFF | 12K | User PROM, Memory mapped I/O |
| DØØØØ-DFFFF | 64K | User PROM, normal LIM Location for Expanded Memory |
| EØØØØ-EFFFF | 64K | ROM expansion, I/O for XT |
| FØØØØ-FDFFF | 56K | ROM BASIC |
| FEØØØ-FFFD9 | 8K | BIOS |
| FFFFØ-FFFF4 | 4 | 1st Code run after system power on |
| FFFF5-FFFFC | 8 | BIOS Release Date |
| FFFFE-FFFFF | 2 | Machine ID (Top of 1 Meg RAM) |
| 1ØØØØØ-FFFFFF | 15Meg | AT Extended Memory |

# IBM PC HARDWARE INTERRUPTS

NMI ........... Non-Maskable Interrupt (Parity)

**Interrupt Controller 1:**

- IRQØ ....... Timer Output
- IRQ1 ....... Keyboard
- IRQ2 ....... XT – Reserved
  AT – Route to Interrupt Controller 2, IRQ8 to 15
- IRQ3 ....... Serial Port COM2: or SDLC
- IRQ4 ....... Serial Port COM1: or SDLC
- IRQ5 ....... XT – Hard Disk Controller
  AT – Parallel Printer Port 2
- IRQ6 ....... Floppy Disk Controller
- IRQ7 ....... Parallel Printer Port LPT1:

**Interrupt Controller 2 (AT Only):**

- IRQ8 ....... Real Time Clock
- IRQ9 ....... Software redirect to IRQ2 (Int ØA Hex)
- IRQ10 ...... Reserved
- IRQ11 ...... Reserved
- IRQ12 ...... Reserved
- IRQ13 ...... 80287 Math Coprocessor
- IRQ14 ...... Hard Disk Controller
- IRQ15 ...... Reserved

# IBM PC HARDWARE I/O MAP

**8088 Class Systems**

| Address | Function |
|---|---|
| ØØØ–ØØF | DMA Controller (8237A) |
| Ø2Ø–Ø21 | Interrupt controller (8259A) |
| Ø4Ø–Ø43 | Timer (8253) |
| Ø6Ø–Ø63 | PPI (8255A) |
| Ø8Ø–Ø83 | DMA page register (74LS612) |
| ØAØ–ØAF | NMI – Non Maskable Interrupt |
| 2ØØ–2ØF | Game Port Joystick controller |
| 21Ø–217 | Expansion Unit |
| 2E8–2EF | COM4: Serial Port |
| 2F8–2FF | COM2: Serial Port |
| 3ØØ–31F | Prototype Card |
| 32Ø–32F | Hard Disk |
| 378–37F | Parallel Printer Port 1 |
| 38Ø–38F | SDLC |
| 3BØ–3BF | MDA – Monochrome Adapter and printer |
| 3DØ–3D7 | CGA – Color Graphics Adapter |
| 3E8–3EF | COM3: Serial Port |
| 3FØ–3F7 | Floppy Diskette Controller |
| 3F8–3FF | COM1: Serial Port |

**80286 Class Systems**

| Address | Function |
|---|---|
| ØØØ–Ø1F | DMA Controller #1 (8237A-5) |
| Ø2Ø–Ø3F | Interrupt controller #1 (8259A) |
| Ø4Ø–Ø5F | Timer (8254) |
| Ø6Ø–Ø6F | Keyboard (8Ø42) |
| Ø7Ø–Ø7F | NMI – Non Maskable Interrupt & CMOS RAM |
| Ø8Ø–Ø9F | DMA page register (74LS612) |
| ØAØ–ØBF | Interrupt controller #2 (8259A) |
| ØCØ–ØDF | DMA Controller #2 (8237A) |
| ØFØ–ØFF | 8Ø287 Math Coprocessor |
| 1FØ–1F8 | Hard Disk |
| 2ØØ–2ØF | Game Port Joystick controller |
| 258–25F | Intel Above Board |
| 278–27F | Parallel Printer Port 2 |
| 2E8–2EF | COM4: Serial Port |
| 2F8–2FF | COM2: Serial Port |
| 3ØØ–31F | Prototype Card |
| 378–37F | Parallel Printer Port 1 |
| 38Ø–38F | SDLC or Bisynchronous Comm Port 2 |
| 3AØ–3AF | Bisynchronous Comm Port 1 |
| 3BØ–3BF | MDA – Monochrome Adapter |
| 3BC–3BE | Parallel Printer on Monochrome Adapter |
| 3CØ–3CF | EGA – Reserved |
| 3DØ–3D7 | CGA – Color Graphics Adapter |
| 3E8–3EF | COM3: Serial Port |
| 3FØ–3F7 | Floppy Diskette Controller |
| 3F8–3FF | COM1: Serial Port |

# IBM PC SOFTWARE INTERRUPTS

| Address | Int # | Interrupt Name |
|---|---|---|
| ØØØ–ØØ3 | Ø | Divide by zero |
| ØØ4–ØØ7 | 1 | Single Step IRET |
| ØØ8–ØØB | 2 | NMI Non Maskable Interrupt |
| ØØC–ØØF | 3 | Breakpoint |
| Ø1Ø–Ø13 | 4 | Overflow IRET |
| Ø14–Ø17 | 5 | Print Screen |
| Ø18–Ø1F | 6 | Reserved Ø18–Ø1B and Ø1C–Ø1F |
| Ø2Ø–Ø23 | 8 | Time of Day Ticker IRQØ |
| Ø24–Ø27 | 9 | Keyboard IRQ1 |
| Ø28–Ø2B | A | XT Reserved, AT IRQ2 direct to IRQ9 |
| Ø2C–Ø2F | B | COM2 communications, IRQ3 |
| Ø3Ø–Ø33 | C | COM1 communications, IRQ4 |
| Ø34–Ø37 | D | XT Hard disk, AT Parallel Printer, IRQ5 |
| Ø38–Ø3B | E | Floppy Diskette, IRQ6 |
| Ø3C–Ø3F | F | Parallel Printer 1, IRQ7, slave 8259, IRET |
| Ø4Ø–Ø43 | 1Ø | ROM Handler – Video |
| Ø44–Ø47 | 11 | ROM Handler – Equipment Check |
| Ø48–Ø4B | 12 | ROM Handler – Memory Check |
| Ø4C–Ø4F | 13 | ROM Handler – Diskette I/O |
| Ø5Ø–Ø53 | 14 | ROM Handler – COMM I/O |
| Ø54–Ø57 | 15 | XT Cassette, AT ROM Catchall Handlers |
| Ø58–Ø5B | 16 | ROM Handler – Keyboard I/O |
| Ø5C–Ø5F | 17 | ROM Handler – Printer I/O |
| Ø6Ø–Ø63 | 18 | ROM Handler – Basic Startup |
| Ø64–Ø67 | 19 | ROM Handler – Bootstrap |
| Ø68–Ø6B | 1A | ROM Handler – Time of Day |
| Ø6C–Ø6F | 1B | ROM Handler – Keyboard Break |
| Ø7Ø–Ø73 | 1C | ROM Handler – User Ticker |
| Ø74–Ø77 | 1D | ROM Pointer, Video Initialization |
| Ø78–Ø7B | 1E | ROM Pointer, Diskette Parameters |
| Ø7C–Ø7F | 1F | ROM Pointer, Graphics Characters Set 2 |
| Ø8Ø–Ø83 | 2Ø | DOS – Terminate Program |
| Ø84–Ø87 | 21 | DOS – Function Call |
| Ø88–Ø8B | 22 | DOS – Program's Terminate Address |
| Ø8C–Ø8F | 23 | DOS – Program's Control–Break Address |
| Ø9Ø–Ø93 | 24 | DOS – Critical Error Handler |
| Ø94–Ø97 | 25 | DOS – Absolute Disk Read |
| Ø98–Ø9B | 26 | DOS – Absolute Disk Write |
| Ø9C–Ø9F | 27 | DOS – TSR Terminate & Stay Ready |
| ØAØ–ØFF | 28–3F | DOS – Idle Loop, IRET |
| 1ØØ–1Ø3 | 4Ø | Hard Disk Pointer – Original Floppy Handler |
| 1Ø4–1Ø7 | 41 | ROM Pointer, XT Hard Disk Parameters |
| 1Ø8–1ØB | 42–45 | Reserved |
| 1ØC–1ØF | 46 | ROM Pointer, AT Hard Disk Parameters |
| 11Ø–17F | 47–5F | Reserved |
| 18Ø–19F | 6Ø–67 | Reserved for User (67 is Expanded Mem) |
| 1AØ–1BF | 68–6F | Not Used |
| 1CØ–1C3 | 7Ø | AT Real Time Clock, IRQ8 |
| 1C4–1C7 | 71 | AT Redirect to IRQ2, IRQ9, LAN Adapter 1 |
| 1C8–1CB | 72 | AT Reserved, IRQ1Ø |
| 1CC–1CF | 73 | AT Reserved, IRQ11 |
| 1DØ–1D3 | 74 | AT Reserved, IRQ12 |
| 1D4–1D7 | 75 | AT 8Ø287 Error to NMI, IRQ13 |
| 1D8–1DB | 76 | AT Hard Disk, IRQ14 |
| 1DC–1DF | 77 | AT Reserved, IRQ15 |
| 1EØ–1FF | 78–7F | Not Used |
| 2ØØ–217 | 8Ø–85 | Reserved for BASIC |
| 218–21B | 86 | NetBIOS, Relocated Interrupt 18H |
| 218–3C3 | 87–FØ | Reserved for BASIC Interpreter |
| 3C4–3FF | F1–FF | Not Used |

# IBM XT/AT CLASS ERROR CODES

A variety of tests are executed automatically when XT/AT class computers are first turned on. Initially, the "Power-On Self Test" (POST) is run. It provides error or warning messages whenever a faulty component is encountered. Typically, two types of messages are issued: **audio codes** and **display screen** messages or codes.

Audio codes consist of a series of beeps that identify a faulty component. If your computer is functioning normally, you will hear one short beep when the system is turned on. However, if a problem is detected, a series of beeps will occur. These audio codes define the problem and are typically the following:

| Beep Code | Problem |
|---|---|
| No beep, continuous beep, or repeating short beeps | Power Supply |
| 1 long beep and 1 short beep | System Board |
| 1 long beep and 2 short beeps, or 1 short beep and blank | Monitor adapter card and/or monitor cable and/or wrong display. |
| 1 short beep and either the red drive LED staying on or Personal Computer BASIC statement | Drive and/or drive adapter card |

If the system completes the POST process, then additional errors are reported in the form of display error messages. See the next page for typical codes and their descriptions.

IBM is a registered trademark of the International Business Machine Corporation.

# MEGABYTES AND KILOBYTES

1 kilobyte = $2^{10}$ bytes = exactly 1,024 bytes

1 megabyte = $2^{20}$ bytes = exactly 1,048,576 bytes

1 gigabyte = 1 billion bytes   1 terabyte = 1 trillion bytes

1 byte = 8 bits (bit is short for binary digit)

8 bit computers (such as the 8088) move data in 1 byte chunks

16 bit computers (such as the 80286) move data in 2 byte chunks

32 bit computers (such as the 80386) move data in 4 byte chunks

| Code | Description |
|---|---|
| 01x | Undetermined problem errors |
| 02x | Power supply errors |
| **1xx** | **System board error** |
| 101 | Interrupt failure |
| 102 | Timer failure |
| 103 | Timer interrupt failure |
| 104 | Protected mode failure |
| 105 | Last 8042 command not accepted |
| 106 | Converting logic test |
| 107 | Hot NMI test |
| 108 | Timer bus test |
| 109 | Direct memory access test error |
| 121 | Unexpected hardware interrupts occurred |
| 131 | Cassette wrap test failed |
| 152 | |
| 161 | System Options Error-(Run SETUP) [Battery failure] |
| 162 | System options not set correctly-(Run SETUP) |
| 163 | Time and date not set-(Run SETUP) |
| 164 | Memory size error-(Run SETUP) |
| 199 | User indicated configuration not correct |
| **2xx** | **Memory (RAM) errors** |
| 201 | Memory test failed |
| 202 | Memory address error |
| 203 | Memory address error |
| **3xx** | **Keyboard errors** |
| 301 | Keyboard did not respond to software reset correctly or a stuck key failure was detected. If a stuck key was detected, the scan code for the key is displayed in hexadecimal. For example, the error code 49 301 indicates that key 73, the PgUp key has failed (49 Hex = 73 decimal) |
| 302 | User indicated error from the keyboard test or AT system unit keylock is locked |
| 303 | Keyboard or system unit error |
| 304 | Keyboard or system unit error; CMOS does not match system |
| **4xx** | **Monochrome monitor errors** |
| 401 | Monochrome memory test, horizontal sync frequency test, or video test failed |
| 408 | User indicated display attributes failure |
| 416 | User indicated character set failure |
| 424 | User indicated 80X25 mode failure |
| 432 | Parallel port test failed (monochrome adapter) |

# IBM XT/AT CLASS ERROR CODES

| Code | Description |
|---|---|
| **5xx** | **Color monitor errors** |
| 501 | Color memory test failed, horizontal sync frequency test, or video test failed |
| 508 | User indicated display attribute failure |
| 516 | User indicated character set failure |
| 524 | User indicated 80X25 mode failure |
| 532 | User indicated 40X25 mode failure |
| 540 | User indicated 320X200 graphics mode failure |
| 548 | User indicated 640X200 graphics mode failure |
| **6xx** | **Diskette drive errors** |
| 601 | Diskette power on diagnostics test failed |
| 602 | Diskette test failed; boot record is not valid |
| 606 | Diskette verify function failed |
| 607 | Write protected diskette |
| 608 | Bad command diskette status returned |
| 610 | Diskette initialization failed |
| 611 | Time-out - diskette status returned |
| 612 | Bad NEC - diskette status returned |
| 613 | Bad DMA - diskette status returned |
| 621 | Bad seek - diskette status returned |
| 622 | Bad CRC - diskette status returned |
| 623 | Record not found - diskette status returned |
| 624 | Bad address mark - diskette status returned |
| 625 | Bad NEC seek - diskette status returned |
| 626 | Diskette data compare error |
| **7xx** | **8087 or 80287 math coprocessor errors** |
| **9xx** | **Parallel printer adapter errors** |
| 901 | Parallel printer adapter test failed |
| **10xx** | **Reserved for parallel printer adapter** |
| **11xx** | **Asynchronous communications adapter errors** |
| 1101 | Async communications adapter test failed |
| **12xx** | **Alternate asynchronous communications adapter errors** |
| 1201 | Alternateasynchronous communications adapter test failed |
| **13xx** | **Game control adapter errors** |
| 1301 | Game control adapter test failed |
| 1302 | Joystick test failed |
| **14xx** | **Printer errors** |
| 1401 | Printer test failed |
| 1404 | Matrix printer failed |
| **15xx** | **Synchronous data link control (SDLC) comm adapter errors** |
| 1510 | 8255 port B failure |
| 1511 | 8255 port A failure |

| Code | Description |
|---|---|
| 1512 | 8255 port C failure |
| 1513 | 8253 timer 1 did not reach terminal count |
| 1514 | 8253 timer 1 stuck on |
| 1515 | 8253 timer 0 did not reach terminal count |
| 1516 | 8253 timer 0 stuck on |
| 1517 | 8253 timer 2 did not reach terminal count |
| 1518 | 8253 timer 2 stuck on |
| 1519 | 8273 port B error |
| 1520 | 8273 port A error |
| 1521 | 8273 command/read time-out |
| 1522 | Interrupt level 4 failure |
| 1523 | Ring Indicate stuck on |
| 1524 | Receive clock stuck on |
| 1525 | Transmit clock stuck on |
| 1526 | Test indicate stuck on |
| 1527 | Ring indicate not on |
| 1528 | Receive clock not on |
| 1529 | Transmit clock not on |
| 1530 | Test indicate not on |
| 1531 | Data set ready not on |
| 1532 | Carrier detect not on |
| 1533 | Clear to send not on |
| 1534 | Data set ready stuck on |
| 1536 | Clear to send stuck on |
| 1537 | Level 3 interrupt failure |
| 1538 | Receive interrupt results error |
| 1539 | Wrap data mis-compare |
| 1540 | DMA channel 1 error |
| 1541 | DMA channel 1 error |
| 1542 | Error in 8273 error checking or status reporting |
| 1547 | Stray interrupt level 4 |
| 1548 | Stray interrupt level 3 |
| 1549 | Interrupt presentation sequence time-out |
| **16xx** | **Display emulation errors (327x, 5520, 525x)** |
| **17xx** | **Fixed disk errors** |
| 1701 | Fixed disk POST error |
| 1702 | Fixed disk adapter error |
| 1703 | Fixed disk drive error |
| 1704 | Fixed disk adapter or drive error |
| 1780 | Fixed disk 0 failure |
| 1781 | Fixed disk 1 failure |
| 1782 | Fixed disk controller failure |
| 1790 | Fixed disk 0 error |
| 1791 | Fixed disk 1 error |
| **18xx** | **I/O expansion unit errors** |

# IBM XT/AT CLASS ERROR CODES

| Code | Description |
|---|---|
| 1801 | I/O expansion unit POST error |
| 1810 | Enable/Disable failure |
| 1811 | Extender card warp test failed (disabled) |
| 1812 | High order address lines failure (disabled) |
| 1813 | Wait state failure (disabled) |
| 1814 | Enable/Disable could not be set on |
| 1815 | Wait state failure (disabled) |
| 1816 | Extender card warp test failed (enabled) |
| 1817 | High order address lines failure (enabled) |
| 1818 | Disable not functioning |
| 1819 | Wait request switch not set correctly |
| 1820 | Receiver card wrap test failure |
| 1821 | Receiver high order address lines failure |
| **19xx** | **3270 PC attachment card errors** |
| **20xx** | **Binary synchronous communications (BSC)** adapter errors |
| 2010 | 8255 port A failure |
| 2011 | 8255 port B failure |
| 2012 | 8255 port C failure |
| 2013 | 8253 timer 1 did not reach terminal count |
| 2014 | 8253 timer 1 stuck on |
| 2016 | 8253 timer 2 did not reach terminal count or timer 2 stuck on |
| 2017 | 8251 Data set ready failed to come on |
| 2018 | 8251 Clear to send not sensed |
| 2019 | 8251 Data set ready stuck on |
| 2020 | 8251 Clear to send stuck on |
| 2021 | 8251 hardware reset failed |
| 2022 | 8251 software reset failed |
| 2023 | 8251 software "error reset" failed |
| 2024 | 8251 transmit ready did not come on |
| 2025 | 8251 receive ready did not come on |
| 2026 | 8251 could not force "overrun" error status |
| 2027 | Interrupt failure - no timer interrupt |
| 2028 | Interrupt failure - transmit, replace card or planar |
| 2029 | Interrupt failure - transmit, replace card |
| 2030 | Interrupt failure - receive, replace card or planar |
| 2031 | Interrupt failure - receive, replace card |
| 2033 | Ring indicate stuck on |
| 2034 | Receive clock stuck on |
| 2035 | Transmit clock stuck on |
| 2036 | Test indicate stuck on |
| 2037 | Ring indicate stuck on |
| 2038 | Receive clock not on |
| 2039 | Transmit clock not on |

| Code | Description |
|---|---|
| 2040 | ...... Test indicate not on |
| 2041 | ...... Data set ready not on |
| 2042 | ...... Carrier detect not on |
| 2043 | ...... Clear to send not on |
| 2044 | ...... Data set ready stuck on |
| 2045 | ...... Carrier detect stuck on |
| 2046 | ...... Clear to send stuck on |
| 2047 | ...... Unexpected transmit interrupt |
| 2048 | ...... Unexpected receive interrupt |
| 2049 | ...... Transmit data did not equal receive data |
| 2050 | ...... 8251 detected overrun error |
| 2051 | ...... Lost data set ready during data wrap |
| 2052 | ...... Receive time-out during data wrap |
| **21xx** | ...... **Alternate binary synchronous communications adapter errors** |
| 2110 | ...... 8255 port A failure |
| 2111 | ...... 8255 port B failure |
| 2112 | ...... 8255 port C failure |
| 2113 | ...... 8253 timer 1 did not reach terminal count |
| 2114 | ...... 8253 timer 1 stuck on |
| 2115 | ...... 8253 timer 2 did not reach terminal count or timer 2 stuck on |
| 2116 | ...... 8251 Data set ready failed to come on |
| 2117 | ...... 8251 Clear to send not sensed |
| 2118 | ...... 8251 Data set ready stuck on |
| 2119 | ...... 8251 Clear to send stuck on |
| 2120 | ...... 8251 hardware reset failed |
| 2121 | ...... 8251 software reset failed |
| 2122 | ...... 8251 software "error reset" failed |
| 2123 | ...... 8251 transmit ready did not come on |
| 2124 | ...... 8251 receive ready did not come on |
| 2125 | ...... 8251 could not force "overrun" error status |
| 2126 | ...... Interrupt failure - no timer interrupt |
| 2128 | ...... Interrupt failure - transmit, replace card or planar |
| 2129 | ...... Interrupt failure - transmit, replace card |
| 2130 | ...... Interrupt failure - receive, replace card or planar |
| 2131 | ...... Interrupt failure - receive, replace card |
| 2133 | ...... Ring indicate stuck on |
| 2134 | ...... Receive clock stuck on |
| 2135 | ...... Transmit clock stuck on |
| 2136 | ...... Test indicate stuck on |
| 2137 | ...... Ring indicate stuck on |
| 2138 | ...... Receive clock not on |
| 2139 | ...... Transmit clock not on |

# IBM XT/AT CLASS ERROR CODES

| Code | Description |
|---|---|
| 2140 | Test indicate not on |
| 2141 | Data set ready not on |
| 2142 | Carrier detect not on |
| 2143 | Clear to send not on |
| 2144 | Data set ready stuck on |
| 2145 | Carrier detect stuck on |
| 2146 | Clear to send stuck on |
| 2147 | Unexpected transmit interrupt |
| 2148 | Unexpected receive interrupt |
| 2149 | Transmit data did not equal receive data |
| 2150 | 8251 detected overrun error |
| 2151 | Lost data set ready during data wrap |
| 2152 | Receive time-out during data wrap |
| **22xx** | **Cluster adapter errors** |
| **24xx** | **Enhanced graphics adapter errors** |
| **29xx** | **Color matrix printer errors** |
| 2901 | |
| 2902 | |
| 2904 | |
| **33xx** | **Compact printer errors** |



# IBM HARDWARE RELEASES

| Date | Code | Hardware Release |
|---|---|---|
| 04-24-81 | FF | PC (the original!) |
| 10-19-81 | FF | PC (fixed bugs) |
| 08-16-81 | FE | XT |
| 10-27-82 | FF | PC with hard drive support and 640k |
| 11-08-82 | FE | PC-XT portable |
| 06-01-83 | FD | PC jr |
| 01-10-84 | FC | AT |
| 06-10-85 | FC | AT revision 1 |
| 09-13-85 | F9 | PC Convertible |
| 11-15-85 | FC | AT w/speed control, 30 meg hard disk |
| 01-10-86 | FB | XT revision 1 |
| 04-21-86 | FC | XT-286 model 2 |
| 05-09-86 | FB | XT revision 2 |
| 09-02-86 | FA | PS/2 Model 30 |
| 02-13-87 | FC | PS/2 Model 50 model 4 |
| 02-13-87 | FC | PS/2 Model 60 model 5 |
| 03-30-87 | F8 | PS/2 Model 80 16 mhz |
| 10-07-87 | F8 | PS/2 Model 80 20 mhz |

# STANDARD 80286 HARD DISK TYPES

| Drive Type | # of Cylinders | # of Heads | Write Precomp | Land Zone | Size in Megabytes |
|---|---|---|---|---|---|
| 1 | 306 | 4 | 128 | 305 | 10 |
| 2 | 615 | 4 | 300 | 615 | 21 |
| 3 | 615 | 6 | 300 | 615 | 31 |
| 4 | 940 | 8 | 512 | 940 | 63 |
| 5 | 940 | 6 | 512 | 940 | 47 |
| 6 | 615 | 4 | 65535 | 615 | 21 |
| 7 | 462 | 8 | 256 | 511 | 31 |
| 8 | 733 | 5 | 65535 | 733 | 31 |
| 9 | 900 | 15 | 65535 | 901 | 112 |
| 10 | 820 | 3 | 65535 | 820 | 21 |
| 11 | 855 | 5 | 65535 | 855 | 36 |
| 12 | 855 | 7 | 65535 | 855 | 50 |
| 13 | 306 | 8 | 128 | 319 | 21 |
| 14 | 733 | 7 | 65535 | 733 | 43 |
| 15 | 0 | 0 | 0 | 0 | 0 |
| 16 | 612 | 4 | 0 | 663 | 21 |
| 17 | 977 | 5 | 300 | 977 | 41 |
| 18 | 977 | 7 | 65535 | 977 | 57 |
| 19 | 1024 | 7 | 512 | 1023 | 60 |
| 20 | 733 | 5 | 300 | 732 | 31 |
| 21 | 733 | 7 | 300 | 732 | 43 |
| 22 | 733 | 5 | 300 | 733 | 31 |
| 23 | 306 | 4 | 0 | 336 | 10 |
| 24 | 698 | 7 | 300 | 732 | 42 |
| 25 | 615 | 4 | 0 | 615 | 21 |
| 26 | 1024 | 4 | 65535 | 1023 | 34 |
| 27 | 1024 | 5 | 65535 | 1023 | 43 |
| 28 | 1024 | 8 | 65535 | 1023 | 68 |
| 29 | 512 | 8 | 256 | 512 | 34 |
| 30 | 615 | 2 | 615 | 615 | 10 |
| 31 | 732 | 7 | 300 | 732 | 44 |
| 32 | 1023 | 5 | 65535 | 1023 | 44 |
| 33 | 306 | 4 | 0 | 340 | 10 |
| 34 | 976 | 5 | 488 | 977 | 42 |
| 35 | 1024 | 9 | 1024 | 1024 | 77 |
| 36 | 1024 | 5 | 512 | 1024 | 43 |
| 37 | 830 | 10 | 65535 | 830 | 69 |
| 38 | 823 | 10 | 256 | 824 | 68 |
| 39 | 615 | 4 | 128 | 664 | 21 |
| 40 | 615 | 8 | 128 | 664 | 41 |
| 41 | 917 | 15 | 65535 | 918 | 114 |
| 42 | 1023 | 15 | 65535 | 1024 | 127 |
| 43 | 823 | 10 | 512 | 823 | 68 |
| 44 | 820 | 6 | 65535 | 820 | 41 |
| 45 | 1024 | 8 | 65535 | 1024 | 68 |
| 46 | 925 | 9 | 65535 | 925 | 69 |
| 47 | 699 | 7 | 256 | 700 | 41 |

Note: Drive types over #24 vary between computer manufacturers

# DIABLO 630 PRINTER CODES

| Code | Hex | Decimal | Command |
|---|---|---|---|
| **Page Format Control:** | | | |
| ESC 9 | 1B 39 | 27 57 | Set left margin at current position |
| ESC Ø | 1B 3Ø | 27 48 | Set right margin at current position |
| ESC T | 1B 54 | 27 84 | Set top margin at current position |
| ESC L | 1B 4C | 27 76 | Set bottom margin at current posit. |
| ESC C | 1B 43 | 27 67 | Clear top and bottom margins |
| ESC FF # | 1B ØC # | 27 12 # | Set lines/page, # is 1 to 126 lines |
| **Horizontal Movement and Spacing Control:** | | | |
| CR | ØD | 13 | Carriage return |
| ESC M | 1B 4D | 27 77 | Enable auto justify |
| ESC = | 1B 3D | 27 61 | Enable auto center |
| ESC ? | 1B 3F | 27 63 | Enable auto carriage return |
| ESC ! | 1B 21 | 27 33 | Disable auto carriage return |
| ESC / | 1B 2F | 27 47 | Enable auto backward printing |
| ESC \ | 1B 5C | 27 92 | Disable auto backward printing |
| ESC < | 1B 3C | 27 6Ø | Enable reverse printing |
| ESC > | 1B 3E | 27 62 | Disable reverse printing |
| ESC 5 | 1B 35 | 27 53 | Enable forward printing |
| ESC 6 | 1B 36 | 27 54 | Enable backward printing |
| SP | 2Ø | 32 | Space |
| BS | Ø8 | Ø8 | Backspace |
| ESC BS | 1B Ø8 | 27 Ø8 | Backspace 1/12Ø inch |
| HT | Ø9 | Ø9 | Horizontal tab |
| ESC HT # | 1B Ø9 # | 27 Ø9 # | Absolute horizontal tab, # is column 1 to 126 |
| ESC DC1 # | 1B 11 # | 27 17 # | Spacing offset, # is 1 to 126 (1/12Ø" units), where # 1 = offset 1 to # 63 = offset 63, # 64 = offset Ø, # 65 =offset –1 to # 126 = offset –62 |
| ESC 1 | 1B 31 | 27 49 | Set horizontal tab stop at current position |
| ESC 8 | 1B 38 | 27 56 | Clear horizontal tab at current position |

# DIABLO 630 PRINTER CODES

| Code | Hex | Decimal | Command |
|---|---|---|---|
| **Horizontal Movement and Spacing Control: (Continued)** | | | |
| ESC 2 | 1B 32 | 27 5Ø | Clear all vertical and horizontal tab stops |
| ESC US # | 1B 1F # | 27 31 # | Set horizontal motion index, # is 1 to 126, where (#–1)/12Ø inch is the column spacing. |
| ESC S | 1B 53 | 27 83 | Return HMI control to spacing switch |
| **Vertical Movement and Spacing Control:** | | | |
| LF | ØA | 1Ø | Line feed |
| ESC LF | 1B ØA | 27 1Ø | Reverse line feed |
| ESC U | 1B 55 | 27 85 | Half line feed |
| ESC D | 1B 44 | 27 68 | Reverse half line feed |
| FF | ØC | 12 | Form feed |
| VT | ØB | 11 | Vertical tab |
| ESC VT # | 1B ØB # | 27 11 # | Absolute vertical tab, # is line 1 to 126 |
| ESC _ | 1B 2D | 27 45 | Set vertical tab stop at current position |
| ESC 2 | 1B 32 | 27 5Ø | Clear all vertical and horizontal tab stops |
| ESC RS # | 1B 1E # | 27 3Ø # | Set vertical motion index, # is 1 to 126, where (#-1)/48 inch is the line spacing. |
| **Character Selection:** | | | |
| ESC P | 1B 5Ø | 27 8Ø | Enable proportional print spacing |
| ESC Q | 1B 51 | 27 81 | Disable proportional print spacing |
| ESC SO DC2 | 1B ØE 12 | 27 14 18 | Enable printwheel down–load mode |
| DC4 | 14 | 28 | Exit printwheel down-load |
| SO | ØE | 14 | Enable ESC mode, supplementary characters |
| SI | ØF | 15 | Disable ESC mode, primary characters |
| ESC A | 1B 41 | 27 65 | Select red ribbon (secondary font) |
| ESC B | 1B 42 | 27 66 | Select black ribbon (primary font) |
| ESC X | 1B 58 | 27 88 | Cancel all WP modes except Proportional |
| ESC Y | 1B 59 | 27 89 | Printwheel Spoke Ø char. |

# DIABLO 630 PRINTER CODES

| Code | Hex | Decimal | Command |
|---|---|---|---|
| **Character Selection: (Continued)** | | | |
| ESC Z | 1B 5A | 27 9Ø | Printwheel Spoke 95 char. |
| **Character Highlight Selection:** | | | |
| ESC E | 1B 45 | 27 69 | Enable underscore print |
| ESC R | 1B 52 | 27 82 | Disable underscore print |
| ESC O | 1B 4F | 27 79 | Enable bold printing |
| ESC W | 1B 57 | 27 87 | Enable shadow printing |
| ESC & | 1B 26 | 27 38 | Disable bold and shadow printing |
| **Graphics:** | | | |
| ESC 3 | 1B 33 | 27 51 | Enable graphics mode |
| ESC 4 | 1B 34 | 27 52 | Disable graphics mode |
| ESC G | 1B 47 | 27 71 | Enable HyPLOT mode |
| **Miscellaneous:** | | | |
| ESC CR P | 1B ØD 5Ø | 27 13 8Ø | Reset all modes to default |
| ESC SUB I | 1B 1A 49 | 27 27 73 | Reset all modes to default |
| ESC EM | 1B 19 | 27 25 | Enable auto sheet feeder |
| ESC SUB | 1B 1A | 27 26 | Enable remote diagnostics |
| ESC N | 1B 4E | 27 78 | Restore normal carriage settling time |
| ESC % | 1B 25 | 27 37 | Increase carriage settling time |
| ESC 7 | 1B 37 | 27 55 | Enable print suppression |
| ESC SO M | 1B ØE 4D | 27 14 77 | Enable program mode |

# EPSON FX-80 PRINTER CODES

| Code | Hex | Decimal | Command |
|---|---|---|---|
| Page Format Control: | | | |
| ESC l # | 1B 6C # | 27 1Ø8 # | Set Left Margin at Col # |
| ESC Q # | 1B 51 # | 27 81 # | Set Right Margin at Col # |
| ESC C # | 1B 43 # | 27 67 # | Set Form Length to # Lines (or n inches) |
| ESC C Ø # | 1B 43 ØØ # | 27 67 Ø # | Set Form Length to # inches |
| ESC N # | 1B 4E # | 27 78 # | Set Skip–over Perforation to # lines |
| ESC O | 1B 4F | 27 79 | Turn Skip–over Perforation Off |

# EPSON FX-80 PRINTER CODES

| Code | Hex | Decimal | Command |
|---|---|---|---|
| **Horizontal Movement and Spacing Control:** | | | |
| CR | ØD | 13 | Carriage return |
| BS | Ø8 | Ø8 | Backspace |
| HT | Ø9 | Ø9 | Horizontal tab |
| ESC a Ø | 1B 61 ØØ | 1B 61 Ø | Alignment Left Justified |
| ESC a 1 | 1B 61 Ø1 | 1B 61 1 | Alignment Auto Centering |
| ESC a 2 | 1B 61 Ø2 | 1B 61 2 | Alignment Right Justified |
| ESC a 3 | 1B 61 Ø3 | 1B 61 3 | Alignment Auto Justified |
| ESC D#Ø | 1B 44 # Ø | 27 68 # ØØ | Set Horizontal Tab(s), # can be 1 or a series of tabs (columns) |
| ESC D Ø | 1B 44 Ø | 27 68 ØØ | Release Horizontal Tab |
| ESC eØ# | 1B 44 Ø # | 27 68 ØØ # | Set Horizontal Unit Tab(s), # is repeating Tab distance in columns |
| ESC e ØØ | 1B 44 Ø Ø | 27 68 ØØ ØØ | Release Horiz Tab Unit |
| ESC f Ø # | 1B 66 ØØ # | 27 1Ø2 Ø # | Move print position # cols |
| ESC \ #1#2 | 1B 5C #1#2 | 27 92 #1#2 | Move print position in increments of 1/12Ø inch |
| ESC $ #1#2 | 1B 24 #1#2 | 27 36 #1#2 | Move print position in 1/6Ø inch increments from left margin |
| ESC SP # | 1B 2Ø # | 27 32 # | Add space after each character in units of 1/24Ø inch where # is from 1 to 63 |
| ESC < | 1B 3C | 27 6Ø | One Line Unidirectional Printing Mode On |
| ESC U | 1B 55 | 27 85 | Select Continuous Print Unidirectional Mode |
| **Vertical Movement and Spacing Control:** | | | |
| LF | ØA | 1Ø | Line feed |
| ESC j # | 1B 6A # | 27 1Ø6 # | Reverse Line Feed of #/216 Inch |
| ESC J # | 1B 4A # | 27 74 # | Forward Line Feed of #/216 inches |
| ESC f 1 # | 1B 66 Ø1 # | 27 1Ø2 1 # | Forward Line Feed # lines |
| FF | ØC | 12 | Form feed |
| ESC Ø | 1B 3Ø | 27 48 | Set Line Spacing to 1/8" (9 points or 8 lpi) |
| ESC 1 | 1B 31 | 27 49 | Set Line Spacing to 7/72" (7 points) |
| ESC 2 | 1B 32 | 27 5Ø | Set Line Spacing to 1/6" (12 points, 6 lpi) |
| ESC 3 # | 1B 33 # | 27 51 # | Set Line Spacing to #/216" |
| ESC A # | 1B 41 # | 27 65 # | Set Line Spacing to # Points (#/72 inch) |

| Code | Hex | Decimal | Command |
|---|---|---|---|
| **Vertical Movement and Spacing Control: (Continued)** | | | |
| VT | ØB | 11 | Vertical tab |
| ESC b#1#2#3 Ø | 1B 62 #1#2#3 ØØ | 27 98 #1#2#3 Ø | Set Vertical Tabs Format Units in Specific Channel, see the manual for details |
| ESC b #1 Ø | 1B 62 #1 ØØ | 27 98 #1 Ø | Release Vertical Tab Format Unit |
| ESC / # | 1B 2F # | 27 47 # | Select Vertical Tab Channel # |
| ESC B#1#2Ø | 1B 42 #1#2 Ø | 27 66 #1 #2 Ø | Set Vertical Tabs for Channel #1, #2 etc |
| ESC B Ø | 1B 42 Ø | 27 66 Ø | Release Vertical Tabs for Channels |
| ESC e 1 # | 1B 65 Ø1 # | 27 1Ø1 1 # | Set Vertical Tab Unit at # of equal space intervals |
| ESC e 1 1 | 1B 65 Ø1 Ø1 | 27 1Ø1 1 1 | Release Vertical Tab Unit of equal space intervals |
| **Character Selection:** | | | |
| ESC I 1 | 1B 49 Ø1 | 27 73 1 | Select Characters (Ø–31, 128–159) to Print |
| ESC I Ø | 1B 49 ØØ | 27 73 Ø | Disable Characters (Ø–31, 128–159) from Printing |
| ESC M | 1B 4D | 27 77 | Enable Elite Pitch Mode |
| ESC P | 1B 5Ø | 27 8Ø | Enable Pica Pitch Mode |
| ESC o | 1B 6F | 27 111 | Enable Elite Pitch Mode |
| ESC n | 1B 6E | 27 11Ø | Enable Pica Pitch Mode |
| ESC w # | 1B 77 # | 27 119 # | Direct Pitch Selection, #=Ø is 1Øcpi, #=1 is 12cpi, #=2 is 15cpi, #=3 is 17cpi, #=4 is proport. |
| ESC p 1 | 1B 7Ø Ø1 | 27 112 1 | Select Proportional Spac |
| ESC p Ø | 1B 7Ø ØØ | 27 112 Ø | Release Proportional Spa |
| ESC W 1 | 1B 57 Ø1 | 27 87 1 | Select Expanded Pitch |
| ESC W Ø | 1B 57 ØØ | 27 87 Ø | Release Expanded Pitch |
| SO or ESC SO | ØE | 14 | Enable 1–line Expanded Print Mode |
| DC4 | 14 | 28 | Disable one–line Expanded Print Mode |
| SI or ESC SI | ØF | 15 | Enable Compressed Print |
| DC2 | 12 | 18 | Disable Compressed Print |
| ESC : | 1B 3A | 27 58 | Duplicate Internal Font |
| ESC ! # | 1B 21 # | 27 33 # | Print Mode Selection, # determines mode, #=128 is underline, #=64 is italic, #=32 is double wide, #=16 is double strike, |

# EPSON FX-80 PRINTER CODES

| Code | Hex | Decimal | Command |
|---|---|---|---|
| **Character Selection: (Continued)** | | | |
| | #=8 is bold, #=4 is compressed, #=2 is proportional, #=1 is Elite, #=Ø is Pica. Add numbers for multiples, eg, 129 is Underlined Elite | | |
| ESC % | 1B 25 | 27 37 | Select Character Set Bank |
| ESC & | 1B 26 | 27 38 | Define User Font |
| ESC 6 | 1B 36 | 27 54 | Enable printing High Bit Symbols (Dec128–Dec159) |
| ESC 7 | 1B 37 | 27 55 | Disable printing High Bit Symbols (Dec128–Dec159) |
| ESC 4 | 1B 34 | 27 52 | Enable Italics printing |
| ESC 5 | 1B 35 | 27 53 | Disable Italics printing |
| ESC R # | 1B 52 # | 27 82 # | Select International Character Set, #=Ø is USA, 1 is France, 2 is Germany, 3 is England, 4 is Denmark A, 5 is Sweden, 6 is Italy, 7 is Spain, 8 is Japan, 9 is Norway, 1Ø is Denmark B |
| ESC S 1 | 1B 53 Ø1 | 27 83 1 | Select Subscripting |
| ESC S Ø | 1B 53 ØØ | 27 83 Ø | Select Superscripting |
| ESC T | 1B 54 | 27 84 | Release Super or Subscripting |
| **Character Highlight Selection:** | | | |
| ESC – 1 | 1B 2D Ø1 | 27 45 1 | Turn underline mode on |
| ESC – Ø | 1B 2D ØØ | 27 45 Ø | Turn underline mode off |
| ESC E | 1B 45 | 27 69 | Enable Bold Print Mode |
| ESC F | 1B 46 | 27 7Ø | Disable Bold Print Mode |
| ESC G | 1B 47 | 27 71 | Enable Double–strike |
| ESC H | 1B 48 | 27 72 | Disable Double–strike |
| **Graphics:** | | | |
| | For values for #1 and #2 below, see printer manuals | | |
| ESC K#1#2 | 1B 4B #1#2 | 27 75 #1#2 | Enable Single–density Graphics Mode, 6Ø dpi |
| ESC L#1#2 | 1B 4C #1#2 | 27 76 #1#2 | Enable Double–density Graphics Mode, 12Ø dpi |
| ESC Y#1#2 | 1B 59 #1#2 | 27 89 #1#2 | Enable Double–density, 12Ø dpi, High–speed Graphics Mode |
| ESC Z#1#2 | 1B 5A #1#2 | 27 9Ø #1#2 | Enable Quadruple – density Graphics Mode, 24Ø dpi |
| ESC *#1#2#3 | 1B 2A #1#2#3 | 27 42 #1#2#3 | Set Graphics Mode |
| ESC ^#1#2#3 | 1B 5E #1 #2 #3 | 27 94 #1#2#3 | 9 pin Graphics Mode |
| ESC ?#1#2 | 1B 3F #1#2 | 27 63 #1#2 | Bit Image Mode Reassignment |

## EPSON FX-80 PRINTER CODES

| Code | Hex | Decimal | Command |
|---|---|---|---|
| **Miscellaneous:** | | | |
| CAN | 18 | 24 | Cancel |
| DC1 | 11 | 17 | Remote Printer Select |
| DC3 | 13 | 19 | Remote Printer Deselect |
| DEL | 7F | 127 | Delete |
| ESC @ | 1B 4Ø | 27 64 | Master Reset |
| ESC # | 1B 23 | 27 35 | Read Bit 7 of Received Word Normally |
| ESC = | 1B 3D | 27 61 | Set Received Bit 7 to Ø |
| ESC > | 1B 3E | 27 62 | Set Received Bit 7 to 1 |
| ESC 8 | 1B 38 | 27 56 | Out of Paper Sensor Off |
| ESC 9 | 1B 39 | 27 57 | Out of Paper Sensor On |
| ESC i | 1B 69 | 27 1Ø5 | Enable Immediate Printing |
| ESC s | 1B 73 | 27 115 | Half Speed Printing |
| ESC EM # | 1B 19 # | 27 25 # | Paper Cassette Selection, #=E is envelope, #=1 is Lower Cassette, #=2 is Upper Cassette, #=R is eject page |

## HP LASERJET PRINTER CODES

| Code | Hex | Decimal | Command |
|---|---|---|---|
| Page Format Control: | | | |
| ESC & l ØO | 1B 26 6C 3Ø 4F | 27 38 1Ø8 48 79 | Portrait Orient. |
| ESC & l 1O | 1B 26 6C 31 4F | 27 38 1Ø8 49 79 | Landscape Orient. |
| ESC & l #P | 1B 26 6C # 5Ø | 27 38 1Ø8 # 8Ø | Page length, # of lines |
| ESC & l #E | 1B 26 6C # 45 | 27 38 1Ø8 # 69 | Top Margin, # of lines |
| ESC & l #F | 1B 26 6C # 46 | 27 38 1Ø8 # 7Ø | Text Length, # of lines |
| ESC & l 1L | 1B 26 6C 31 4C | 27 38 1Ø8 49 76 | Skip Perforation, Set on |
| ESC & l ØL | 1B 26 6C 3Ø 4C | 27 38 1Ø8 48 76 | Skip Perforation, Set off |
| ESC & l #D | 1B 26 6C # 44 | 27 38 1Ø8 # 68 | Lines Per Inch, # of lines/inch |
| ESC & l #C | 1B 26 6C # 43 | 27 38 1Ø8 # 67 | Vertical Motion Index, # of 1/48 inch |
| ESC &k#H | 1B 26 6B # 48 | 27 38 1Ø7 # 72 | Horizontal Motion Index, # of 1/12Ø inch |
| ESC &a#L | 1B 26 61 # 4C | 27 38 97 # 76 | Left Margin, Left column # |

# HP LASERJET PRINTER CODES

| Code | Hex | Decimal | Command |
|---|---|---|---|
| **Page Format Control: (Continued)** | | | |
| ESC &a#M | 1B 26 61 # 4D | 27 38 97 # 77 | Right Margin, Right column # |
| ESC 9 | 1B 39 | 27 57 | Clear Margins |
| **Horizontal Movement and Spacing Control:** | | | |
| BS | Ø8 | 8 | Backspace |
| CR | ØD | 13 | Carriage Return |
| ESC &k#G | 1B 26 6B # 47 | 27 38 1Ø7 # 71 | CR/LF/FF Line Termination Action |

Line Termination Action

| # | CR | LF | FF |
|---|---|---|---|
| Ø | CR | LF | FF |
| 1 | CR+LF | LF | FF |
| 2 | CR | CR+LF | CR+FF |
| 3 | CR+LF | CR+LF | CR+FF |

| Code | Hex | Decimal | Command |
|---|---|---|---|
| ESC &sØC | 1B 26 73 3Ø 43 | 27 38 115 48 67 | Set Wrap Around |
| ESC &s1C | 1B 26 73 31 43 | 27 38 115 49 67 | Release Wrap Around |
| ESC &a#C | 1B 26 61 # 43 | 27 38 97 # 67 | Move Print Position to Column # |
| ESC &a#H | 1B 26 61 # 48 | 27 38 97 # 72 | Move Print Position Horizontal # of Decipoints |
| ESC *p#X | 1B 2A 7Ø # 58 | 27 42 112 # 88 | Move Print Position Horizontal # of Dots |
| **Vertical Movement and Spacing Control:** | | | |
| LF | ØA | 1Ø | Line Feed |
| FF | ØC | 12 | Formfeed |
| ESC = | 1B 3D | 27 61 | Half Line Feed |
| ESC &a#R | 1B 26 61 # 52 | 27 38 97 # 82 | Move Print Position to Row # |
| ESC &a#V | 1B 26 61 # 56 | 27 38 97 # 86 | Move Print Position Vertical # of Decipoints |
| ESC *p#Y | 1B 2A 7Ø # 59 | 27 42 112 # 89 | Move Print Position Vertical # of Dots |
| **Font Selection:** | | | |
| ESC (#ID | 1B 28 # ID | 27 4Ø # ID | Symbol Set, Primary, # is Character ID |
| ESC )#ID | 1B 29 # ID | 27 41 # ID | Symbol Set, Secondary, # is Character ID |

Character ID's: Roman-8bit = 8U, Kana-8bit = 8K,

| Code | Hex | Decimal | Command |
|---|---|---|---|
| **Font Selection: (Continued)** | | | |
| | Math-8bit = 8M, ANSI-8bit = 9U, USASCII = ØU, Line Draw = ØB, Math Symbols =ØA, US Legal = 1U, Roman Ext =ØE, ISO Denmark = ØD, ISO Italy = ØI, ISO United Kingdom = 1E, ISO France = ØF, ISO Germany = ØG, ISO Sweden = ØS, ISO Spain = 1S | | |
| ESC (sØP | 1B 28 73 3Ø 5Ø | 27 4Ø 115 48 8Ø | Spacing, Primary Fixed |
| ESC (s1P | 1B 28 73 31 5Ø | 27 4Ø 115 49 8Ø | Spacing, Primary Proportional |
| ESC )sØP | 1B 29 73 3Ø 5Ø | 27 41 115 48 8Ø | Spacing, Secondary Fixed |
| ESC )s1P | 1B 29 73 31 5Ø | 27 41 115 49 8Ø | Spacing, Secondary Proportional |
| ESC (s#H | 1B 28 73 # 48 | 27 4Ø 115 # 72 | Print Pitch, Primary, # is characters/inch |
| ESC )s#H | 1B 29 73 # 48 | 27 41 115 # 72 | Print Pitch, Secondary, # is characters/inch |
| ESC &k#S | 1B 26 6B # 53 | 27 38 1Ø7 # 83 | Print Pitch, Prim. & Secondary, #=Ø is 1Ø cpi, #=1 is 16.66 cpi |
| ESC (s#V | 1B 28 73 # 56 | 27 4Ø 115 # 86 | Print Point Size, Primary, # is points |
| ESC )s#V | 1B 29 73 # 56 | 27 41 115 # 86 | Print Point Size, Secondary, # is points |
| ESC (sØS | 1B 28 73 3Ø 53 | 27 4Ø 115 48 83 | Print Style, Primary, Upright |
| ESC (s1S | 1B 28 73 31 53 | 27 4Ø 115 49 83 | Print Style, Primary, Italic |
| ESC )sØS | 1B 29 73 3Ø 53 | 27 41 115 48 83 | Print Style, Secondary, Upright |
| ESC )s1S | 1B 29 73 31 53 | 27 41 115 49 83 | Print Style, Secondary, Italic |
| ESC (s#B | 1B 28 73 # 42 | 27 4Ø 115 # 66 | Stroke Weight, Primary, # is –7 to +7 |
| ESC )s#B | 1B 29 73 # 42 | 27 41 115 # 66 | Stroke Weight, Secondary, # is –7 to +7, –1 to –7 =light, Ø =Medium, 1 to 7 =Bold |
| ESC (s#T | 1B 28 73 # 54 | 27 4Ø 115 # 84 | Typeface, Primary, # is typeface |
| ESC )s#T | 1B 29 73 # 54 | 27 41 115 # 84 | Typeface, Secondary, # is typeface: Ø=Line printer, 1=Pica, 2=Elite, 3=Courier, 4=Swiss 721, 5=Dutch, 6=Gothic, 7=Script, 8=Prestige, 9=Caslon, 1Ø=Orator, 23=Century 7Ø. |

# HP LASERJET PRINTER CODES

| Code | Hex | Decimal | Command |
|---|---|---|---|
| **Font Control:** | | | |
| SI | ØF | 15 | Shift In Primary |
| SO | ØE | 14 | Shift In Secondary |
| ESC (#X | 1B 28 # 58 | 27 4Ø # 88 | Define Font, Primary, # is the Font ID number |
| ESC )#X | 1B 29 # 58 | 27 41 # 88 | Define Font, Secondary, # is the Font ID numbr |
| ESC *c#F | 1B 2A 63 # 46 | 27 42 99 # 7Ø | Font/Character Control, see printer manual |
| ESC (#@ | 1B 28 # 4Ø | 27 4Ø # 64 | Primary Font Default, see printer manual |
| ESC )#@ | 1B 29 # 4Ø | 27 41 # 64 | Secondary Font Default, see printer manual |
| ESC *c#D | 1B 2A 63 # 44 | 27 42 99 # 68 | Define Font ID, # is the ID |
| ESC )s#W | 1B 29 73 # 57 | 27 41 115 # 87 | Font Header, # is byte number of font attribute |
| ESC *c#E | 1B 2A 63 # 45 | 27 42 99 # 69 | Define Character Code to downloaded, # is Ø to 255 |
| ESC (s#W | 1B 28 73 # 57 | 27 4Ø 115 # 87 | Produce Down–load Character, see printer manual |
| **Character Highlight Selection:** | | | |
| ESC &dD | 1B 26 64 44 | 27 38 1ØØ 68 | Turn underline on |
| ESC &d@ | 1B 26 64 4Ø | 27 38 1ØØ 64 | Turn underline off |
| **Graphics:** | | | |
| ESC *t#R | 1B 2A 74 # 52 | 27 42 116 # 82 | Resolution, # is 75, 1ØØ, 15Ø, or 3ØØ Dots/inch |
| ESC *r#A | 1B 2A 72 # 41 | 27 42 114 # 65 | Graphics Start, #=Ø is start vertical from left end of print area, #=1 is start from present position. |
| ESC *b#W | 1B 2A 62 # 57 | 27 42 98 # 87 | Sending Graphics data, # is number of bytes of bit image data. |
| ESC *rB | 1B 2A 72 42 | 27 42 114 66 | End Raster Graphics Mode |
| ESC *c#A | 1B 2A 63 # 41 | 27 42 99 # 65 | Set Horizontal Rule Width to # dots (1 dot=1/3ØØ inch) |
| ESC *c#H | 1B 2A 63 # 48 | 27 42 99 # 72 | Set Horizontal Rule Width to # decipoints (1 decipoint=1/72Ø inch) |
| ESC *c#B | 1B 2A 63 # 42 | 27 42 99 # 66 | Set Vertical Rule Width to # dots (1 dot=1/3ØØ inch) |
| ESC *c#V | 1B 2A 63 # 56 | 27 42 99 # 86 | Set Vertical Rule Width to # decipoints (1 decipoint=1/72Ø inch) |

| Code | Hex | Decimal | Command |
|---|---|---|---|
| **Graphics: (Continued)** | | | |
| ESC *c#G | 1B 2A 63 # 47 | 27 42 99 # 71 | Set Gray Scale or Hatch Pattern ID #, see printer manual for a sample of each pattern/hatch and its associated ID # |
| ESC *c#P | 1B 2A 63 # 5Ø | 27 42 99 # 8Ø | Set Print Pattern # |
| **Macro's:** | | | |
| ESC &f#Y | 1B 26 66 # 59 | 27 38 1Ø2 # 89 | Set Macro ID # |
| ESC &fØX | 1B 26 66 3Ø 58 | 27 38 1Ø2 48 88 | Start Macro |
| ESC &f1X | 1B 26 66 31 58 | 27 38 1Ø2 49 88 | End Macro |
| ESC &f2X | 1B 26 66 32 58 | 27 38 1Ø2 5Ø 88 | Jump to Macro |
| ESC &f3X | 1B 26 66 33 58 | 27 38 1Ø2 51 88 | Call Macro |
| ESC &f4X | 1B 26 66 34 58 | 27 38 1Ø2 52 88 | Set Overlay Macro |
| ESC &f5X | 1B 26 66 35 58 | 27 38 1Ø2 53 88 | Release Overlay Macro |
| ESC &f6X | 1B 26 66 36 58 | 27 38 1Ø2 54 88 | Release all Macro |
| ESC &f7X | 1B 26 66 37 58 | 27 38 1Ø2 55 88 | Release all temporary Macro |
| ESC &f8X | 1B 26 66 38 58 | 27 38 1Ø2 56 88 | Release current Macro |
| ESC &f9X | 1B 26 66 39 58 | 27 38 1Ø2 57 88 | Assign temporary attribute to Macro |
| ESC &f1ØX | 1B 26 66 31 3Ø 58 | 27 38 1Ø2 49 48 88 | Assign permanent attribute to Macro |
| **Miscellaneous:** | | | |
| ESC Y | 1B 59 | 27 89 | Set Display Function of control codes |
| ESC Z | 1B 5A | 27 9Ø | Release Display Function of control codes |
| ESC & p # X | 1B 26 7Ø # 58 | 27 38 112 # 88 | Transparent Print Data (no ESC commands exist) |
| ESC &fØS | 1B 26 66 3Ø 53 | 27 38 1Ø2 48 83 | Push Printing Position. Puts present printing position on the top of the stack |
| ESC &f1S | 1B 26 66 31 53 | 27 38 1Ø2 49 83 | Pop Printing Position. Recall stored printing position and put on the top of the stack |
| ESC & l #X | 1B 26 6C # 58 | 27 38 1Ø8 # 88 | Set Number of Copies to # |
| ESC & l #H | 1B 26 6C # 48 | 27 38 1Ø8 # 72 | Paper Input Control. #=Ø is Feed out current page, #=1 is Lower Cassette supplies paper, #=3 is Envelope |

## HP LASERJET PRINTER CODES

| Code | Hex | Decimal | Command |
|---|---|---|---|
| **Miscellaneous: (Continued)** | | | |
| | feeder supplies envelope, #=4 is Upper Cassette supplies paper | | |
| ESC E | 1B 45 | 27 69 | Reset Printer |
| ESC z | 1B 7A | 27 122 | Start Printer Self Test |

## IBM PROPRINTER PRINTER CODES

| Code | Hex | Decimal | Command |
|---|---|---|---|
| Page Format Control: | | | |
| ESC C Ø # | 1B 43 ØØ # | 27 67 Ø # | Page Length, # is in Inch |
| ESC C # | 1B 43 # | 27 67 # | Page Length, # is in Lines |
| ESC X #1#2 | 1B 58 #1#2 | 27 88 #1#2 | Left/Right Margins Set, #1 is left inches, #2 is right inches |
| ESC N # | 1B 4E # | 27 78 # | Skip Perforation Set, # is Top + Bottom |
| ESC O | 1B 4F | 27 79 | Skip Perforation Release |
| ESC 4 | 1B 34 | 27 52 | Top of Page Set |
| **Horizontal Movement and Spacing Control:** | | | |
| BS | Ø8 | 8 | Backspace |
| CR | ØD | 13 | Carriage Return |
| ESC D # Ø | 1B 44 # ØØ | 27 68 # Ø | Horizontal Tab Set, # is the column, can use more than one # |
| ESC D Ø | 1B 44 ØØ | 27 68 Ø | Horizontal Tab Release |
| HT | Ø9 | 9 | Horizontal Tab, moves to next preset tab |
| ESC R | 1B 52 | 27 82 | Reset all Tabs |
| **Vertical Movement and Spacing Control:** | | | |
| ESC Ø | 1B 3Ø | 27 48 | Set Line Spacing to 1/8 inch (9 points or 8 lpi) |
| ESC 1 | 1B 31 | 27 49 | Set Line Spacing to 7/72 inch (7 points) |
| ESC 2 | 1B 32 | 27 5Ø | Execute a Line Feed, must follow ESC A # command |
| ESC 3 # | 1B 33 # | 27 51 # | Set Line Spacing to #/216 inch |
| ESC A # | 1B 41 # | 27 65 # | Set Line Spacing to # Points (#/72 inch) |
| LF | ØA | 1Ø | Line feed |
| ESC 5 1 | 1B 35 Ø1 | 27 53 1 | Set Auto Line Feed |

| Code | Hex | Decimal | Command |
|---|---|---|---|
| **Vertical Movement and Spacing Control (Continued):** | | | |
| ESC 5 Ø | 1B 35 ØØ | 27 53 Ø | Release Auto Line Feed |
| ESC j # | 1B 6A # | 27 1Ø6 # | Reverse Line Feed of #/216 Inches |
| ESC J # | 1B 4A # | 27 74 # | Forward Line Feed of #/216 Inches |
| FF | ØC | 12 | Form feed |
| ESC B # Ø | 1B 42 # ØØ | 27 66 # Ø | Vertical Tab Set, # is the line, can use more than one # |
| ESC B Ø | 1B 42 ØØ | 27 66 Ø | Vertical Tab Release |
| VT | ØB | 11 | Vertical Tab, moves to next preset tab |
| ESC R | 1B 52 | 27 82 | Reset all Tabs |
| **Character Selection:** | | | |
| DC2 | 12 | 18 | Pica Pitch (12 pt,1Ø cpi) |
| ESC : | 1B 3A | 27 58 | Elite Pitch (1Ø pt, 12 cpi) |
| SI | ØF | 15 | Compressed Print |
| ESC SI | 1B ØF | 27 15 | Compressed Print |
| SO | ØE | 14 | Set Double Width for a single line |
| ESC SO | 1B ØE | 27 14 | Set Double Width for a single line |
| DC4 | 14 | 2Ø | Release Double Width for a single line |
| ESC WØ | 1B 57 ØØ | 27 87 Ø | Release Double Wide Line |
| ESC W1 | 1B 57 Ø1 | 27 87 1 | Set Double Width Line |
| ESC SØ | 1B 53 ØØ | 27 83 Ø | Set Superscript Mode On |
| ESC S1 | 1B 53 Ø1 | 27 83 1 | Set Subscript Mode On |
| ESC T | 1B 54 | 27 84 | Release Superscript and Subscript |
| ESC 7 | 1B 37 | 27 55 | Set IBM Character Set 1 |
| ESC 6 | 1B 36 | 27 54 | Set IBM Character Set 2 |
| ESC ^ | 1B 5E | 27 94 | Select 1 Character from the All Character Chart |
| ESC \ #1 #2 | 1B 5C | 27 92 | Select Print Continuously from All Character Chart for a total of (#2 X 256) + #1 |
| **Character Highlight Selection:** | | | |
| ESC – 1 | 1B 2D Ø1 | 27 45 1 | Turn Underline Mode On |
| ESC – Ø | 1B 2D ØØ | 27 45 Ø | Turn Underline Mode Off |
| ESC _ 1 | 1B 5F Ø1 | 27 95 1 | Enable Overline Mode |
| ESC _ Ø | 1B 5F ØØ | 27 95 Ø | Disable Overline Mode |
| ESC E | 1B 45 | 27 69 | Enable Bold Print Mode |
| ESC F | 1B 46 | 27 7Ø | Disable Bold Print Mode |

# IBM PROPRINTER PRINTER CODES

| Code | Hex | Decimal | Command |
|---|---|---|---|
| **Character Highlight Selection: (Continued)** | | | |
| ESC G | 1B 47 | 27 71 | Enable Double–strike |
| ESC H | 1B 48 | 27 72 | Disable Double–strike |
| **Graphics:** | | | |
| For values of #1 and #2 below, see printer manuals | | | |
| ESC K#1#2 | 1B 4B #1#2 | 27 75 #1#2 | Enable Single–density Graphics Mode, 6Ø dpi |
| ESC L#1#2 | 1B 4C #1#2 | 27 76 #1#2 | Enable Double–density Graphics Mode, 12Ø dpi |
| ESC Y#1#2 | 1B 59 #1#2 | 27 89 #1#2 | Enable Double–density, 12Ø dpi, High–speed Graphics Mode |
| ESC Z#1#2 | 1B 5A #1#2 | 27 9Ø #1#2 | Enable Quad–density Graphics Mode, 24Ø dpi |
| **Miscellaneous:** | | | |
| CAN | 18 | 24 | Cancel |
| DC1 | 11 | 17 | Remote Printer Select |
| ESC Q3 | 1B 51 Ø3 | 27 8 3 | Remote Printer Deselect |
| ESC EM # | 1B 19 # | 27 25 # | Paper Cassette Selection, #=E is envelope, #=1 is Lower Cassette, #=2 is Upper Cassette, #=R is eject page |
| NUL | ØØ | Ø | Null |
| BEL | Ø7 | 7 | Sound Beeper |

# NEC PINWRITER PRINTER CODES

| Code | Hex | Decimal | Command |
|---|---|---|---|
| Pinwriters use most of the same codes as the Epson LQ1500 except for the following FS Codes: | | | |
| FS 3 # | 1C | 28 | Line space Ø-255 #/360 |
| FS C # | 1C | 28 | Set Font Cartridge, #=Ø is resident font, #=1 is slot 1, #=2 is slot 2 |
| FS E # | 1C | 28 | Ø=Cancel horiz enlarge., 1=2X horiz enlargement, 2=3X horiz enlargement |
| FS F | 1C | 28 | Release Enhanced Print |
| FS I # | 1C | 28 | Ø=Italic Set, 1=IBM Set |
| FS R | 1C | 28 | Set Reverse Line Feed |
| FS S # | 1C | 28 | Ø=Draft 12,1=high speed |
| FS V 1 | 1C | 28 | Set double vertical enlarge |
| FS V Ø | 1C | 28 | Release double vertical enlargement |
| FS Z #1 #2 | 1C | 28 | Set 360 dpi graphics |
| FS @ | 1C | 28 | Initialize except user buffer |

# Constants

| CONSTANT | SYMBOL | VALUE |
|---|---|---|
| Acceleration due to gravity | $g$ | 32.174 ft/sec$^2$ |
| | | 980.665 cm/sec$^2$ |
| | | 386.09 in/sec$^2$ |
| | | 21.94 miles/hr–sec |
| Air, density @ 0°C, 760mm Hg | | 1.2929 gm/liter |
| AMU | | $1.6605 \times 10^{-27}$ kg |
| Astronomic unit | $AU$ | $149.669 \times 10^{6}$ km |
| Atomic mass constant, unified | $m_u$ | $1.66043 \times 10^{-27}$ kg |
| Atomic specific heat constant | $h/k$ | $4.79928 \times 10^{-11}$ sec deg |
| Avogadro constant | $N_A$ | $6.02252 \times 10^{23}$ mol$^{-1}$ |
| Barn Cross Section | | $10^{-24}$ cm$^2$ |
| Bohr Magneton | $\mu_B$ | $9.2732 \times 10^{-24}$ Am$^2$ |
| Bohr radius | $a_o$ | $5.29177 \times 10^{-9}$ cm |
| Equation for above | | $\propto / 4 \pi R_\infty$ |
| Boltzmann entropy constant | $k$ | $1.38054 \times 10^{-23}$ JK$^{-1}$ |
| Earth, equatorial radius | | 6378.39 km, 3969.34 mi |
| Earth, polar radius | | 6356.91km, 3949.99 mi |
| Earth, mass | M | $5.983 \times 10^{24}$ kg |
| | | $6.595 \times 10^{21}$ tons |
| Earth, mean density | | 344.7 lbs / ft$^3$ |
| Electric field constant | $\varepsilon_o$ | $8.8541853 \times 10^{-12}$Fm$^{-1}$ |
| Equation for above | | $1 / \mu_o c^2$ |
| Electron, Atomic mass | Nm | $5.48598 \times 10^{-4}$ amu |
| Electron charge to mass ratio | $e/m_e$ | $1.758796 \times 10^{-11}$Ckg$^{-1}$ |
| Electron radius | $\gamma_e$ | $2.81777 \times 10^{-15}$ m |
| Electron, magnetic moment | $\mu_e$ | $9.28389 \times 10^{-21}$ erg gauss$^{-1}$ |
| Electron mass at rest | $m_e$ | $9.1091 \times 10^{-31}$ kg |
| Electron, Compton wave length | $\lambda_c$ | $2.42621 \times 10^{-12}$ m |
| Electron–volt | $ev$ | $1.60210 \times 10^{-19}$ J |

| CONSTANT | SYMBOL | VALUE |
|---|---|---|
| Euler's constant | $\gamma$ | 0.57721566490153386O |
| Faraday constant | $F$ | $9.64870 \times 10^{4}$ Cmol$^{-1}$ |
| Gas constant, Universal | $R$ | 8.3143 JK$^{-1}$mol$^{-1}$ |
| Gas constant, Universal | $R_o$ | $8.20545 \times 10^{-2}$ lit-Atm/°K-mole |
| Gas constant, Universal | $R$ | 1546 lbf–ft / lb mole-°$_R$ |
| Golden Ratio | $\phi$ | 1.618033988749894848 |
| Gravitational constant | G | $6.670 \times 10^{-11}$ Nm$^2$kg$^{-2}$ |
| Hydrogen, Atomic mass | H | 1.00782522 amu |
| Hydrogen atom, mass | | $1.67339 \times 10^{-24}$ gm |
| Ice–point temperature | $T_{Ice}$ | $2.731500 \times 10^{2}$ K |
| Impedance of free space | $Z_o$ | 376.731 ohms |
| Light, speed of in a vacuum | $C_o$ | $2.997925 \times 10^{10}$ cm/sec |
| Logarithmic constant | $e$ | 2.718281828459045235 |
| Loschmidt's number | $n_o$ | $2.68702 \times \times 10^{19}$ cm$^{-3}$ |
| Magnetic field constant | $\mu_o$ | $1.256637 \times 10^{-6}$ Hm$^{-1}$ |
| Equation for above | | $4\pi 10^{-7}$ Hm$^{-1}$ |
| Mercury, density @ 0°C | | 13.5955 gm/cm$^3$ |
| Neutron rest mass | $m_n$ | $1.67482 \times 10^{-27}$ kg |
| Parsec | pc | 206.265 AU |
| Permeability of free space | $\mu o$ | $12.5664 \times 10^{-7}$ henry/m |
| Permittivity of free space | $\varepsilon_o$ | $8.8542 \times 10^{-12}$ farad/m |
| Pi (ratio circle circum/diameter) | $\pi$ | 3.141592653589793238 |
| | $2\pi$ | 6.283185 |
| | $\pi^2$ | 9.869604 |
| | $\sqrt{\pi}$ | 1.772454 |
| | $\pi/4$ | 0.785398 |
| Planck constant | $h$ | $6.6256 \times 10^{-34}$ Js |

| CONSTANT | SYMBOL | VALUE |
|---|---|---|
| Planck constant | | $6.6252 \times 10^{-27}$ erg–sec |
| Planck – 1st radiation constant | $c_1$ | $3.7418 \times 10^{-5}$ erg $cm^2 s^{-1}$ |
| Planck – 2nd radiation constant | $c_2$ | 1.4388 cm K |
| Proton, Atomic mass | $Nm_p$ | 1.00727663 amu |
| Proton rest mass | $m_p$ | $1.67252 \times 10^{-27}$ kg |
| Proton, Compton wave length | $\lambda_{c \bullet p}$ | $1.32140 \times 10^{-15}$ $m$ |
| Quantum charge ratio | $h/e$ | $4.13556 \times 10^{15}$ J/C |
| Radiation constant, 1st | c1 | $4.99208 \times 10^{-15}$ erg-cm |
| Radiation constant, 2nd | c2 | 1.43879 cm–deg |
| Rydberg constant, infinite mass | $R_\infty$ | $1.0973731 \times 10^{7}$ $m^{-1}$ |
| Sound, velocity in air @ STP | | 331.7 m/sec |
| | | 1087.1 ft/sec |
| Sound, velocity in water @ 20°C | | 1470 m/sec |
| | | 4823 ft/sec |
| Sefan-Boltzmann Constant | $\sigma$ | $5.6686 \times 10^{-5}$ erg/$cm^2$–sec–$(°K)^4$ |
| Vacuum permeability | $\mu_o$ | $4\pi \times 10^{-7}$ $kgms^{-2}A^{-2}$ |
| Water, density @ 3.98°C | | 1.000000 gm/ml |
| | | 0.03613 lb/$in^3$ |
| | | 62.43 lb/$ft^3$ |
| Water, heat of fusion @ 0°C | | 79.71 cal/gm |
| Water, heat of vaporization @ 100°C | | 539.55 cal/gm |
| Water, viscosity @ 20°C | | 1.002 centipoise |
| | | 0.01002 dyne–sec/$cm^2$ |
| Wave length, krypton 86 orange-red line | | 6057.802Å |
| Wien displacement constant | $\lambda_{max}T$ | 0.289779 cm deg |
| Zeeman displacement | $\mu_1$/hc | (see next line) |
| | | $4.668583 \times 10^{-5} cm^{-1} gauss^{-1}$ |

# POCKET REF

# Electrical



(See also ELECTRONICS Chapter on page 129)

# COPPER WIRE CURRENT CAPACITY

## Single wire in open air, ambient temp

Ampacities of Wire Types (w/ Temp Rating) @ 0–2000 Volts

| Wire Size AWG | RUW, T, TW (140°F) | FEPW, RH, RHW RUH, THW, THWN XHHW, ZW (167°F) | V, MI (185°F) | TA, TBS, SA, AVB, SIS, FEP, FEPB, RHH (194°F) |
|---|---|---|---|---|
| 0000 | 300 | 360 | 390 | 405 |
| 000 | 260 | 310 | 335 | 350 |
| 00 | 225 | 265 | 290 | 300 |
| 0 | 195 | 230 | 250 | 260 |
| 1 | 165 | 195 | 215 | 220 |
| 2 | 140 | 170 | 185 | 190 |
| 3 | 120 | 145 | 160 | 165 |
| 4 | 105 | 125 | 135 | 140 |
| 6 | 80 | 95 | 100 | 105 |
| 8 | 60 | 70 | 75 | 80 |
| 10 | 40 | 50 | 55 | 55 |
| 12 | 30 | 35 | 40 | 40 |
| 14 | 25 | 30 | 30 | 35 |
| 16 | – | – | 23 | 24 |
| 18 | – | – | – | 18 |

Note: Types T and TW are the most common for house wiring. T is for dry conditions only and TW is for dry or wet conditions; both are single layer plastic covered.

If the ambient temperature[1] is over 86°F (30°C), then the following corrections should be applied by multiplying the above ampacities by the correction factor below.

| Ambient[1] Temp °F | Ampacity Correction for above Wire types 140°F | 167°F | 185°F | 194°F |
|---|---|---|---|---|
| 87-104 | 0.82 | 0.88 | 0.90 | 0.91 |
| 105-113 | 0.71 | 0.82 | 0.85 | 0.87 |
| 114-122 | 0.58 | 0.75 | 0.80 | 0.82 |
| 123-141 | ... | 0.58 | 0.67 | 0.71 |
| 142-158 | ... | 0.35 | 0.52 | 0.58 |
| 159-176 | ... | ... | 0.30 | 0.41 |

**NOTE:** The information on pages 108 and 109 has been extracted from the National Electrical Code ®, National Fire Protection Association, Quincy, Massachusetts 02269, Copyright 1983 and does not represent the complete code.

[1] Ambient temperature is the temperature of the material (air, earth, etc) surrounding the wire.

# COPPER WIRE CURRENT CAPACITY
## Three wires in cable, ambient temp 86°F

| | Ampacities of Wire Types (w/ Temp Rating) @ 0–2000 Volts | | | |
|---|---|---|---|---|
| Wire Size AWG | RUW, T, TW UF (140°F) | FEPW, RH, RHW RUH, THW, THWN XHHW, ZW, USE (167°F) | V, MI (185°F) | TA, TBS, SA, AVB, SIS, FEP, FEPB, RHH (194°F) |
| 0000 | 195 | 230 | 250 | 260 |
| 000 | 165 | 200 | 215 | 225 |
| 00 | 145 | 175 | 190 | 195 |
| 0 | 125 | 150 | 165 | 170 |
| 1 | 110 | 130 | 145 | 150 |
| 2 | 95 | 115 | 125 | 130 |
| 3 | 85 | 100 | 110 | 110 |
| 4 | 70 | 85 | 95 | 95 |
| 6 | 55 | 65 | 70 | 75 |
| 8 | 40 | 50 | 55 | 55 |
| 10 | 30 | 35 | 40 | 40 |
| 12 | 25 | 25 | 30 | 30 |
| 14 | 20 | 20 | 25 | 25 |
| 16 | – | – | 18 | 18 |
| 18 | – | – | – | 14 |

Note: All notes on ambient temperature and T and TW types on the previous page also apply to this Three Wire section.

# STANDARD LAMP & EXTENSION CORD CURRENT CAPACITIES

| Wire Size AWG | Wire Types SP, SPT, S, SJ, SV, ST, SJT, SVT 2 Conductor | 3 Conductor | 4 Conductor |
|---|---|---|---|
| 10 | 28 | 25 | 20 |
| 12 | 23 | 20 | 15 |
| 14 | 18 | 15 | 10 |
| 16 | 13 | 10 | 15 |
| 18 | 10 | 7 | 20 |

**NOTE:** In all Copper Wire Types listed on pages 108 and 109 (except Types V, MI, TA, TBS, SA, and SIS) overcurrent protection should not exceed 15 amps for 14 AWG, 20 amps for 12 AWG, and 30 amps for 10 AWG. This is not true if specifically permitted elsewhere in the Code.

# ALUMINUM WIRE AMP CAPACITY

## Single wire in open air, ambient temp

| Wire Size AWG | RUW, T, TW (140°F) | RH, RHW RUH, THW, THWN XHHW (167°F) | V, MI (185°F) | TA, TBS, SA, AVB, SIS, RHH, THHN, XHHW (194°F) |
|---|---|---|---|---|
| | Ampacities of Wire Types (w/ Temp Rating) @ 0–2000 Volts | | | |
| 500MCM | 405 | 485 | 525 | 545 |
| 400MCM | 355 | 425 | 465 | 480 |
| 300MCM | 290 | 350 | 380 | 395 |
| 0000 | 235 | 280 | 305 | 315 |
| 000 | 200 | 240 | 265 | 275 |
| 00 | 175 | 210 | 225 | 235 |
| 0 | 150 | 180 | 195 | 205 |
| 1 | 130 | 155 | 165 | 175 |
| 2 | 110 | 135 | 145 | 150 |
| 3 | 95 | 115 | 125 | 130 |
| 4 | 80 | 100 | 105 | 110 |
| 6 | 60 | 75 | 80 | 80 |
| 8 | 45 | 55 | 60 | 60 |
| 10 | 35 | 40 | 40 | 40 |
| 12 | 25 | 35 | 30 | 35 |

Note: Types T and TW are the most common for house wiring. T is for dry conditions only and TW is for dry or wet conditions; both are single layer plastic covered.

If the ambient[(1)] temperature is over 86°F (30°C), then the following corrections should be applied by multiplying the above ampacities by the correction factor below.

| Ambient[(1)] Temp °F | 140°F | 167°F | 185°F | 194°F |
|---|---|---|---|---|
| | *Ampacity Correction for above Wire Types* | | | |
| 87-104 | 0.82 | 0.88 | 0.90 | 0.91 |
| 105-113 | 0.71 | 0.82 | 0.85 | 0.87 |
| 114-122 | 0.58 | 0.75 | 0.80 | 0.82 |
| 123-141 | ... | 0.58 | 0.67 | 0.71 |
| 142-158 | ... | 0.35 | 0.52 | 0.58 |
| 159-176 | ... | ... | 0.30 | 0.41 |

**NOTE:** The information on pages 110 and 111 has been extracted from the National Electrical Code ®, National Fire Protection Association, Quincy, Massachusetts 02269, Copyright 1983 and does not represent the complete code.

[(1)] Ambient temperature is the temperature of the material (air, earth, etc) surrounding the wire.

# ALUMINUM WIRE AMP CAPACITY

## Three wires in cable, ambient temp 86°F

| Ampacities of Wire Types (w/ Temp Rating) @ 0–2000 Volts | | | | |
|---|---|---|---|---|
| Wire Size AWG | RUW, T, TW UF (140°F) | RH, RHW RUH, THW, THWN XHHW, USE (167°F) | V, MI (185°F) | TA, TBS, SA, AVB, SIS, RHH, THHN, XHHW (194°F) |
| 500MCM | 260 | 310 | 335 | 350 |
| 400MCM | 225 | 270 | 295 | 305 |
| 300MCM | 190 | 230 | 250 | 255 |
| 0000 | 150 | 180 | 195 | 205 |
| 000 | 130 | 155 | 170 | 175 |
| 00 | 115 | 135 | 145 | 150 |
| 0 | 100 | 120 | 130 | 135 |
| 1 | 85 | 100 | 110 | 115 |
| 2 | 75 | 90 | 100 | 100 |
| 3 | 65 | 75 | 85 | 85 |
| 4 | 55 | 65 | 75 | 75 |
| 6 | 40 | 50 | 55 | 60 |
| 8 | 30 | 40 | 40 | 45 |
| 10 | 25 | 30 | 30 | 35 |
| 12 | 20 | 20 | 25 | 25 |

Note: All notes on ambient temperature and T and TW types on the previous page also apply to this Three Wire section.

# CURRENT ADJUSTMENT FOR MORE THAN 3 WIRES IN A

| Number of Conductors | Percentage of amperage value listed in amperage tables on the previous 4 pages. |
|---|---|
| 4 to 6 | 80% |
| 7 to 24 | 70% |
| 25 to 42 | 60% |
| over 43 | 50% |

Basically, the above table reflects the rule that the higher the temperature (more wires=higher temperature) the lower the current carrying capacity of the wire.

**NOTE:** In all Aluminum and Copper Clad Aluminum Wire Types listed on pages 110 and 111 (except Types V, MI, TA, TBS, SA, and SIS) overcurrent protection should not exceed 15 amps for 12 AWG and 25 amps for 10 AWG. This is not true if specifically permitted elsewhere in the Code.

# COPPER WIRE RESISTANCE

| Gauge A.W.G.* | Feet per Ohm @ 77°F | Ohms per 1000 ft @ 77°F | Feet per Ohm @ 149°F | Ohms per 1000 ft @ 149°F |
|---|---|---|---|---|
| 0000 | 20000 | 0.050 | 17544 | 0.057 |
| 000 | 15873 | 0.063 | 13699 | 0.073 |
| 00 | 12658 | 0.079 | 10870 | 0.092 |
| 0 | 10000 | 0.100 | 8621 | 0.116 |
| 1 | 7936 | 0.126 | 6849 | 0.146 |
| 2 | 6289 | 0.159 | 5435 | 0.184 |
| 3 | 4975 | 0.201 | 4310 | 0.232 |
| 4 | 3953 | 0.253 | 3425 | 0.292 |
| 5 | 3135 | 0.319 | 2710 | 0.369 |
| 6 | 2481 | 0.403 | 2151 | 0.465 |
| 7 | 1968 | 0.508 | 1706 | 0.586 |
| 8 | 1560 | 0.641 | 1353 | 0.739 |
| 9 | 1238 | 0.808 | 1073 | 0.932 |
| 10 | 980.4 | 1.02 | 847.5 | 1.18 |
| 11 | 781.3 | 1.28 | 675.7 | 1.48 |
| 12 | 617.3 | 1.62 | 534.8 | 1.87 |
| 13 | 490.2 | 2.04 | 423.7 | 2.36 |
| 14 | 387.6 | 2.58 | 336.7 | 2.97 |
| 15 | 307.7 | 3.25 | 266.7 | 3.75 |
| 16 | 244.5 | 4.09 | 211.4 | 4.73 |
| 17 | 193.8 | 5.16 | 167.8 | 5.96 |
| 18 | 153.6 | 6.51 | 133.2 | 7.51 |
| 19 | 121.8 | 8.21 | 105.5 | 9.48 |
| 20 | 96.2 | 10.4 | 84.0 | 11.9 |
| 21 | 76.3 | 13.1 | 66.2 | 15.1 |
| 22 | 60.6 | 16.5 | 52.6 | 19.0 |
| 23 | 48.1 | 20.8 | 41.7 | 24.0 |
| 24 | 38.2 | 26.2 | 33.1 | 30.2 |
| 25 | 30.3 | 33.0 | 26.2 | 38.1 |
| 26 | 24.0 | 41.6 | 20.8 | 48.0 |
| 27 | 19.0 | 52.5 | 16.5 | 60.6 |
| 28 | 15.1 | 66.2 | 13.1 | 76.4 |
| 29 | 12.0 | 83.4 | 10.4 | 96.3 |
| 30 | 9.5 | 105 | 8.3 | 121 |
| 31 | 7.5 | 133 | 6.5 | 153 |
| 32 | 6.0 | 167 | 5.2 | 193 |
| 33 | 4.7 | 211 | 4.1 | 243 |
| 34 | 3.8 | 266 | 3.3 | 307 |
| 35 | 3.0 | 335 | 2.6 | 387 |
| 36 | 2.4 | 423 | 2.0 | 488 |
| 37 | 1.9 | 533 | 1.6 | 616 |
| 38 | 1.5 | 673 | 1.3 | 776 |
| 39 | 1.2 | 848 | 1.0 | 979 |
| 40 | 0.93 | 1070 | 0.81 | 1230 |

* *American Wire Gauge (formerly Brown & Sharp)*

# STANDARD COPPER WIRE SPECS

| Gauge A.W.G | Diameter in mils (1000th in) | Diameter Millimeters | Area in Circular Mils | Weight Lbs per 1000 feet | Turns / inch Enamel |
|---|---|---|---|---|---|
| 0000 | 460.0 | 11.684 | 212000 | 641.0 | 2.2 |
| 000 | 410.0 | 10.414 | 168000 | 508.0 | 2.4 |
| 00 | 365.0 | 9.271 | 133000 | 403.0 | 2.7 |
| 0 | 325.0 | 8.255 | 106000 | 319.0 | 3.0 |
| 1 | 289.0 | 7.348 | 83700 | 253.0 | 3.3 |
| 2 | 258.0 | 6.544 | 66400 | 201.0 | 3.8 |
| 3 | 229.0 | 5.827 | 52600 | 159.0 | 4.2 |
| 4 | 204.0 | 5.189 | 41700 | 126.0 | 4.7 |
| 5 | 182.0 | 4.621 | 33100 | 100.0 | 5.2 |
| 6 | 162.0 | 4.115 | 26300 | 79.5 | 5.9 |
| 7 | 144.0 | 3.665 | 20800 | 63.0 | 6.5 |
| 8 | 128.0 | 3.264 | 16500 | 50.0 | 7.6 |
| 9 | 114.0 | 2.906 | 13100 | 39.6 | 8.6 |
| 10 | 102.0 | 2.588 | 10400 | 31.4 | 9.6 |
| 11 | 91.0 | 2.305 | 8230 | 24.9 | 10.7 |
| 12 | 81.0 | 2.053 | 6530 | 19.8 | 12.0 |
| 13 | 72.0 | 1.828 | 5180 | 15.7 | 13.5 |
| 14 | 64.0 | 1.628 | 4110 | 12.4 | 15.0 |
| 15 | 57.0 | 1.450 | 3260 | 9.86 | 16.8 |
| 16 | 51.0 | 1.291 | 2580 | 7.82 | 18.9 |
| 17 | 45.0 | 1.150 | 2050 | 6.2 | 21.2 |
| 18 | 40.0 | 1.024 | 1620 | 4.92 | 23.6 |
| 19 | 36.0 | 0.912 | 1290 | 3.90 | 26.4 |
| 20 | 32.0 | 0.812 | 1020 | 3.09 | 29.4 |
| 21 | 28.5 | 0.723 | 810 | 2.45 | 33.1 |
| 22 | 25.3 | 0.644 | 642 | 1.94 | 37.0 |
| 23 | 22.6 | 0.573 | 509 | 1.54 | 41.3 |
| 24 | 20.1 | 0.511 | 404 | 1.22 | 46.3 |
| 25 | 17.9 | 0.455 | 320 | 0.970 | 51.7 |
| 26 | 15.9 | 0.405 | 254 | 0.769 | 58.0 |
| 27 | 14.2 | 0.361 | 202 | 0.610 | 64.9 |
| 28 | 12.6 | 0.321 | 160 | 0.484 | 72.7 |
| 29 | 11.3 | 0.286 | 127 | 0.384 | 81.6 |
| 30 | 10.0 | 0.255 | 101 | 0.304 | 90.5 |
| 31 | 8.9 | 0.227 | 79.7 | 0.241 | 101 |
| 32 | 8.0 | 0.202 | 63.2 | 0.191 | 113 |
| 33 | 7.1 | 0.180 | 50.1 | 0.152 | 127 |
| 34 | 6.3 | 0.160 | 39.8 | 0.120 | 143 |
| 35 | 5.6 | 0.143 | 31.5 | 0.095 | 158 |
| 36 | 5.0 | 0.127 | 25.0 | 0.0757 | 175 |
| 37 | 4.5 | 0.113 | 19.8 | 0.0600 | 198 |
| 38 | 4.0 | 0.101 | 15.7 | 0.0476 | 224 |
| 39 | 3.5 | 0.090 | 12.5 | 0.0377 | 248 |
| 40 | 3.1 | 0.080 | 9.9 | 0.0200 | 282 |

** American Wire Gauge (formerly Brown & Sharp)*

# WIRE CLASSES & INSULATION

Standard cable, as used in home and general construction, is classified by the wire size, number of wires, insulation type and dampness condition of the wire environment. Example: a cable with the code "12/2 with Ground – Type UF – 600V – (UL)" has the following specifications:

- Wire size is 12 gauge (minimum required size for homes today; see the National Electric Code).
- The " / 2 " indicates there are two wires in the cable.
- "Ground" indicates there is a third wire in the cable to be used as a grounding wire.
- "Type UF" indicates the insulation type and acceptable dampness rating.
- "600V" means the wire is rated at 600 volts maximum.
- "UL" indicates the wire has been certified by Underwriters Laboratory to be safe.

Cables are dampness rated as follows:

- DRY: No dampness normally encountered. Indoor location above ground level.
- DAMP: Partially protected locations. Moderate amount of moisture. Indoor location below ground level.
- WET: Water saturation probable, such as underground or in concrete slabs or outside locations exposed to weather.

There are literally hundreds of different types of insulation used in wire and cable. To make things simple, the following descriptions are for wires commonly used in home wiring:

"BX" Armor covered with flexible, galvanized steel. Normally used in dry locations. Not legal to use in some states such as California.

"ROMEX" Although actually a trade name, it is used to describe a general class of plastic coated cable. Each wire is plastic wrapped except possibly the ground wire, which is sometimes bare or paper covered. Very flexible. There are three general types:
"NM" – Dry only, 2 or 3 wire, ground wire plastic wrapped.
"NMC" – Dry, 2 or 3 wire, all wires in solid plastic.
"UF" – Wet, 2 or 3 wire, all wires in solid, water resistant plastic. Use also instead of conduit.

NEVER put ROMEX or BX inside conduit.

# WIRE CLASSES & INSULATION

Wire types are typically coded by the type of insulation, temperature range, dampness rating, and type and composition of the jacket. The following are some of the "Type Codes":

| | |
|---|---|
| "T..." | Very common, dry only, full current load temperature must be less than 60°C (140°F). |
| "F" | Fixture wire. CF has cotton insulation (90°C), AF has asbestos insulation (150°C), SF has silicone insulation (200°C). |
| "R..." | Rubber (natural, neoprene, etc) covered |
| "S..." | Appliance cord, stranded conductors, cotton layer between wire and insulation, jute fillers, rubber outer jacket. S is extra hard service, SJ lighter service, SV light service. |
| "SP..." | Lamp cord, rubber insulation. |
| "SPT..." | Lamp cord, plastic insulation. |
| "X..." | Insulation is a cross linked synthetic polymer. Very tough and heat and moisture resistant. |
| "FEP..." | Fluorinated ethylene propylene insulation. Rated over 90°C ( 194°F). Dry only. |
| "...B" | Suffix indicating a outer braid is used, such as glass. |
| "...H" | Suffix indicating Higher loaded current temperatures may be used, up to 75°C (167°F). |
| "...HH" | Suffix indicating much higher loaded current temperatures may be used, up to 90°C (194°F). |
| "...L" | Suffix indicating a seamless lead jacket. |
| "...N" | Suffix indicating the jacket is extruded nylon or thermoplastic polyester and is very resistant to gas and oil and is very tough. |
| "...O" | Suffix indicating neoprene jacket. |
| "...W" | Suffix indicating WET use type. |

Examples of some of the more common wire types are "T", "TW", "THWN", "THHN", "XHHW", "RHH", and "RHW".

# STANDARD WIRING COLOR CODES

Standard wire color codes are very different between electronic circuitry and household 110 Volt AC wiring.

**Household wiring** (or other AC applications in the 100 + volt range) uses the following color codes:

| Wire Color | Circuit type |
|---|---|
| Black | "Hot" wire. In an outlet, it is always wired to the narrow spade or brass colored terminal. |
| Green | "Ground" wire, always wired to the green terminal. Also called chassis ground. This wire is also sometimes green w/ yellow stripe. |
| Red | "Traveler" wire used in connecting 3–way switches. Connects power between the 3–way switches. |
| White | "Neutral" wire. In an outlet, it is always wired to the wide spade or silver colored terminal. |

Typically, the following color codes are used for **electronic applications** (as established by the Electronic Industries Association – EIA):

| Wire Color (solid) | Circuit type |
|---|---|
| Black | Chassis grounds, returns, primary leads |
| Blue | Plate leads, transistor collectors, FET drain |
| Brown | Filaments, plate start lead |
| Gray | AC main power leads |
| Green | Transistor base, finish grid, diodes, FET gate |
| Orange | Transistor base 2, screen grid |
| Red | B plus dc power supply |
| Violet | Power supply minus |
| White | B–C minus of bias supply, AVC–AGC return |
| Yellow | Emitters-cathode and transistor, FET source |

**Stereo Audio Channels** are color coded as follows:

| Wire Color (solid) | Circuit type |
|---|---|
| White | Left channel high side |
| Blue | Left channel low side |
| Red | Right channel high side |
| Green | Right channel low side |

# STANDARD WIRING COLOR CODES

**Power Transformers** are color coded as follows:

| Wire Color (solid) | Circuit type |
|---|---|
| Black | If a transformer does not have a tapped primary, both leads are black. |
| Black | If a transformer does have a tapped primary, the black is the common lead. |
| Black & Yellow | Tap for a tapped primary. |
| Black & Red | End for a tapped primary. |

**AF Transformers** (audio) are color coded as follows:

| Wire Color (solid) | Circuit type |
|---|---|
| Black | Ground line. |
| Blue | Plate, collector, or drain lead. End of primary winding. |
| Brown | Start primary loop, Opposite to blue lead. |
| Green | High side, end secondary loop. |
| Red | B plus, center tap push–pull loop. |
| Yellow | Secondary center tap. |

**IF Transformers** (Intermediate Frequency) are color coded as follows:

| Wire Color (solid) | Circuit type |
|---|---|
| Blue | Primary high side of plate, collector, or drain lead. |
| Green | Secondary high side for output. |
| Red | Low side of primary returning B plus. |
| Violet | Secondary outputs. |
| White | Secondary low side. |

# WIRE SIZE vs VOLTAGE DROP

Voltage drop is the amount of voltage lost over the length of a piece of wire. Voltage drop changes as a function of the resistance of the wire and should be less than 2% if possible. If the drop is greater than 2%, efficiency of the appliance is severely decreased and life of the equipment will be decreased. As an example, if the voltage drop on an incandescent light bulb is 10%, the light output of the bulb decreases over 30%!

Voltage drop can be calculated using Ohm's Law, which is Voltage Drop = Current in amps x Resistance in ohms. For example, the voltage drop over a 200 foot long, 14 gauge power line supplying a 1000 watt floodlight is calculated as follows:

Current = 1000 watts / 120 volts = 8.4 amps
Resistance of #14 wire = 2.58 ohms / 1000 feet @ 77$^{\circ}$F
Resistance of power line = 200 feet X 0.00258 ohms/foot
= 0.516 ohms
Voltage drop = 8.4 amps X 0.516 ohms = 4.33 volts
Percent voltage drop = 4.33 volts / 120 volts = 3.6 %

The 4.2% drop is over the maximum 2% so either the wattage of the bulbs must be decreased or the diameter of the wire must be increased (a decrease in wire gauge number). If #12 wire were used in the above example, the voltage drop would have only been 2.2%. The wire resistance values for various size wire are contained in the Copper Wire table on page 112.

An interesting corollary to the above example is that if the line voltage doubles ( 240 volts instead of 120 volts) the voltage drop decreases by 50%. That means that a line can carry the same power 2 times further! Higher voltage lines are more efficient.

A more commonly used method of calculating voltage drop is as follows:

$$\text{Voltage drop} = \frac{22 \times \text{Wire length in feet} \times \text{current in amps}}{\text{Circular Mils}}$$

Circular mils are given in the Standard Copper Wire Specs table on page 113. Note that the 22 value applies to copper wire only, if aluminum is used, change the value to 36.

# WIRE SIZE vs VOLTAGE DROP

Max Wire Feet @ 120 Volts, 1 Phase, 2% Max Voltage Drop

| Amps | Volt-Amps | #14 | #12 | #10 | #8 | #6 |
|---|---|---|---|---|---|---|
| 1 | 120 | 450 | 700 | 1100 | 1800 | 2800 |
| 5 | 600 | 90 | 140 | 225 | 360 | 575 |
| 10 | 1200 | 45 | 70 | 115 | 180 | 285 |
| 15 | 1800 | 30 | 47 | 75 | 120 | 190 |
| 20 | 2400 | ... | 36 | 57 | 90 | 140 |
| 25 | 3000 | ... | ... | 45 | 72 | 115 |
| 30 | 3600 | ... | ... | 38 | 60 | 95 |
| 40 | 4800 | ... | ... | ... | 45 | 72 |
| 50 | 6000 | ... | ... | ... | ... | 57 |

| Amps | Volt-Amps | #4 | #2 | 1/0 | 2/0 | 3/0 |
|---|---|---|---|---|---|---|
| 1 | 120 | 4500 | 7000 | ... | ... | ... |
| 5 | 600 | 910 | 1400 | 2250 | 2800 | ... |
| 10 | 1200 | 455 | 705 | 1100 | 1400 | 1800 |
| 15 | 1800 | 305 | 485 | 770 | 965 | 1200 |
| 20 | 2400 | 230 | 365 | 575 | 725 | 900 |
| 25 | 3000 | 180 | 290 | 460 | 580 | 720 |
| 30 | 3600 | 150 | 240 | 385 | 490 | 600 |
| 40 | 4800 | 115 | 175 | 290 | 360 | 440 |
| 50 | 6000 | 90 | 145 | 230 | 290 | 360 |
| 60 | 7200 | 76 | 120 | 190 | 240 | 305 |
| 70 | 8400 | 65 | 105 | 165 | 205 | 260 |
| 80 | 9600 | ... | 90 | 144 | 180 | 230 |

Max Wire Feet @ 240 Volts, 1 Phase, 2% Max Voltage Drop

| Amps | Volt-Amps | #14 | #12 | #10 | #8 | #6 |
|---|---|---|---|---|---|---|
| 1 | 240 | 900 | 1400 | 2200 | 3600 | 5600 |
| 5 | 1200 | 180 | 285 | 455 | 720 | 1020 |
| 10 | 2400 | 90 | 140 | 225 | 360 | 525 |
| 15 | 3600 | 60 | 95 | 150 | 240 | 350 |
| 20 | 4800 | ... | 70 | 110 | 180 | 265 |
| 25 | 6000 | ... | ... | 90 | 144 | 210 |
| 30 | 7200 | ... | ... | 75 | 120 | 175 |
| 40 | 5600 | ... | ... | ... | 90 | 130 |
| 50 | 12000 | ... | ... | ... | ... | 105 |

| Amps | Volt-Amps | #4 | #2 | 1/0 | 2/0 | 3/0 |
|---|---|---|---|---|---|---|
| 1 | 240 | 9000 | ... | ... | ... | ... |
| 5 | 1200 | 1750 | 2800 | 4500 | 5600 | 7000 |
| 10 | 2400 | 910 | 1400 | 2200 | 2800 | 3600 |
| 15 | 3600 | 605 | 965 | 1500 | 1900 | 2400 |
| 20 | 4800 | 455 | 725 | 1100 | 1400 | 1800 |
| 25 | 6000 | 365 | 580 | 920 | 1100 | 1440 |
| 30 | 7200 | 300 | 485 | 770 | 970 | 1200 |
| 40 | 5600 | 230 | 360 | 575 | 725 | 880 |
| 50 | 12000 | 180 | 290 | 460 | 580 | 720 |
| 60 | 14400 | 150 | 240 | 385 | 485 | 600 |
| 70 | 16800 | 130 | 205 | 330 | 415 | 520 |
| 80 | 19200 | ... | 180 | 290 | 365 | 440 |
| 100 | 24000 | ... | ... | 230 | 280 | 360 |
| 150 | 36000 | ... | ... | 185 | 190 | 240 |
| 200 | 48000 | ... | ... | ... | ... | 180 |

# CONDUIT SIZE vs WIRE SIZE

| Wire Size AWG | Minimum Conduit Size (inches) per Number of Type T & TW Wires. Number of Wires Inside Conduit 2 | 3 | 4 | 5 | 6 |
|---|---|---|---|---|---|
| 14 | 1/2 | 1/2 | 1/2 | 1/2 | 1/2 |
| 12 | 1/2 | 1/2 | 1/2 | 1/2 | 1/2 |
| 10 | 1/2 | 1/2 | 1/2 | 1/2 | 3/4 |
| 8 | 1/2 | 1/2 | 3/4 | 3/4 | 1 |
| 6 | 3/4 | 1 | 1 | 1-1/4 | 1-1/4 |
| 4 | 1 | 1 | 1-1/4 | 1-1/4 | 1-1/2 |
| 2 | 1 | 1-1/4 | 1-1/4 | 1-1/2 | 2 |
| 1/0 | 1-1/4 | 1-1/2 | 2 | 2 | 2-1/2 |
| 2/0 | 1-1/2 | 1-1/2 | 2 | 2 | 2-1/2 |
| 3/0 | 1-1/2 | 2 | 2 | 2-1/2 | 2-1/2 |

See the National Electric Code for conduit sizes when using wire types other than Type T & TW.

# BOX SIZE vs NUMBER OF WIRES

| Box Size in inches | Maximum Number of Wires in a Junction Box, Wire Size AWG #14 | #12 | #10 | #8 |
|---|---|---|---|---|
| **Outlet Boxes** | | | | |
| 3-1/4 X 1-1/2 octagon | 5 | 4 | 4 | 3 |
| 3-1/2 X 1-1/2 octagon | 5 | 5 | 4 | 3 |
| 4 X 1-1/2 octagon | 8 | 7 | 6 | 5 |
| 4 X 2-1/8 octagon | 11 | 10 | 9 | 7 |
| 4 X 4 X 1-1/2 | 11 | 10 | 9 | 7 |
| 4 X 4 X 2-1/8 | 15 | 14 | 12 | 10 |
| 4 X 2-1/8 X 1-1/2 | 3 | 3 | 3 | 2 |
| 4 X 2-1/8 X 1-7/8 | 5 | 4 | 4 | 3 |
| 4 X 2-1/8 X 2-1/8 | 5 | 5 | 4 | 3 |
| **Switch Boxes** | | | | |
| 3 X 2 X 1-1/2 | 6 | 5 | 5 | 4 |
| 3 X 2 X 2 | 7 | 6 | 5 | 4 |
| 3 X 2 X 2-1/4 | 9 | 8 | 7 | 6 |
| 3 X 2 X 2-1/2 | 5 | 4 | 4 | 3 |
| 3 X 2 X 2-3/4 | 6 | 6 | 5 | 4 |
| 3 X 2 X 3-1/2 | 7 | 6 | 6 | 5 |

The above number are maximums and you should deduct 1 wire if an outlet, switch, cable clamp, fixture stud, or similar part is also installed in the box.

# AVERAGE ELECTRIC MOTOR SPECS

NOTE: Use the following table as a general guide only! These numbers are for normal fan, furnace, appliance, pump, and normal duty applications. The exact specifications for any given motor can vary greatly from those listed below. For 230V motors simply divide the indicated amps by 2.

Specs for 115 volt, 60 Hz, 1 Phase, AC Electric Motors

| Motor Horsepower | RPM | Full Load Amps |
|---|---|---|
| 1/20 | 1550 | 2.5 |
| 1/15 | 1550 | 2.8 |
| 1/12 | 1725 | 2.2–2.8 |
| | 1550 | 4.1 |
| | 850 | 3.2 |
| 1/10 | 1550 | 3.5 |
| | 1050 | 3.4–4.2 |
| 1/8 | 1725 | 1.8–2.7 |
| | 1140 | 3.8 |
| | 1075 | 1.8–5.0 |
| 1/6 | 1725 | 3.3–4.7 |
| | 1550 | 4.0–4.8 |
| | 1140 | 4.0–4.9 |
| | 1075 | 2.4–5.0 |
| 1/4 | 1725 | 4.4–6.3 |
| | 1625 | 3.1–3.6 |
| | 1140 | 5.6–6.8 |
| | 1075 | 3.4–6.8 |
| | 850 | 6.9 |
| 1/3 | 3450 | 5.6–6.5 |
| | 1725 | 5.3–6.8 |
| | 1140 | 5.0–7.2 |
| | 1075 | 5.1 |
| 1/2 | 3450 | 9.8 |
| | 1725 | 7.0–9.2 |
| | 1075 | 7.3 |
| 3/4 | 3450 | 11.8 |
| | 1725 | 11.6 |
| | 1075 | 9.5 |
| 1 | 3450 | 13.0–15.0 |
| | 1725 | 13.6–16.0 |
| 1-1/2 | 3450 | 16.4–19.6 |
| | 1725 | 19.6 |
| 2 | 3450 | 19–23 |

The above general specifications are based on motor data from the *1988 Graingers Catalog, Chicago, Illinois.*

# NEMA ELECTRIC MOTOR FRAMES

| Motor Frame | NEMA Frame Dimension - Inches | | | | | | |
|---|---|---|---|---|---|---|---|
| | D | E | F | U | V | M+N | Keyway |
| 42 | 2-5/8 | 1-3/4 | 27/32 | 3/8 | ... | 4-1/32 | ... |
| 48 | 3 | 2-1/8 | 1-3/8 | 1/2 | ... | 5-3/8 | ... |
| 56 | 3-1/2 | 2-7/16 | 1-1/2 | 5/8 | ... | 6-1/8 | 3/16x3/32 |
| 66 | 4-1/8 | 2-15/16 | 2-1/2 | 3/4 | ... | 7-7/8 | 3/16x3/32 |
| 143T | 3-1/2 | 2-3/4 | 2 | 7/8 | 2 | 6-1/2 | 3/16x3/32 |
| 145T | 3-1/2 | 2-3/4 | 2-1/2 | 7/8 | 2 | 7 | 3/16x3/32 |
| 182 | 4-1/2 | 3-3/4 | 2-1/4 | 7/8 | 2 | 7-1/4 | 3/16x3/32 |
| 182T | 4-1/2 | 3-3/4 | 2-1/4 | 1-1/8 | 2-1/2 | 7-3/4 | 1/4x1/8 |
| 184 | 4-1/2 | 3-3/4 | 2-3/4 | 7/8 | 2 | 7-3/4 | 3/16x3/32 |
| 184T | 4-1/2 | 3-3/4 | 2-3/4 | 1-1/8 | 2-1/2 | 8-1/4 | 1/4x1/8 |
| 213 | 5-1/4 | 4-1/4 | 2-3/4 | 1-1/8 | 2-3/4 | 9-1/4 | 1/4x1/8 |
| 213T | 5-1/4 | 4-1/4 | 2-3/4 | 1-3/8 | 3-1/8 | 9-5/8 | 5/16x5/32 |
| 215 | 5-1/4 | 4-1/4 | 3-1/2 | 1-1/8 | 2-3/4 | 10 | 1/4x1/8 |
| 215T | 5-1/4 | 4-1/4 | 3-1/2 | 1-3/8 | 3-1/8 | 10-3/8 | 5/16x5/32 |
| 254T | 6-1/4 | 5 | 4-1/8 | 1-5/8 | 3-3/4 | 12-3/8 | 3/8x3/16 |
| 254U | 6-1/4 | 5 | 4-1/8 | 1-3/8 | 3-1/2 | 12-1/8 | 5/16x5/32 |
| 256T | 6-1/4 | 5 | 5 | 1-5/8 | 3-3/4 | 13-1/4 | 3/8x3/16 |
| 256U | 6-1/4 | 5 | 5 | 1-3/8 | 3-1/2 | 13 | 5/16x5/32 |
| 284T | 7 | 5-1/2 | 4-3/4 | 1-7/8 | 4-3/8 | 14-1/8 | 1/2x1/4 |
| 284TS | 7 | 5-1/2 | 4-1/4 | 1-5/8 | 3 | 13-1/2 | 3/8x3/16 |
| 284U | 7 | 5-1/2 | 4-3/4 | 1-5/8 | 4-5/8 | 14-2/8 | 3/8x3/16 |
| 286T | 7 | 5-1/2 | 5-1/2 | 1-7/8 | 4-3/8 | 14-7/8 | 1/2x1/4 |
| 286U | 7 | 5-1/2 | 5-1/2 | 1-5/8 | 4-5/8 | 15-1/8 | 3/8x3/16 |
| 324T | 8 | 6-1/4 | 5-1/4 | 2-1/8 | 5 | 15-3/4 | 1/2x1/4 |
| 324U | 8 | 6-1/4 | 5-1/4 | 1-7/8 | 5-3/8 | 16-1/8 | 1/2x1/4 |
| 326T | 8 | 6-1/4 | 6 | 2-1/8 | 5 | 16-1/2 | 1/2x1/4 |
| 326TS | 8 | 6-1/4 | 6 | 1-7/8 | 3-1/2 | 15 | 1/2x1/4 |
| 326U | 8 | 6-1/4 | 6 | 1-7/8 | 5-3/8 | 16-7/8 | 1/2x1/4 |
| 364T | 9 | 7 | 5-5/8 | 2-3/8 | 5-5/8 | 17-3/8 | 5/8x5/16 |
| 364U | 9 | 7 | 5-5/8 | 2-1/8 | 6-1/8 | 17-7/8 | 1/2x1/4 |
| 365T | 9 | 7 | 6-1/8 | 2-3/8 | 5-5/8 | 17-7/8 | 5/8x5/16 |
| 365U | 9 | 7 | 6-1/8 | 2-1/8 | 6-1/8 | 1-3/8 | 1/2x1/4 |
| 404T | 10 | 8 | 6-1/8 | 2-7/8 | 7 | 20 | 3/4x3/8 |
| 404U | 10 | 8 | 6-1/8 | 2-3/8 | 6-7/8 | 19-7/8 | 5/8x5/16 |
| 405T | 10 | 8 | 6-7/8 | 2-7/8 | 7 | 20-3/4 | 3/4x3/8 |
| 405U | 10 | 8 | 6-7/8 | 2-3/8 | 6-7/8 | 20-5/8 | 5/8x5/16 |
| 444T | 11 | 9 | 7-1/4 | 3-3/8 | 8-1/4 | 23-1/4 | 7/8x7/16 |
| 444U | 11 | 9 | 7-1/4 | 2-7/8 | 8-3/8 | 23-3/8 | 3/4x3/8 |
| 445T | 11 | 9 | 8-1/4 | 3-3/8 | 8-1/4 | 24-1/4 | 7/8x7/16 |
| 445U | 11 | 9 | 8-1/4 | 2-7/8 | 8-3/8 | 24-3/8 | 3/4x3/8 |

The above standards were established by the *National Electrical Manufacturers Association (NEMA)*

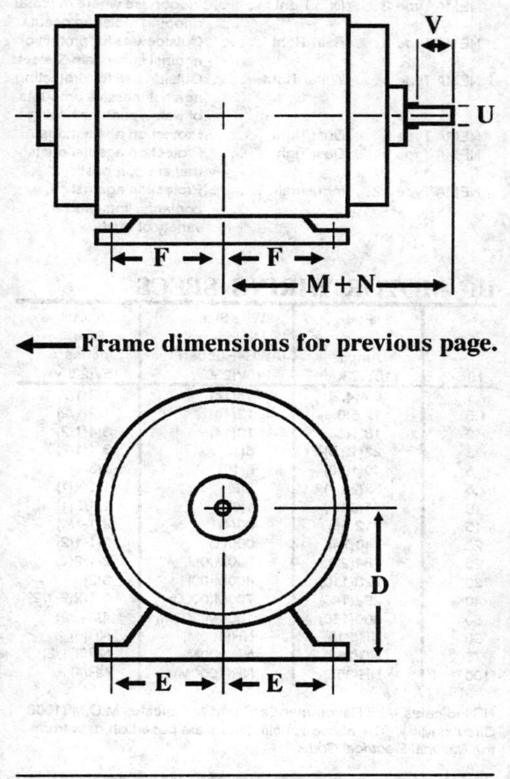


Frame dimensions for previous page.

# NEMA ELECTRIC ENCLOSURES

| Enclosure Type | Class | Description |
|---|---|---|
| NEMA Type 1 | General Purpose | Indoor use where no oil, water or dust is present. |
| NEMA Type 2 | Drip Tight | Indoor use where minimal dripping moisture occurs. |
| NEMA Type 3 | Rain Tight | Outside use for protection against snow, rain & sleet. |
| NEMA Type 4 | Water Tight | Outside use for protection against massive amounts of water, such as hosing. |
| NEMA Type 5 | Dust Tight | Protection against dust. |
| NEMA Type 9 | Dust Tight | Protection against dusts that are combustible. |
| NEMA Type 12 | Industrial | Protection against oil, coolants, lints, and a variety of dusts. |

# DC MOTOR WIRING SPECS

| HP | Full Load Amps 115V(230V) | Wire Size Minimum (AWG–Rubber) 115V(230V) | Conduit Size Inches 115V(230V) |
|---|---|---|---|
| 1 | 8.4(4.2) | 14(14) | 1/2(1/2) |
| 1.5 | 12.5(6.3) | 12(14) | 1/2(1/2) |
| 2 | 16.1(8.3) | 10(14) | 3/4(1/2) |
| 3 | 23(12.3) | 8(12) | 3/4(1/2) |
| 5 | 40(19.8) | 6(10) | 1(3/4) |
| 7.5 | 58(28.7) | 3(6) | 1-1/4(1) |
| 10 | 75(38) | 1(6) | 1-1/2(1) |
| 15 | 112(56) | 00(4) | 2(1-1/4) |
| 20 | 140(74) | 000(1) | 2(1-1/2) |
| 25 | 184(92) | 300M(0) | 2-1/2(2) |
| 30 | 220(110) | 400M(00) | 3(2) |
| 40 | 292(146) | 700M(0000) | 3-1/2(2-1/2) |
| 50 | 360(180) | 1000M(300M) | 4(2-1/2) |
| 60 | NR(215) | NR(400M) | NR(3) |
| 75 | NR(268) | NR(600M) | NR(3-1/2) |
| 100 | NR(355) | NR(1000M) | NR(4) |

NR indicates "Not Recommended" and M indicates M.C.M (1000 Circular Mils). The above specifications are based on data from the *National Electrical Code*.

# 3 PHASE ELECTRIC MOTOR SPECS

| HP | Full Load Amps 230V(460V) | Wire Size Minimum (AWG–Rubber) 230V(460V) | Conduit Size Inches 230V(460V) |
|---|---|---|---|
| 1 | 3.3(1.7) | 14(14) | 1/2(1/2) |
| 1.5 | 4.7(2.4) | 14(14) | 1/2(1/2) |
| 2 | 6(3.0) | 14(14) | 1/2(1/2) |
| 3 | 9(4.5) | 14(14) | 1/2(1/2) |
| 5 | 15(7.5) | 12(14) | 1/2(1/2) |
| 7.5 | 22(11) | 8(14) | 3/4(1/2) |
| 10 | 27(14) | 8(12) | 3/4(1/2) |
| 15 | 38(19) | 6(10) | 1-1/4(3/4) |
| 20 | 52(26) | 4(8) | 1-1/4(3/4) |
| 25 | 64(32) | 3(6) | 1-1/4(1-1/4) |
| 30 | 77(39) | 1(6) | 1-1/2(1-1/4) |
| 40 | 101(51) | 00(4) | 2(1-1/4) |
| 50 | 125(63) | 000(3) | 2(1-1/4) |
| 60 | 149(75) | 200M(1) | 2-1/2(1-1/2) |
| 75 | 180(90) | 0000(0) | 2-1/2(2) |
| 100 | 245(123) | 500M(000) | 3(2) |
| 125 | 310(155) | 750M(0000) | 3-1/2(2-1/2) |
| 150 | 360(180) | 1000M(300M) | 4(2-1/2) |
| 200 | 480(240) | NR(500M) | NR(3) |
| 250 | 580(290) | NR(NR) | NR(NR) |
| 300 | 696(348) | NR(NR) | NR(NR) |

NR indicates "Not Recommended" and "M" indicates M.C.M (1000 Circular Mils).

Note that starting currents for the above motors can be many times the Full Load Amps and fuses must be adjusted accordingly. If the powerline becomes too long, voltage drop will exceed safe limits and the wire size should be adjusted to the next larger (smaller AWG number) gauge wire. See the Copper Wire Specifications table in this chapter for more specific information on wire.

The above specifications are from the *National Electrical Code.*

# HP vs TORQUE vs RPM – MOTORS

| | Torque in Inch Pounds @ Motor R.P.M. | | | | | |
|---|---|---|---|---|---|---|
| HP | 3450 | 2000 | 1725 | 1550 | 1140 | 1050 |
| 1 | 18 | 32 | 37 | 41 | 55 | 60 |
| 1.5 | 27 | 47 | 55 | 61 | 83 | 90 |
| 2 | 37 | 63 | 73 | 81 | 111 | 120 |
| 3 | 55 | 95 | 110 | 122 | 166 | 180 |
| 5 | 91 | 158 | 183 | 203 | 276 | 300 |
| 7.5 | 137 | 236 | 274 | 305 | 415 | 450 |
| 10 | 183 | 315 | 365 | 407 | 553 | 600 |
| 15 | 274 | 473 | 548 | 610 | 829 | 900 |
| 20 | 365 | 630 | 731 | 813 | 1106 | 1200 |
| 25 | 457 | 788 | 913 | 1017 | 1382 | 1501 |
| 30 | 548 | 945 | 1096 | 1220 | 1659 | 1801 |
| 40 | 731 | 1261 | 1461 | 1626 | 2211 | 2401 |
| 50 | 913 | 1576 | 1827 | 2033 | 2764 | 3001 |
| 60 | 1096 | 1891 | 2192 | 2440 | 3317 | 3601 |
| 70 | 1279 | 2206 | 2558 | 2846 | 3870 | 4202 |
| 80 | 1461 | 2521 | 2923 | 3253 | 4423 | 4802 |
| 90 | 1644 | 2836 | 3288 | 3660 | 4976 | 5402 |
| 100 | 1827 | 3151 | 3654 | 4066 | 5529 | 6002 |
| 125 | 2284 | 3939 | 4567 | 5083 | 6911 | 7503 |
| 150 | 2740 | 4727 | 5480 | 6099 | 8293 | 9004 |
| 175 | 3197 | 5515 | 6394 | 7116 | 9675 | 10504 |
| 200 | 3654 | 6303 | 7307 | 8132 | 11057 | 12005 |
| 225 | 4110 | 7090 | 8221 | 9149 | 12439 | 13505 |
| 250 | 4567 | 7878 | 9134 | 10165 | 13821 | 15006 |
| 275 | 5024 | 8666 | 10047 | 11182 | 15203 | 16507 |
| 300 | 5480 | 9454 | 10961 | 12198 | 16586 | 18007 |
| 350 | 6394 | 11029 | 12788 | 14231 | 19350 | 21008 |
| 400 | 7307 | 12605 | 14614 | 16265 | 22114 | 24010 |
| 450 | 8221 | 14181 | 16441 | 18298 | 24878 | 27011 |
| 500 | 9134 | 15756 | 18268 | 20331 | 27643 | 30012 |
| 550 | 10047 | 17332 | 20095 | 22364 | 30407 | 33013 |
| 600 | 10961 | 18908 | 21922 | 24397 | 33171 | 36014 |

$$\text{Torque in Inch Pounds} = \frac{\text{Horsepower} \times 63025}{\text{Motor RPM}}$$

To convert to Foot Pounds, divide the torque by 12.

# HP vs TORQUE vs RPM - MOTORS

| HP | Torque in Inch Pounds @ Motor R.P.M. 1000 | 850 | 750 | 600 | 500 | 230 |
|---|---|---|---|---|---|---|
| 1 | 63 | 74 | 84 | 105 | 126 | 274 |
| 1.5 | 95 | 111 | 126 | 158 | 189 | 411 |
| 2 | 126 | 148 | 168 | 210 | 252 | 548 |
| 3 | 189 | 222 | 252 | 315 | 378 | 822 |
| 5 | 315 | 371 | 420 | 525 | 630 | 1370 |
| 7.5 | 473 | 556 | 630 | 788 | 945 | 2055 |
| 10 | 630 | 741 | 840 | 1050 | 1261 | 2740 |
| 15 | 945 | 1112 | 1261 | 1576 | 1891 | 4110 |
| 20 | 1261 | 1483 | 1681 | 2101 | 2521 | 5480 |
| 25 | 1576 | 1854 | 2101 | 2626 | 3151 | 6851 |
| 30 | 1891 | 2224 | 2521 | 3151 | 3782 | 8221 |
| 40 | 2521 | 2966 | 3361 | 4202 | 5042 | 10961 |
| 50 | 3151 | 3707 | 4202 | 5252 | 6303 | 13701 |
| 60 | 3782 | 4449 | 5042 | 6303 | 7563 | 16441 |
| 70 | 4412 | 5190 | 5882 | 7353 | 8824 | 19182 |
| 80 | 5042 | 5932 | 6723 | 8403 | 10084 | 21922 |
| 90 | 5672 | 6673 | 7563 | 9454 | 11345 | 24662 |
| 100 | 6303 | 7415 | 8403 | 10504 | 12605 | 27402 |
| 125 | 7878 | 9268 | 10504 | 13130 | 15756 | 34253 |
| 150 | 9454 | 11122 | 12605 | 15756 | 18908 | 41103 |
| 175 | 11029 | 12976 | 14706 | 18382 | 22059 | 47954 |
| 200 | 12605 | 14829 | 16807 | 21008 | 25210 | 54804 |
| 225 | 14181 | 16683 | 18908 | 23634 | 28361 | 61655 |
| 250 | 15756 | 18537 | 21008 | 26260 | 31513 | 68505 |
| 275 | 17332 | 20390 | 23109 | 28886 | 34664 | 75356 |
| 300 | 18908 | 22244 | 25210 | 31513 | 37815 | 82207 |
| 350 | 22059 | 25951 | 29412 | 36765 | 44118 | 95908 |
| 400 | 25210 | 29659 | 33613 | 42017 | 50420 | 109609 |
| 450 | 28361 | 33366 | 37815 | 47269 | 56723 | 123310 |
| 500 | 31513 | 37074 | 42017 | 52521 | 63025 | 137011 |
| 550 | 34664 | 40781 | 46218 | 57773 | 69328 | 150712 |
| 600 | 37815 | 44488 | 50420 | 63025 | 75630 | 164413 |

$$\text{Torque in Inch Pounds} = \frac{\text{Horsepower} \times 63025}{\text{Motor RPM}}$$

NOTE: Ratings below 500 RPM are for gear motors.
To convert to Foot Pounds, divide the torque by 12.

# HP vs TORQUE vs RPM – MOTORS

| | Torque in Inch Pounds @ Motor R.P.M. | | | | | |
|---|---|---|---|---|---|---|
| HP | 190 | 155 | 125 | 100 | 84 | 68 |
| 1 | 332 | 407 | 504 | 630 | 750 | 927 |
| 1.5 | 498 | 610 | 756 | 945 | 1125 | 1390 |
| 2 | 663 | 813 | 1008 | 1261 | 1501 | 1854 |
| 3 | 995 | 1220 | 1513 | 1891 | 2251 | 2781 |
| 5 | 1659 | 2033 | 2521 | 3151 | 3751 | 4634 |
| 7.5 | 2488 | 3050 | 3782 | 4727 | 5627 | 6951 |
| 10 | 3317 | 4066 | 5042 | 6303 | 7503 | 9268 |
| 15 | 4976 | 6099 | 7563 | 9454 | 11254 | 13903 |
| 20 | 6634 | 8132 | 10084 | 12605 | 15006 | 18537 |
| 25 | 8293 | 10165 | 12605 | 15756 | 18757 | 23171 |
| 30 | 9951 | 12198 | 15126 | 18908 | 22509 | 27805 |
| 40 | 13268 | 16265 | 20168 | 25210 | 30012 | 37074 |
| 50 | 16586 | 20331 | 25210 | 31513 | 37515 | 46342 |
| 60 | 19903 | 24397 | 30252 | 37815 | 45018 | 55610 |
| 70 | 23220 | 28463 | 35294 | 44118 | 52521 | 64879 |
| 80 | 26537 | 32529 | 40336 | 50420 | 60024 | 74147 |
| 90 | 29854 | 36595 | 45378 | 56723 | 67527 | 83415 |
| 100 | 33171 | 40661 | 50420 | 63025 | 75030 | 92684 |
| 125 | 41464 | 50827 | 63025 | 78781 | 93787 | 115855 |
| 150 | 49757 | 60992 | 75630 | 94538 | 112545 | 139026 |
| 175 | 58049 | 71157 | 88235 | 110294 | 131302 | 162197 |
| 200 | 66342 | 81323 | 100840 | 126050 | 150060 | 185368 |
| 225 | 74635 | 91488 | 113445 | 141806 | 168817 | 208539 |
| 250 | 82928 | 101653 | 126050 | 157563 | 187574 | 231710 |
| 275 | 91220 | 111819 | 138655 | 173319 | 206332 | 254881 |
| 300 | 99513 | 121984 | 151260 | 189075 | 225089 | 278051 |
| 350 | 116099 | 142315 | 176470 | 220588 | 262604 | 324393 |
| 400 | 132684 | 162645 | 201680 | 252100 | 300119 | 370735 |
| 450 | 149270 | 182976 | 226890 | 283613 | 337634 | 417077 |
| 500 | 165855 | 203306 | 252100 | 315125 | 375149 | 463419 |
| 550 | 182441 | 223637 | 277310 | 346638 | 412664 | 509761 |
| 600 | 199026 | 243968 | 302520 | 378150 | 450179 | 556103 |

$$\text{Torque in Inch Pounds} = \frac{\text{Horsepower} \times 63025}{\text{Motor RPM}}$$

NOTE: Ratings below 500 RPM are for gear motors.
To convert to Foot Pounds, divide the torque by 12.

# POCKET REF

# Electronics



(See also Frequency Spectrum on page 192)
(See also ELECTRIC Chapter on page 107)

# RESISTOR COLOR CODES

| Color | 1st Digit(A) | 2nd Digit(B) | Multiplier(C) | Tolerance(D) |
|---|---|---|---|---|
| Black | 0 | 0 | 1 | |
| Brown | 1 | 1 | 10 | 1% |
| Red | 2 | 2 | 100 | 2% |
| Orange | 3 | 3 | 1,000 | 3% |
| Yellow | 4 | 4 | 10,000 | 4% |
| Green | 5 | 5 | 100,000 | |
| Blue | 6 | 6 | 1,000,000 | |
| Violet | 7 | 7 | 10,000,000 | |
| Gray | 8 | 8 | 100,000,000 | |
| White | 9 | 9 | $10^9$ | |
| Gold | | | 0.1 (EIA) | 5% |
| Silver | | | 0.01 (EIA) | 10% |
| No Color | | | | 20% |

Example: Red–Red–Orange = 22,000 ohms, 20%

Additional information concerning the Axial Lead resistor can be obtained if Band A is a wide band. Case 1: If only Band A is wide, it indicates that the resistor is wirewound. Case 2: If Band A is wide and there is also a blue fifth band to the right of Band D on the Axial Lead Resistor, it indicates the resistor is wirewound and flame proof.

Axial Lead Resistor

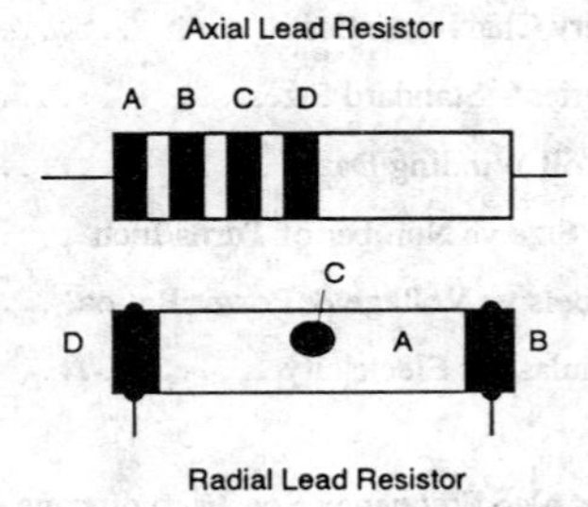


Radial Lead Resistor

# RESISTOR STANDARD VALUES

Standard Resistor Values for 5% class

| | | | |
|---|---|---|---|
| 1 | 62 | 3.9k | 240k |
| 1.1 | 68 | 4.3k | 270k |
| 1.2 | 75 | 4.7k | 300k |
| 1.3 | 82 | 5.1k | 330k |
| 1.5 | 91 | 5.6k | 360k |
| 1.6 | 100 | 6.2k | 390k |
| 1.8 | 110 | 6.8k | 430k |
| 2.0 | 120 | 7.5k | 470k |
| 2.2 | 130 | 8.2k | 510k |
| 2.4 | 150 | 9.1k | 560k |
| 2.7 | 160 | 10k | 620k |
| 3.0 | 180 | 11k | 680k |
| 3.3 | 200 | 12k | 750k |
| 3.6 | 220 | 13k | 820k |
| 3.9 | 240 | 15k | 910k |
| 4.3 | 270 | 16k | 1.0M |
| 4.7 | 300 | 18k | 1.1M |
| 5.1 | 330 | 20k | 1.2M |
| 5.6 | 360 | 22k | 1.3M |
| 6.2 | 390 | 24k | 1.5M |
| 6.8 | 430 | 27k | 1.6M |
| 7.5 | 470 | 30k | 1.8M |
| 8.2 | 510 | 33k | 2.0M |
| 9.1 | 560 | 36k | 2.2M |
| 10 | 620 | 39k | 2.4M |
| 11 | 680 | 43k | 2.7M |
| 12 | 750 | 47k | 3.0M |
| 13 | 820 | 51k | 3.3M |
| 15 | 910 | 56k | 3.6M |
| 16 | 1.0k | 62k | 3.9M |
| 18 | 1.1k | 68k | 4.3M |
| 20 | 1.2k | 75k | 4.7M |
| 22 | 1.3k | 82k | 5.1M |
| 24 | 1.5k | 91k | 5.6M |
| 27 | 1.6k | 100k | 6.2M |
| 30 | 1.8k | 110k | 6.8M |
| 33 | 2.0k | 120k | 7.5M |
| 36 | 2.2k | 130k | 8.2M |
| 39 | 2.4k | 150k | 9.1M |
| 43 | 2.7k | 160k | 10.0M |
| 47 | 3.0k | 180k | |
| 51 | 3.3k | 200k | |
| 56 | 3.6k | 220k | |

k = kohms = 1,000 ohms     M = megohms = 1,000,000 ohms

# CAPACITOR COLOR CODES

| Color | 1st Digit(A) | 2nd Digit(B) | Multiplier(C) | Tolerance(D) |
|---|---|---|---|---|
| Black | 0 | 0 | 1 | 20% |
| Brown | 1 | 1 | 10 | 1% |
| Red | 2 | 2 | 100 | 2% |
| Orange | 3 | 3 | 1,000 | 3% |
| Yellow | 4 | 4 | 10,000 | 4% |
| Green | 5 | 5 | 100,000 | 5% |
| Blue | 6 | 6 | 1,000,000 | 6% |
| Violet | 7 | 7 | 10,000,000 | 7% |
| Gray | 8 | 8 | 100,000,000 | 8% |
| White | 9 | 9 | $10^9$ | 9% |
| Gold | | | 0.1 (EIA) | 5% |
| Silver | | | 0.01 (EIA) | 10% |
| No Color | | | | 20% |

# Color Codes for Ceramic Capacitors

| Color | Decimal Multiplier(C) | Tolerance (D) Above 10pf | Tolerance (D) Below 10pf | Temp Coef ppm/°C (E) |
|---|---|---|---|---|
| Black | 1 | 20 | 2.0 | 0 |
| Brown | 10 | 1 | | –30 |
| Red | 100 | 2 | | –80 |
| Orange | 1000 | | | –150 |
| Yellow | | | | –220 |
| Green | | 5 | 0.5 | –330 |
| Blue | | | | –470 |
| Violet | | | | –750 |
| Gray | 0.01 | | 0.25 | 30 |
| White | 0.1 | 10 | 1.0 | 500 |

Ceramic disc capacitors are usually labeled. If the number is < 1 then the value is picofarads, if > 1 the value is microfarads. The letter R is sometimes used as a decimal, eg, 4R7 is 4.7.

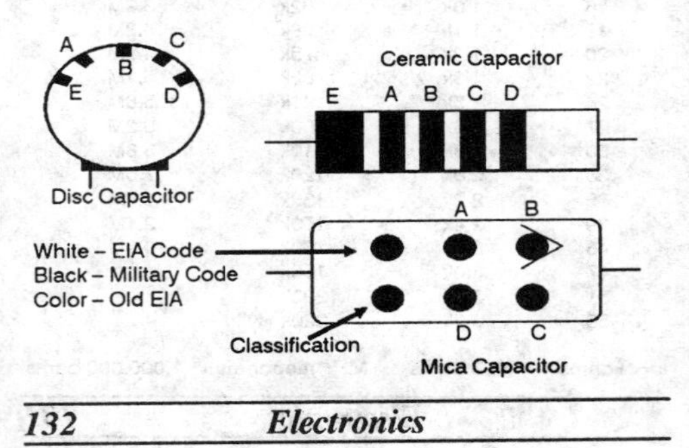

# CAPACITOR STANDARD VALUES

| pF | mF | mF | mF | mF |
|---|---|---|---|---|
| 10 | 0.001 | 0.1 | 10 | 1000 |
| 12 | 0.0012 | | | |
| 13 | 0.0013 | | | |
| 15 | 0.0015 | 0.15 | 15 | |
| 18 | 0.0018 | | | |
| 20 | 0.002 | | | |
| 22 | 0.0022 | 0.22 | 22 | 2200 |
| 24 | | | | |
| 27 | | | | |
| 30 | | | | |
| 33 | 0.0033 | 0.33 | 33 | 3300 |
| 36 | | | | |
| 43 | | | | |
| 47 | 0.0047 | 0.47 | 47 | 4700 |
| 51 | | | | |
| 56 | | | | |
| 62 | | | | |
| 68 | 0.0068 | 0.68 | 68 | 6800 |
| 75 | | | | |
| 82 | | | | |
| 100 | 0.01 | 1.0 | 100 | 10,000 |
| 110 | | | | |
| 120 | | | | |
| 130 | | | | |
| 150 | 0.015 | 1.5 | | |
| 180 | | | | |
| 200 | | | | |
| 220 | 0.022 | 2.2 | 220 | 22,000 |
| 240 | | | | |
| 270 | | | | |
| 300 | | | | |
| 330 | 0.033 | 3.3 | 330 | |
| 360 | | | | |
| 390 | | | | |
| 430 | | | | |
| 470 | 0.047 | 4.7 | 470 | 47,000 |
| 510 | | | | |
| 560 | | | | |
| 620 | | | | |
| 680 | 0.068 | 6.8 | | |
| 750 | | | | |
| 820 | | | | 82,000 |
| 910 | | | | |

pf = picofarads = $1 \times 10^{-12}$ farads
mf = micro farads = $1 \times 10^{-6}$ farads

# PILOT LAMPS

| Lamp Number | Bead Color | Base Type | Bulb Volts | Amps | Type |
|---|---|---|---|---|---|
| 12 | | 2 Pin | 6.3 | 0.15 | G3-1/2 |
| 12PSB5 | | Slide | 12 | 0.17 | T2 |
| 13 | | Screw | 3.7 | 0.30 | G3-1/2 |
| 14 | | Screw | 2.47 | 0.30 | G3-1/2 |
| 19 | | 2 Pin | 14.4 | 0.10 | G3-1/2 |
| 24PSB5 | | Slide | 24 | 0.073 | T2 |
| 27 | | Screw | 4.9 | 0.30 | G4-1/2 |
| 28PSB5 | | Slide | 28 | 0.04 | T2 |
| 31 | | Screw | 6.15 | 0.30 | G4-1/2 |
| 40 | Brown | Screw | 6-8 | 0.15 | T3-1/4 |
| 40A | Brown | Bayonet | 6-8 | 0.15 | T3-1/4 |
| 41 | White | Screw | 2.5 | 0.5 | T3-1/4 |
| 42 | Green | Screw | 3.2 | 0.35/0.5 | T3-1/4 |
| 43 | White | Bayonet | 2.5 | 0.5 | T3-1/4 |
| 44 | Blue | Bayonet | 6-8 | 0.25 | T3-1/4 |
| 45 | White/Grn | Bayonet | 3.2 | 0.35/0.5 | T3-1/4 |
| 46 | Blue | Screw | 6-8 | 0.25 | T3-1/4 |
| 47 | Brown | Bayonet | 6-9 | 0.15 | T3-1/4 |
| 48 | Pink | Screw | 2.0 | 0.06 | T3-1/4 |
| 49 | Pink | Bayonet | 2.0 | 0.06 | T3-1/4 |
| 49A | White | Bayonet | 2.1 | 0.12 | T3-1/4 |
| 50 | White | Screw | 6-8 | 0.2 | G3-1/2 |
| 51 | White | Bayonet | 6-8 | 0.2 | G3-1/2 |
| 53 | | Bayonet | 14.4 | 0.12 | G3-1/2 |
| 55 | White | Bayonet | 6-8 | 0.4 | G4-1/2 |
| 57 | White | Bayonet | 14 | 0.24 | G4-1/2 |
| 63 | | Bayonet | 7.0 | 0.63 | G6 |
| 67 | | Bayonet | 13.5 | 0.59 | G6 |
| 73 | | Wedge | 14 | 0.08 | T1-3/4 |
| 81 | | Bayonet | 6.5 | 1.02 | G6 |
| 82 | | Dbl Bayonet | 6.5 | 1.02 | G6 |
| 85 | | Wedge | 28 | 0.04 | T1-3/4 |
| 86 | | Wedge | 6.3 | 0.20 | T1-3/4 |
| 87 | | Bayonet | 6.8 | 1.91 | S8 |
| 88 | | Dbl Bayonet | 6.8 | 1.90 | S8 |
| 89 | | Bayonet | 13.0 | 0.58 | G6 |
| 93 | | Bayonet | 12.8 | 1.04 | S8 |
| 112 | Pink | Screw | 1.1 | 0.22 | TL-3 |
| 120MB | | Min. Bayonet | 120.0 | 0.025 | T2 |
| 123 | | Screw | 1.25 | 0.30 | G3-1/2 |
| 136 | | Screw | 1.25 | 0.60 | G4-1/2 |
| 158 | | Wedge | 14.0 | 0.24 | T3-1/4 |
| 161 | | Wedge | 14.0 | 0.19 | T3-1/4 |
| 168 | | Wedge | 14.0 | 0.35 | T3-1/4 |
| 194 | | Wedge | 14.0 | 0.27 | T3-1/4 |
| 222 | White | Screw | 2.2 | 0.25 | GTL-3 |
| 292 | White | Screw | 2.9 | 0.17 | T3-1/4 |
| 292A | White | Bayonet | 2.9 | 0.17 | T3-1/4 |
| 301 | | Bayonet | 28.0 | 0.17 | G5 |
| 302 | | Dbl Bayonet | 28.0 | 0.17 | G5 |
| 303 | | Bayonet | 28.0 | 0.30 | G6 |
| 305 | | Bayonet | 28.0 | 0.51 | S8 |
| 307 | | Bayonet | 28.0 | 0.66 | S8 |
| 308 | | Dbl Bayonet | 28.0 | 0.67 | S8 |
| 309 | | Bayonet | 28.0 | 0.90 | S11 |
| 313 | | Min. Bayonet | 28.0 | 0.17 | T3-1/4 |
| 327 | | Mig. Flanged | 28.0 | 0.04 | T1-3/4 |

# PILOT LAMPS

| Lamp Number | Bead Color | Base Type | Bulb Volts | Amps | Type |
|---|---|---|---|---|---|
| 328 | | Mig. Flanged | 6.0 | 0.20 | T1-3/4 |
| 330 | | Mig. Flanged | 14.0 | 0.08 | T1-3/4 |
| 331 | | Mig. Flanged | 1.35 | 0.06 | T1-3/4 |
| 334 | | Mig. Grooved | 28.0 | 0.04 | T1-3/4 |
| 335 | | Mig. Screw | 28.0 | 0.04 | T1-3/4 |
| 344 | | Mig. Flanged | 10.0 | 0.014 | T1-3/4 |
| 381 | | Mig. Flanged | 6.3 | 0.02 | T1-3/4 |
| 382 | | Mig. Flanged | 14.0 | 0.08 | T1-3/4 |
| 385 | | Mig. Flanged | 28.0 | 0.04 | T1-3/4 |
| 387 | | Mig. Flanged | 28.0 | 0.04 | T1-3/4 |
| 388 | | Mig. Grooved | 28.0 | 0.04 | T1-3/4 |
| 656 | | Wedge | 28.0 | 0.06 | T3-1/4 |
| 680 | | Wires | 5.0 | 0.06 | T1 |
| 682 | | Mig. Flange | 5.0 | 0.06 | T1 |
| 683 | | Wires | 5.0 | 0.06 | T1 |
| 683AS15 | | Wires | 5.0 | 0.06 | T1 |
| 685 | | Mig. Flange | 5.0 | 0.06 | T1 |
| 713 | | Wires | 5.0 | 0.075 | T1 |
| 714 | | Mig. Flange | 5.0 | 0.075 | T1 |
| 715 | | Wires | 5.0 | 0.115 | T1 |
| 715AS15 | | Wires | 5.0 | 0.115 | T1 |
| 718 | | Mig. Flange | 5.0 | 0.115 | T1 |
| 755 | | Min. Bayonet | 6.3 | 0.15 | T3-1/4 |
| 756 | | Min. Bayonet | 14.0 | 0.08 | T3-1/4 |
| 757 | | Min. Bayonet | 28.0 | 0.08 | T3-1/4 |
| 1003 | | Bayonet | 12.8 | 0.94 | B6 |
| 1004 | | Dbl Bayonet | 12.8 | 0.94 | B6 |
| 1034 | | Bayonet | 12.8 | 1.80 | S8 |
| 1076 | | Dbl Bayonet | 12.8 | 1.80 | S8 |
| 1133 | | | 6.2 | 3.91 | RP11 |
| 1156 | | Bayonet | 12.8 | 2.10 | S8 |
| 1157 | | DC Bayonet | 12.8 | 2.10 | S8 |
| 1176 | | Dbl Bayonet | 12.8 | 1.34 | S8 |
| 1195 | | | 12.5 | 3.0 | RP11 |
| 1251 | | Bayonet | 28.0 | 0.23 | G6 |
| 1445 | | Bayonet | 14.4 | 0.135 | G3-1/2 |
| 1447 | | Screw | 18 | 0.15 | G3-1/2 |
| 1455 | Brown | Screw | 18.0 | 0.25 | G5 |
| 1455A | Brown | Bayonet | 18.0 | 0.25 | G5 |
| 1458 | | Bayonet | 20.0 | 0.25 | G5 |
| 1487 | | Screw | 12-16 | 0.2 | T3-1/4 |
| 1488 | | Bayonet | 14 | 0.15 | T3-1/4 |
| 1490 | White | Bayonet | 3.2 | 0.15 | T3-1/4 |
| 1495 | | Bayonet | 28.0 | 0.30 | T4-1/2 |
| 1705 | | Wires | 14.0 | 0.08 | T1-3/4 |
| 1764 | | Wires | 28.0 | 0.04 | T1-3/4 |
| 1784 | | Wires | 6.0 | 0.20 | T1-3/4 |
| 1813 | | Bayonet | 14.4 | 0.10 | T3-1/4 |
| 1815 | | Bayonet | 12-16 | 0.20 | T3-1/4 |
| 1816 | | Bayonet | 13.0 | 0.33 | T1-3/4 |
| 1819 | | Min. Bayonet | 28.0 | 0.04 | T3-1/4 |
| 1820 | | Min. Bayonet | 28.0 | 0.10 | T3-1/4 |
| 1822 | | Min. Bayonet | 36.0 | 0.10 | T3-1/4 |
| 1829 | | Min. Bayonet | 28.0 | 0.07 | T3-1/4 |
| 1847 | | Bayonet | 6.3 | 0.15 | T1-3/4 |
| 1864 | | Min. Bayonet | 28.0 | 0.17 | T3-1/4 |
| 1891 | Pink | Bayonet | 14 | 0.23 | T3-1/2 |

# PILOT LAMPS

| Lamp Number | Bead Color | Base Type | Bulb Volts | Amps | Type |
|---|---|---|---|---|---|
| 1892 | White | Screw | 14 | 0.12 | T3-1/2 |
| 1895 | | Min. Bayonet | 14.0 | 0.27 | G4-1/2 |
| 2181 | | Wires | 6.3 | 0.20 | T1-3/4 |
| 2182 | | Wires | 6.3 | 0.20 | T1-3/4 |
| 2187 | | Wires | 28.0 | 0.04 | T1-3/4 |
| 3150 | | Mig. Flange | 5.0 | 0.06 | T1-3/4 |
| 6838 | | Wires | 28.0 | 0.024 | T1 |
| 6839 | | Mig. Flange | 28.0 | 0.024 | T1 |
| 7327 | | Bi Pin | 28.0 | 0.04 | T1-3/4 |
| 7333 | | Mig. Flange | 5.0 | 0.06 | T1-3/4 |
| 7361 | | Bi Pin | 5.0 | 0.06 | T1-3/4 |
| 7381 | | Bi Pin | 6.3 | 0.20 | T1-3/4 |
| 7382 | | Bi Pin | 14.0 | 0.08 | T1-3/4 |
| 7387 | | Bi Pin | 28.0 | 0.04 | T1-3/4 |
| 7632 | | Bi Pin | 28.0 | 0.04 | T1-3/4 |
| 7839 | | Bi Pin | 28.0 | 0.024 | T1 |
| 8623 | | Thr. Knurled | 28.0 | 0.04 | T1-1/4 |
| 8627 | | Wires | 28.0 | 0.04 | T1-1/4 |
| PR-2 | Blue | Flange | 2.4 | 0.50 | B3-1/2 |
| PR-3 | Green | Flange | 3.6 | 0.50 | B3-1/2 |
| PR-4 | Yellow | Flange | 2.3 | 0.27 | B3-1/2 |
| PR-6 | Brown | Flange | 2.5 | 0.30 | B3-1/2 |
| PF-7 | | Flange | 3.7 | 0.30 | B3-1/2 |
| PR-12 | White | Flange | 5.95 | 0.50 | B3-1/2 |
| PR-13 | | Flange | 4.75 | 0.50 | B3-1/2 |
| PR-18 | | Flange | 7.2 | 0.55 | B3-1/2 |

| Neon Number | Resistor Required | Base Type | Bulb Volts | Milli-amps | Type |
|---|---|---|---|---|---|
| NE-2 (A1A) | 150K | Wire | 110VAC | 0.5 | T2 |
| NE-2A (A2A) | 220K | Wire | 110VAC | 0.3 | T2 |
| A1B | 220K | Wire | 110VAC | 0.3 | T2 |
| A1C | 47K | Wire | 110VAC | 1.2 | T2 |
| NE-2D (C7A) | 100K | Midg. Flange | 110VAC | 0.6 | T2 |
| NE-2E (A9A) | 100K | Wire | 110VAC | 0.6 | T2 |
| NE-2H (C2A) | 30K | Wire | 110VAC | 1.7 | T2 |
| NE-2J (C9A) | 30K | Flange | 110VAC | 1.7 | T2 |
| NE-2M | 150K | Wire | 110VAC | 0.5 | T2 |
| NE-2P | 30K | Wire | 110VAC | 1.7 | T2 |
| NE-7 (B4A) | 30K | Wire | 110VAC | 2.0 | |
| NE-17 (B5A) | 30K | DC Bayonet | 110VAC | 2.0 | T4-1/2 |
| NE-21 (B6A) | 30K | | 110VAC | 2.0 | |
| NE-30 | None | Screw | 110VAC | 12.0 | S11 |
| NE-34 | | Screw | 110VAC | 18.0 | S14 |
| NE-42 | | | 110VAC | 30.0 | |
| NE-45 (B7A) | 30K | Candelabra S | 110VAC | 2.0 | T4-1/2 |
| NE-47 (B8A) | 30K | SC Bayonet | 110VAC | 2.0 | T4-1.2 |
| NE-48 | 30K | DC Bayonet | 110VAC | 2.0 | T4-1/2 |
| NE-51 (B1A) | 200K | Min. Bayonet | 110VAC | 0.3 | T3-1/4 |
| NE-51H (B2A) | 45K | Min. Bayonet | 110VAC | 1.2 | T3-1/4 |
| NE-56 | None | Screw | 220VAC | 5.0 | S11 |
| NE-57 | None | Candelabra S | 110VAC | 2.0 | T4-1/4 |
| NE-58 (F4A) | 100K | Candelabra S | 220VAC | 2.0 | T4-1/4 |
| NE-79 | 7.5K | DC Bayonet | 110VAC | 12.0 | S7 |
| 6S6DC | | Dbl Bayonet | 120V | 6 watt | S6 |
| 7C7 | | Candelabra S | 115-0125 | 7 watt | S7 |

# FUSES – SMALL TUBE TYPE

| TYPE | Description | Diameter Inches | Length Inches |
|---|---|---|---|
| 3AB | Ceramic body, normal, 200% 15sec | 1/4 | 1-1/4 |
| 1AG | Auto Glass, fast blow, 200% 5sec | 1/4 | 5/8 |
| 2AG | Auto Glass, fast blow, 200% 10sec | 0.177 | 0.57 |
| 3AG | Auto Glass, fast blow, 200% 5sec | 1/4 | 1-1/4 |
| 4AG | Auto Glass, fast blow, 200% 5sec | 9/32 | 1-1/4 |
| 5AG | Auto Glass, fast blow, 200% 5sec | 13/32 | 1-1/2 |
| 7AG | Auto Glass, fast blow, 200% 5sec | 1/4 | 7/8 |
| 8AG | Auto Glass, fast blow, 200% 5sec | 1/4 | 1 |
| 9AG | Auto Glass, fast blow, 200% 5sec | 1/4 | 1-7/16 |
| 216 | Metric, fast blow, high int.,210% 30m | 5mm | 20mm |
| 217 | Glass, Metric, fast blow, 210% 30m | 5mm | 20mm |
| 218 | Glass, Metric, slow blow, 210% 2 min | 5mm | 20mm |
| ABC | No Delay, Ceramic, 110% rating, Will blow at 135% load in one hour | 1/4 | 1-1/4 |
| AGC | Fast Acting, glass tube, 110% rating, Will blow at 135% load in one hour | 1/4 | 1-1/4 |
| AGX | Fast Acting, glass tube | 1/4 | 1 |
| BLF | No delay, 200% 15sec | 13/32 | 1-1/2 |
| BLN | No delay, military, 200% 15sec | 13/32 | 1-1/2 |
| BLS | Fast clearing, 600V, 135% 1hr | 13/32 | 1-3/8 |
| FLA | Time delay, indicator pin, 135% 1hr | 13/32 | 1-1/2 |
| FLM | Dual element, delay, 200% 12 sec | 13/32 | 1-1/2 |
| FLQ | Dual element, delay,500V,200%12sec | 13/32 | 1-1/2 |
| FNM | Slow Blow Time Delay | 13/32 | 1-1/4 |
| FNA | Slow Blow, Indicator, silver pin pops out when blown, Dual Element | 13/32 | 1-1/2 |
| GBB | Rectifier Fuse, Fast, low let through | 1/4 | 1-1/4 |
| GLD | Indicator Fuse, silver pin pops out to show blown fuse. 110% rating | 1/4 | 1-1/4 |
| GGS | Metric, fast acting | 5mm | 20mm |
| KLK | Fast, current limiting, 600V, 135% 1hr | 13/32 | 1-1/2 |
| KLW | Fast, protect solid state, 250% 1sec | 13/32 | 1-1/2 |
| MDL | Dual Element, Time Delay, glass tube | 1/4 | 1-1/4 |
| MDX | Dual Element, glass tube | 1/4 | 1-1/4 |
| MDV | Dual Element, glass tube, Pigtail | 1/4 | 1-1/4 |
| SC | Slow Blow, Time Delay Size rejection also | 13/32 | 1-5/16to 2-1/4 |
| 218000 | Slow blow, glass body, 200% 5sec | 0.197 | 0.787 |
| 251000 | Pico II™ Subminiature, fast blow | Wire lead | |
| 273000 | Microfuse, fast blow, 200% 5sec | Wire lead | |
| 313000 | Slow Blow, glass body, 200% 5sec | 1/4 | 1-1/4 |
| 326000 | Slow Blow, ceramic, 200% 5sec | 1/4 | 1-1/4 |

**Note:** The 200% 10 sec figures above indicate that a 200% overload will blow the fuse in 10 seconds.

# BATTERY CHARACTERISTICS

| Battery (1) | Anode | Cathode | Voltage(2) | Amp-hrs/kg |
|---|---|---|---|---|
| Ammonia | Mg | m–DNB | 2.2 (1.7) | 1,400 |
| Cadmium-Air (C) | Cd | $O_2$ | 1.2 (0.8) | 475 |
| Cuprous chloride | Mg | CuCl | 1.5 (1.4) | 240 |
| Edison (C) | Fe | NiO | 1.5 (1.2) | 195 |
| $H_2$–$O_2$ (C) | $H_2$ | $O_2$ | 1.23 (0.8) | 3,000 |
| Lead-Acid (C) | Pb | $PbO_2$ | 2.1 (2.0) | 55 |
| Leclance (NC) | Zn | $MnO_2$ | 1.6 (1.2) | 230 |
| Lithium–High Temp, 350°C, with fused salt | | | | |
| | Li | S | 2.1 (1.8) | 685 |
| Magnesium (NC) | Mg | $MnO_2$ | 2.0 (1.5) | 270 |
| Mercury (NC) | Zn | HgO | 1.34 (1.2) | 185 |
| Mercad (NC) | Cd | HgO | 0.9 (0.85) | 165 |
| $MnO_2$ alkaline (NC) | Zn | $MnO_2$ | 1.5 (1.15) | 230 |
| NiCad (C) | Cd | NiO | 1.35 (1.2) | 165 |
| Organic Cath.(NC) | Mg | m–DNB | 1.8 (1.15) | 1,400 |
| Silver Cadmium (C) | Cd | AgO | 1.4 (1.05) | 230 |
| Silver Chloride | Mg | AgCl | 1.6 (1.5) | 170 |
| Silver Oxide | Zn | AgO | 1.85 (1.5) | 285 |
| Silver–Poly | Ag | Polyiodide | 0.66 (0.6) | 180? |
| Sodium – High Temp, 300°C, with $\beta$–alumina electrolyte | | | | |
| | Na | S | 2.2 (1.8) | 1,150 |
| Thermal | Ca | Fuel | 2.8 (2.6) | 240 |
| Zinc-Air (NC) | Zn | $O_2$ | 1.6 (1.1) | 815 |
| Zinc–Nickel (C) | Zn | Ni oxides | 1.75 (1.6) | 185 |
| Zinc-Silver Ox | Zn | AgO | 1.85 (1.5) | 285 |

**Fuel Cells:**

| Fuel Cell | Anode | Cathode | Voltage | Amp-hrs/kg |
|---|---|---|---|---|
| Hydrogen | $H_2$ | $O_2$ | 1.23 (0.7) | 26,000 |
| Hydrazine | $N_2H_4$ | $O_2$ | 1.5 (0.7) | 2,100 |
| Methanol | $CH_2OH$ | $O_2$ | 1.3 (0.9) | 1,400 |

(1) (NC) after the name indicates the cell is a Primary Cell and cannot be recharged. (C) indicates the cell is a Secondary Cell and can be recharged.

(2) The first voltage is the theoretical voltage developed by the cell and the value in parenthesis is the typical voltage generated by a working cell. Amp–hrs/kg is the theoretical capacity of the cell.

Battery data listed above was obtained from the *Electronic Engineers Master Catalog, Hearst Business Communications Inc., 1986–1987.*

# BATTERIES - STANDARD SIZES

| Size | Eveready # | NEDA # | Voltage | Capacity |
|---|---|---|---|---|
| **Carbon Zinc Cells:** | | | | |
| AAA | 912 | 24F | 1.5 | 20 ma @ 21 hrs |
| AA | 915 | 15F | 1.5 | 54 ma @ 20 hrs |
| C | 935 | 14F | 1.5 | 20 ma @ 140 hrs |
| C | 1235 | 14D | 1.5 | 37.5 ma @ 97 hrs |
| D | 950 | 13F | 1.5 | 20 ma @ 360 hrs |
| D | 1150 | 13C | 1.5 | 375 ma @ 15.8 hrs |
| D | 1250 | 13D | 1.5 | 60 ma @ 139 hrs |
| N | 904 | 910F | 1.5 | 20 ma @ 22 hrs |
| WO | 201 | | 1.5 | 0.1 ma @ 650 hrs |
| | 750 | 704 | 3.0 | 20 ma @ 37 hrs |
| | 715 | 903 | 4.5 | 120 ma @ 90 hrs |
| | 724 | 2 | 6.0 | 60 ma @ 175 hrs |
| | 509 | 908 | 6.0 | 187 ma @ 40 |
| 109 | 206 | 1611 | 9 | 12 ma @ 40 hrs |
| 127 | 226 | 1600 | 9 | 12 ma @ 61 hrs |
| | 276 | 1603 | 9 | 20 ma @ 350 hrs |
| 117 | 216 | 1604 | 9 | 9 ma @ 50 hrs |
| | 228 | 1810 | 12 | 12 ma @ 59 hrs |
| | 420 | 225 | 22.5 | 5 ma @ 60 hrs |
| | 482 | 207 | 45 | 40 ma @ 125 hrs |
| | 490 | 204 | 90 | 10 ma @ 63 |
| **Alkaline–Manganese:** | | | | |
| AAA | E92 | 24A | 1.5 | 37.5 ma @ 25 hrs |
| AA | E91 | 15A | 1.5 | 20 ma @ 107 hrs |
| C | E93 | 14A | 1.5 | 37.5 ma @ 160 hrs |
| D | E95 | 13A | 1.5 | 50 ma @ 270 hrs |
| G | 520 | 930A | 6.0 | 375 ma @ 59 hrs |
| N | E90 | 910A | 1.5 | 9 ma @ 90 hrs |
| | 532 | 1308AP | 3.0 | 20 ma @ 35 hrs |
| | 531 | 1307AP | 4.5 | 20 ma @ 35 hrs |
| | 522 | 1604A | 9.0 | 18 ma @ 33 hrs |
| | 539 | | 6.0 | 18 MA @ 30.5 |
| **Ni Cad Rechargeable:** | | | | |
| AAA | CH12ABP–2 | 10024 | 1.2 | 180 milliamp–hours |
| AA | CH15 | 10015 | 1.2 | 500 milliamp–hours |
| C | CH35 | 10014 | 1.2 | 1.2 ampere–hours |
| Sub C | CH1.2 | 10022 | 1.2 | 1.2 ampere–hours |
| D | CH50 | 10013 | 1.2 | 1.2 ampere–hours |
| D | CH4 | 10013HC | 1.2 | 4 ampere–hours |
| N | CH150 | 10910 | 1.2 | 150 milliamp–hours |
| | CH22 | | 8.4 | 80 milliamp–hours |

(CF series rechargeable is Fast Charge, CH is Standard Charge)

Note: CH should be charged at a rate of 1/10 amp-hour for 10 hours.

# RF COIL WINDING DATA

The inductance ( I ), in microhenrys, of air–core coil can be calculated to within 1% or 2% with the following formulas:

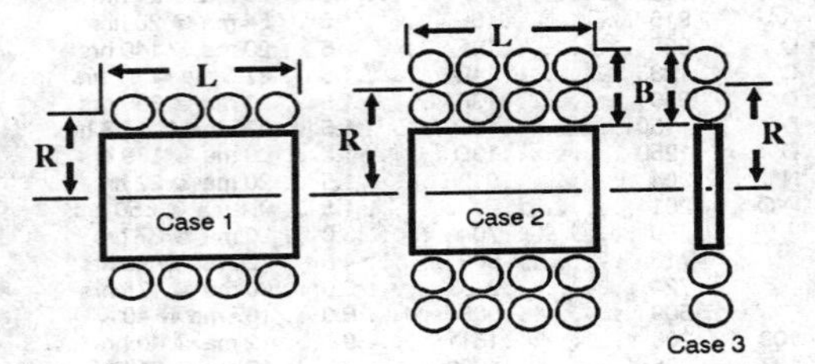


**CASE 1: Single Layer Coil**

$$I = \frac{R^2N^2}{9R + 10L}$$

**CASE 2: Multiple Layer Coil**

$$I = \frac{0.8\,(R^2N^2)}{6R + 9L + 10B}$$

**CASE 3: Single Layer, Single Row Coil**

$$I = \frac{R^2N^2}{8R + 11B}$$

In all of the above equations, N = number of turns and I is the inductance in microhenrys. L and R are distances in inches.

# WIRE SIZE vs TURNS/INCH

| Gauge AWG | Number of Turns Per Inch of Length Enamel | S.S.C. | D.C.C. |
|---|---|---|---|
| 1 | ... | ... | 3.3 |
| 2 | ... | ... | 3.6 |
| 3 | ... | ... | 4.0 |
| 4 | ... | ... | 4.5 |
| 5 | ... | ... | 5.0 |
| 6 | ... | ... | 5.6 |
| 7 | ... | ... | 6.2 |
| 8 | 7.6 | ... | 7.1 |
| 9 | 8.6 | ... | 7.8 |
| 10 | 9.6 | ... | 8.9 |
| 11 | 10.7 | ... | 9.8 |
| 12 | 12.0 | ... | 10.9 |
| 13 | 13.5 | ... | 12.0 |
| 14 | 15.0 | ... | 13.8 |
| 15 | 16.8 | ... | 14.7 |
| 16 | 18.9 | 18.9 | 16.4 |
| 17 | 21.2 | 21.2 | 18.1 |
| 18 | 23.6 | 23.6 | 19.8 |
| 19 | 26.4 | 26.4 | 21.8 |
| 20 | 29.4 | 29.4 | 23.8 |
| 21 | 33.1 | 32.7 | 26.0 |
| 22 | 37.0 | 36.5 | 30.0 |
| 23 | 41.3 | 40.6 | 31.6 |
| 24 | 46.3 | 45.3 | 35.6 |
| 25 | 51.7 | 50.4 | 38.6 |
| 26 | 58.0 | 55.6 | 41.8 |
| 27 | 64.9 | 61.5 | 45.0 |
| 28 | 72.7 | 68.6 | 48.5 |
| 29 | 81.6 | 74.8 | 51.8 |
| 30 | 90.5 | 83.3 | 55.5 |
| 31 | 101.0 | 92.0 | 59.2 |
| 32 | 113.0 | 101.0 | 62.6 |
| 33 | 127.0 | 110.0 | 66.3 |
| 34 | 143.0 | 120.0 | 70.0 |
| 35 | 158.0 | 132.0 | 73.5 |
| 36 | 175.0 | 143.0 | 77.0 |
| 37 | 198.0 | 154.0 | 80.3 |
| 38 | 224.0 | 166.0 | 83.6 |
| 39 | 248.0 | 181.0 | 86.6 |
| 40 | 282.0 | 194.0 | 89.7 |

The above values will vary slightly depending the manufacturer of the wire and thickness of enamel.

# DECIBELS vs VOLT & POWER RATIOS

| Voltage | Power | + DB − | Voltage | Power |
|---|---|---|---|---|
| 1.000 | 1.000 | 0.0 | 1.000 | 1.000 |
| 1.059 | 1.122 | 0.5 | 0.891 | 0.944 |
| 1.122 | 1.259 | 1.0 | 0.794 | 0.891 |
| 1.189 | 1.413 | 1.5 | 0.708 | 0.841 |
| 1.259 | 1.585 | 2.0 | 0.631 | 0.794 |
| 1.334 | 1.778 | 2.5 | 0.562 | 0.750 |
| 1.413 | 1.995 | 3.0 | 0.501 | 0.708 |
| 1.496 | 2.239 | 3.5 | 0.447 | 0.668 |
| 1.585 | 2.512 | 4.0 | 0.398 | 0.631 |
| 1.679 | 2.818 | 4.5 | 0.355 | 0.596 |
| 1.778 | 3.162 | 5.0 | 0.316 | 0.562 |
| 1.884 | 3.548 | 5.5 | 0.282 | 0.531 |
| 1.995 | 3.981 | 6.0 | 0.251 | 0.501 |
| 2.113 | 4.467 | 6.5 | 0.224 | 0.473 |
| 2.239 | 5.012 | 7.0 | 0.200 | 0.447 |
| 2.371 | 5.623 | 7.5 | 0.178 | 0.422 |
| 2.512 | 6.310 | 8.0 | 0.158 | 0.398 |
| 2.661 | 7.079 | 8.5 | 0.141 | 0.376 |
| 2.818 | 7.943 | 9.0 | 0.126 | 0.355 |
| 2.985 | 8.913 | 9.5 | 0.112 | 0.335 |
| 3.162 | 10.000 | 10.0 | 0.100 | 0.316 |
| 3.350 | 11.220 | 10.5 | 0.089 | 0.299 |
| 3.548 | 12.589 | 11.0 | 0.079 | 0.282 |
| 3.758 | 14.125 | 11.5 | 0.071 | 0.266 |
| 3.981 | 15.849 | 12.0 | 0.063 | 0.251 |
| 4.217 | 17.783 | 12.5 | 0.056 | 0.237 |
| 4.467 | 19.953 | 13.0 | 0.050 | 0.224 |
| 4.732 | 22.387 | 13.5 | 0.045 | 0.211 |
| 5.012 | 25.119 | 14.0 | 0.040 | 0.200 |
| 5.309 | 28.184 | 14.5 | 0.035 | 0.188 |
| 5.623 | 31.623 | 15.0 | 0.032 | 0.178 |
| 5.957 | 35.481 | 15.5 | 0.028 | 0.168 |
| 6.310 | 39.811 | 16.0 | 0.025 | 0.158 |
| 6.683 | 44.668 | 16.5 | 0.022 | 0.150 |
| 7.079 | 50.119 | 17.0 | 0.020 | 0.141 |
| 7.499 | 56.234 | 17.5 | 0.018 | 0.133 |
| 7.943 | 63.096 | 18.0 | 0.016 | 0.126 |
| 8.414 | 70.795 | 18.5 | 0.014 | 0.119 |
| 8.913 | 79.433 | 19.0 | 0.013 | 0.112 |
| 9.441 | 89.125 | 19.5 | 0.011 | 0.106 |
| 10.0 | 100 | 20.0 | 0.010 | 0.100 |
| 31.6 | 1000 | 30.0 | 0.001 | 0.0316 |
| 100.0 | 10000 | 40.0 | 0.0001 | 0.01 |
| 316.2 | $10^5$ | 50.0 | 0.00001 | 0.00316 |
| 1000 | $10^6$ | 60.0 | $10^{-6}$ | 0.001 |
| 3162 | $10^7$ | 70.0 | $10^{-7}$ | 0.000316 |
| 10000 | $10^8$ | 80.0 | $10^{-8}$ | 0.001 |
| 31620 | $10^9$ | 90.0 | $10^{-9}$ | 0.0000316 |
| $10^5$ | $10^{10}$ | 100.0 | $10^{-10}$ | $10^{-5}$ |
| 316200 | $10^{11}$ | 110.0 | $10^{-11}$ | 0.0000031 |
| $10^6$ | $10^{12}$ | 120.0 | $10^{-12}$ | $10^{-6}$ |

# FORMULAS FOR ELECTRICITY

## (1) Ohms Law (DC Current):

$$\text{Current in amps} = \frac{\text{Voltage in volts}}{\text{Resistance in ohms}} = \frac{\text{Power in watts}}{\text{Voltage in volts}}$$

$$\text{Current in amps} = \sqrt{\frac{\text{Power in watts}}{\text{Resistance in ohms}}}$$

Voltage in volts = Current in amps x Resistance in ohms

Voltage in volts = Power in watts / Current in amps

$\text{Voltage in volts} = \sqrt{\text{Power in watts} \times \text{Resistance in ohms}}$

$\text{Power in watts} = (\text{Current in watts})^2 \times \text{Resistance in ohms}$

Power in watts = Voltage in volts x Current in amps

$\text{Power in watts} = (\text{Voltage in volts})^2 / \text{Resistance in Ohms}$

Resistance in ohms = Voltage in volts / Current in amps

$\text{Resistance in ohms} = \text{Power in watts} / (\text{Current in amps})^2$

## (2) Resistors in Series (values in Ohms):

$\text{Total Resistance} = \text{Resistance}_1 + \text{Resistance}_2 + \ldots\text{Resistance}_n$

## (3) Two Resistors in Parallel (values in Ohms):

$$\text{Total Resistance} = \frac{\text{Resistance}_1 \times \text{Resistance}_2}{\text{Resistance}_1 + \text{Resistance}_2}$$

## (4) Multiple Resistors in Parallel (values in Ohms):

$$\text{Total Resistance} = \frac{1}{1/\text{Resistance}_1 + 1/\text{Resistance}_2 + \ldots\ 1/\text{Resistance}_n}$$

## (5) Ohms Law (AC Current):

In the following AC Ohms Law formulas, $\theta$ is the phase angle in degrees by which current lags voltage (in an inductive circuit) or by which current leads voltage (in a capacitive circuit). In a resonant circuit (such as normal household 120VAC) the phase angle is 0° and Impedance = Resistance.

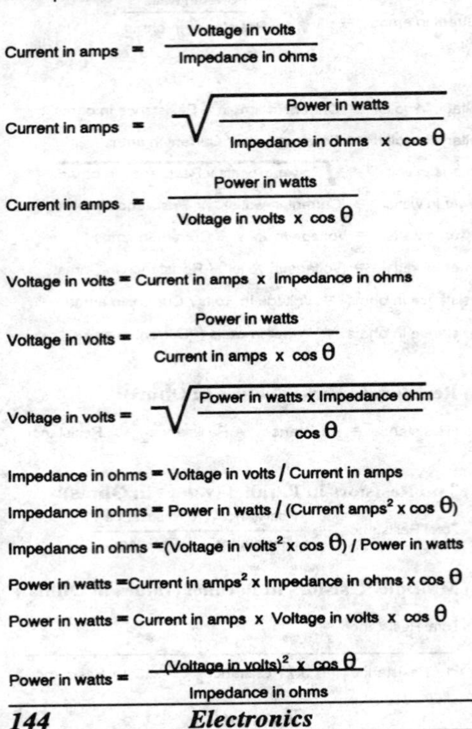

$$\text{Current in amps} = \frac{\text{Voltage in volts}}{\text{Impedance in ohms}}$$

$$\text{Current in amps} = \sqrt{\frac{\text{Power in watts}}{\text{Impedance in ohms} \times \cos\theta}}$$

$$\text{Current in amps} = \frac{\text{Power in watts}}{\text{Voltage in volts} \times \cos\theta}$$

Voltage in volts = Current in amps x Impedance in ohms

$$\text{Voltage in volts} = \frac{\text{Power in watts}}{\text{Current in amps} \times \cos\theta}$$

$$\text{Voltage in volts} = \sqrt{\frac{\text{Power in watts} \times \text{Impedance ohm}}{\cos\theta}}$$

Impedance in ohms = Voltage in volts / Current in amps

Impedance in ohms = Power in watts / (Current amps$^2$ x cos $\theta$)

Impedance in ohms =(Voltage in volts$^2$ x cos $\theta$) / Power in watts

Power in watts =Current in amps$^2$ x Impedance in ohms x cos $\theta$

Power in watts = Current in amps x Voltage in volts x cos $\theta$

$$\text{Power in watts} = \frac{(\text{Voltage in volts})^2 \times \cos\theta}{\text{Impedance in ohms}}$$

**(6) Resonance: – *f***

Resonant frequency in hertz (where $XL = X_C$) =

$$\frac{1}{2\pi\sqrt{\text{Inductance in henrys} \times \text{Capacitance in farads}}}$$

**(7) Reactance: – X**

Reactance in ohms of an inductance is $X_L$
Reactance in ohms of a capacitance is $X_C$

$X_L = 2\ \pi$(frequency in hertz x Inductance in henrys)

XC = 1 / (2 $\pi$(frequency in hertz x Capacitance in farads)

**(8) Impedance: – Z**

Impedance in ohms (series) = $\sqrt{\text{Resistance in ohms}^2 + (X_L - X_C)^2}$

Impedance in ohms (parallel) = $$\frac{\text{Resistance in ohms} \times \text{Reactance}}{\sqrt{\text{Resistance in ohms}^2 + \text{Reactance}^2}}$$

**(9) Susceptance: – B**

Susceptance in mhos =

$$\frac{\text{Reactance in ohms}}{\text{Resistance in ohms}^2 + \text{Reactance in ohms}^2}$$

**(10) Admittance: – Y**

Admittance in mhos =

$$\frac{1}{\sqrt{\text{Resistance in ohms}^2 \times \text{Reactance in ohms}^2}}$$

Admittance in mhos = 1 / Impedance in ohms

## (11) Power Factor: – pf

Power Factor = cos (Phase Angle)
Power Factor = True Power / Apparent Power
Power Factor = Power in watts / (volts x current in amps)
Power Factor = Resistance in ohms / Impedance in ohms

## (12) Q or Figure of Merit: – Q

Q = Inductive Reactance in ohms / Series Resistance in ohms
Q = Capacitive Reactance in ohms / Series Resistance in ohms

## (13) Efficiency of any Device:

Efficiency = Output / Input

## (14) Sine Wave Voltage and Current:

Effective (RMS) value = 0.707 x Peak value
Effective (RMS) value = 1.11 x Average value
Average value = 0.637 x Peak value
Average value = 0.9 x Effective (RMS) value
Peak Value = 1.414 x Effective (RMS) value
Peak Value = 1.57 x Average value

## (15) Decibels: – db

db = 10 Log10 (Power in Watts #1 / Power in Watts #2)
db = 10 Log10 (Power Ratio)
db = 20 Log10 (Volts or Amps #1 / Volts or Amps #2)
db = 20 Log10 (Voltage or Current Ratio)
Power Ratio = $10^{(db/10)}$
Voltage or Current Ratio = $10^{(db/20)}$

If impedances are not equal:

$$db = 20 \, Log_{10} \left[ (Volt1\sqrt{Z_2}) \, / \, (Volt2\sqrt{Z_1}) \right]$$

### (16) Capacitors in Parallel (values in any farad):

Total Capacitance = $Capacitance_1 + Capacitance_2 + \ldots Capacitance_n$

### (17) Two Capacitors in Serial (values in any farad):

$$\text{Total Capacitance} = \frac{Capacitance_1 \times Capacitance_2}{Capacitance_1 + Capacitance_2}$$

### (18) Multiple Capacitors in Series (values in farads):

$$\text{Total Capacitance} = \frac{1}{1/Capacitance_1 + 1/Capacitance_2 + \ldots 1/Capacitance_n}$$

### (19) Quantity of Electricity in a Capacitor: – Q

Q in coulombs = Capacitance in farads x Volts

### (20) Capacitance of a Capacitor: – C

Capacitance in picofarads =

$$0.0885 \times \frac{\text{Dielectric constant} \times \text{area in cm}^2 \times (\text{\# of plates} - 1)}{\text{thickness of dielectric in cm}}$$

### (21) Self Inductance:

Use the same formulas as those for Resistance, substituting inductance for resistance. When including the effects of coupling, add 2 x mutual inductance if fields are adding and subtract 2 x mutual inductance if the fields are opposing. e.g.

Series: Lt = L1 + L2 + 2M or Lt = L1 + L2 – 2M

Parallel: Lt = 1 / [ (1/L1 + M) + (1/L2 + M) ]

## (22) Frequency and Wavelength: $f$ and $\lambda$

Frequency in kilohertz = $(3 \times 10^5)$ / wavelength in cm
Frequency in megahertz = $(3 \times 10^4)$ / wavelength in cm
Frequency in megahertz = (984) / wavelength in feet
Wavelength in cm = $(3 \times 10^4)$ / frequency in megahertz
Wavelength in meters = $(3 \times 10^5)$ / frequency in kilohertz
Wavelength in feet = (984) / frequency in megahertz

## (23) Length of an Antenna:

Quarter–wave antenna:
Length in feet = 234 / frequency in megahertz

Half–wave antenna:
Length in feet = 468 / frequency in megahertz

## (24) LCR Series Time Circuits:

Time in seconds =
Inductance in henrys / Resistance in ohms

Time in seconds =
Capacitance in microfarads x Resistance in ohms

## (25) 70 Volt Loud Speaker Matching Transformer:

Transformer Primary Impedance =
(Amplifier output volts)$^2$ / Speaker Power

## (26) Time Duration of One Cycle:

10 megahertz = 100 nanoseconds cycle
4 megahertz = 250 nanoseconds cycle
1 megahertz = 1 microsecond cycle
250 kilohertz = 4 microsecond cycle
100 kilohertz = 10 microsecond cycle

# POCKET REF

## General Information



(See also GENERAL SCIENCE on page 177)

# HOLIDAYS

| Holiday | Date (listed in chronological order) |
|---|---|
| New Years Day | January 1 |
| Epiphany | Sunday on or before January 6 |
| Martin Luther King Day | 3rd Monday January or January 15 |
| Robert E. Lee Day | January 18 |
| National Freedom Day | February 1 |
| Groundhog Day | February 2 |
| Lincoln's Birthday | February 12 |
| Presidents Day | 3rd Monday February |
| Valentine's Day | February 14 |
| Susan B. Anthony Day | February 15 |
| Ash Wednesday | 47 days before Easter |
| St Patrick's Day | March 17 |
| St. Joseph's Day | March 19 |
| Juarez' Birthday | March 21 (Mexico) |
| Palm Sunday | Sunday before Easter |
| Maundy Thursday | Thursday before Easter |
| Good Friday | Friday before Easter |
| Easter | 1st Sunday after 1st full moon after the Spring equinox. |
| Pan American Day | April 14 |
| Secretaries Day | 4th Wednesday in April |
| Arbor Day | Last Friday in April |
| Loyalty Day | May 1 |
| Cinco de Mayo | May 5 (Mexico) |
| Ascension | 40 days after Easter |
| Pentecost | 50 days after Easter |
| Mother's Day | 2nd Sunday in May |
| Armed Forces Day | 3rd Saturday in May |
| National Maritime Day | May 22 |
| Victoria Day | 1st Monday before May 25 |
| Memorial Day | Last Monday in May |
| Flag Day | June14 |
| Father's Day | 3rd Sunday in June |
| Independence Day(US) | July 4 |
| Assumption Day | August 15 |
| Labor Day | 1st Monday in September |
| Grandparent's Day | 1st Sunday after Labor Day |
| Citizenship Day | September 17 |
| Child Health Day | 1st Monday in October |
| Columbus Day | 2nd Monday in October |
| World Poetry Day | October 15 |
| Boss Day | October 16 |
| United Nations Day | October 24 |
| Mother-in-Law's Day | 4th Sunday in October |
| Halloween | October 31 |

# HOLIDAYS

| Holiday | Date (listed in chronological order) |
| --- | --- |
| Reformation Day | October 31 (Protestant) |
| All Saints' Day | November 1 |
| Election Day | 1st Tuesday in November |
| Veterans' Day | November 11 |
| Sadie Hawkins Day | 1st Saturday after November 11 |
| Thanksgiving Day | 4th Thursday in November |
| Immaculate Conception | December 8 |
| Bill of Rights Day | December 15 |
| Wright Brothers Day | December 17 |
| Christmas Eve | December 24 (In some states) |
| Christmas Day | December 25 |
| National Day of Prayer | Set by president, any day but Sunday |

*Most Jewish holidays are not included because they are difficult to calculate (not on the same date) and require the Jewish calendar.*

## STATE SPECIFIC HOLIDAYS

| | |
| --- | --- |
| Three Kings Day, Puerto Rico | January 6 |
| Confederate Heroes Day, in the South | January 19 |
| Kentucky, F. D. Roosevelt's Bday | January 30 |
| Texas Independence Day | March 2 |
| Alabama, Thomas Jefferson Bday | March 12 |
| Louisiana & Alabama, Mardi Gras | Tuesday before Ash Wed. |
| Alaska, Seward's Day | March 28 |
| Alabama, Confederate Memorial Day | April 1 |
| San Jacinto Day, Texas | April 21 |
| Arbor Day, Nebraska | April 22 |
| Mississippi, Confederate Memorial Day | April 24 |
| Maine & Massachusetts, Patriots Day | 3rd Monday in April |
| Missouri, Harry S. Truman's Bday | May 8 |
| Mississippi, Jefferson Davis' Bday | May 29 |
| Alabama, Jefferson Davis' Bday | 1st Monday in June |
| Kentucky, Jefferson Davis' Bday | June 3 |
| Kentucky, Confederate Memorial Day | June 3 |
| Hawaii, King Kamehameha I Day | June 11 |
| Texas, Emancipation Day | June 19 |
| West Virginia Day | June 20 |
| Vermont, Bennington Battle Day | June 16 |
| Utah, Pioneer Day | July 24 |
| Puerto Rico Constitution Day | July 25 |
| Colorado, Colorado Day | August 1 |
| Victory Day, Rhode Island | August 14 |
| Defender's Day, Maryland | September 12 |
| Alaska, Alaska Day | October 18 |
| Nevada, Nevada Day | October 30 |
| New York, Verrazano Day | April 7 |

## SEASON & CLOCK DATES

| Season | Date |
|---|---|
| **Spring Equinox** (Spring begins & day and night are equal lengths) | March 20 |
| **Daylight Savings Time, Start**, move 1 hour ahead, 2 a.m. on the first Sunday in April | |
| **Summer Solstice** (Summer begins & sun is farthest north of the equator | June 20 |
| **Autumn Equinox** (Fall begins & day and night are equal lengths) | September 22 |
| **Daylight Savings Time, End**, move 1 hour back, 2 a.m. on the last Sunday in October | |
| **Winter Solstice** (Winter begins & the sun is farthest south of the equator | December 21 |

## SIGNS OF THE ZODIAC

| Name | Symbol | Dates |
|---|---|---|
| Aries | Ram | March 21–April 19 |
| Taurus | Bull | April 20–May 20 |
| Gemini | Twins | May 21–June 20 |
| Cancer | Crab | June 21–July 22 |
| Leo | Lion | July 23–Aug 22 |
| Virgo | Virgin | Aug 23–Sept 22 |
| Libra | Balance | Sept 23–Oct 22 |
| Scorpio | Scorpion | Oct 23–Nov 21 |
| Sagittarius | Archer | Nov 22–Dec 21 |
| Capricorn | Goat | Dec 22–Jan 19 |
| Aquarius | Water Bearer | Jan 20–Feb 18 |
| Pisces | Fish | Feb 19–March 20 |

## FLOWERS FOR EACH MONTH

| Month | Flower |
|---|---|
| January | Carnation |
| February | Violet |
| March | Jonquil |
| April | Sweet Pea |
| May | Lily of the Valley |
| June | Rose |
| July | Larkspur |
| August | Gladiola |
| September | Aster |
| October | Calendula |
| November | Chrysanthemum |
| December | Narcissus |

# BIRTHSTONES

| Month | Stone | Significance |
|---|---|---|
| January | Garnet | Constancy |
| February | Amethyst | Sincerity |
| March | Jasper, bloodstone, aquamarine | Wisdom |
| April | Diamond | Innocence |
| May | Emerald, chrysoprase | Love |
| June | Pearl, moonstone, alexandrite | Wealth |
| July | Ruby, carnelian | Freedom |
| August | Sardonyx, peridot | Friendship |
| September | Sapphire, lapis lazuli | Truth |
| October | Opal, tourmaline | Hope |
| November | Topaz | Loyalty |
| December | Turquoise, zircon, lapis lazuli | Success |

# ANNIVERSARY NAMES

| Anniversary Year | Traditional | Modern |
|---|---|---|
| 1 | paper | clocks |
| 2 | cotton, straw, calico | china |
| 3 | leather | crystal or glass |
| 4 | flowers, fruit, books | appliances |
| 5 | wood | silverware |
| 6 | iron or sugar (sweets) | wood |
| 7 | copper, wool, brass | desk sets |
| 8 | bronze, rubber | linens & laces |
| 9 | pottery, | leather |
| 10 | tin or aluminum | diamond jewelry |
| 11 | steel | fashion jewelry |
| 12 | silk or fine linen | pearls |
| 13 | lace | textiles or furs |
| 14 | ivory or agate | gold jewelry |
| 15 | crystal, glass | watches |
| 20 | china | platinum |
| 25 | silver | silver |
| 30 | pearl | diamond |
| 35 | coral | jade |
| 40 | ruby or garnet | ruby |
| 45 | sapphire | sapphire |
| 50 | gold | gold |
| 55 | emerald, turquoise | emerald |
| 60 | diamond | diamond |
| 75 | diamond | diamond |

# ENGLISH – GREEK ALPHABET

| English | Greek | Greek Name |
|---|---|---|
| A, a | Α, α | alpha |
| B, b | Β, β | beta |
| G, g | Γ, γ | gamma |
| D, d | Δ, δ | delta |
| E, e | Ε, ε | epsilon |
| Z, z | Ζ, ζ | zeta |
| E, e | Η, η | eta |
| Th, th | Θ, θ | theta |
| I, i | Ι, ι | iota |
| K, k | Κ, κ | kappa |
| L, l | Λ, λ | lambda |
| M, m | Μ, μ | mu |
| N, n | Ν, ν | nu |
| X, x | Ξ, ξ | xi |
| O, o | Ο, ο | omicron |
| P, p | Π, π | pi |
| R, r | Ρ, ρ | rho |
| S, s | Σ, σ | sigma |
| T, t | Τ, τ | tau |
| U, u | Υ, υ | upsilon |
| Ph, ph | Φ, φ | phi |
| Ch, ch | Χ, χ | chi |
| Ps, ps | Ψ, ψ | psi |
| O, o | Ω, ω | omega |

# RADIO ALPHABET

| Letter | Word | Pronunciation |
|---|---|---|
| A | Alfa | Al Fah |
| B | Bravo | Bra Voh |
| C | Charlie | Char Lee |
| D | Delta | Del Tah |
| E | Echo | Ek Oh |
| F | Foxtrot | Foks Trot |
| G | Golf | Golf |
| H | Hotel | Ho Tell |
| I | India | In Dee Ah |
| J | Juliett | Jew Lee Ett |
| K | Kilo | Key Loh |
| L | Lima | Lee Mah |
| M | Mike | Mike |
| N | November | No Vem Ber |
| O | Oscar | Oss Cahr |
| P | Papa | Pah Pah |
| Q | Quebec | Ke Beck |
| R | Romeo | Row Me Oh |
| S | Sierra | See Air Rah |
| T | Tango | Tang Go |
| U | Uniform | You Nee Form |
| V | Victor | Vick Ter |
| W | Whiskey | Wiss Key |
| X | X-Ray | Ecks Ray |
| Y | Yankee | Yang Key |
| Z | Zulu | Zoo Loo |

# MORSE CODE

| Letter | Code | Letter | Code | Letter | Code |
|---|---|---|---|---|---|
| A | •– | Q | ––•– | 1 | •–––– |
| B | –••• | R | •–• | 2 | ••––– |
| C | –•–• | S | ••• | 3 | •••–– |
| D | –•• | T | – | 4 | ••••– |
| E | • | U | ••– | 5 | ••••• |
| F | ••–• | V | •••– | 6 | –•••• |
| G | ––• | W | •–– | 7 | ––••• |
| H | •••• | X | –••– | 8 | –––•• |
| I | •• | Y | –•–– | 9 | ––––• |
| J | •––– | Z | ––•• | 0 | ––––– |
| K | –•– | Error | •••••••• | . | •–•–•– |
| L | •–•• | Wait | •–••• | , | ––••–– |
| M | –– | End Msg | •–•–• | : | –––•• |
| N | –• | End Wrk | •••–•– | – | –•••– |
| O | ––– | Inv Xmit | –•– | ( | –•––•– |
| P | •––• | / | –••–• | ? | ••––•• |

## "TEN" RADIO CODES

| | |
|---|---|
| 10–1 | Receiving poorly, bad signal |
| 10–2 | Receiving OK, signal strong |
| 10–3 | Stop transmitting |
| 10–4 | Message received |
| 10–5 | Relay message |
| 10–6 | Busy, please stand by |
| 10–7 | Out of service |
| 10–8 | In service |
| 10–9 | Repeat message |
| 10–10 | Finished, standing by |
| 10–11 | Talk slower |
| 10–12 | Visitors present |
| 10–13 | Need weather or road conditions |
| 10–16 | Pickup needed at ________ |
| 10–17 | Urgent Business |
| 10–18 | Is there anything for us |
| 10–19 | Nothing for you, return to base |
| 10–20 | My location is ________ |
| 10–21 | Use a telephone |
| 10–22 | Report in person to ________ |
| 10–23 | Stand by |
| 10–24 | Finished last assignment |
| 10–25 | Can you contact ________? |
| 10–26 | Disregard last information |
| 10–27 | I'm changing to channel ________ |
| 10–28 | Identify your station |
| 10–29 | Your time is up for contact |
| 10–30 | Does not conform to FCC rules |
| 10–32 | I'll give you a radio check |
| 10–33 | Emergency traffic at this station |
| 10–34 | Help needed at this station |
| 10–35 | Confidential information |
| 10–36 | The correct time is ________ |
| 10-37 | Wrecker needed at ________ |
| 10-38 | Ambulance needed at ________ |
| 10-39 | Your message has been delivered |
| 10-41 | Please change to channel ________ |
| 10-42 | Traffic accident at ________ |
| 10-43 | Traffic congestion at ________ |
| 10-44 | I have a message for ________ |
| 10-45 | All units within range please report in |
| 10-50 | Break channel |
| 10-60 | What is the next message number |
| 10-62 | Unable to copy, please call on the phone |
| 10-63 | Net directed to ________ |
| 10-64 | Net clear |
| 10-65 | Standing by, awaiting your next message |
| 10-67 | All units comply |
| 10-70 | Fire at ________ |
| 10-71 | Proceed with transmission in sequence |
| 10-73 | Speed trap at ________ |
| 10-75 | Your transmission is causing interference |
| 10-77 | Negative contact |
| 10-81 | Reserve hotel room for ________ |
| 10-82 | Reserve room for ________ |
| 10-84 | My telephone number is ________ |
| 10-85 | My address is ________ |
| 10-89 | Radio repairman is needed at ________ |
| 10-90 | I have TVI |
| 10-91 | Talk closer to the microphone |
| 10-92 | Your transmitter needs adjustment |
| 10-93 | Check my frequency on this channel |
| 10-94 | Please give me a long count |
| 10-95 | Transmit dead carrier for 5 seconds |
| 10-99 | Mission completed, all units secure |
| 10-200 | Police needed at ________ |

# PAPER SIZES

| Paper Size | Standard | Millimeters | Inches |
|---|---|---|---|
| Eight Crown | IMP | 1461 x 1060 | 57-1/2 x 41-3/4 |
| Antiquarian | IMP | 1346 x 533 | 53 x 21 |
| Quad Demy | IMP | 1118 x 826 | 44 x 32-1/2 |
| Double Princess | IMP | 1118 x 711 | 44 x 28 |
| Quad Crown | IMP | 1016 x 762 | 40 x 30 |
| Double Elephant | IMP | 1016 x 686 | 40 x 27 |
| B0 | ISO | 1000 x 1414 | 39.37 x 55.67 |
| Arch-E | USA | 914 x 1219 | 36 x 48 |
| Double Demy | IMP | 889 x 572 | 35 x 22-1/2 |
| E | ANSI | 864 x 1118 | 34 x 44 |
| A0 | ISO | 841 x 1189 | 33.11 x 46.81 |
| Imperial | IMP | 762 x 559 | 30 x 22 |
| Princess | IMP | 711 x 546 | 28 x 21-1/2 |
| B1 | ISO | 707 x 1000 | 27.83 x 39.37 |
| Arch-D | USA | 610 x 914 | 24 x 36 |
| A1 | ISO | 594 x 841 | 23.39 x 33.11 |
| Demy | IMP | 584 x 470 | 23 x 18-1/2 |
| D | ANSI | 559 x 864 | 22 x 34 |
| B2 | ISO | 500 x 707 | 19.68 x 27.83 |
| Arch-C | USA | 457 x 610 | 18 x 24 |
| C | ANSI | 432 x 559 | 17 x 22 |
| A2 | ISO | 420 x 594 | 16.54 x 23.39 |
| B3 | ISO | 353 x 500 | 13.90 x 19.68 |
| Brief | IMP | 333 x 470 | 13-1/8 x 18-1/2 |
| Foolscap folio | IMP | 333 x 210 | 13-1/8 x 8-1/4 |
| Arch-B | USA | 305 x 457 | 12 x 18 |
| A3 | ISO | 297 x 420 | 11.69 x 16.54 |
| B | ANSI | 279 x 432 | 11 x 17 |
| Demy quarto | IMP | 273 x 216 | 10-3/4 x 8-1/2 |
| B4 | ISO | 250 x 353 | 9.84 x 13.90 |
| Crown quarto | IMP | 241 x 184 | 9-1/2 x 7-1/4 |
| Royal octavo | IMP | 241 x 152 | 9-1/2 x 6 |
| Arch-A | USA | 229 x 305 | 9 x 12 |
| Demy octavo | IMP | 222 x 137 | 8-3/4 X 5-3/8 |
| A | ANSI | 216 x 279 | 8.5 x 11 |
| A4 | ISO | 210 x 297 | 8.27 x 11.69 |
| Foolscap quarto | IMP | 206 x 165 | 8-1/8 x 6-1/2 |
| Crown Octavo | IMP | 181 x 121 | 7-1/8 x 4-3/4 |
| B5 | ISO | 176 x 250 | 6.93 x 9.84 |
| A5 | ISO | 148 x 210 | 5.83 x 8.27 |
| | USA | 140 x 216 | 5.5 x 8.5 |
| | USA | 127 x 178 | 5 x 7 |
| A6 | ISO | 105 x 148 | 4.13 x 5.83 |
| | USA | 102 x 127 | 4 x 5 |
| | USA | 76 x 102 | 3 x 5 |
| A7 | ISO | 74 x 105 | 2.91 x 4.13 |
| A8 | ISO | 52 x 74 | 2.05 x 2.91 |
| A9 | ISO | 37 x 52 | 1.46 x 2.05 |
| A10 | ISO | 26 x 37 | 1.02 x 1.46 |

Abbreviations for the above table are:

| | |
|---|---|
| ISO | International Standards Organization |
| ANSI | American National Standards Institute |
| USA | United States |
| IMP | Imperial paper and plan sizes |
| Arch | United States architectural standards |

# MILITARY RANK & GRADE

| Grade | Air Force | Navy & Coast Guard |
|---|---|---|
| E1 | Airman Basic (AB) | Seaman Recruit |
| E2 | Airman (Amn) | Seaman Apprentice |
| E3 | Airman 1st Class (A1C) | Seaman |
| E4 | Senior Airman (SrA) | Petty Officer 3rd Class |
| E5 | Staff Sergeant | Petty Off 2nd Class |
| E6 | Technical Sergeant | Petty Off 1st Class |
| E7 | Master Sergeant | Chief Petty Officer |
| E8 | Senior Master Sergeant | Sr Chief Petty Officer |
| E9 | Chief Master Sergeant | Mst Chief Petty Officer |
| W1 | Warrant Officer | Warrant Officer |
| W234 | Chief Warrant Officers | Chief Warrant Officers |
| O1 | 2nd Lieutenant | Ensign |
| O2 | 1st Lieutenant | Lieutenant Jr Grade |
| O3 | Captain | Lieutenant |
| O4 | Major | Lieutenant Commander |
| O5 | Lieutenant Colonel | Commander |
| O6 | Colonel | Captain |
| O7 | Brigadier General * | Commodore * |
| O8l | Major General ** | Rear Admiral ** |
| O8 | Lieutenant General *** | Vice Admiral *** |
| O8 | General **** | Admiral **** |
| O8 | General of the Air Force 5* | Fleet Admiral 5* |

| Grade | Army | Marines |
|---|---|---|
| E1 | Private | Private |
| E2 | Private | Private 1st Class |
| E3 | Private 1st Class | Lance Corporal |
| E4 | Corporal Specialist 4 | Corporal |
| E5 | Sergeant Specialist 5 | Sergeant |
| E6 | Staff Sergeant Specialist 6 | Staff Sergeant |
| E7 | Sergeant 1st Class Specialist 7 | Gunnery Sergeant |
| E8 | 1st/Mst Sergeant Specialist 8 | 1st/Master Sergeant |
| E9 | Sergeant Major Specialist 9 | Sgt Major/ Mgy Sergeant |
| W1 | Warrant Officer | Warrant Officer |
| W234 | Chief Warrant Officer | Chief Warrant Officer |
| O1 | 2nd Lieutenant | 2nd Lieutenant |
| O2 | 1st Lieutenant | 1st Lieutenant |
| O3 | Captain | Captain |
| O4 | Major | Major |
| O5 | Lieutenant Colonel | Lieutenant Colonel |
| O6 | Colonel | Colonel |
| O7 | Brigadier General * | Brigadier General * |
| O8l | Major General ** | Major General ** |
| O8 | Lieutenant General *** | Lieutenant General *** |
| O8 | General **** | General **** |
| O8 | Gen of Army 5* | – – |

# STATE INFORMATION

| State | Abbreviation | Population (1990) | Capital |
|---|---|---|---|
| United States | USA | 248,709,873 | Washington, DC |
| Alabama | AL | 4,040,587 | Montgomery |
| Alaska | AK | 550,043 | Juneau |
| Arizona | AZ | 3,655,228 | Phoenix |
| Arkansas | AR | 2,350,725 | Little Rock |
| California | CA | 29,760,021 | Sacramento |
| Colorado | CO | 3,294,394 | Denver |
| Connecticut | CT | 3,287,116 | Hartford |
| Delaware | DE | 666,168 | Dover |
| Dist. of Columbia | DC | 606,900 | |
| Florida | FL | 12,937,926 | Tallahassee |
| Georgia | GA | 6,478,216 | Atlanta |
| Hawaii | HI | 1,108,229 | Honolulu |
| Idaho | ID | 1,006,749 | Boise |
| Illinois | IL | 11,430,602 | Springfield |
| Indiana | IN | 5,544,159 | Indianapolis |
| Iowa | IA | 2,776,755 | Des Moines |
| Kansas | KS | 2,477,574 | Topeka |
| Kentucky | KY | 3,685,296 | Frankfort |
| Louisiana | LA | 4,219,973 | Baton Rouge |
| Maine | ME | 1,227,928 | Augusta |
| Maryland | MD | 4,781,468 | Annapolis |
| Massachusetts | MA | 6,016,425 | Boston |
| Michigan | MI | 9,295,297 | Lansing |
| Minnesota | MN | 4,375,099 | St. Paul |
| Mississippi | MS | 2,573,216 | Jackson |
| Missouri | MO | 5,117,073 | Jefferson City |
| Montana | MT | 799,065 | Helena |
| Nebraska | NE | 1,578,385 | Lincoln |
| Nevada | NV | 1,201,833 | Carson City |
| New Hampshire | NH | 1,109,353 | Concord |
| New Jersey | NJ | 7,730188 | Trenton |
| New Mexico | NM | 1,515,069 | Santa Fe |
| New York | NY | 17,990,455 | Albany |
| North Carolina | NC | 6,628,637 | Raleigh |
| North Dakota | ND | 638,800 | Bismarck |
| Ohio | OH | 10,847,115 | Columbus |
| Oklahoma | OK | 3,486,703 | Oklahoma City |
| Oregon | OR | 2,842,321 | Salem |
| Pennsylvania | PA | 11,881,643 | Harrisburg |
| Rhode Island | RI | 1,003,464 | Providence |
| South Carolina | SC | 3,486,703 | Columbia |
| South Dakota | SD | 696,004 | Pierre |
| Tennessee | TN | 4,877,185 | Nashville |
| Texas | TX | 16,986,510 | Austin |
| Utah | UT | 1,722,850 | Salt Lake City |
| Vermont | VT | 562,758 | Montpelier |
| Virginia | VA | 6,187,358 | Richmond |
| Washington | WA | 4,866,692 | Olympia |
| West Virginia | WV | 1,793,477 | Charleston |
| Wisconsin | WI | 4,891,769 | Madison |
| Wyoming | WY | 453,588 | Cheyenne |

# CLIMATE DATA IN U.S. CITIES

| State, City | Temperature (°F) Winter | Summer | Avg Precipitation (in.) Rain | Snow |
|---|---|---|---|---|
| AL, Mobile | 52.5 | 81.5 | 64.6 | 0.1 |
| AK, Juneau | 25.5 | 54.3 | 53.1 | 102.3 |
| AZ, Phoenix | 53.9 | 89.6 | 7.1 | trace |
| AR, Little Rock | 42.4 | 80.5 | 49.2 | 5.6 |
| CA, Los Angeles | 52.1 | 68.3 | 12.1 | trace |
| CO, Denver | 31.9 | 70.6 | 15.3 | 59.9 |
| CT, Hartford | 27.3 | 71.1 | 44.4 | 49.4 |
| DC, Washington | 37.2 | 77.0 | 39.0 | 16.7 |
| DE, Wilmington | 33.3 | 74.0 | 41.4 | 20.9 |
| FL, Miami | 67.8 | 82.0 | 57.5 | 0 |
| GA, Atlanta | 43.8 | 77.5 | 48.6 | 2.0 |
| HI, Honolulu | 73.2 | 80.0 | 23.5 | 0 |
| ID, Boise | 32.7 | 70.8 | 11.7 | 21.8 |
| IL, Chicago | 25.0 | 71.2 | 33.3 | 40.1 |
| IN, Indianapolis | 29.1 | 73.4 | 39.1 | 23.3 |
| IA, Des Moines | 22.9 | 73.9 | 30.8 | 35.0 |
| KS, Wichita | 33.0 | 79.0 | 28.6 | 16.3 |
| KY, Louisville | 35.2 | 75.9 | 43.6 | 17.3 |
| LA, New Orleans | 53.9 | 81.3 | 59.7 | 0.2 |
| ME, Portland | 23.4 | 65.6 | 43.5 | 72.0 |
| MD, Baltimore | 34.6 | 74.8 | 41.8 | 21.6 |
| MA, Boston | 31.3 | 71.1 | 43.8 | 41.6 |
| MI, Detroit | 25.9 | 70.0 | 31.0 | 41.3 |
| MN, Minneapolis-St.Paul | 16.0 | 70.6 | 26.4 | 49.9 |
| MS, Jackson | 47.8 | 80.7 | 52.8 | 1.1 |
| MO, St. Louis | 32.3 | 76.9 | 33.9 | 19.7 |
| MT, Great Falls | 23.7 | 66.2 | 15.2 | 59.1 |
| NE, Omaha | 24.9 | 75.3 | 30.3 | 31.1 |
| NV, Reno | 34.0 | 66.2 | 7.5 | 25.3 |
| NH, Concord | 22.2 | 67.1 | 36.5 | 64.4 |
| NJ, Atlantic City | 33.6 | 72.2 | 41.9 | 16.4 |
| NM, Albuquerque | 36.6 | 76.4 | 8.1 | 10.6 |
| NY, New York | 33.8 | 74.5 | 44.1 | 28.8 |
| NC, Raleigh | 41.1 | 76.2 | 41.8 | 7.5 |
| ND, Bismarck | 12.2 | 67.8 | 15.4 | 40.7 |
| OH, Cleveland | 28.0 | 69.8 | 35.4 | 54.0 |
| OK, Oklahoma City | 38.9 | 80.0 | 30.9 | 9.0 |
| OR, Portland | 41.0 | 63.8 | 37.4 | 7.0 |
| PA, Pittsburgh | 29.0 | 70.2 | 36.3 | 44.7 |
| RI, Providence | 29.9 | 70.1 | 45.3 | 36.6 |
| SC, Columbia | 46.2 | 79.6 | 49.1 | 1.9 |
| SD, Sioux Falls | 17.1 | 71.4 | 24.1 | 40.3 |
| TN, Nashville | 39.5 | 77.8 | 48.5 | 11.3 |
| TX, Houston | 53.3 | 82.1 | 44.8 | 0.4 |
| UT, Salt Lake City | 31.0 | 73.5 | 15.3 | 59.4 |
| VT, Burlington | 19.1 | 67.3 | 33.7 | 78.3 |
| VA, Richmond | 38.5 | 76.0 | 44.1 | 14.5 |
| WA, Seattle | 41.0 | 63.0 | 38.6 | 12.9 |
| WV, Charleston | 35.1 | 73.0 | 42.4 | 32.2 |
| WI, Milwaukee | 22.3 | 68.2 | 30.9 | 47.3 |
| WY, Cheyenne | 28.2 | 65.9 | 13.3 | 54.4 |

**Temperature** is average daily temperature (°F) in Dec, Jan, & Feb (Winter) and June, July, & Aug (Summer).
**Rain** is average annual rain plus snow, etc water equivalent.
**Snow** is the average depth of unmelted snow.
*Data from U.S. NOAA, Climatography of the United States*

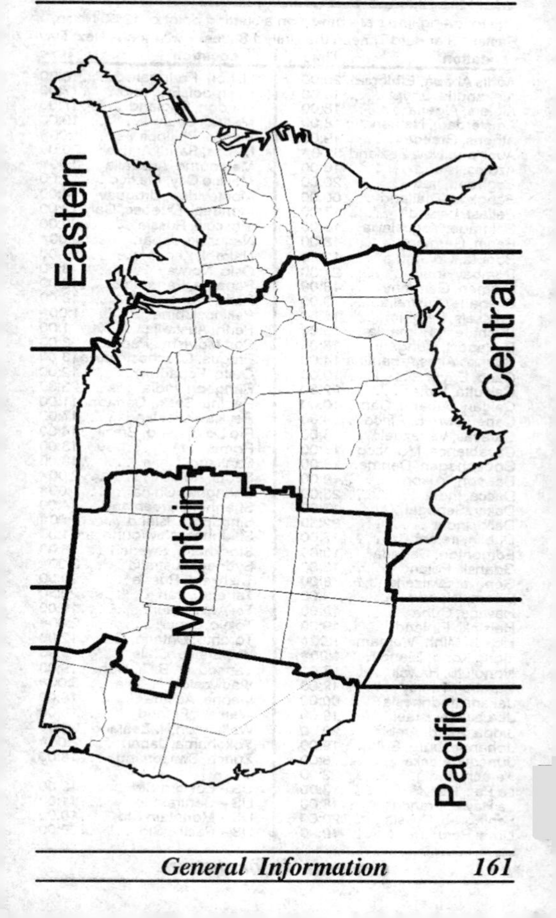
Eastern
Central
Mountain
Pacific

# TIME ZONES IN THE WORLD

The following times are based on a starting point of 12:00 Noon, Eastern Standard Time in the United States. "*" means Next Day

| Location | Time |
|---|---|
| Addis Ababa, Ethiopia | 20:00 |
| Alexandria, Egypt | 19:00 |
| Algiers, Algeria | 18:00 |
| Amsterdam, Netherland | 18:00 |
| Athens, Greece | 19:00 |
| Auckland,New Zealand | 5:00* |
| Azores, Island | 16:00 |
| Baghdad, Iraq | 20:00 |
| Bangkok, Thailand | 00:00 |
| Belfast, Ireland | 17:00 |
| Belgrade, Yugoslavia | 18:00 |
| Berlin, Germany | 18:00 |
| Bogota, Columbia | 12:00 |
| Bombay, India | 22:30 |
| Bremen, Germany | 18:00 |
| Brisbane, Australia | 3:00 |
| Brussels, Belgium | 18:00 |
| Bucharest, Rumania | 19:00 |
| Budapest, Hungary | 18:00 |
| Buenos Aires,Argentina | 14:00 |
| Cairo, Egypt | 19:00 |
| Calcutta, India | 22:30 |
| Calgary, Alberta, Can. | 10:00 |
| Cape Town, S. Africa | 19:00 |
| Caracas, Venezuela | 13:00 |
| Casablanca, Morocco | 17:00 |
| Copenhagen, Denmark | 18:00 |
| Dawson, Yukon | 9:00 |
| Dacca, India | 23:00 |
| Dakar, Senegal | 17:00 |
| Delhi, India | 22:30 |
| Dublin, Ireland | 17:00 |
| Edmonton, Canada | 10:00 |
| Gdansk, Poland | 18:00 |
| Geneva, Switzerland | 18:00 |
| Guam, Island | 3:00 |
| Havana, Cuba | 12:00 |
| Helsinki, Finland | 19:00 |
| Ho Chi Minh, Vietnam | 1:00* |
| Hong Kong, Island | 1:00* |
| Honolulu, Hawaii | 7:00 |
| Istanbul, Turkey | 19:00 |
| Jakarta, Indonesia | 00:00 |
| Jerusalem, Israel | 19:00 |
| Jidda, Saudi Arabia | 20:00 |
| Johannesburg, S.Africa | 19:00 |
| Juneau, Alaska | 8:00 |
| Karachi, India | 22:00 |
| La Paz, Bolivia | 13:00 |
| Le Havre, France | 18:00 |
| Leningrad, Russia | 20:00 |
| Lima, Peru | 12:00 |
| Lisbon, Portugal | 18:00 |
| Liverpool, England | 17:00 |
| London, England | 17:00 |
| Madrid, Spain | 19:00 |
| Manila, Philippines | 1:00* |
| Mecca, Saudi Arabia | 20:00 |
| Melbourne, Australia | 3:00* |
| Mexico City, Mexico | 11:00 |
| Montevideo, Uruguay | 14:00 |
| Montreal, Quebec, Can. | 12:00 |
| Moscow, Russia | 20:00 |
| Nagasaki, Japan | 2:00* |
| Nairobi, Kenya | 20:00 |
| Oslo, Norway | 18:00 |
| Panama, Panama | 12:00 |
| Paris, France | 18:00 |
| Peking, China | 1:00* |
| Perth, Australia | 1:00 |
| Port Moresby, Papua | 3:00 |
| Prague, Czechoslovakia | 18:00 |
| Quito, Ecuador | 12:00 |
| Rangoon, India | 23:30 |
| Regina, Sask., Canada | 11:00 |
| Reykjavik, Iceland | 17:00 |
| Rio De Janeiro, Brazil | 14:00 |
| Rome, Italy | 18:00 |
| Santiago, Chile | 13:00 |
| Seoul, Korea | 2:00* |
| Shanghai, China | 1:00* |
| Shannon, Greenland | 14:00 |
| Singapore, Island | 1:00* |
| St. Johns, Newfoundland | 1:30 |
| Stockholm, Sweden | 18:00 |
| Sydney, Australia | 3:00* |
| Tashkent, Russia | 23:00 |
| Teheran, Iran | 20:30 |
| Tel Aviv, Israel | 19:00 |
| Tokyo, Japan | 2:00* |
| Toronto, Ontario, Can. | 12:00 |
| Valparaiso, Chile | 13:00 |
| Vancouver, B.C., Can. | 9:00 |
| Vladivostok, Russia | 3:00* |
| Vienna, Austria | 18:00 |
| Warsaw, Poland | 18:00 |
| Wellington,N. Zealand | 5:00* |
| Yokohama, Japan | 2:00* |
| Zurich, Switzerland | 18:00 |
| US – Eastern Std | 12:00 |
| US – Central Std | 11:00 |
| US – Mountain Std | 10:00 |
| US – Pacific Std | 9:00 |

# TELEPHONE AREA CODES by STATE

| State | City | Code |
|---|---|---|
| Alabama | | 205 |
| Alaska | | 907 |
| Anguilla | | 809 |
| Antigua | | 809 |
| Arizona | | 602 |
| Arkansas | | 501 |
| Bahamas | | 809 |
| Barbados | | 809 |
| Bequia | | 809 |
| Bermuda | | 809 |
| California | Alameda County | 510 |
| | Anaheim | 714 |
| | Bakersfield | 805 |
| | Barstow | 619 |
| | Bishop | 619 |
| | El Centro | 619 |
| | Eureka | 707 |
| | Fresno | 209 |
| | Los Angeles | 213 |
| | Los Angeles, Malabu | 310 |
| | Modesto | 209 |
| | Monterey | 408 |
| | Oakland | 415 |
| | Orange | 714 |
| | Palm Springs | 619 |
| | Pasadena | 818 |
| | Redding | 916 |
| | Riverside | 714 |
| | Sacramento | 916 |
| | San Diego | 619 |
| | San Francisco | 415 |
| | San Jose | 408 |
| | Santa Barbara | 805 |
| | Santa Rosa | 707 |
| Canada | Alberta | 403 |
| | Brit Columbia | 604 |
| | London | 519 |
| | Manitoba | 204 |
| | New Brunswick | 506 |
| | Newfoundland | 709 |
| | NW Territories | 403 |
| | Nova Scotia | 902 |
| | Ontario, West | 807 |
| | Ontario, East | 705 |
| | Ottawa | 613 |
| | Prince Edward I | 902 |
| | Toronto | 416 |
| | Quebec, Montreal | 514 |
| | Quebec, Quebec | 418 |
| | Quebec, Sherbrooke | 819 |
| | Saskatchewan | 306 |
| | Yukon | 403 |
| Cayman Islands | | 809 |
| Colorado | Northern Colorado | 303 |
| | Southern Colorado | 719 |
| Connecticut | | 203 |
| Delaware | | 302 |

| State | City | Code |
|---|---|---|
| District of Columbia | Washington | 202 |
| Dominica | | 809 |
| Dominican Republic | | 809 |
| Florida | Ft. Lauderdale | 305 |
| | Ft. Myers | 813 |
| | Jacksonville | 904 |
| | Miami | 305 |
| | Orlando | 407 |
| | Pensacola | 904 |
| | Tallahassee | 904 |
| Georgia | Atlanta | 404 |
| | Savannah | 912 |
| Hawaii | | 808 |
| Idaho | | 208 |
| Illinois | Centralia | 618 |
| | Champaign | 217 |
| | Chicago, Downtown | 312 |
| | Chicago | 708 |
| | Peoria | 309 |
| | Rockford | 815 |
| | Springfield | 217 |
| | Waukegan | 312 |
| Indiana | Evansville | 812 |
| | Indianapolis | 317 |
| | South Bend | 219 |
| Iowa | Council Bluffs | 712 |
| | Dubuque | 319 |
| | Des Moines | 515 |
| Jamaica | | 809 |
| Kansas | Dodge City | 316 |
| | Topeka | 913 |
| | Wichita | 316 |
| Kentucky | Covington | 606 |
| | Frankfort | 502 |
| | Louisville | 502 |
| Louisiana | Baton Rouge | 504 |
| | Lake Charles | 318 |
| | New Orleans | 504 |
| | Shreveport | 318 |
| Maine | | 207 |
| Maryland | East side of state | 410 |
| | West side of state | 301 |
| Massachusetts | East State | 508 |
| | West state | 413 |
| | Boston | 617 |
| Mexico | Northwest Mexico | 706 |
| | Mexico City | 905 |
| Michigan | Ann Arbor | 313 |
| | Battle Creek | 616 |
| | Detroit | 313 |
| | Escanaba | 906 |
| | Flint | 313 |
| | Grand Rapids | 616 |
| | Lansing | 517 |
| Minnesota | Duluth | 218 |
| | Minneapolis | 612 |
| | Rochester | 507 |

# TELEPHONE AREA CODES by STATE

| State | City | Code |
|---|---|---|
| | St. Paul | 612 |
| Mississippi | | 601 |
| Missouri | Jefferson City | 314 |
| | Kansas City | 816 |
| | Springfield | 417 |
| | St. Louis | 314 |
| Montana | | 406 |
| Montserrat | | 809 |
| Mustique | | 809 |
| Nebraska | Lincoln | 402 |
| | North Platte | 308 |
| | Omaha | 402 |
| Nevada | | 702 |
| Nevis | | 809 |
| New Hampshire | | 603 |
| New Jersey | Newark, north state | 201 |
| | Trenton, south state | 609 |
| | Middle of state | 908 |
| New Mexico | | 505 |
| New York | Albany | 518 |
| | Binghamton | 607 |
| | Bronx | 212 |
| | Brooklyn | 718 |
| | Buffalo | 716 |
| | Elmira | 607 |
| | Hempstead | 516 |
| | Long Island | 516 |
| | Manhattan | 212 |
| | New York City | 212 |
| | Niagra Falls | 716 |
| | Queens | 718 |
| | Rochester | 716 |
| | Stanton Island | 718 |
| | Syracuse | 315 |
| | White Plains | 914 |
| | Yonkers | 914 |
| North Carolina | Charlotte | 704 |
| | Raleigh | 919 |
| North Dakota | | 701 |
| Ohio | Cincinnati | 513 |
| | Cleveland | 216 |
| | Columbus | 614 |
| | Toledo | 419 |
| Oklahoma | Oklahoma City | 405 |
| | Tulsa | 918 |
| Oregon | | 503 |
| Palm Island | | 809 |
| Pennsylvania | Altoona | 814 |
| | Harrisburg | 717 |
| | Philadelphia | 215 |
| | Pittsburgh | 412 |
| Puerto Rico | | 809 |
| Rhode Island | | 401 |
| South Carolina | | 803 |
| South Dakota | | 605 |
| Sts. Kitts & Lucia | | 809 |
| St. Vincent | | 809 |

## TELEPHONE AREA CODES by STATE

| State | City | Code |
|---|---|---|
| Tennessee | Memphis | 901 |
| | Nashville | 615 |
| Texas | Abilene | 915 |
| | Amarillo | 806 |
| | Austin | 512 |
| | Corpus Christi | 512 |
| | Dallas | 214 |
| Texas | El Paso | 915 |
| | Fort Worth | 817 |
| | Galveston | 409 |
| | Houston | 713 |
| | Lubbock | 806 |
| | San Antonio | 512 |
| | Sweetwater | 915 |
| | Waco | 817 |
| Trinidad | | 809 |
| Tobago | | 809 |
| Union Island | | 809 |
| Utah | | 801 |
| Vermont | | 802 |
| Virgin Islands | | 809 |
| Virginia | Arlington | 703 |
| | Norfolk | 804 |
| | Richmond | 804 |
| Washington | Olympia | 206 |
| | Seattle | 206 |
| | Spokane | 509 |
| West Virginia | | 304 |
| Wisconsin | Eau Claire | 715 |
| | Green Bay | 414 |
| | Madison | 608 |
| | Milwaukee | 414 |
| Wyoming | | 307 |

## TELEPHONE AREA CODES by CODE

| State | City | Code |
|---|---|---|
| New Jersey | Newark, north state | 201 |
| District of Columbia | Washington | 202 |
| Connecticut | | 203 |
| Canada | Manitoba | 204 |
| Alabama | | 205 |
| Washington | Olympia & Seattle | 206 |
| Maine | | 207 |
| Idaho | | 208 |
| California | Fresno & Modesto | 209 |
| New York | Bronx & Manhattan | 212 |
| New York | New York City | 212 |
| California | Los Angeles | 213 |
| Texas | Dallas | 214 |
| Pennsylvania | Philadelphia | 215 |
| Ohio | Cleveland | 216 |
| Illinois | Champaign&Springfield | 217 |
| Minnesota | Duluth | 218 |
| Indiana | South Bend | 219 |
| Maryland | West side of state | 301 |

# TELEPHONE AREA CODES by CODE

| State | City | Code |
|---|---|---|
| Delaware | | 302 |
| Colorado | Northern Colorado | 303 |
| West Virginia | | 304 |
| Florida | Fort Lauderdale | 305 |
| Florida | Miami | 305 |
| Canada | Saskatchewan | 306 |
| Wyoming | | 307 |
| Nebraska | North Platte | 308 |
| Illinois | Peoria | 309 |
| California | Los Angeles, Malibu | 310 |
| Illinois | Chicago, Downtown | 312 |
| Illinois | Waukegan | 312 |
| Michigan | Ann Arbor | 313 |
| Michigan | Detroit & Flint | 313 |
| Missouri | Jefferson City | 314 |
| Missouri | St. Louis | 314 |
| New York | Syracuse | 315 |
| Kansas | Dodge City & Wichita | 316 |
| Indiana | Indianapolis | 317 |
| Louisiana | Lake Charles | 318 |
| Louisiana | Shreveport | 318 |
| Iowa | Dubuque | 319 |
| Rhode Island | | 401 |
| Nebraska | Lincoln & Omaha | 402 |
| Canada | Alberta & Yukon | 403 |
| Canada | Northwest Territories | 403 |
| Georgia | Atlanta | 404 |
| Oklahoma | Oklahoma City | 405 |
| Montana | | 406 |
| Florida | Orlando | 407 |
| California | Monterey & San Jose | 408 |
| Texas | Galveston | 409 |
| Maryland | East side of state | 410 |
| Pennsylvania | Pittsburgh | 412 |
| Massachusetts | West side of state | 413 |
| Wisconsin | Green Bay & Milwaukee | 414 |
| California | Oakland | 415 |
| California | San Francisco | 415 |
| Canada | Toronto | 416 |
| Missouri | Springfield | 417 |
| Canada | Quebec, Quebec | 418 |
| Ohio | Toledo | 419 |
| Arkansas | | 501 |
| Kentucky | Frankfort & Louisville | 502 |
| Oregon | | 503 |
| Louisiana | Baton Rouge | 504 |
| Louisiana | New Orleans | 504 |
| New Mexico | | 505 |
| Canada | New Brunswick | 506 |
| Minnesota | Rochester | 507 |
| Massachusetts | East side of state | 508 |
| Washington | Spokane | 509 |
| California | Alameda County | 510 |
| Texas | Austin &Corpus Christi | 512 |
| Texas | San Antonio | 512 |
| Ohio | Cincinnati | 513 |
| Canada | Quebec, Montreal | 514 |

# TELEPHONE AREA CODES by CODE

| State | City | Code |
|---|---|---|
| Iowa | Des Moines | 515 |
| New York | Hempstead | 516 |
| New York | Long Island | 516 |
| Michigan | Lansing | 517 |
| New York | Albany | 518 |
| Canada | London | 519 |
| Mississippi | | 601 |
| Arizona | | 602 |
| New Hampshire | | 603 |
| Canada | Brit Columbia | 604 |
| South Dakota | | 605 |
| Kentucky | Covington | 606 |
| New York | Binghamton & Elmira | 607 |
| Wisconsin | Madison | 608 |
| New Jersey | Trenton, south state | 609 |
| Minnesota | Minneapolis & St. Paul | 612 |
| Canada | Ottawa | 613 |
| Ohio | Columbus | 614 |
| Tennessee | Nashville | 615 |
| Michigan | Battle Creek | 616 |
| Michigan | Grand Rapids | 616 |
| Massachusetts | Boston | 617 |
| Illinois | Centralia | 618 |
| California | Barstow & Bishop | 619 |
| California | El Centro | 619 |
| California | Palm Springs | 619 |
| California | San Diego | 619 |
| North Dakota | | 701 |
| Nevada | | 702 |
| Virginia | Arlington | 703 |
| North Carolina | Charlotte | 704 |
| Canada | Ontario, East | 705 |
| Mexico | Northwest Mexico | 706 |
| California | Eureka & Santa Rosa | 707 |
| Illinois | Chicago | 708 |
| Canada | Newfoundland | 709 |
| Iowa | Council Bluffs | 712 |
| Texas | Houston | 713 |
| California | Anaheim & Orange | 714 |
| California | Riverside | 714 |
| Wisconsin | Eau Claire | 715 |
| New York | Buffalo & Niagra Falls | 716 |
| New York | Rochester | 716 |
| Pennsylvania | Harrisburg | 717 |
| New York | Brooklyn & Queens | 718 |
| New York | Staten Island | 718 |
| Colorado | Southern Colorado | 719 |
| Utah | | 801 |
| Vermont | | 802 |
| South Carolina | | 803 |
| Virginia | Norfolk& Richman | 804 |
| California | Bakersfield | 805 |
| California | Santa Barbara | 805 |
| Texas | Amarillo & Lubbock | 806 |
| Canada | Ontario, West | 807 |
| Hawaii | | 808 |

# TELEPHONE AREA CODES by CODE

| State | City | Code |
| --- | --- | --- |
| Anguilla, Antigua, Bahamas, Barbados, Bequia, Bermuda, Cayman Islands, Dominica, Domin. Republic, Jamaica, Montserrat, Mustique, Nevis, Palm Island, Puerto Rico, Sts. Kitts & Lucia, St. Vincent, Trinidad, Tobago, Union Island, Virgin Islands | | 809 |
| Indiana | Evansville | 812 |
| Florida | Ft. Myers | 813 |
| Pennsylvania | Altoona | 814 |
| Illinois | Rockford | 815 |
| Missouri | Kansas City | 816 |
| Texas | Fort Worth & Waco | 817 |
| California | Pasadena | 818 |
| Canada | Quebec, Sherbrooke | 819 |
| Tennessee | Memphis | 901 |
| Canada | Nova Scotia | 902 |
| Canada | Prince Edward Island | 902 |
| Florida | Jacksonville | 904 |
| Florida | Pensacola | 904 |
| Florida | Tallahassee | 904 |
| Mexico | Mexico City | 905 |
| Michigan | Escanaba | 906 |
| Alaska | | 907 |
| New Jersey | Middle of state | 908 |
| Georgia | Savannah | 912 |
| Kansas | Topeka | 913 |
| New York | White Plains & Yonkers | 914 |
| Texas | Abilene & El Paso | 915 |
| Texas | Sweetwater | 915 |
| California | Redding | 916 |
| California | Sacramento | 916 |
| Oklahoma | Tulsa | 918 |
| North Carolina | Raleigh | 919 |

# MAJOR WORLD AIRPORTS

| Airport City | Airport Name | Elevation, Feet |
|---|---|---|
| Addis–Ababa, Ethiopia | Haile Selassi II Intl. | 7625 |
| Algiers, Algeria | Dar el Beida | 82 |
| Amsterdam, Netherlands | Schiphol | –13 |
| Anchorage, Alaska | Anchorage Intl. | 124 |
| Athens, Greece | Athens Central | 90 |
| Atlanta, Georgia | Hartsfield Atlanta Intl. | 1026 |
| Auckland, New Zealand | Auckland Intl. | 23 |
| Azores, Island | Santa Maria | 305 |
| Baghdad, Iraq | Baghdad Intl. | 113 |
| Bangkok, Thailand | Bangkok | 12 |
| Beirut, Lebanon | Beirut Intl. | 85 |
| Belgrade, Yugoslavia | Belgrade Intl. | 331 |
| Berlin, Germany | Tegel | 121 |
| Berlin, Germany | Tempelhof | 164 |
| Bermuda, Island | Kindley AFB | 11 |
| Bogota, Columbia | El Dorado | 8355 |
| Bombay, India | Bombay | 27 |
| Boston, Massachusetts | Logan Intl. | 20 |
| Brisbane, Australia | Brisbane | 7 |
| Bucharest, Rumania | Otopeni | 31 |
| Budapest, Hungary | Ferihegy | 440 |
| Buenos Aires, Argentina | Ezeiza | 66 |
| Cairo, Egypt | Cairo Intl. | 366 |
| Calcutta, India | Calcutta | 17 |
| Calgary, Canada | Calgary Intl. | 3557 |
| Capetown, South Africa | D.F. Malan | 151 |
| Caracas, Venezuela | Maiquetia | 230 |
| Casablanca, Morocco | Nouasser | 656 |
| Chicago, Illinois | O'Hare Intl. | 667 |
| Copenhagen, Denmark | Kastrup | 17 |
| Dakar, Senegal W. Africa | Yoff | 89 |
| Dallas, Texas | Dallas/Ft. Worth | 596 |
| Damascus, Syria | Damascus Intl. | 2020 |
| Darwin, Australia | Darwin | 94 |
| Denver, Colorado | Stapleton Intl. | 5330 |
| Dublin, Ireland | Dublin | 222 |
| Edmonton, Canada | Edmonton Intl. | 2373 |
| Fairbanks, Alaska | Fairbanks Intl. | 434 |
| Frankfurt, Germany | Rhein/Main | 368 |
| Geneva, Switzerland | Geneva–Cointrin | 1411 |
| Guam, Island | Agana | 298 |
| Halifax, Canada | Halifax Intl. | 477 |
| Hamburg, Germany | Hamburg | 53 |
| Hartford, Connecticut | Bradley Intl. | 173 |
| Helsinki, Finland | Helsinki Airport | 167 |
| Ho Chi Minh City, Viet Nam | Tan Son Nhut | 33 |
| Hong Kong, Island | Hong Kong Intl. | 15 |
| Honolulu, Hawaii | Honolulu Intl. | 13 |
| Houston, Texas | Intercontinental | 98 |
| Istanbul, Turkey | Yesilkoy | 92 |
| Jakarta, Indonesia | Halim | 86 |
| Jidda, Saudi Arabia | Jidda Intl. | 157 |
| Johannesburg, South Africa | Jan Smuts | 5557 |
| Kansas City, Missouri | Kansas City Intl. | 1025 |
| Karachi, Pakistan | Karachi | 100 |

# MAJOR WORLD AIRPORTS

| Airport City | Airport Name | Elevation, Feet |
|---|---|---|
| Khartoum, Sudan | Khartoum | 1256 |
| Kinshasa, Zaire | Ndjili | 1014 |
| La Paz, Bolivia | Kennedy | 13354 |
| Lima, Peru | Lima–Callao Intl. | 105 |
| Lisbon, Portugal | Lisbon | 374 |
| London, England | Heathrow Intl. | 80 |
| Los Angeles, California | Los Angeles Intl. | 126 |
| Madrid, Spain | Barajas | 1998 |
| Manila, Philippines | Manila Intl. | 74 |
| Melbourne, Australia | Tullamarine | 392 |
| Mexico City, Mexico | Mexico City Intl. | 7341 |
| Miami, Florida | Miami Intl. | 10 |
| Montreal, Canada | Montreal Intl. | 117 |
| Moscow, Russia | Sheremetyevo | 623 |
| Moscow, Russia | Vnukovo | 669 |
| Nairobi, Kenya | Nairobi | 5327 |
| New Delhi, India | Palam | 776 |
| New Orleans, Louisiana | Moisant Intl. | 4 |
| New York, New York | Kennedy Intl. | 12 |
| Osaka, Japan | Osaka Intl. | 39 |
| Panama, Panama | Tocumen National | 135 |
| Paris, France | Charles B. deGaulle | 387 |
| Paris, France | Le Bourget | 217 |
| Paris, France | Orly | 292 |
| Peking, China | Peking Intl. | 15 |
| Perth, Australia | Perth | 53 |
| Port Moresby, Papua | Jacksons Aero | 125 |
| Quito, Ecuador | Mariscal Sucre | 9228 |
| Rangoon, India | Mingaladon | 109 |
| Recife, Brazil | Guararepes | 36 |
| Reykjavik, Iceland | Keflavik NAS | 169 |
| Rio de Janeiro, Brazil | Galeao | 16 |
| Rome, Italy | Leonardo da Vinci | 7 |
| San Francisco, California | San Francisco Intl. | 12 |
| Santiago, Chili | Pudahuel | 1554 |
| Seattle, Washington | Seattle–Tacoma Intl. | 428 |
| Seoul, Korea | Kimpo Intl. | 58 |
| Shanghai, China | Shanghai Intl. | 15 |
| Shannon, Greenland | Shannon | 47 |
| Singapore, Island in Asia | Singapore | 65 |
| Stockholm, Sweden | Arlanda | 123 |
| Sydney, Australia | Kingsford–Smith | 6 |
| Taipei, Taiwan | Taipei Intl. | 21 |
| Teheran, Iran | Mehrabad | 3949 |
| Tel Aviv, Israel | Ben Gurion Intl. | 135 |
| Tokyo, Japan | Tokyo Intl. | 8 |
| Toronto, Canada | Toronto Intl. | 569 |
| Tunis, Algeria | Carthage | 20 |
| Vancouver, Canada | Vancouver Intl. | 8 |
| Wake, Island | Wake Island | 14 |
| Warsaw, Poland | Okecie | 361 |
| Washington, D.C. | Dulles Intl. | 313 |
| Zurich, Switzerland | Zurich | 1416 |

# AIRLINE TWO LETTER CODES

| Code | Airline |
|---|---|
| AA | American |
| AC | Air Canada |
| AE | Air Ceylon |
| AF | Air France |
| AH | Air Algerie |
| AI | Air India |
| AL | Allegheny |
| AM | Aeromexico |
| AN | Ansett-Australia |
| AO | Aviaco |
| AQ | Air Anglia |
| AR | Aerolineas Argentinas |
| AS | Alaska |
| AT | Royal Air Maroc |
| AU | Austral |
| AV | Avianca |
| AW | Air Niger |
| AX | Air Togo |
| AY | Finnair |
| AZ | Alitalia |
| BA | British Airways BAOD |
| BB | Balair |
| BD | British Midland |
| BE | British Airways BAED |
| BG | Bangladesh Biman |
| BH | Turks & Caicos |
| BI | Royal Brunei |
| BJ | Bakhtar Afghan |
| BL | Air BVI |
| BM | Aero Tras. Italiani |
| BN | Braniff Intl. |
| BO | Bouraq Indonesia |
| BP | Air Botswana |
| BQ | Business Jets |
| BR | British Caledonian |
| BS | Burnett |
| BU | Braathens |
| BV | Bavaria |
| BW | British W. Indian |
| BX | Spantax |
| BY | Britannia |
| CA | Caac |
| CE | Central Australian |
| CF | Faucett |
| CG | Ciba-Pilatus |
| CI | China Airlines |
| CK | Connair |
| CL | Capitol Intl. |
| CM | Copa |
| CO | Continental |
| CP | CP Air |
| CS | Cambrian (British) |
| CU | Cubana |
| CV | Cargolux |
| CX | Cathay Pacific |
| CY | Cyprus |
| DA | Dan-Air |
| DE | Delta Intl. |
| DF | Condor |
| DG | Affretair |
| DJ | Air Djibouti |
| DK | Scanair |
| DL | Delta |
| DM | Maersk |
| DO | Dominicana |
| DS | Air Senegal |
| DT | Taag-Angola |
| DV | Germanair |
| DX | Danair |
| DY | Alyemda |
| EA | Eastern |
| EC | East African |
| EF | Far Eastern |
| EI | Aer Lingus-Irish |
| EQ | Tame |
| ET | Ethiopian |
| EU | Ecuatoriana |
| EW | East-West |
| EX | Air Champagne Ardennes |
| EY | Europe Aero |
| FF | Intl Aviation |
| FG | Ariana Afghan |
| FI | Flugfelag Iceland |
| FJ | Air Pacific |
| FL | Frontier |
| FT | Flying Tiger |
| FU | Air Littoral |
| GA | Garuda Indonesia |
| GB | Air Inter Gabon |
| GC | Linacongo |
| GD | Air North |
| GF | Gulf Air |
| GH | Ghana |
| GI | Air Guinee |
| GJ | Ansett-S. Australia |
| GK | Laker |
| GL | Greenland Air |
| GN | Air Gabon |
| GP | Hadag General |
| GQ | General Air |
| GR | Aurigny Air |
| GS | Air Vosges |
| GT | Gibraltar |
| GU | Aviateca |
| GV | Territory |
| GW | Gambia |
| GX | Great Lakes |
| GY | Guyana |
| HA | Hawaiian |
| HB | Air Melanesiae |
| HE | Trans European |
| HF | Hapag-Lloyd-Flug |
| HH | Somali |

# AIRLINE TWO LETTER CODES

| Code | Airline |
|---|---|
| HI | Hong Kong Air |
| HJ | Air Haiti |
| HN | NLM–Dutch |
| HT | Air Tchad |
| HV | Transavia |
| IA | Iraqi |
| IB | Iberia |
| IC | Indian |
| IE | Solomon Islands |
| IF | Interflug |
| IG | Alisarda |
| IH | Itavia |
| IJ | Touraine Air |
| IM | Invicta |
| IN | Aerlinte Eireann |
| IO | Air Paris |
| IQ | Intl. Caribbean |
| IR | Iran Air |
| IT | Air Inter |
| IV | Guinea Ecuatorial |
| IW | Intl. Air Bahama |
| IX | In Air |
| IY | Yemen |
| IZ | Arkia–Israel Inland |
| JA | Air Spain |
| JD | Toa Domestic |
| JH | Pan Adria |
| JJ | Aviogenex |
| JK | Tae |
| JL | Japan Air Lines |
| JM | Air Jamaica |
| JP | Inex–Adria |
| JR | Air Yugoslavia |
| JU | Yugoslav –JAT |
| JW | Trek |
| KA | Kalinga |
| KB | Kenya Air Charters |
| KD | Kendell |
| KE | Korean |
| KH | Cook Island |
| KL | KLM |
| KM | Air Malta |
| KR | Kar Air |
| KS | Saturn |
| KU | Kuwait |
| KW | KLM Air Charter |
| LA | LAN |
| LB | LAB |
| LC | Logan Air |
| LD | Lade–Lineas Aereas Del Estado |
| LE | Air Lowveld |
| LF | Linjeflyg |
| LG | Lux Air |
| LH | Lufthansa |
| LI | Leeward Islands |
| LJ | Sierra Leone |
| LL | Loftleidir Icelandic |
| LM | Alm–Dutch Antil. |
| LO | Lot Polish |
| LP | Air Alpes |
| LR | Lacsa |
| LT | Lufttransport Unter. |
| LU | Saeta |
| LV | Lav |
| LY | El Al |
| LZ | Balkan Bulgarian |
| MA | Malev Hungarian |
| MD | Air Madagascar |
| ME | Middle East |
| MG | Melanesian |
| MH | Malaysian |
| MK | Air Mauritius |
| MM | Sociedad Aeronautica Medellin |
| MN | Commercial |
| MP | Martin Air |
| MR | Air Mauritanie |
| MS | Egypt Air |
| MU | Misr Air |
| MV | Macrobertson Miller |
| MW | Maya |
| MX | Mexicana |
| MY | Air Mali |
| MZ | Merpati Nusantara |
| NA | National |
| NB | Sterling |
| NC | North Central |
| ND | Nord Air |
| NE | Air New England |
| NH | All Nippon |
| NI | Lancia |
| NJ | Namakwaland Lugdiens |
| NL | Air Liberia |
| NM | Mt. Cook |
| NS | Northeast (Brit. Airways) |
| NU | Southwest |
| NW | Northwest Orient |
| NZ | New Zealand National |
| OA | Olympic |
| OB | Austrian Air Transport |
| OC | Air California |
| OD | Aero Condor |
| OG | Air Guadeloupe |
| OH | SFO Helicopters |
| OI | Slov Air |
| OK | Czechoslovak |
| OL | Ostfriesische Luft Transcontinental |
| OM | Monarch |

# AIRLINE TWO LETTER CODES

| Code | Airline |
|---|---|
| ON | Air Nauru |
| OO | Sobel Air |
| OP | Air Panama |
| OR | Air Comores |
| OS | Austrain |
| OU | Aerial Tours |
| OV | Overseas National |
| OY | Conair |
| OZ | Ozark |
| PA | Pan American |
| PC | Fiji Air |
| PH | Polynesian |
| PI | Piedmont |
| PK | Pakistan Intl. |
| PL | Aero Peru |
| PR | Philippine |
| PS | Pacific Southwest |
| PV | Eastern Provincial |
| PW | Pacific Western |
| PX | Air Niugini |
| PY | Surinam |
| PZ | Lap |
| QB | Quebec Air |
| QC | Air Zaire |
| QD | Transbrazil |
| QE | Air Tahiti |
| QF | Qantas |
| QM | Air Malawi |
| QN | Bush Pilots |
| QP | Casp Air |
| QQ | Aerovias Quisqueyana |
| QR | Air Centrafique |
| QS | African Safari |
| QU | Air Limousin |
| QZ | Zambia |
| RA | Royal Nepal |
| RB | Syrian Arab |
| RC | Air Cambodge |
| RD | Airlift Intl |
| RF | Air Samoa |
| RG | Varig |
| RH | Air Rhodesia |
| RJ | Alia-Royal Jordan. |
| RK | Air Afrique |
| RM | Aerolineas Tao |
| RN | Royal Air Inter |
| RO | Tarom |
| RT | Transportes Aer-eos De Timor |
| RU | Rousseau |
| RW | Hughes Airwest |
| RY | Royal Air Laos |
| RZ | Tarca |
| SA | South African |
| SB | Seaboard World |
| SC | Cruzeiro |
| SD | Sudan |
| SG | Sabah Air |
| SH | Sahsa |
| SK | SAS |
| SL | Southeast |
| SN | Sabena |
| SO | Southern |
| SP | Sata |
| SQ | Singapore |
| SR | Swiss Air |
| SU | Aeroflot |
| SV | Saudi Arabian |
| SW | Suidwes Lugdiens |
| TA | Taca |
| TB | Trans Air (Sweden) |
| TD | Transportes Aereos De Carga |
| TE | Air New Zealand |
| TF | Avia Taxi France |
| TG | Thai Airways Intl |
| TH | Thai Airways Co |
| TK | Thy-Turk Hava Yollari |
| TL | Trans Mediterranean |
| TM | Deta |
| TN | Trans Australia |
| TO | Tempair |
| TP | Tap |
| TQ | Trans Oceanic |
| TR | Trans Europa |
| TS | Aloha |
| TT | Texas Intl |
| TU | Tunis Air |
| TV | Trans International |
| TW | Transworld |
| TX | TAN |
| TY | Air Caledonie |
| TZ | Transair Ltd |
| UA | United |
| UB | Burma |
| UC | Ladeco |
| UE | United Air Serivces |
| UF | Tonga |
| UK | British Island |
| UL | Lansa-Honduras |
| UM | Air Manila |
| UO | Trans Union |
| UP | Bahamas Air |
| UR | Trans Africa Air |
| UT | UTA |
| UY | Cameroon |
| UZ | Air Rouergue |
| VA | Viasa |
| VC | Laco |
| VE | Avensa |
| VF | British Air Ferries |
| VG | Air Siam |

# AIRLINE TWO LETTER CODES

| Code | Airline |
|---|---|
| VH | Air Volta |
| VK | Crowley |
| VP | Vasp |
| VQ | Air Pacific |
| VS | Sata |
| VT | Air Polynesie |
| VU | Air Ivoire |
| VY | Alas Del Caribe |
| VX | Transval Air |
| WA | Western |
| WB | San |
| WC | Wien Air Alaska |
| WD | Ward Air |
| WF | Wideroes Flyvesel. |
| WG | ALAG–Alpine Lufttransport |
| WK | Western Alaska |
| WM | Windward Islands |
| WN | Southwest |
| WO | World |
| WQ | Bahamas World |
| WT | Nigeria |
| WU | AVNA |
| WW | Westwing |
| WX | Ansett–New South Wales |
| WZ | Swazi Air |
| YD | Pyren Air Gaspe |
| YK | Cyprus Turkish |
| YO | Brit Airways Helico. |
| YP | Pagas |
| ZT | Satena |

# LOST CREDIT CARD PHONE #'S

| Carrier | USA Number | World Wide Number |
|---|---|---|
| American Express | 800-227-2639 | 402-392-2429 |
| Amoco | 800-548-6482 | 515-226-4100 |
| Chevron Oil Co | 800-243-8766 | 415-827-6000 |
| Conoco | Call collect | 405-767-3456 |
| Diamond Shamrock | 800-333-3560 | 806-378-3601 |
| Diners Club Int./Carte Bl. | 800-525-9135 | 303-790-2433 |
| Discovery Card | 800-347-2683 | 602-481-2300 |
| Exxon Oil Co | 800-231-4674 | 713-680-6500 |
| Joslins | 800-333-2878 | 303-781-1111 |
| May D & F/ May Co | 800-828-4120 | 303-620-7500 |
| Mastercard | 800-826-2181 | 416-232-8020 |
| Montgomery Ward | 800-367-0468 | 913-676-4025 |
| J.C. Penney | Any Store | 303-779-6900 |
| Phillips 66 | 800-331-0961 | 918-661-5000 |
| Sears & Roebuck | 800-877-8691 | 303-989-9410 |
| Shell Oil Co | 800-331-3703 | 918-496-4300 |
| Texaco Oil Co | 800-552-7827 | 713-666-8000 |
| Visa Worldwide | 800-336-8472 | 415-574-7700 |

NOTE: MOST of the above carriers require that you have your credit card number in order to report it lost or stolen. You should carry a list of all you card numbers somewhere other than your wallet or purse (use the inside cover of POCKET REF if you want).

When calling the world wide number, try calling collect first. Many carriers, such as Visa, will accept a collect call to report a lost or stolen card. Some of the world wide numbers listed above are regional numbers, but they will be able to direct you to the correct number if they can't help.

# AIRLINE 1-800 PHONE NUMBERS

| Airline | Phone # |
|---|---|
| Aeroperu | 255-7378 |
| Air Canada | 776-3000 |
| Air France | 237-2747 |
| Air India | 233-7776 |
| Air Jamaica | 523-5585 |
| Air Midwest | 835-2953 |
| Air Nevada | 634-6377 |
| Air New Zealand | 262-1234 |
| Alaska Airlines | 426-0333 |
| Alitalia Airlines | 223-5730 |
| Aloha Airlines | 367-5250 |
| Alm Antillean | 327-7230 |
| Amer. Airlines | 433-7300 |
| Amer. Eagle Air. | 446-7834 |
| American West | 247-5692 |
| Australian Air. | 922-5122 |
| Austrian Air | 843-0002 |
| Avianca | 327-1330 |
| Big Sky Air. | 882-4475 |
| British Airways | 247-9297 |
| British Caledonian | 538-2225 |
| Canadian Air. | 669-3377 |
| Cannonball Air | 323-6850 |
| Capitol Air Cargo | 221-4468 |
| Cathay Pacific | 233-2742 |
| Cayman Air. Ltd | 422-9626 |
| China Airlines | 227-5118 |
| Christman Air Sys. | 999-8359 |
| Continental Air. | 525-0280 |
| CSA Czech. Air. | 223-2365 |
| Data Air Courier | 323-6808 |
| Delta Air. Dash | 638-7333 |
| Delta Air. Inc. | 221-1212 |
| DHL Worldwide | 225-5345 |
| El Al Israel Air. | 223-6700 |
| Faucett Peruvian | 334-3356 |
| Finnair | 223-5700 |
| Gulf Air/Arabian | 223-1740 |
| Guyana Airways | 242-4210 |
| Hawaiian Air. | 367-5320 |
| Horizon Air Frt. | 547-9308 |
| Iberia Air. | 772-4642 |
| Icelandair | 223-5500 |
| Japan Air. | 525-3663 |
| KLM Royal Dutch | 777-5553 |
| Korean Air. | 421-8200 |
| Lacsa Air. | 225-2272 |
| LAN Chile Air. | 735-5526 |
| Lloyd Aero Boliv. | 327-7407 |
| Lufthansa Air. | 645-3880 |
| Malaysian Air. | 421-8641 |
| Malev Hungarian | 223-6884 |
| Mesa Airlines | 637-2247 |
| Mexicana Air. | 531-7921 |
| Midway Air. | 621-5700 |
| Midwest Express | 452-2022 |
| New York Air (see Continental) | |
| Northwest Air. | 225-2525 |
| Olympic Airways | 223-1226 |
| Pacific SW Air. | 345-9772 |
| Pan Am World Aw. | 221-1111 |
| Philippine Air. | 435-9725 |
| Piedmont Air | (see US Air) |
| Priority Air Express | 257-4777 |
| Quantas Airways | 227-4500 |
| Sabena Belgian | 382-7567 |
| Saeta Air | 338-3975 |
| SAS Scandinavian | 221-2350 |
| Saudi Arabian Air. | 472-8342 |
| Sequoia Publishing | 873-7126 |
| Singapore Air. | 742-3333 |
| Skywest Air. | 453-9417 |
| S. African Air. | 722-9675 |
| Southwest Air. | 531-5601 |
| Swissair | 221-4750 |
| TAP Air Portugal | 221-7890 |
| Telerad Air Cour. | 221-1430 |
| Texas Air. | 713-834-2950 |
| Thai Airways Int | 426-5204 |
| Trans World Air. | 221-2000 |
| United Air. | 241-6522 |
| United Air. Cargo | 722-5243 |
| USAir | 428-4322 |
| UTA French Air | 282-4484 |
| Varig Brasilian | 468-2744 |
| Viasa-Venezuelan | 327-5454 |
| Western Air | (see Delta) |

# POCKET REF

# General Science



(See also GEOLOGY for Richter Earthquake scales)

# TEMPERATURE CONVERSIONS

| °C | °F | °C | °F | °C | °F |
|---|---|---|---|---|---|
| 10000 | 18032 | 430 | 806 | 200 | 392.0 |
| 9500 | 17132 | 420 | 788 | 195 | 383.0 |
| 9000 | 16232 | 410 | 770 | 190 | 374.0 |
| 8500 | 15332 | 400 | 752 | 185 | 365.0 |
| 8000 | 14432 | 395 | 743 | 180 | 356.0 |
| 7500 | 13532 | 390 | 734 | 175 | 347.0 |
| 7000 | 12632 | 385 | 725 | 170 | 338.0 |
| 6500 | 11732 | 380 | 716 | 165 | 329.0 |
| 6000 | 10832 | 375 | 707 | 160 | 320.0 |
| 5500 | 9932 | 370 | 698 | 155 | 311.0 |
| 5000 | 9032 | 365 | 689 | 150 | 302.0 |
| 4500 | 8132 | 360 | 680 | 145 | 293.0 |
| 4000 | 7232 | 355 | 671 | 140 | 284.0 |
| 3500 | 6332 | 350 | 662 | 135 | 275.0 |
| 3000 | 5432 | 345 | 653 | 130 | 266.0 |
| 2500 | 4532 | 340 | 644 | 125 | 257.0 |
| 2000 | 3632 | 335 | 635 | 120 | 248.0 |
| 1500 | 2732 | 330 | 626 | 115 | 239.0 |
| 1000 | 1832 | 325 | 617 | 110 | 230.0 |
| 950 | 1742 | 320 | 608 | 105 | 221.0 |
| 900 | 1652 | 315 | 599 | 100 | 212.0 |
| 850 | 1562 | 310 | 590 | 99 | 210.2 |
| 800 | 1472 | 305 | 581 | 98 | 208.4 |
| 750 | 1382 | 300 | 572 | 97 | 206.6 |
| 700 | 1292 | 295 | 563 | 96 | 204.8 |
| 650 | 1202 | 290 | 554 | 95 | 203.0 |
| 600 | 1112 | 285 | 545 | 94 | 201.2 |
| 590 | 1094 | 280 | 536 | 93 | 199.4 |
| 580 | 1076 | 275 | 527 | 92 | 197.6 |
| 570 | 1058 | 270 | 518 | 91 | 195.8 |
| 560 | 1040 | 265 | 509 | 90 | 194.0 |
| 550 | 1022 | 260 | 500 | 89 | 192.2 |
| 540 | 1004 | 255 | 491 | 88 | 190.4 |
| 530 | 986 | 250 | 482 | 87 | 188.6 |
| 520 | 968 | 245 | 473 | 86 | 186.8 |
| 510 | 950 | 240 | 464 | 85 | 185.0 |
| 500 | 932 | 235 | 455 | 84 | 183.2 |
| 490 | 914 | 230 | 446 | 83 | 181.4 |
| 480 | 896 | 225 | 437 | 82 | 179.6 |
| 470 | 878 | 220 | 428 | 81 | 177.8 |
| 460 | 860 | 215 | 419 | 80 | 176.0 |
| 450 | 842 | 210 | 410 | 79 | 174.2 |
| 440 | 824 | 205 | 401 | 78 | 172.4 |

°C = Degrees Centigrade (Celsius scale). 1 unit is 1/100 of the difference between the temperature of melting ice and boiling water at standard temperature and pressure.

°F = Degrees Fahrenheit. 1 unit is 1/180 of the difference between the temperature of melting ice and boiling water at standard temperature and pressure.

# TEMPERATURE CONVERSIONS

| °C | °F | °C | °F | °C | °F |
|---|---|---|---|---|---|
| 77 | 170.6 | 34 | 93.2 | −9 | 15.8 |
| 76 | 168.8 | 33 | 91.4 | −10 | 14.0 |
| 75 | 167.0 | 32 | 89.6 | −11 | 12.2 |
| 74 | 165.2 | 31 | 87.8 | −12 | 10.4 |
| 73 | 163.4 | 30 | 86.0 | −13 | 8.6 |
| 72 | 161.6 | 29 | 84.2 | −14 | 6.8 |
| 71 | 159.8 | 28 | 82.4 | −15 | 5.0 |
| 70 | 158.0 | 27 | 80.6 | −16 | 3.2 |
| 69 | 156.2 | 26 | 78.8 | −17 | 1.4 |
| 68 | 154.4 | 25 | 77.0 | −18 | −0.4 |
| 67 | 152.6 | 24 | 75.2 | −19 | −2.2 |
| 66 | 150.8 | 23 | 73.4 | −20 | −4.0 |
| 65 | 149.0 | 22 | 71.6 | −21 | −5.8 |
| 64 | 147.2 | 21 | 69.8 | −22 | −7.6 |
| 63 | 145.4 | 20 | 68.0 | −23 | −9.4 |
| 62 | 143.6 | 19 | 66.2 | −24 | −11.2 |
| 61 | 141.8 | 18 | 64.4 | −25 | −13.0 |
| 60 | 140.0 | 17 | 62.6 | −26 | −14.8 |
| 59 | 138.2 | 16 | 60.8 | −27 | −16.6 |
| 58 | 136.4 | 15 | 59.0 | −28 | −18.4 |
| 57 | 134.6 | 14 | 57.2 | −29 | −20.2 |
| 56 | 132.8 | 13 | 55.4 | −30 | −22.0 |
| 55 | 131.0 | 12 | 53.6 | −31 | −23.8 |
| 54 | 129.2 | 11 | 51.8 | −32 | −25.6 |
| 53 | 127.4 | 10 | 50.0 | −33 | −27.4 |
| 52 | 125.6 | 9 | 48.2 | −34 | −29.2 |
| 51 | 123.8 | 8 | 46.4 | −35 | −31.0 |
| 50 | 122.0 | 7 | 44.6 | −36 | −32.8 |
| 49 | 120.2 | 6 | 42.8 | −37 | −34.6 |
| 48 | 118.4 | 5 | 41.0 | −38 | −36.4 |
| 47 | 116.6 | 4 | 39.2 | −39 | −38.2 |
| 46 | 114.8 | 3 | 37.4 | −40 | −40.0 |
| 45 | 113.0 | 2 | 35.6 | −50 | −58.0 |
| 44 | 111.2 | 1 | 33.8 | −60 | −76.0 |
| 43 | 109.4 | 0 | 32.0 | −70 | −94.0 |
| 42 | 107.6 | −1 | 30.2 | −80 | −112.0 |
| 41 | 105.8 | −2 | 28.4 | −90 | −130.0 |
| 40 | 104.0 | −3 | 26.6 | −100 | −148.0 |
| 39 | 102.2 | −4 | 24.8 | −125 | −193.0 |
| 38 | 100.4 | −5 | 23.0 | −150 | −238.0 |
| 37 | 98.6 | −6 | 21.2 | −200 | −328.0 |
| 36 | 96.8 | −7 | 19.4 | −250 | −418.0 |
| 35 | 95.0 | −8 | 17.6 | −273 | −459.4 |

$^\circ C = 5/9\ (^\circ F - 32)$ $\quad ^\circ F = 9/5\ ^\circ C + 32$

Absolute Zero = $0^\circ K = -273.16^\circ C = -459.69^\circ F$

°K = Degrees Kelvin (Absolute temperature). This scale is based on the average kinetic energy per molecule of a perfect gas and uses the same size degrees as the Centigrade scale. Zero (0°K) on the scale is the temperature at which a perfect gas has lost all of its energy.

# SOUND INTENSITIES

| Decibels | Degree | Loudness or Feeling |
|---|---|---|
| 225 | Deafening | 12" cannon @ 12 ft, in front & below |
| 140 | | Jet Aircraft |
| | | Artillery fire |
| 130 | | Threshold of Pain |
| | | >130 causes immediate ear damage |
| | | Propeller aircraft at 5 meters |
| | | Hydraulic press, pneumatic rock drill |
| 120 | | Thunder, Diesel engine room |
| | | Nearby riviter |
| 110 | | Close to a train, ball mill |
| 100 | Very Loud | Boiler factory, home lawn mower |
| | | Car horn at 5 meters, wood saw |
| 90 | | Symphony or a band |
| | | >90 regularly can cause ear damage |
| | | Noisy factory |
| | | Truck with out muffler |
| 80 | Loud | Inside a high speed auto |
| | | Police whistle, electric shaver |
| | | Noisy office, alarm clock |
| 70 | | Average radio |
| | | Normal street noise |
| 60 | Moderate | Normal conversation, close up |
| 50 | | Normal office noise, quiet stream |
| 45 | | To awaken a sleeping person |
| 40 | Faint | Normal private office noise |
| | | Residential neighborhood, no cars |
| 30 | | Quiet conversation |
| 20 | Very Faint | Inside an empty theater |
| | | Ticking of a watch |
| | | Rustle of leaves |
| | | Whisper |
| 10 | | Sound proof room |
| | | Threshold of Hearing |
| 0 | | Absolute silence |

Sound intensities are typically measured in decibels (db). A decibel is defined as 10 times the logarithm of the power ratio (power ratio is the ratio of the intensity of the sound to the intensity of an arbitrary standard point.) Normally a change of 1 db is the smallest volume change detectable by the human ear.

Sound intensity is also defined in terms of energy (ergs) transmitted per second over a 1 square centimeter surface. This energy is proportional to the velocity of propagation of the sound.

$$\text{Ergs/cm3} = 2\pi^2 \times \text{density in gm/cm}^3 \times \text{frequency}^2 \text{ in hz} \times \text{amplitude in cm}$$

# SOUND INTENSITIES

## Permissible Noise Exposures

| Hours Duration per Day | Sound Level in Decibels (Slow Response) |
|---|---|
| 8 | 90 |
| 6 | 92 |
| 4 | 95 |
| 3 | 97 |
| 2 | 100 |
| 1.5 | 102 |
| 1 | 105 |
| 0.5 | 110 |
| 0.25 | 115 |

The above restrictions are based on the *Occupational Safety and Health Act of 1970.* That Code basically states that if the above exposures are exceeded, then hearing protection must be worn. Note that these are based on the "A scale" of a standard sound level meter at slow response and will change if some other standard is used. See the *OSHA Section 1910.95* for additional details on the differences.

## Perception of Changes in Sound

| Sound Level Change in Decibels | Perception |
|---|---|
| 3 | Barely perceptible |
| 5 | Clearly perceptible |
| 10 | Twice as loud |

Note that the sound level scale in decibels is a logarithmic rather than linear scale. This means that as the sound level goes up, the effective increase is much more. For example, by changing from 5 up to 10 decibels, the sound is twice as loud but a change from 90 up to 100 yields a sound that is three times as intense.

The human ear can hear sounds in the frequency range from 20 hz (cycles/second) up to 20,000 hz (cycles/second).

# HUMAN BODY COMPOSITION

| Element | Percent (1) |
|---|---|
| Oxygen | .65 |
| Carbon | .18 |
| Hydrogen | .10 |
| Nitrogen | .3 |
| Calcium | 1.5 |
| Phosphorus | 1.0 |
| Sulphur | 0.25 |
| Potassium | 0.20 |
| Chlorine | 0.15 |
| Sodium | 0.15 |
| Magnesium | 0.05 |
| Fluorine | 0.02 |
| Iron | 0.006 |
| Zinc | 0.0033 |
| Silicon | 0.0020 |
| Rubidium | 0.00170 |
| Zirconium | 0.00035 |
| Strontium | 0.00020 |
| Aluminum | 0.00014 |
| Niobium | 0.00014 |
| Copper | 0.00014 |
| Antimony | < 0.00013 |
| Lead | 0.00011 |
| Cadmium | 0.000043 |
| Tin | 0.000043 |
| Iodine | 0.00004 |
| Manganese | 0.00003 |
| Vanadium | 0.00003 |
| Barium | 0.000023 |
| Arsenic | 0.00002 |
| Titanium | < 0.00002 |
| Boron | 0.000014 |
| Nickel | < 0.000014 |
| Chromium | < 0.000009 |
| Cobalt | < 0.000004 |
| Molybdenum | < 0.000007 |
| Silver | < 0.000001 |
| Gold | < 0.000001 |
| Uranium | $3 \times 10^{-8}$ |
| Cesium | $< 1.4 \times 10^{-8}$ |
| Radium | $1.4 \times 10^{-13}$ |

*(1) Scientific Tables. Ciba–Geigy Ltd*
*Basle. Switzerland 1971*

# BODY WEIGHT vs HEIGHT

## MEN

| Height in Feet–Inches | Small Frame (lbs) | Medium Frame (lbs) | Large Frame (lbs) |
|---|---|---|---|
| 5–2 | 128–134 | 131–141 | 138–150 |
| 5–3 | 130–136 | 133–143 | 140–153 |
| 5–4 | 132–138 | 135–145 | 142–156 |
| 5–5 | 134–140 | 137–148 | 144–160 |
| 5–6 | 136–142 | 139–151 | 146–164 |
| 5–7 | 138–145 | 142–154 | 149–168 |
| 5–8 | 140–148 | 145–157 | 152–172 |
| 5–9 | 142–151 | 148–160 | 155–176 |
| 5–10 | 144–154 | 151–163 | 158–180 |
| 5–11 | 146–157 | 154–166 | 161–184 |
| 6–0 | 149–160 | 157–170 | 164–188 |
| 6–1 | 152–164 | 160–174 | 168–192 |
| 6–2 | 155–168 | 164–178 | 172–197 |
| 6–3 | 158–172 | 167–182 | 176–202 |
| 6–4 | 162–176 | 171–187 | 181–207 |

## WOMEN

| Height in Feet–Inches | Small Frame (lbs) | Medium Frame (lbs) | Large Frame (lbs) |
|---|---|---|---|
| 4–10 | 102–111 | 109–121 | 118–131 |
| 4–11 | 103–113 | 111–123 | 120–134 |
| 5–0 | 104–115 | 113–126 | 122–137 |
| 5–1 | 106–118 | 115–129 | 125–140 |
| 5–2 | 108–121 | 118–132 | 128–143 |
| 5–3 | 111–124 | 121–135 | 131–147 |
| 5–4 | 114–127 | 124–138 | 134–151 |
| 5–5 | 117–130 | 127–141 | 137–155 |
| 5–6 | 120–133 | 130–144 | 140–159 |
| 5–7 | 123–136 | 133–147 | 143–163 |
| 5–8 | 126–139 | 136–150 | 146–167 |
| 5–9 | 129–142 | 139–153 | 149–170 |
| 5–10 | 132–145 | 142–156 | 152–173 |
| 5–11 | 135–148 | 145–159 | 155–176 |
| 6–0 | 138–151 | 148–162 | 158–179 |

Based on data from *Metropolitan Life Insurance Company*

# PHYSICAL GROWTH % – BOYS

Select the age in years, read the weight/height on the same row and then read the top line for the Percentile category.

| Age in Years | Boys Weight (Pounds) Percentile | | | | | | |
|---|---|---|---|---|---|---|---|
| | 5% | 10% | 25% | 50% | 75% | 90% | 95% |
| 2 | 24 | 25 | 27 | 29 | 31 | 33 | 35 |
| 3 | 27 | 28 | 30 | 33 | 35 | 38 | 39 |
| 4 | 31 | 32 | 34 | 37 | 40 | 43 | 45 |
| 5 | 34 | 35 | 38 | 41 | 45 | 48 | 51 |
| 6 | 37 | 39 | 42 | 46 | 50 | 54 | 58 |
| 7 | 41 | 43 | 46 | 51 | 55 | 61 | 66 |
| 8 | 45 | 47 | 51 | 56 | 62 | 68 | 76 |
| 9 | 49 | 52 | 56 | 62 | 69 | 79 | 87 |
| 10 | 54 | 56 | 62 | 70 | 78 | 90 | 99 |
| 11 | 59 | 62 | 69 | 78 | 89 | 103 | 113 |
| 12 | 66 | 69 | 77 | 88 | 101 | 116 | 127 |
| 13 | 74 | 79 | 88 | 99 | 114 | 130 | 143 |
| 14 | 84 | 89 | 99 | 112 | 128 | 144 | 158 |
| 15 | 95 | 102 | 113 | 126 | 143 | 159 | 174 |
| 16 | 105 | 112 | 123 | 136 | 154 | 171 | 188 |
| 17 | 113 | 121 | 132 | 145 | 162 | 183 | 200 |
| 18 | 118 | 126 | 136 | 150 | 166 | 192 | 208 |

| Age in Years | Boys Height (Inches) Percentile | | | | | | |
|---|---|---|---|---|---|---|---|
| | 5% | 10% | 25% | 50% | 75% | 90% | 95% |
| 2 | 32.2 | 32.5 | 33.2 | 33.8 | 34.6 | 35.8 | 36.6 |
| 3 | 35.0 | 35.5 | 36.5 | 37.4 | 38.4 | 39.5 | 40.2 |
| 4 | 37.7 | 38.2 | 39.4 | 40.6 | 41.6 | 42.6 | 43.4 |
| 5 | 40.2 | 40.9 | 42.0 | 43.4 | 44.4 | 45.5 | 46.1 |
| 6 | 42.5 | 43.2 | 44.4 | 45.7 | 47.0 | 48.0 | 48.6 |
| 7 | 44.5 | 45.4 | 46.5 | 47.9 | 49.2 | 50.4 | 51.0 |
| 8 | 46.5 | 47.3 | 48.5 | 50.0 | 51.4 | 52.6 | 53.4 |
| 9 | 48.4 | 49.3 | 50.5 | 52.0 | 53.5 | 54.8 | 55.7 |
| 10 | 50.3 | 51.2 | 52.5 | 54.0 | 55.6 | 57.2 | 58.2 |
| 11 | 52.1 | 53.1 | 54.5 | 56.4 | 58.0 | 59.6 | 60.8 |
| 12 | 54.1 | 55.2 | 56.7 | 58.8 | 60.6 | 62.6 | 63.8 |
| 13 | 56.0 | 57.2 | 59.1 | 61.5 | 63.4 | 65.5 | 66.5 |
| 14 | 58.5 | 59.6 | 61.5 | 64.1 | 66.2 | 68.2 | 69.2 |
| 15 | 61.0 | 62.0 | 64.1 | 66.4 | 68.4 | 70.3 | 71.5 |
| 16 | 63.4 | 64.4 | 66.2 | 68.2 | 70.0 | 71.6 | 72.9 |
| 17 | 64.8 | 66.0 | 67.6 | 69.4 | 70.9 | 72.5 | 73.6 |
| 18 | 65.2 | 66.4 | 67.8 | 69.5 | 71.2 | 73.0 | 73.8 |

Data from the *National Center for Health Statistics (NCHS) Hyattsville, Maryland.*

# PHYSICAL GROWTH % – GIRLS

Select the age in years, read the weight/height on the same row and then read the top line for the Percentile category.

| Age in | Girls Weight (Pounds) Percentile | | | | | | |
|---|---|---|---|---|---|---|---|
| Years | 5% | 10% | 25% | 50% | 75% | 90% | 95% |
| 2 | 22 | 23 | 25 | 27 | 29 | 31 | 32 |
| 3 | 26 | 27 | 29 | 31 | 34 | 37 | 38 |
| 4 | 29 | 31 | 33 | 36 | 39 | 42 | 44 |
| 5 | 32 | 33 | 36 | 39 | 43 | 47 | 50 |
| 6 | 36 | 37 | 39 | 43 | 47 | 53 | 57 |
| 7 | 39 | 41 | 44 | 48 | 53 | 60 | 71 |
| 8 | 43 | 45 | 49 | 55 | 61 | 71 | 76 |
| 9 | 48 | 50 | 55 | 63 | 71 | 82 | 89 |
| 10 | 53 | 57 | 63 | 72 | 82 | 96 | 103 |
| 11 | 60 | 64 | 71 | 81 | 94 | 109 | 118 |
| 12 | 67 | 71 | 80 | 91 | 105 | 123 | 134 |
| 13 | 75 | 80 | 89 | 101 | 116 | 134 | 147 |
| 14 | 83 | 88 | 97 | 110 | 125 | 144 | 160 |
| 15 | 90 | 95 | 105 | 118 | 132 | 152 | 170 |
| 16 | 95 | 100 | 110 | 123 | 137 | 157 | 177 |
| 17 | 98 | 103 | 112 | 124 | 138 | 158 | 180 |
| 18 | 99 | 104 | 113 | 124 | 138 | 159 | 181 |

| Age in | Girls Height (Inches) Percentile | | | | | | |
|---|---|---|---|---|---|---|---|
| Years | 5% | 10% | 25% | 50% | 75% | 90% | 95% |
| 2 | 31.6 | 31.7 | 32.5 | 33.8 | 34.8 | 36.0 | 36.5 |
| 3 | 34.7 | 35.1 | 36.0 | 37.0 | 38.0 | 39.0 | 39.5 |
| 4 | 37.4 | 38.0 | 38.9 | 40.0 | 41.0 | 42.0 | 42.6 |
| 5 | 39.8 | 40.4 | 41.5 | 42.7 | 43.9 | 44.8 | 45.5 |
| 6 | 42.0 | 42.7 | 43.9 | 45.2 | 46.5 | 47.6 | 48.4 |
| 7 | 44.0 | 44.8 | 46.0 | 47.5 | 49.0 | 50.2 | 51.0 |
| 8 | 46.0 | 46.7 | 48.1 | 49.8 | 51.4 | 52.8 | 53.6 |
| 9 | 48.0 | 48.9 | 50.3 | 52.1 | 53.7 | 55.3 | 56.2 |
| 10 | 50.1 | 51.0 | 52.5 | 54.5 | 56.1 | 57.9 | 61.4 |
| 11 | 52.5 | 53.3 | 55.1 | 57.0 | 58.6 | 60.5 | 62.8 |
| 12 | 55.0 | 56.0 | 57.8 | 59.6 | 61.3 | 63.0 | 64.0 |
| 13 | 57.1 | 58.2 | 60.0 | 61.8 | 63.4 | 65.0 | 66.0 |
| 14 | 58.5 | 59.5 | 61.4 | 63.0 | 64.8 | 66.4 | 67.3 |
| 15 | 59.2 | 60.3 | 61.8 | 63.6 | 65.4 | 67.0 | 68.0 |
| 16 | 59.6 | 60.6 | 62.0 | 64.0 | 65.6 | 67.3 | 68.1 |
| 17 | 60.2 | 61.0 | 62.4 | 64.2 | 65.8 | 67.3 | 68.2 |
| 18 | 60.4 | 61.4 | 62.7 | 64.4 | 65.9 | 67.3 | 68.2 |

Data from the *National Center for Health Statistics (NCHS) Hyattsville, Maryland.*

# ACCELERATION DUE TO GRAVITY

| Degrees Latitude | Acceleration Due to Gravity at Sea Level | |
|---|---|---|
| | Feet/second$^2$ | Cm/second$^2$ |
| 0 | 32.0878 | 978.039 |
| 5 | 32.0891 | 978.078 |
| 10 | 32.0929 | 978.195 |
| 15 | 32.0991 | 978.384 |
| 20 | 32.1076 | 978.641 |
| 25 | 32.1180 | 978.960 |
| 30 | 32.1302 | 979.329 |
| 31 | 32.1327 | 979.407 |
| 32 | 32.1353 | 979.487 |
| 33 | 32.1380 | 979.569 |
| 34 | 32.1407 | 979.652 |
| 35 | 32.1435 | 979.737 |
| 36 | 32.1463 | 979.822 |
| 37 | 32.1491 | 979.908 |
| 38 | 32.1520 | 979.995 |
| 39 | 32.1549 | 979.083 |
| 40 | 32.1578 | 980.171 |
| 41 | 32.1607 | 980.261 |
| 42 | 32.1636 | 980.350 |
| 43 | 32.1666 | 980.440 |
| 44 | 32.1696 | 980.531 |
| 45 | 32.1725 | 980.621 |
| 46 | 32.1755 | 980.711 |
| 47 | 32.1785 | 980.802 |
| 48 | 32.1814 | 980.892 |
| 49 | 32.1844 | 980.981 |
| 50 | 32.1873 | 981.071 |
| 51 | 32.1902 | 981.159 |
| 52 | 32.1931 | 981.247 |
| 53 | 32.1960 | 981.336 |
| 54 | 32.1988 | 981.422 |
| 55 | 32.2016 | 981.507 |
| 56 | 32.2044 | 981.592 |
| 57 | 32.2071 | 981.675 |
| 58 | 32.2098 | 981.757 |
| 59 | 32.2125 | 981.839 |
| 60 | 32.2151 | 981.918 |
| 65 | 32.2272 | 982.288 |
| 70 | 32.2377 | 982.608 |
| 75 | 32.2463 | 982.868 |
| 80 | 32.2525 | 983.059 |
| 85 | 32.2564 | 983.178 |
| 90 | 32.2577 | 983.217 |

# BEAUFORT WIND STRENGTH SCALE

| Beaufort Number | Wind Speed Miles/hour | Description |
|---|---|---|
| Ø | < 1 | **Calm:** Still: Smoke will rise vertically. |
| 1 | 1–5 | **Light Air:** Rising smoke drifts, weather vane is inactive. |
| 2 | 6–11 | **Light Breeze:** Leaves rustle, can feel wind on your face, weather vane is active. |
| 3 | 12–19 | **Gentle Breeze:** Leaves and twigs move around. Light weight flags extend. |
| 4 | 20–28 | **Moderate Breeze:** Moves thin branches, raises dust and paper. |
| 5 | 29–38 | **Fresh Breeze:** Small trees sway. |
| 6 | 39–49 | **Strong Breeze:** Large tree branches move, open wires (such as telegraph wires) begin to "whistle", umbrellas are difficult to keep under control. |
| 7 | 50–61 | **Moderate Gale:** Large trees begin to sway, noticeably difficult to walk. |
| 8 | 62–74 | **Fresh Gale:** Twigs and small branches are broken from trees, walking into the wind is very difficult. |
| 9 | 75–88 | **Strong Gale:** Slight damage occurs to buildings, shingles are blown off of roofs. |
| 10 | 89–102 | **Whole Gale:** Large trees are uprooted, building damage is considerable. |
| 11 | 103–117 | **Storm:** Extensive widespread damage, These typically occur only at sea, and rarely inland. |
| 12 | >117 | **Hurricane:** Extreme destruction. |

NOTE: The Beaufort Number is also referred to as a "Force" number, for example, "Force 10 Gale".

# WIND CHILL FACTORS

In order to determine a "Wind Chill Factor", locate the measured outside temperature column and then the wind speed row and then read the corresponding "Wind Chill Factor" at the intersection of the row and column. "Wind Chill Factor" is the combined effect of actual temperature and wind speed that increases heat loss in the body and makes the measured outside temperature "feel" colder.

| Wind Speed | Measured Outside Temperature °F | | | | |
|---|---|---|---|---|---|
| Miles/hour | 50 | 45 | 40 | 35 | 30 |
| 4 | 50 | 45 | 40 | 35 | 30 |
| 6 | 46 | 41 | 35 | 30 | 24 |
| 8 | 43 | 37 | 31 | 25 | 20 |
| 10 | 40 | 34 | 28 | 22 | 16 |
| 12 | 38 | 32 | 26 | 19 | 13 |
| 14 | 37 | 30 | 23 | 17 | 10 |
| 16 | 35 | 28 | 21 | 15 | 8 |
| 18 | 34 | 27 | 20 | 13 | 6 |
| 20 | 32 | 25 | 18 | 11 | 4 |
| 22 | 31 | 24 | 17 | 10 | 2 |
| 24 | 30 | 23 | 16 | 8 | 1 |
| 26 | 30 | 22 | 15 | 7 | 0 |
| 28 | 29 | 21 | 14 | 6 | −1 |
| 30 | 28 | 21 | 13 | 5 | −2 |
| 35 | 27 | 19 | 11 | 3 | −4 |
| 40 | 26 | 18 | 10 | 2 | −6 |
| 45 | 25 | 17 | 9 | 1 | −7 |

| Wind Speed | Measured Outside Temperature °F | | | | |
|---|---|---|---|---|---|
| Miles/hour | 25 | 20 | 15 | 10 | 5 |
| 4 | 25 | 20 | 15 | 10 | 5 |
| 6 | 19 | 13 | 8 | 2 | −3 |
| 8 | 14 | 8 | 2 | −4 | −10 |
| 10 | 10 | 4 | −3 | −9 | −15 |
| 12 | 6 | 0 | −7 | −13 | −19 |
| 14 | 3 | −3 | −10 | −17 | −23 |
| 16 | 1 | −6 | −13 | −20 | −26 |
| 18 | −1 | −8 | −15 | −22 | −29 |
| 20 | −3 | −10 | −18 | −25 | −32 |
| 22 | −5 | −12 | −19 | −27 | −34 |
| 24 | −6 | −14 | −21 | −29 | −36 |
| 26 | −8 | −15 | −23 | −30 | −38 |
| 28 | −9 | −17 | −24 | −32 | −39 |
| 30 | −10 | −18 | −25 | −33 | −41 |
| 35 | −12 | −20 | −28 | −36 | −43 |
| 40 | −14 | −22 | −29 | −37 | −45 |
| 45 | −15 | −23 | −31 | −39 | −47 |

# WIND CHILL FACTORS

| Wind Speed Miles/hour | Measured Outside Temperature °F 0 | -5 | -10 | -15 | -20 |
|---|---|---|---|---|---|
| 4 | 0 | -5 | -10 | -15 | -20 |
| 6 | -9 | -14 | -20 | -25 | -31 |
| 8 | -16 | -21 | -27 | -33 | -39 |
| 10 | -21 | -27 | -33 | -40 | -46 |
| 12 | -26 | -32 | -39 | -45 | -51 |
| 14 | -30 | -36 | -43 | -50 | -56 |
| 16 | -33 | -40 | -47 | -54 | -60 |
| 18 | -36 | -43 | -50 | -57 | -64 |
| 20 | -39 | -46 | -53 | -60 | -67 |
| 22 | -41 | -48 | -56 | -63 | -70 |
| 24 | -43 | -51 | -58 | -65 | -73 |
| 26 | -45 | -53 | -60 | -68 | -75 |
| 28 | -47 | -54 | -62 | -69 | -77 |
| 30 | -48 | -56 | -64 | -71 | -79 |
| 35 | -51 | -59 | -67 | -75 | -82 |
| 40 | -53 | -61 | -69 | -77 | -85 |
| 45 | -55 | -63 | -71 | -79 | -86 |

| Wind Speed Miles/hour | Measured Outside Temperature °F -25 | -30 | -35 | -40 | -45 |
|---|---|---|---|---|---|
| 4 | -25 | -30 | -35 | -40 | -45 |
| 6 | -36 | -42 | -47 | -53 | -58 |
| 8 | -45 | -51 | -57 | -62 | -68 |
| 10 | -52 | -58 | -64 | -70 | -77 |
| 12 | -58 | -64 | -71 | -77 | -83 |
| 14 | -63 | -70 | -76 | -83 | -89 |
| 16 | -67 | -74 | -81 | -88 | -95 |
| 18 | -71 | -78 | -85 | -92 | -99 |
| 20 | -75 | -82 | -89 | -96 | -103 |
| 22 | -78 | -85 | -92 | -99 | -107 |
| 24 | -80 | -88 | -95 | -102 | -110 |
| 26 | -83 | -90 | -97 | -105 | -112 |
| 28 | -85 | -92 | -100 | -107 | -115 |
| 30 | -86 | -94 | -102 | -109 | -117 |
| 35 | -90 | -98 | -106 | -114 | -121 |
| 40 | -93 | -101 | -109 | -116 | -124 |
| 45 | -94 | -102 | -110 | -118 | -126 |

**NOTE: Wind speeds greater than 45 miles/hour have little additional effect on the Wind Chill Factor. The following formula can be used to determine Wind Chill Factors not listed in the above table:**

$$\text{Wind Chill Factor} = (\{(10.45 + (6.686112 \times \sqrt{\text{Wind speed}}) - (0.447041 \times \text{Wind speed}))/22.034\} \times (\text{Temperature} - 91.4)) + 91.4$$

# HEAT – HUMIDITY FACTOR

In order to determine a "Heat Factor", locate the measured outside temperature row and then the humidity column and then read the corresponding apparent temperature at the intersection of the row and column. This "Heat Factor" is the combined effect of actual temperature and humidity that makes the measured outside temperature "feel" hotter. Heat exhaustion danger occurs when the "Heat Factor" is greater than 105°F.

| Measured Temp °F | Percent Relative Humidity 0 | 10 | 20 | 30 |
|---|---|---|---|---|
| 70 | 64 | 65 | 66 | 67 |
| 75 | 69 | 70 | 72 | 73 |
| 80 | 73 | 75 | 77 | 78 |
| 85 | 78 | 80 | 82 | 84 |
| 90 | 83 | 85 | 87 | 90 |
| 95 | 87 | 90 | 93 | 96 |
| 100 | 91 | 95 | 99 | 104 |
| 105 | 95 | 100 | 105 | 113 |
| 110 | 99 | 105 | 112 | 123 |
| 115 | 103 | 111 | 120 | 135 |
| 120 | 107 | 116 | 130 | 148 |

| Measured Temp °F | Percent Relative Humidity 40 | 50 | 60 | 70 |
|---|---|---|---|---|
| 70 | 68 | 69 | 70 | 70 |
| 75 | 74 | 75 | 76 | 77 |
| 80 | 79 | 81 | 82 | 85 |
| 85 | 86 | 88 | 90 | 93 |
| 90 | 93 | 96 | 100 | 106 |
| 95 | 101 | 107 | 114 | 124 |
| 100 | 110 | 120 | 132 | 144 |
| 105 | 123 | 135 | 149 | |
| 110 | 137 | 150 | | |
| 115 | 151 | | | |

| Measured Temp °F | Percent Relative Humidity 80 | 90 | 100 |
|---|---|---|---|
| 70 | 71 | 71 | 72 |
| 75 | 78 | 79 | 80 |
| 80 | 86 | 88 | 91 |
| 85 | 97 | 102 | 108 |
| 90 | 113 | 122 | |
| 95 | 136 | | |

# FIREWOOD / FUEL COMPARISONS

| Fuel Type | Million BTU /Unit (1) | Available Units /million BTU (2) | Comment |
|---|---|---|---|
| #2 Fuel Oil | 0.135/gallon | 11.5 | 65% efficient |
| Charcoal | 0.013/pound | 128 | 60% efficient |
| Coal: | | | 60% efficient |
| Anthracite | 15.2/ton | 0.07 | |
| Bituminous | 22.0/ton | 0.08 | Low Volatile |
| Bituminous | 28.6/ton | 0.06 | High Volatile |
| Lignite | 13.8/ton | 0.12 | |
| Electricity | 0.003/KWH | 293 | 100% efficient |
| Kerosene | 0.135/gallon | 11.5 | 65% efficient |
| Natural Gas | 700/MCF | 1.43 | 70% efficient |
| Propane | 0.09/gallon | 15.7 | 70% efficient |
| Wood: | | | 50% to 60% efficient |
| Apple | 30/cord | 0.047 | L–smoke, L–spark |
| Aspen | 18/cord | 0.077 | M–smoke, H–spark |
| Cottonwood | 17/cord | 0.082 | M–smoke, L–spark |
| Elm, Red | 29/cord | 0.048 | M–smoke, M–spark |
| Fir, Douglas | 24/cord | 0.058 | H–smoke, M–spark |
| Hickory | 27/cord | 0.052 | L–smoke, L–spark |
| Juniper | 15/cord | 0.093 | M–smoke, M–spark |
| Maple, Silver | 20/cord | 0.070 | L–smoke, L–spark |
| Oak, Red | 30/cord | 0.047 | L–smoke, L–spark |
| Oak, White | 32/cord | 0.044 | L–smoke, L–spark |
| Pine, Lodgepole | 21/cord | 0.066 | M–smoke, M–spark |
| Pine, Pinon | 27/cord | 0.052 | M–smoke, M–spark |
| Pine, Ponderosa | 20/cord | 0.070 | M–smoke, M–spark |
| Spruce, Englem. | 18/cord | 0.077 | M–smoke, H–spark |

"L–" is Low, "M–" is Medium, and "H–" is High
In order to calculate the actual cost of heat for each type, simply multiply the "Available Units/million BTU" by the current cost per unit. For example, if natural gas is currently $4 per MCF, the cost of 1 million BTU is $4 x 1.43 = $5.72. In the case of White Oak, the cost of 1 million BTU is $150/cord x 0.072 = $10.80. Note that the wood efficiency can vary greatly, depending on moisture and efficiency of the furnace you are using.

**(1)** Million BTU/Unit defines the average amount of heat per unit that is available for that fuel type, assuming 100% burning efficiency. For example, Aspen wood contains 18,000,000 BTU per dry cord.

**(2)** Available Units/million BTU defines the actual number of units required to produce 1,000,000 BTU. The efficiency of burning (shown in the Comment column) is considered, as well as the moisture content of woods (average 20% moisture for dry wood).

# FREQUENCY SPECTRUM

| Frequency (Wavelength) | Name |
|---|---|
| 0 Hertz | Steady direct current |
| 16–16,000 Hz | Audio frequencies |
| 10–30 kHz (30,000–10,000m) | v.l.f. – very low frequency |
| 10–16 kHz | ultrasonic |
| 30 kHz to 30,000 megahertz | Radio Frequencies |
| 30–300 kHz(10,000–1000m) | l.f. – low frequencies |
| 30–535 kHz | Marine com & navigation, aero nav. |
| 300–3000 kHz(1000–100m) | m.f. – medium frequencies |
| 535–1605 kHz | AM broadcast bands |
| 1800–2000 kHz | 160 meter band |
| 3–30 MHz(100–10m) | h.f. – high frequencies |
| 3.5–4 MHz | 80 meter band |
| 7–7.3 MHz | 40 meter band |
| 14–14.35 MHz | 20 meter band |
| 21–21.45 MHz | 15 meter band |
| 26.95–27.54 MHz | Industrial, scientific, & medical |
| 28–29.7 MHz | 10 meter band |
| 26.965–27.455 MHz | Citizens Band Class D |
| 30–300 MHz(10–1m) | v.h.f. – very high frequencies |
| 30–50 MHz | Police, fire, forest, highway, railroad |
| 50–54 MHz | 6 meter band |
| 54–72 MHz | TV channels 2 to 4 |
| 72–76 MHz | Government, Aero. Marker 75MHz |
| 76–88 MHz | TV channels 5 and 6 |
| 88–108 MHz | FM broadcast band |
| 108–118 MHz | Aeronautical navigation |
| 118–136 MHz | Civil Communication Band |
| 148–174 MHz | Government |
| 144–148 MHz | 2 meter band |
| 174–216 MHz | TV channels 7 to 13 |
| 216–470 MHz | Amateur, government, CB Band, non–government, fixed or mobile aeronautical navigation |
| 220–225 MHz | Amateur band, 1-1/4 meter |
| 225–400 MHz | Military |
| 420–450 MHz | Amateur band, 0.7 meter |
| 462.5–465 MHz | Citizens Band |

## FREQUENCY SPECTRUM

| Frequency (Wavelength) | Name |
|---|---|
| 300–3000 MHz(100–10cm) | u.h.f. – ultra high frequencies |
| 470–890 MHz | TV channels 14 to 83 |
| 890–3000 MHz | Aero navigation, amateur bands, government & non–government, fixed and mobile |
| 1300–1600 | Radar band |
| 3000–30,000 MHz(10–1cm) | s.h.f. – super high frequencies Government and non–government, amateur bands, radio navigation |
| 30,000 MHz to 300 GHz (1–0.1cm) | Extra–high frequencies (weather radar, experimental, government) |
| 30–0.76 μm | Infrared light and heat |
| 0.76–0.39 μm | Visible light |
| 6470–7000 angstroms | Red light |
| 5850–6740 angstroms | Orange light |
| 5750–5850 angstroms | Yellow light |
| 5560–5750 angstroms | Maximum visibility |
| 4912–5560 angstroms | Green light |
| 4240–4912 angstroms | Blue light |
| 4000–4240 angstroms | Violet light |
| 0.39–0.032 μm | Ultraviolet light |
| 0.032–0.00001 μm | X–rays |
| 0.00001–0.0000006 μm | Gamma rays |
| 0.0005 angstroms | Cosmic rays |

μm = micrometer($10^{-6}$m): m = meter: cm = centimeter
Hz = hertz: MHz = megahertz ($10^6$ Hz): kHz = kilohertz($10^3$ Hz)
GHz = gigahertz($10^9$ Hz): 1 angstrom = $10^{-10}$ meters

# PLANETARY DATA

## SUN:

. Mass ........................ $4.381 \times 10^{30}$ pounds
. Density ........................ 88.0 lbs/cubic foot
. Mean Radius ........................ 434,959 miles
. Gravity relative to earth ........ 27.9
. Rotation period ................ 24 days, 16 hours, 48 minutes
. Number of moons ............. 0

## MERCURY:

. Mass ........................ $6.982 \times 10^{23}$ pounds
. Density ........................ 340.9 lbs/cubic foot
. Mean Radius ................. 1516 miles
. Max distance from the sun ....... 43,770,000 miles
. Min distance from the sun ....... 28,580,000 miles
. Gravity relative to earth ........ 0.37
. Rotation period ................ 58 days, 21 hours, 58 minutes
. Revolution time around sun ...... 88 days
. Orbital velocity ................ 29.75 miles/second
. Number of moons ............. 0

## VENUS:

. Mass ........................ $1.074 \times 10^{25}$ pounds
. Density ........................ 309.6 lbs/cubic foot
. Mean Radius ................. 3759 miles
. Max distance from the sun ....... 67,730,000 miles
. Min distance from the sun ....... 66,490,000 miles
. Gravity relative to earth ........ 0.88
. Rotation period ................ 243 days
. Revolution time around sun ...... 224.7 days
. Orbital velocity ................ 21.76 miles/second
. Number of moons ............. 0

## EARTH:

. Mass ........................ $1.317 \times 10^{25}$ pounds
. Density ........................ 347.7 lbs/cubic foot
. Mean Radius ................. 3963 miles
. Max distance from the sun ....... 94,510,000 miles
. Min distance from the sun ....... 91,400,000 miles
. Gravity relative to earth ........ 1.00
. Rotation period ................ 23 hours, 56 minutes
. Revolution time around sun ...... 365.26 days
. Orbital velocity ................ 18.51 miles/second
. Number of moons ............. 1

# PLANETARY DATA

## EARTH'S MOON:

Mass ........................... $1.619 \times 10^{23}$ pounds
Density ........................ 207.9 lbs/cubic foot
Mean Radius .................... 1080 miles
Max distance from earth ........ 252,900 miles
Min distance from earth ........ 221,800 miles
Gravity relative to earth ...... 0.17
Rotation period ................ 27 days, 7 hours, 43 minutes
Orbital velocity ............... 0.101 miles/second

## MARS:

Mass ........................... $1.409 \times 10^{24}$ pounds
Density ........................ 246.6 lbs/cubic foot
Mean Radius .................... 2109 miles
Max distance from the sun ...... 154,700,000 miles
Min distance from the sun ...... 127,400,000 miles
Gravity relative to earth ...... 0.38
Rotation period ................ 24 hours, 37 minutes
Revolution time around sun ..... 687 days
Orbital velocity ............... 14.99 miles/second
Number of moons ................ 2, Deimos, Phobos

## JUPITER:

Mass ........................... $4.189 \times 10^{27}$ pounds
Density ........................ 83.0 lbs/cubic foot
Mean Radius .................... 44,365 miles
Max distance from the sun ...... 507,000,000 miles
Min distance from the sun ...... 459,800,000 miles
Gravity relative to earth ...... 2.64
Rotation period ................ 9 hours, 3 minutes
Revolution time around sun ..... 12 years (4332.6 days)
Orbital velocity ............... 8.12 miles/second
Number of moons,17+........Adrastea, Metis, Amalthea, Thebe, Io, Europa, Ganymede, Callisto, Leda, Himalia, Lysithea, Elara, Ananke, Carme, Pasiphae, Sinope.

## SATURN:

Mass ........................... $1.254 \times 10^{27}$ pounds
Density ........................ 43.7 lbs/cubic foot
Mean Radius .................... 37,468 miles
Max distance from the sun ...... 938,300,000 miles
Min distance from the sun ...... 838,900,000 miles
Gravity relative to earth ...... 1.15
Rotation period ................ 10 hours, 30 minutes
Revolution time around sun ..... 29 years (10759.2 days)
Orbital velocity ............... 5.99 miles/second
Number of moons, 22+..........Atlas, Prometheus, Pandora, Janus, Epimetheus, Mimas, Enceladus, Tethys, Telesto, Calypso, Dione, Helene, Rhea, Titan, Hyperion, Iapetus, Phoebe.

# PLANETARY DATA

## URANUS:

. Mass .......................... $1.916 \times 10^{26}$ pounds
. Density .......................... 97.4 lbs/cubic foot
. Mean Radius .................... 15,938 miles
. Max distance from the sun ....... 1,870,000,000 miles
. Min distance from the sun ....... 1,696,000,000 miles
. Gravity relative to earth ......... 1.15
. Rotation period ................. 15 hours, 36 minutes
. Revolution time around sun ...... 84 years (30685.9 days)
. Orbital velocity ................. 4.23 miles/second
. Number of moons, 15........ Cordelia, Ophelia, Bianca, Cressida,Desdemona, Juliet, Portia, Rosalind, Belinda, Puck, Miranda, Ariel, Umbriel, Titania, Oberon.

## NEPTUNE:

. Mass .......................... $2.271 \times 10^{26}$ pounds
. Density .......................... 142.3 lbs/cubic foot
. Mean Radius .................... 15,255 miles
. Max distance from sun ........... 2,881,000,000 miles
. Min distance from sun ........... 2,771,000,000 miles
. Gravity relative to earth ......... 1.12
. Rotation period ................. 18 hours, 26 minutes
. Revolution time around sun ...... 164 years (60187.6 days)
. Orbital velocity ................. 3.38 miles/second
. Number of moons ............... 3, Triton, Nereid, ?

## PLUTO:

. Mass .......................... $1.184 \times 10^{25}$ pounds
. Density .......................... 280.9 lbs/cubic foot
. Mean Radius .................... 715 miles
. Max distance from sun ........... 4,580,000,000 miles
. Min distance from sun ........... 2,765,000,000 miles
. Gravity relative to earth ......... 0.04
. Rotation period ................. 6 days, 9 hours, 17 minutes
. Revolution time around sun ...... 247.7 years (90885 days)
. Orbital velocity ................. 2.95 miles/second
. Number of moons ............... 1, Charon

# Geology



***Dana's Manual of Mineralogy, Field Geologists' Manual and A Field Guide to Rocks and Minerals*** **were used as source material for much of the Geology chapter, see page 2 for the reference.**


(See also GENERAL SCIENCE on page 177)
(See alsoWEIGHTS OF MATERIALS on page 389)

# MINERAL TABLE ABBREVIATIONS

Abbreviations used in the "Name" column are:

| | | |
|---|---|---|
| (A) | = | Amphibole group |
| (B) | = | Bauxite component |
| (C) | = | Clay group or clay like |
| (D) | = | Diopside series |
| (E) | = | Enstatite group |
| (F) | = | Feldspar group |
| (Fp) | = | Feldspathoid group |
| (G) | = | Garnet group |
| (H) | = | Hornblende |
| (J) | = | Jamesonite group |
| (M) | = | Mica group |
| (O) | = | Orthoclase |
| (Ov) | = | Olivine group |
| (P) | = | Pyroxene group |
| (R) | = | Rare Earth Oxide group |
| (S) | = | Spinel group |
| (Sc) | = | Scapolite series |
| (W) | = | Wolframite series |
| (Z) | = | Zeolite group |

The "Hard" column lists hardness as defined by Mhos scale of hardness (see Mhos table, page 231, also in this geology section).

"Sys" column lists the crystal system of each mineral:

| | | |
|---|---|---|
| Is | = | Isometric |
| Hx | = | Hexagonal–Hexagonal |
| Rh | = | Hexagonal–rhombohedral |
| Te | = | Tetragonal |
| Or | = | Orthorhombic |
| Mo | = | Monoclinic |
| Tr | = | Triclinic |

## – A –

| Name | Composition | Density | Hard | Sys |
|---|---|---|---|---|
| Acanthite | $Ag_2S$ | 7.2-7.3 | 2-2.5 | Mo |
| Achroite | Colorless tourmaline | | | |
| Acmite (P) | $NaFe(SiO_3)_2$ | 3.4-3.6 | 6-6.5 | Mo |
| Actinolite (A) | $Ca_2(Mg,Fe)_5(Si_8O_{22})(OH)_2$ | 3.0-3.2 | 5-6 | Mo |
| Adularia (O) | Clear orthoclase | | | |
| Aegirte | Acmite, Aegirine | | | |
| Agate | Banded chalcedony | | | |
| Alabandite | MnS | 4.0 | 3.4-4 | Is |
| Alabaster | Fine grained gypsum | | | |
| Albite (P) | $Na(AlSi_3O_8)$ | 2.62 | 6 | Tr |
| Alexandrite | Chrysoberyl - gemstone | | | |
| Allanite | $(Ce,Ca,Y)(Al,Fe)_3(SiO_4)_3(OH)$ | 3.5-4.2 | 5.5-6 | Mo |
| Allemontite | AsSb | 5.8-6.2 | 3-4 | Hx |
| Allophane (C) | $Al_2O_3 \bullet SiO_2 \bullet nH_2O$ | 1.8-1.9 | 3 | Am |
| Almandite (G) | $Fe_3Al_2(SiO_4)_3$ - red | 4.25 | 7 | Is |
| Altaite | PbTe | 8.16 | 3 | Is |
| Alunite | $KAl_3(SO_4)_2(OH)_6$ | 2.6-2.8 | 4 | Rh |
| Amazonstone (F) | Green microcline | | | |
| Amblygonite | $(Li,Na)AlPO_4(F,OH)$ | 3.0-3.1 | 6 | Tr |
| Amethyst | Purple quartz | | | |
| Amphibole | A group of minerals | | | |
| Analcime (Fp) | $Na(AlSi_2O_6) \bullet H_2O$ | 2.27 | 5-5.5 | Is |
| Anatase | $TiO_2$ | 3.9 | 5.5-6 | Te |
| Anauxite | Silicon rich Kaolinite | | | |
| Andalusite | $Al_2SiO_5$ | 3.1-3.2 | 7.5 | Or |
| Andesine (P) | $Ab_{70}An_{30}$-$Ab_{50}An_{50}$ | 2.69 | 6 | Tr |
| Andradite (G) | $Ca_3Fe_2(SiO_4)_3$ | 3.75 | 7 | Is |
| Anglesite | $PbSO_4$ | 6.2-6.4 | 3 | Or |
| Anhydrite | $CaSO_4$ | 2.8-3.0 | 3-3.5 | Or |
| Ankerite | $Ca(Fe,Mg,Mn)(CO_3)_2$ | 2.9-3 | 3.5 | Rh |
| Annabergite | $(Ni,Co)_3(AsO_4)_2 \bullet 8H_2O$ | 3.0 | 3.5-3 | Mo |
| Anorthite (P) | $CaAl_2Si_2O_8$ | 2.76 | 6 | Tr |
| Anorthoclase (O) | $(Na,K)AlSi_3O_8$ | 2.58 | 6 | Tr |
| Anthophyllite (A) | $(Mg,Fe)_7(Si_8O_{22})(OH)_2$ | 2.8-3.2 | 5.5-6 | Or |
| Antigorite | Serpentine | | | |
| Antimony | Sb | 6.7 | 3 | Rh |
| Antlerite | $Cu_3SO_4(OH)_4$ | 3.9 | 3.5-4 | Or |
| Apatite | $Ca_5(PO_4,CO_3)_3(F,OH,Cl)$ | 3.1-3.2 | 5 | He |
| Apophyllite | $KCa_4Si_8O_{20}(F,OH) \bullet H_2O)$ | 2.3-2.4 | 4.5-5 | Te |
| Aquamarine | Green-blue beryl - gemstone | | | |
| Aragonite | $CaCO_3$ | 2.95 | 3.5-4 | Or |
| Arfvedsonite (A) | $Na_{2-3}(Fe,Mg,Al)_5Si_8O_{22}(OH)_2$ | 3.45 | 6 | Mo |
| Argentite | $Ag_2S$ | 7.3 | 2-2.5 | Is |
| Arsenic | As | 5.7 | 3.5 | Rh |
| Arsenopyrite | FeAsS | 5.9-6.2 | 5.5-6 | Mo |
| Asbestos | A group of minerals | | | |

| Name | Composition | Density | Hard | Sys |
|---|---|---|---|---|
| Atacamite | Cu2Cl(OH)3 | 3.7-3.8 | 3-3.5 | Or |
| Augite (P) | (Ca,Na)(Mg,Fe,Al)(Si,Al)2(O)6 | 3.2-3.4 | 5-6 | Mo |
| Aurichalcite | (Zn,Cu)5(CO3)2(OH)6 | 3.2-3.7 | 2 | Mo |
| Autunite | Ca(UO2)2((PO4)2 • 10H2O | 3.1-3.2 | 2-2.5 | Te |
| Awaruite | FeNi2 | 7.7-8.1 | 5 | Is |
| Axinite | (Ca,Mn,Fe)3Al2BSi4O15(OH) | 3.2-3.4 | 6.5-7 | Tr |
| Azurite | Cu3(CO3)2(OH)2 | 3.77 | 3.5-4 | Mo |

## – B –

| Name | Composition | Density | Hard | Sys |
|---|---|---|---|---|
| Balas ruby | Red spinel - gemstone | | | |
| Barite | BaSO4 | 4.5 | 3-3.5 | Or |
| Barytes | Barite | | | |
| Bastnaesite(R) | (Ce,La)(CO3)(F,OH) | 4.9-5.2 | 4-4.5 | He |
| Bauxite | Aluminum hydroxide mixture... | | | |
| Beidellite (C) | Al8(Si4O10)3(OH)12 • 12H2O | 2.6 | 1.5 | Or |
| Bentonite (C) | Montmorillonite clay | | | |
| Beryl | Be3Al2(Si6O18) | 2.7-2.8 | 7.5-8 | He |
| Biotite (M) | K(Mg,$Fe^{2+}$)3(Al,$Fe^{3+}$).... Si3O10(OH)2 | 2.8-3.2 | 2.5-3 | Mo |
| Bismite | Bi2O3 | 8 | 4.5 | Mo |
| Bismuth | Bi | 9.8 | 2-2.5 | Rh |
| Black Jack | Sphalerite | | | |
| Blende | Sphalerite | | | |
| Bloodstone | Heliotrope | | | |
| Blue vitriol | Chalcanthite | | | |
| Boehmite (B) | AlO(OH) | 3.0-3.1 | | Or |
| Boracite | Mg3B7O13Cl | 2.9-3 | 7 | Or |
| Borax | Na2B4O7 • 10H2O | 1.7 | 2-2.5 | Mo |
| Bornite | Cu5FeS4 | 5.0-5.1 | 3 | Is |
| Boulangerite | Pb5Sb4S11 | 6-6.3 | 2.5-3 | Or |
| Bournonite | PbCuSbS3 | 5.8-5.9 | 2.5-3 | Or |
| Brannerite | (U,Ca,Ce)(Ti,Fe)2O6 | 4.5-5.4 | 4.5 | ? |
| Braunite | 3Mn2O3 • MnSiO3 | 4.8 | 6-6.5 | Te |
| Bravoite | (Ni,Fe)S2 | 4.66 | 5.5-6 | Is |
| Brochantite | Cu4(OH)6SO4 | 3.9 | 3.5-4 | Mo |
| Bromyrite | Ag(Br,Cl) - BrCl | 6-6.5 | 2.5 | Is |
| Bronzite (E) | (Mg,Fe)SiO3 | 3.1-3.3 | 5.5 | Or |
| Brookite | TiO2 | 3.9-4.1 | 5.5-6 | Or |
| Brucite | Mg(OH)2 | 2.39 | 2.5 | Rh |
| Bytownite (P) | Ab30An70-Ab10An90 | 2.74 | 6 | Tr |

## – C –

| Name | Composition | Density | Hard | Sys |
|---|---|---|---|---|
| Cairngorm | Quartz - black to smoky | | | |
| Calamine | Hemimorphite | | | |
| Calaverite | AuTe2 | 9.35 | 2.5 | Mo |
| Calcite | CaCO3 | 2.72 | 3 | Rh |
| Californite | Idocrase - gemstone | | | |
| Calomel | Hg2Cl2 | 7.2 | 1.5 | Te |

| Name | Composition | Density | Hard | Sys |
|---|---|---|---|---|
| Cancrinite(Fp) | $(Na_2,Ca)_4(AlSiO_4)_6 \cdot CO_3 \bullet nH_2O$ | 2.45 | 5-6 | He |
| Carnallite | $KMgCl_3 \bullet 6H_2O$ | 1.6 | 1 | Or |
| Carnelian | Chalcedony - red | | | |
| Carnotite | $K(UO_2)_2(VO_4)_2 \bullet 3H_2O$ | 4.1 | soft | Or |
| Cassiterite | $SnO_2$ | 6.8-7.1 | 6-7 | Te |
| Cat's Eye | Chrysoberyl or quartz - gemstone | | | |
| Celestite | $SrSo_4$ | 3.9-4.0 | 3-3.5 | Or |
| Celsian (F) | $BaAl_2Si_2O_8$ | 3.37 | 6 | Mo |
| Cerargyrite | Ag(Cl,Br) - ClBr | 5.5-6 | 2.5 | Is |
| Cerussite | $PbCO_3$ | 6.55 | 3-3.5 | Or |
| Cervanite | $Sb_2O_4$ | 4.0-5.0 | 4-5 | Or |
| Chabazite (Z) | $Ca(Al_2Si_4O_{12}) \bullet 6H_2O$ | 2.0-2.2 | 4-5 | Rh |
| Chalcanthite | $CuSO_4 \bullet 5H_2O$ | 2.1-2.3 | 2.5 | Tr |
| Chalcedony | Cryptocrystalline quartz | 2.6-2.7 | | |
| Chalcocite | $Cu_2S$ | 5.5-5.8 | 2.5-3 | Or |
| Chalcopyrite | $CuFeS_2$ | 4.1-4.3 | 3.5-4 | Te |
| Chalcotrichite | Cuprite - fibrous | | | |
| Chalk | Calcite - fine grained | | | |
| Chalybite | Siderite | | | |
| Chert | $SiO_2$ - cryptocrystalline quartz | 2.65 | 7 | |
| Chessylite | Azurite | | | |
| Chiastolite | Andalusite | | | |
| Chloanthite | Skutterudite - nickel variety | | | |
| Chlorite | $(Mg,Fe^{2+},Fe^{3+})_6 \cdot AlSi_3O_{10}(OH)_8$ | 2.6-2.9 | 2-2.5 | Mo |
| Chloritoid (M) | $Fe_2Al_4Si_2O_{10}(OH)_4$ | 3.5 | 6-7 | Mo |
| Chondrodite | $(Mg,Fe)_3SiO_4(OH,F)_2$ | 3.1-3.2 | 6-6.5 | Mo |
| Chromite | $(Fe,Mg)O \bullet (Fe,Al,Cr)_2O_3$ | 4.3-4.6 | 5.5 | Is |
| Chrysoberyl | $BeAl_2O_4$ | 3.6-3.8 | 8.5 | Or |
| Chrysocolla | $Cu_2H_2(Si_2O_5)(OH)_4$ | 2.0-2.4 | 2-4 | ? |
| Chrysolite | Olivine | | | |
| Chrysoprase | Chalcedony - green | | | |
| Chrysotile | Serpentine asbestos | | | |
| Cinnabar | HgS | 8.10 | 2.5 | Rh |
| Cinnamon stone | Grossularite garnet | | | |
| Citrine | Quartz - pale yellow | | | |
| Clay | A group of minerals | | | |
| Cleavelandite | Albite - white | | | |
| Cliachite | Al hydroxide in bauxite | | | |
| Clinochlore | Chlorite | | | |
| Clinoclase | $Cu_3(AsO_4)(OH)_3$ | 4.38 | 2.5-3 | Mo |
| Clinoenstatite(E) | $(Mg,Fe)SiO_3$ | 3.19 | 6 | Mo |
| Clinoferrosilite(P) | $(Fe,Mg)SiO_3$ | 3.6 | 6 | Mo |
| Clinohumite | $Mg_9Si_4O_{16}(F,OH)_2$ | 3.1-3.2 | 6 | Mo |
| Clinozoisite | $Ca_2Al_3Si_3O_{12}(OH)$ | 3.2-3.4 | 6-6.5 | Mo |
| Cobaltite | CoAsS | 6.33 | 5.5 | Is |
| Colemanite | $Ca_2B_6O_{11} \bullet 5H_2O$ | 2.42 | 4-4.5 | Mo |
| Collophane | Apatite | | | |

| Name | Composition | Density | Hard | Sys |
|---|---|---|---|---|
| Columbite | $(Fe,Mn)(Nb,Ta)_2O_6$ - NbTa | 5.2-6.7 | 6 | Or |
| Copper | Cu | 8.9 | 2.5-3 | Is |
| Copper glance | Chalcocite | | | |
| Copper pyrites | Chalcopyrite | | | |
| Cordierite | $(Mg,Fe)_2Al_4Si_5O_{18}$ | 2.6-2.7 | 7-7.5 | Or |
| Corundum | $Al_2O_3$ | 4.02 | 9 | Rh |
| Covellite | CuS | 4.6-4.7 | 1.5-2 | He |
| Cristobalite | $SiO_2$ - high temp quartz | 2.3 | 7 | |
| Crocidolite | $Na_3Fe^{2+}{}_3Fe^{3+}{}_2(SiO_{23})(OH)$ | 3.2-3.3 | 4 | Mo |
| Crocoite | $PbCrO_4$ | 5.9-6.1 | 2.5-3 | Mo |
| Cryolite | $Na_3AlF_6$ | 2.9-3 | 2.5 | Mo |
| Cubanite | $CuFe_2S_3$ | 4.0-4.2 | 3.5 | Or |
| Cummingtonite (A) | $(Fe,Mg)_7(Si_8O_{22})(OH)_2$ | 3.1-3.6 | 6 | Mo |
| Cuprite | $Cu_2O$ | 6 | 3.5-4 | Is |
| Cyanite | Kyanite | | | |
| Cymophane | Chrysoberyl | | | |

## – D –

| Name | Composition | Density | Hard | Sys |
|---|---|---|---|---|
| Danaite | (Fe,Co)AsS | 5.9-6.2 | 5.5-6 | Mo |
| Danburite | $CaB_2(SiO_4)_2$ | 2.9-3.0 | 7 | Or |
| Datolite | $CaB(SiO_4)(OH)$ | 2.8-3 | 5-5.5 | Mo |
| Davidite | Brannerite - Th variety | | | |
| Demantoid (G) | Andradite garnet - green gemstone | | | |
| Diallage | Diopside | | | |
| Diamond | C | 3.5 | 10 | Is |
| Diaspore | AlO(OH) | 3.3-3.4 | 6.5-7 | Or |
| Diatomite | Diatoms - siliceous | 0.4-0.6 | 2 | |
| Dichroite | Cordierite | | | |
| Dickite (C) | $Al_2Si_2O_5(OH)_4$ - Kaoline | 2.6 | 2-2.5 | Mo |
| Digenite | $Cu_9S_5$ | 5.6 | 2.5-3 | Is |
| Diopside (P) | $CaMg(SiO_3)_2$ | 3.2-3.3 | 5-6 | Mo |
| Dioptase | $CuSiO_2(OH)_2$ | 3.3 | 5 | Rh |
| Disthene | Kyanite | | | |
| Dolomite | $CaMg(CO_3)_2$ | 2.85 | 3.5-4 | Rh |
| Dry bone ore | Smithsonite | | | |
| Dumortierite | $(Al,Fe)_7O_3(BO_3)(SiO_4)_3$ | 3.2-3.4 | 7 | Or |

## – E –

| Name | Composition | Density | Hard | Sys |
|---|---|---|---|---|
| Edenite (H) | $Ca_2NaMg_5(AlSi_7O_{22})(OH)_2$ | 3 | 6 | Mo |
| Electrum | Au,Ag - natural alloy | 13.5-17 | 3 | Is |
| Eleolite | Nepheline | | | |
| Embolite | Ag(Cl,Br) - Cl = Br | 5.6 | 1-1.5 | Is |
| Emerald | Beryl - green gemstone | | | |
| Emery | Corundum with magnetite | | | |
| Enargite | $Cu_3AsS_4$ | 4.4-4.5 | 3 | Or |
| Endlichite | Vanadinite - arsenic variety | | | |
| Enstatite (P) | $MgSiO_3$ | 3.2-3.5 | 5.5 | Or |
| Epidote | $Ca_2(Al,Fe)_3Si_3O_{12}(OH)$ | 3.3-3.5 | 6-7 | Mo |

| Name | Composition | Density | Hard | Sys |
|---|---|---|---|---|
| Epsomite | $MgSO_4 \bullet 7H_2O$ - Epsom salt | 1.75 | 2-2.5 | Or |
| Erythrite | $Co_3(AsO_4)_2 \bullet 8H_2O$ | 2.95 | 1.5-2.5 | Mo |
| Essonite (G) | Grossularite | | | |
| Euclase | $BeAlSiO_4(OH)$ | 3.1 | 7.5 | Mo |
| Eucryptite | $LiAlSiO_4$ - after spodumene | 2.67 | | He |
| Euxenite | $(Y,Ce,Ca,U,Th)_2 \ldots (Ti,Nb,Ta,Fe)_2O_6$ | 5-5.9 | 5.5-6.5 | Or |

## – F –

| Name | Composition | Density | Hard | Sys |
|---|---|---|---|---|
| Fahlore | Tetrahedrite | | | |
| Fayalite (Ov) | $Fe_2SiO_4$ | 4.14 | 6.5 | Or |
| Feather ore | Jamesonite | | | |
| Feldspar (F) | A group of minerals | | | |
| Feldspathoid | A group of minerals | | | |
| Ferberite (W) | $FeWO_4$ | 7.5 | 5 | Mo |
| Fergusonite (R) | $(RE,Fe)(Nb,Ta,Ti)O_4$ | 4.2-5.8 | 5.5-6.5 | Te |
| Ferrimolybdite | $Fe_2(MoO_4)_3 \bullet 8H_2O$ | 3 | 1.5 | Or |
| Ferrosilite (P) | $FeSiO_3$ | 3.6 | 6 | Or |
| Fibrolite | Sillimanite | | | |
| Flint | $SiO_2$ - cryptocrystalline quartz | 2.65 | 7 | |
| Flos ferri | Aragonite - arborescent | | | |
| Fluorite | $CaF_2$ | 3.18 | 4 | Is |
| Fool's gold | Pyrite | | | |
| Formanite (R) | Fergusonite with TaNb | | | |
| Forsterite (Ov) | $Mg_2SiO_4$ | 3.2 | 6.5 | Or |
| Fowlerite | Rhodonite - zinc bearing | | | |
| Franklinite | $(Fe^{2+},Zn,Mn^{2+}) \ldots (Fe^{3+},Mn^{3+})_2O_4$ | 5.15 | 6 | Is |
| Freibergite | Tetrahedrite - silver bearing | | | |

## – G –

| Name | Composition | Density | Hard | Sys |
|---|---|---|---|---|
| Gadolinite (R) | $Be_2FeY_2Si_2O_{10}$ | 4-4.5 | 6.5-7 | Mo |
| Gahnite (S) | $ZnAl_2O_4$ | 4.55 | 7.5-8 | Is |
| Galaxite (S) | $MnAl_2O_4$ | 4.03 | 7.5-8 | Is |
| Galena | PbS | 7.4-7.6 | 2.5 | Is |
| Garnet (G) | A group of minerals | 3.5-4.3 | 6.5-7.5 | Is |
| Garnierite | $(Ni,Mg)_3Si_2O_5(OH)_4$ | 2.2-2.8 | 2-3 | Am |
| Gaylussite | $Na_2Ca(CO_3)_2 \bullet 5H_2O$ | 1.99 | 2-3 | Mo |
| Gedrite (A) | Anthophyllite - Al variety | | | |
| Geocronite | $Pb_5(Sb,As)_2S_8$ | 6.3-6.5 | 2.5 | Or |
| Gersdorffite | NiAsS | 5.9 | 5.5 | Is |
| Geyserite | Opal | | | |
| Gibbsite | $Al(OH)_3$ | 2.3-2.4 | 2.5-3.5 | Mo |
| Glauberite | $Na_2Ca(SO_4)_2$ | 2.7-2.8 | 2.5-3 | Mo |
| Glaucodot | Danaite | | | |
| Glauconite (M) | $(K,Na)(Al,Fe^{3+},Mg)_2 \ldots (Al,Si)_4O_{10}(OH)_2$ | 2.3 | 2 | Mo |

| Name | Composition | Density | Hard | Sys |
|---|---|---|---|---|
| Glaucophane (A) | $Na_2(Mg,Fe^{2+})_3Al_2Si_8O_{22}(OH)_2$ | 3.0-3.2 | 6-6.5 | Mo |
| Gmelinite (Z) | $(Na_2,Ca)Al_2Si_4O_{12} \bullet 6H_2O$ | 2.0-2.2 | 4.5 | Rh |
| Goethite | $FeO(OH)$ | 4.37 | 5-5.5 | Or |
| Gold | Au | 15-19.3 | 2.5-3 | Is |
| Goslarite | $ZnSO_4 \bullet 7H_2O$ | 1.98 | 2-2.5 | Or |
| Graphite | C | 2.3 | 1-2 | He |
| Greenockite | CdS | 4.9 | 3-3.5 | He |
| Grossularite (G) | $Ca_3Al_2(SiO_4)_3$ | 3.53 | 6.5 | Is |
| Gummite | $UO_3 \bullet nH_2O$ | 3.9-6.4 | 2.5-5 | |
| Gypsum | $CaSO_4 \bullet 2H_2O$ | 2.32 | 2 | Mo |

## – H –

| Name | Composition | Density | Hard | Sys |
|---|---|---|---|---|
| Halite | NaCl - common salt | 2.16 | 2.5 | Is |
| Halloysite (C) | $Al_2Si_2O_5(OH) \bullet nH_2O$ | 2.0-2.2 | 1-2 | Am |
| Harmotome (Z) | $(Ba,K)(Al,Si)_2Si_6O_{16} \bullet 6H_2O$ | 2.45 | 4.5 | Mo |
| Hastingsite (H) | $NaCa_2(Fe,Mg)_5Al_2Si_6O_{22}(OH)_2$ | 3.2 | 6.0 | Mo |
| Hausmannite | $MnMn_2O_4$ | 4.84 | 5.5 | Te |
| Hauynite (Fp) | $(Na,Ca)_{4-8}Al_6Si_6O_{24} \bullet (SO_4,S)_{1-2}$ | 2.4-2.5 | 5.5-6 | Is |
| Hectorite (C) | $(Mg,Li)_6Si_8O_{20}(OH)_4$ | 2.5 | 1-1.5 | Mo |
| Hedenbergite (P) | $CaFe(Si_2O_6)$ | 3.55 | 5-6 | Mo |
| Heliotrope | Chalcedony - green and red | | | |
| Helvite | $(Mn,Fe,Zn)_4Be_3(SiO_4)_3S$ | 3.2-3.4 | 6-6.5 | Is |
| Hematite | $Fe_2O_3$ | 5.26 | 5.5-6.5 | Rh |
| Hemimorphite | $Zn_4(Si_2O_7)(OH)_2 \bullet H_2O$ | 3.4-3.5 | 4.5-5 | Or |
| Hercynite (S) | $FeAl_2O_4$ | 4.39 | 7.5-8 | Is |
| Hessite | $Ag_2Te$ | 8.4 | 2.5-3 | Is |
| Heulandite | $(Na,Ca)_{4-6}Al_6(Al,Si)_4Si_{26}O_{72} \bullet 24H_2O$ | 2.2 | 3.5-4 | Mo |
| Hiddenite | Spodumene - green | | | |
| Holmquisite (A) | Glaucophane - Li variety | | | |
| Hornblende | $Ca_2Na(Mg,Fe^{2+})_4(Al,Fe^{3+},Ti)AlSi_8O_{22}(O,OH)_2$ | 3.2 | 5-6 | Mo |
| Horn silver | Cerargyrite | | | |
| Huebnerite (W) | $MnWO_4$ | 7.0 | 5 | Mo |
| Humite | $Mg_7(SiO_4)_3(F,OH)_2$ | 3.1-3.2 | 6 | Or |
| Hyacinth | Zircon - brown to orange | | | |
| Hyalite | Opal - globular and colorless | | | |
| Hyalophane (O) | $(K,Ba)Al(Al,Si)_3O_8$ | 2.8 | 6 | Mo |
| Hydromica (M) | Illite | | | |
| Hydrozincite | $Zn_5(CO_3)_2(OH)_6$ | 3.6-3.8 | 2-2.5 | Mo |
| Hypersthene (P) | $(Mg,Fe)SiO_3$ | 3.4-3.5 | 5-6 | Or |

## – I –

| Name | Composition | Density | Hard | Sys |
|---|---|---|---|---|
| Ice | $H_2O$ | 0.917 | 1.5 | He |
| Iceland spar | Calcite - clear | | | |

| Name | Composition | Density | Hard | Sys |
|---|---|---|---|---|
| Iddingsite | $H_8Mg_9Fe_2Si_3O_{14}$ | 3.5-3.8 | 3 | Or |
| Idocrase | $Ca_{10}(Mg,Fe)_2Al_4(SiO_4)_5(Si_2O_7)_2(OH)_4$ | 3.3-3.4 | 6.5 | Te |
| Illite (C) | Al,K,Ca,Mg | | | |
| Ilmenite | $FeTiO_3$ | 4.7 | 5.5-6 | Rh |
| Ilvaite | $CaFe^{2+}{}_2Fe^{3+}(SiO_4)_2(OH)$ | 4.0 | 5.5-6 | Or |
| Indicolite | Tourmaline - dark blue | | | |
| Iodobromite | Ag(Cl,Br,I) | 5.7 | 1-1.5 | Is |
| Iodyrite | AgI | 5.7 | 1-1.5 | He |
| Iolite | Cordierite - gemstone | | | |
| Iridium | Ir - platinoid | 22.7 | 6-7 | Is |
| Iridosmine | Ir,Os - platinoid | 19.3-21 | 6-7 | Rh |
| Iron pyrite | Pyrite | | | |

## – J –

| Name | Composition | Density | Hard | Sys |
|---|---|---|---|---|
| Jacinth | Hyacinth, zircon | | | |
| Jacobsite (S) | $(Mn^{2+},Fe^{2+},Mg)(Fe^{3+},Mn^{3+})_2O_4$ | 5.1 | 5.5-6.5 | Is |
| Jade | Jadeite or nephrite | | | |
| Jadeite (P) | $Na(Al,Fe)Si_2O_6$ | 3.3-3.5 | 6.5-7 | Mo |
| Jamesonite | $Pb_4FeSb_6S_{14}$ | 5.5-6 | 2-3 | Mo |
| Jargon | Zircon - clear, yellow, or smoky | | | |
| Jarosite | $KFe_3(SO_4)_2(OH)_6$ | 2.9-3.3 | 3 | Rh |
| Jasper | Quartz - red cryptocrystalline | | | |

## – K –

| Name | Composition | Density | Hard | Sys |
|---|---|---|---|---|
| Kainite | $MgSO_4 \bullet KCl \bullet 3H_2O$ | 2.1 | 3 | Mo |
| Kalinite | Alum - potash variety | | | |
| Kaliophilite | $K(AlSiO_4)$ | 2.61 | 6 | He |
| Kalsilite | Nepheline series | | | |
| Kaolin group | Clay mineral family | | | |
| Kaolinite | $Al_2(Si_2O_5)(OH)_4$ | 2.6-2.7 | 2-2.5 | Mo |
| Kernite | $Na_2B_4O_7 \bullet 4H_2O$ | 1.95 | 3 | Mo |
| Krennerite | $AuTe_2$ | 8.62 | 2-3 | Or |
| Kunzite | Spodumene - pink | | | |
| Kyanite | $Al_2SiO_5$ | 3.6-3.7 | 5-7 | Tr |

## – L –

| Name | Composition | Density | Hard | Sys |
|---|---|---|---|---|
| Labradorite (P) | $Ab_{50}An_{50}$-$Ab_{30}An_{70}$ | 2.71 | 6 | Tr |
| Langbeinite | $K_2Mg_2(SO_4)_3$ | 2.83 | 2.5-3.5 | Is |
| Lapis lazuli | Lazurite - impure | | | |
| Larsenite (Ov) | $PbZnSiO_4$ | 5.9 | 3 | Or |
| Laumontite (Z) | $(Ca,Na)Al_2Si_4O_{12} \bullet 4H_2O$ | 2.28 | 4 | Mo |
| Lawsonite | $CaAl_2(Si_2O_7)(OH)_2 \bullet H_2O$ | 3.09 | 8 | Or |
| Lazulite | $(Mg,Fe^{3+})Al_2(PO_4)_2(OH)_2$ | 3.0-3.1 | 5-5.5 | Mo |
| Lazurite | $(Na,Ca)_4(AlSiO_4)_3(SO_4,S,Cl)$ | 2.4-2.5 | 5-5.5 | Is |
| Lechatelierite | $SiO_2$ - fused silica | 2.2 | 6-7 | Am |
| Lepidocrocite | $FeO(OH)$ | 4.09 | 5 | Or |

| Name | Composition | Density | Hard | Sys |
|---|---|---|---|---|
| Lepidolite (M) | $K(Li,Al)_3(Si,Al)_4O_{10}(F,OH)_2$ | 2.8-3 | 2.5-4 | Mo |
| Leucite (Fp) | $K(AlSi_2O_6)$ | 2.4-2.5 | 5.5-6 | Is |
| Libethenite | $Cu_2(PO_4)(OH)$ | 4 | 4 | Or |
| Limonite | $FeO(OH) \bullet nH_2O$ | 3.6-4 | 5-5.5 | Am |
| Linarite | $PbCu(SO_4)(OH)_2$ | 5.3 | 2.5 | Mo |
| Linnaeite | $Co_3S_4$ | 4.8 | 4.5-5.5 | Is |
| Lithium mica | Lepidolite | | | |
| Lithiophilite | $Li(Mn^{2+},Fe^{2+})PO_4$ | 3.5 | 5 | Or |
| Loellingite | $FeAs_2$ | 7.4-7.5 | 5-5.5 | Or |

## – M –

| Name | Composition | Density | Hard | Sys |
|---|---|---|---|---|
| Magnesite | $MgCO_3$ | 3.0-3.2 | 3.5-5 | Rh |
| Magnetite (S) | $(Fe,Mg)Fe_2O_4$ | 5.18 | 6 | Is |
| Malachite | $Cu_2CO_3(OH)_2$ | 3.9-4.0 | 3.5-4 | Mo |
| Manganite | $MnO(OH)$ | 4.3 | 4 | Or |
| Manganosite | $MnO$ | 5.0-5.4 | 5.5 | Is |
| Marcasite | $FeS_2$ - white iron pyrite | 4.89 | 6-6.5 | Or |
| Margarite (M) | $CaAl_2(Al_2Si_2O_{10})(OH)_2$ | 3.0-3.1 | 3.5-5 | Mo |
| Marialite (Sc) | $3NaAlSi_3O_8 \bullet NaCl$ | 2.7 | 5.5-6 | Te |
| Marmatite | Sphalerite - iron bearing | 3.9-4 | | |
| Martite | Hematite after magnetite | | | |
| Meerschaum | Sepiolite | | | |
| Meionite (Sc) | $3CaAl_2Si_2O_8 \bullet CaCO_3$ | 2.7 | 5.5-6 | Te |
| Melaconite | Tenorite | | | |
| Melanite (G) | Andradite garnet - black | | | |
| Melanterite | $FeSO_4 \bullet 7H_2O$ | 1.90 | 2 | Mo |
| Melilite | $(Na,Ca)_2(Mg,Al)(Si,Al)_2O_7$ | 2.9-3.1 | 5 | Te |
| Menaccanite | Ilmenite | | | |
| Menaghinite (J) | $CuPb_{13}Sb_7S_{24}$ | 6.36 | 2.5 | Or |
| Mercury | Hg - fluid, quicksilver | 13.6 | | |
| Miargyrite | $AgSbS_2$ | 5.2-5.3 | 2.5 | Mo |
| Mica (M) | A group of minerals | | | |
| Microcline (F) | $K(AlSi_3O_8)$ - k feldspar | 2.5-2.6 | 6 | Tr |
| Microlite | $(Na,Ca)_2(Ta,Nb)_2O_6$ | 6.33 | 5.5 | Is |
| Microperthite (F) | Microcline and albite | | | |
| Millerite | $NiS$ | 5.3-5.7 | 3-3.5 | Rh |
| Mimetite | $Pb_5Cl(AsO_4)_3$ | 7.0-7.2 | 3.5 | He |
| Minium | $Pb_3O_4$ | 8.9-9.2 | 2.5 | |
| Mispickel | Arsenopyrite | | | |
| Molybdenite | $MoS_2$ | 4.6-4.7 | 1-1.5 | He |
| Monazite | $(Ce,La,Y,Th)(PO_4,SiO_4)$ | 5.0-5.3 | 5-5.5 | Mo |
| Monticellite | $CaMgSiO_4$ - rare olivine | 3.2 | 5 | Or |
| Montmorillonite (C) | $(Al,Mg)_8(Si_4O_{10})_3$ . . . $(OH)_{10} \bullet 10H_2O$ | 2.5 | 1-1.5 | Mo |
| Moonstone (O) | Opalescent albite or orthoclase | | | |
| Morganite | Beryl - rose color | | | |
| Mullite | $Al_6Si_2O_{13}$ | 3.23 | 6-7 | Or |
| Muscovite (M) | $KAl_2(AlSi_3O_{10})(OH)_2$ | 2.7-3.1 | 2-2.5 | Mo |

| Name | Composition | Density | Hard | Sys |
|---|---|---|---|---|
| **–N–** | | | | |
| Nacrite (C) | $Al_2(Si_2O_5)(OH)_2$ -kaolin group | 2.6 | 2-2.5 | Mo |
| Nagyagite | $Pb_5Au(Te,Sb)_4S_{5-8}$ | 7.4 | 1-1.5 | Mo |
| Natroalunite | Alunite with NaK | | | |
| Natrolite (Z) | $Na_2(Al_2Si_3O_{10}) \bullet 2H_2O$ | 2.25 | 5-5.5 | Mo |
| Nepheline (Fp) | $(Na,K)AlSiO_4$ | 2.5-2.7 | 5.5-6 | He |
| Nephrite | Tremolite, similar to jade | | | |
| Niccolite | NiAs | 7.78 | 5-5.5 | Ne |
| Nickel bloom | Annabergite | | | |
| Nickel iron | Ni,Fe - Meteorite alloy | 7.8-8.2 | 5 | Is |
| Ni skutterudite | $(Ni,CO,Fe)As_3$ | 6.1-6.9 | 5.5-6 | Is |
| Nitre | $KNO_3$ - saltpeter | 2-2.1 | 2 | Or |
| Nontronite (C) | $Fe(AlSi)_8O_{20}(OH)_4$ | 2.5 | 1-1.5 | Mo |
| Norbergite | $Mg_3(SiO_4)(F,OH)_2$ | 3.1-3.2 | 6 | Or |
| Noselite (Fp) | $Na_8Al_6Si_6O_{24}(SO_4)$ - | 2.2-2.4 | 6 | Is |
| **–O–** | | | | |
| Octahedrite | Anatase | | | |
| Oligoclase (P) | $Ab_{90}An_{10}$-$Ab_{70}An_{30}$ | 2.65 | 6 | Tr |
| Olivine (Ov) | $(Mg,Fe)_2SiO_4$ | 3.3-4.4 | 6.5-7 | Or |
| Onyx | Chalcedony - layered structure | | | |
| Opal | $SiO_2 \bullet nH_2O$ | 1.9-2.2 | 5-6 | Am |
| Orpiment | $As_2S_3$ | 3.49 | 1.5-2 | Mo |
| Orthite | Allanite | | | |
| Orthoclase (F) | $K(AlSi_3O_8)$ - K feldspar | 2.57 | 6 | Mo |
| Osmiridium | Iridosmine | | | |
| Ottrelite (M) | $(Fe^{2+},Mn)(Al,Fe^{3+})$ . . . . $Si_3O_{10} \bullet H_2O$ | 3.5 | 6-7 | Mo |
| **–P–** | | | | |
| Palladium | Pd | 11.9 | 4.5-5 | Is |
| Paragonite (M) | $NaAl_2(AlSi_3O_{10})(OH)_2$ | 2.85 | 2 | Mo |
| Pargasite (H) | $NaCa_2Mg_4Al_3Si_6O_{22}(OH)_2$ | 3-3.5 | 5.5 | Mo |
| Peacock ore | Bornite | | | |
| Pearceite | $(Ag,Cu)_{16}As_2S_{11}$ | 6.15 | 3 | Mo |
| Pectolite | $NaCa_2Si_3O_8(OH)$ | 2.7-2.8 | 5 | Tr |
| Penninite | Chlorite | | | |
| Pentlandite | $(Fe,Ni)_9S_8$ | 4.6-5 | 3.5-4 | Is |
| Peridot (Ov) | Olivine - gemstone | | | |
| Perovskite | $CaTiO_3$ | 4.03 | 5.5 | Is |
| Perthite (F) | Microcline and albite mix | | | |
| Petalite (Fp) | $Li(AlSi_4O_{10})$ | 2.4 | 6-6.5 | Mo |
| Petzite | $Ag_3AuTe_2$ | 8.7-9 | 2.5-3 | Is |
| Phenacite | $Be_2SiO_4$ | 2.9-3.0 | 7.5-8 | Rh |
| Phillipsite (Z) | $(K_2,Na_2,Ca)Al_2$ . . . . $Si_4O_{12} \bullet 4.5H_2O$ | 2.2 | 4.5-5 | Mo |
| Phlogopite (M) | $K(Mg,Fe)_3AlSi_3O_{10}(OH,F)_2$ | 2.86 | 2.5-3 | Mo |

| Name | Composition | Density | Hard | Sys |
|---|---|---|---|---|
| Phosgenite | $Pb_2Cl_2CO_3$ | 6.0-6.3 | 3 | Te |
| Phosphuranylite | $Ca(UO_2)_4(PO_4)_2 \ldots (OH)_4 \bullet 7H_2O$ | ? | 2.5 | Te |
| Picotite (S) | Spinel - chromium | | | |
| Piedmontite | Epidote - $Mn^{2+}$ | 3.4 | 6.5 | Mo |
| Pigeonite (P) | $(Ca,Mg,Fe)SiO_3$ | 3.2-3.4 | 5-6 | Mo |
| Pinite (M) | Muscovite mica | | | |
| Pitchblende | Uraninite | | | |
| Plagioclase (P) | A group of Al silicate minerals | | | |
| Plagionite (J) | $Pb_5Sb_8S_{17}$ - | 5.56 | 2.5 | Mo |
| Platinum | Platinum metal alloy | 14-19 | 4-4.5 | Is |
| Pleonaste (S) | Spinel - iron | | | |
| Plumbago | Graphite | | | |
| Polianite | $MnO_2$ - pyrolusite | 5.0 | 6-6.5 | Te |
| Pollucite | $(Cs,Na)_2Al_2Si_4O_{12} \bullet H_2O$ | 2.9 | 6.5 | Is |
| Polybasite | $(Ag,Cu)_{16}Sb_2S_{11}$ | 6.0-6.2 | 2-3 | Mo |
| Polycrase (R) | Y,Ce,Ca,U,Th,Ti, . . . . Nb,Ta,Fe oxide | 4.7-5.9 | 5.5-6.5 | Or |
| Polyhalite | $K_2Ca_2Mg(SO_4)_4 \bullet 2H_2O$ | 2.78 | 2.5-3 | Tr |
| Potash Alum | $KAl(SO_5)_2 \bullet 11H_2O$ | 1.75 | 2-2.5 | Is |
| Potassium Feld | $KAlSi_3O_8$ - see orthoclase | | | |
| Potash mica (M) | Muscovite | | | |
| Powellite | $CaMoO_4$ | 4.23 | 3.5-4 | Te |
| Prase | Jasper - green | | | |
| Prehnite | $Ca_2Al_2(Si_3O_{10})(OH)_2$ | 2.8-2.9 | 6-6.5 | Or |
| Prochlorite | Chlorite group | | | |
| Proustite | $Ag_3AsS_3$ | 5.55 | 2-2.5 | Rh |
| Psilomelane | Manganese mineral group - massive | | | |
| Pyrargyrite | $Ag_3SbS_3$ | 5.85 | 2.5 | Rh |
| Pyrite | $FeS_2$ | 5.02 | 6-6.5 | Is |
| Pyrochlore | $(Na,Ca)_2(Nb,Ta)_2O_6(OH,F)$ | 4.2-4.5 | 5 | Is |
| Pyrolusite | $MnO_2$ | 4.75 | 1-2 | Te |
| Pyromorphite | $Pb_5(PO_4)_3Cl$ | 6.5-7.1 | 3.5-4 | He |
| Pyrope (G) | $(Mg,Fe)_3Al_2(SiO_4)_3$ | 3.51 | 7 | Is |
| Pyrophyllite | $AlSi_2)_5(OH)$ | 2.8-2.9 | 1-2 | Mo |
| Pyroxene (P) | A group of minerals | | | |
| Pyrrhotite | $Fe_{1-x}S$ where x = 0 to 0.2 | 4.6 | 4 | He |

## – Q –

| Name | Composition | Density | Hard | Sys |
|---|---|---|---|---|
| Quartz | $SiO_2$ | 2.65 | 7 | Rh |

## – R –

| Name | Composition | Density | Hard | Sys |
|---|---|---|---|---|
| Rammelsbergite | $NiAs_2$ | 7.1 | 5.5-6 | Or |
| Rasorite | Kernite | | | |
| Realgar | AsS | 3.48 | 1.5-2 | Mo |
| Red ochre | Hematite | | | |
| Rhodochrosite | $MnCO_3$ | 3.5-3.6 | 3.5-4.5 | Rh |
| Rhodolite (G) | $3(Mg,Fe)O,Al_2O_3 \bullet 3SiO_2$ | 3.84 | 7 | Is |

| Name | Composition | Density | Hard | Sys |
|---|---|---|---|---|
| Rhodonite | $MnSiO_3$ | 3.6-3.7 | 5.5-6 | Tr |
| Riebeckite (A) | $Na_2(Fe,Mg)_5Si_8O_{22}(OH)_2$ | 3.44 | 4 | Mo |
| Rock salt | Halite | | | |
| Roscoelite (M) | $K(U,Al,Mg)_3Si_3O_{10}(OH)_2$ | 2.97 | 2.5 | Mo |
| Rubellite | Tourmaline - red or pink | | | |
| Ruby | Corundum - red, gemstone | | | |
| Ruby copper | Cuprite | | | |
| Ruby silver | Pyrargyrite or proustite | | | |
| Rutile | $TiO_2$ | 4.2-4.3 | 6-6.5 | Te |

## – S –

| Name | Composition | Density | Hard | Sys |
|---|---|---|---|---|
| Samarskite | $(RE,U,Ca,Fe,Pb,Th,Nb,Ta,Ti,Sn)O_6$ | 4.1-6.2 | 5-6 | Or |
| Sanadine (O) | Orthoclase - high temperature | | | |
| Saponite (C) | $(Mg,Al)_6(Si,Al)_8O_{20}(OH)_4$ | 2.5 | 1-1.5 | Mo |
| Sapphire | Corundum - blue, gemstone | | | |
| Satin spar | Gypsum - fibrous | | | |
| Scapolite | $(Na\ or\ Ca)_4Al_3(Al,Si)_3Si_6O_{24}(Cl,CO_3,SO_4)$ | 2.6-2.7 | 5-6 | Te |
| Scheelite | $CaWO_4$ | 5.9-6.1 | 4.5-5 | Te |
| Schorlite | Tourmaline - black | | | |
| Scolecite (Z) | $Ca(Al_2Si_3O_{10}) \bullet 3H_2O$ | 2.2-2.4 | 5-5.5 | Mo |
| Scorodite | $FeAsO_4 \bullet 2H_2O$ | 3.1-3.3 | 3.5-4 | Or |
| Scorzalite | $(Fe,Mg)Al_2(PO_4)_2(OH)_2$ | 3.35 | 5.5-6 | Mo |
| Selenite | Gypsum - clear, crystalline | | | |
| Semseyite | $Pb_9Sb_8S_{21}$ | 5.8 | 2.5 | Mo |
| Sepiolite | $Mg_4(Si_2O_5)_3(OH)_2 \bullet 6H_2O$ - Meerschaum | 2.0 | 2-2.5 | Mo |
| Sericite (M) | Muscovite mica - fine grained | | | |
| Serpentine | $(Mg,Fe)_3Si_2O_5(OH)_4$ | 2.2 | 2-5 | Mo |
| Siderite | $FeCO_3$ | 3.8-3.9 | 3.5-4 | Rh |
| Siegenite | $(CO,Ni)_3S_4$ - Linnaeite series | 4.8 | 4.5-5.5 | Is |
| Sillimanite | $Al_2SiO_5$ | 3.2 | 6-7 | Or |
| Silver | Ag | 10.5 | 2.5-3 | Is |
| Silver glance | Argentite | | | |
| Sklodowskite | $Mg(UO_2)_2Si_2O_7 \bullet 6H_2O$ | 3.54 | | Or |
| Skutterudite | $(Co,Ni,Fe)As_3$ | 6.1-6.9 | 5 | Is |
| Smaltite | Skutterudite variety | | | |
| Smithsonite | $ZnCO_3$ | 4.3-4.4 | 5 | Rh |
| Soapstone | Talc | | | |
| Sodalite (Fp) | $Na_4Al_3Si_3O_{12}Cl$ | 2.2-2.3 | 5.5-6 | Is |
| Soda nitre | $NaNO_3$ | 2.29 | 1-2 | Rh |
| Specular iron | Hematite - foliated | | | |
| Sperrylite | $PtAs_2$ | 10.5 | 6-7 | Is |
| Spessartite (G) | $Mn_3Al_2(SiO_4)_3$ - red,brown | 4.18 | 7 | Is |
| Sphalerite | $(Zn,Fe)S$ | 3.9-4.1 | 3.5-4 | Is |
| Sphene | $CaTiO(SiO_4)$ | 3.4-3.5 | 5-5.5 | Mo |
| Spinel group | $(Mg,Fe,Zn,Mn)Al_2O_4$ | 3.6-4 | 8 | Is |

| Name | Composition | Density | Hard | Sys |
|---|---|---|---|---|
| Spodumene (P) | $LiAl(Si_2O_6)$ | 3.1-3.2 | 6.5-7 | Mo |
| Stannite | $Cu_2FeSnS_4$ | 4.4 | 4 | Te |
| Staurolite | $(Fe,Mg)_2Al_2Si_4O_{23}(OH)$ | 3.6-3.8 | 7-7.5 | Or |
| Steatite | Talc | | | |
| Stephanite | $Ag_5SbS_4$ | 6.2-6.3 | 2-2.5 | Or |
| Sternbergite | $AgFe_2S_3$ | 4.1-4.2 | 1-1.5 | Or |
| Stibnite | $Sb_2S_3$ | 4.5-4.6 | 2 | Or |
| Stilbite (Z) | $NaCa_2Al_5Si_3O_{36} \bullet 14H_2O$ | 2.1-2.2 | 3.5-4 | Mo |
| Stillwellite | $(Ce,La,Ca)BSiO_5$ | 4.57 | | Rh |
| Stolzite | $PbWO_4$ | 8.3-8.4 | 2.5-3 | Te |
| Stromeyerite | $(Cu,Ag)S$ | 6.2-6.3 | 2.5-3 | Or |
| Strontianite | $SrCO_3$ | 3.7 | 3.5-4 | Or |
| Sulphur | S | 2-2.1 | 1.5-2.5 | Or |
| Sunstone (F) | Oligoclase, translucent | | | |
| Sylvanite | $(Au,Ag)Te_2$ | 8-8.2 | 1.5-2 | Mo |
| Sylvite | KCl | 1.99 | 2 | Is |

## – T –

| Name | Composition | Density | Hard | Sys |
|---|---|---|---|---|
| Talc | $Mg_3(Si_4O_{10})(OH)_2$ | 2.7-2.8 | 1 | Mo |
| Tantalite | $(Fe,Mn)(Ta,Nb)_2O_6$, TaNb | 6.2-8 | 6-6.5 | Or |
| Tennantite | $(Cu,Fe,Zn,Ag)_{12}As_4S_{13}$ | 4.6-5.1 | 3-4.5 | Is |
| Tenorite | CuO | 5.8-6.4 | 3-4 | Tr |
| Tephroite (Ov) | $Mn_2(SiO_4)$ | 4.1 | 6 | Or |
| Tetrahedrite | $(Cu,Fe,Zn,Ag)_{12}Sb_4S_{13}$ | 4.6-5.1 | 3-4.5 | Is |
| Thenardite | $Na_2SO_4$ | 2.68 | 2.5 | Or |
| Thomsonite (Z) | $NaCa_2Al_5Si_5O_{20} \bullet 6H_2O$ | 2.3 | 5 | Or |
| Thorianite | $ThO_2$ | 9.7 | 6.5 | Is |
| Thorite | $Th(SiO_4)$ | 5.3 | 5 | Te |
| Thulite | Zoisite, pink to red | | | |
| Tiger's eye | Quartz after crocidolite, yellow | | | |
| Tin | Sn | 7.3 | 2 | Te |
| Tinstone | Cassiterite | | | |
| Titanite | Sphene | | | |
| Topaz | $Al_2(SiO_4)(F,OH)_2$ | 3.4-3.6 | 8 | Or |
| Torbernite | $Cu(UO_2)_2(PO_4)_2 \bullet 8H_2O$ | 3.22 | 2-2.5 | Te |
| Tourmaline | $(Na,Ca)(Al,Fe,Li,Mg)_3Al_6 \ldots (BO_3)_3(Si_6O_{18})(OH)_4$ | 3-3.2 | 7-7.5 | Rh |
| Tremolite (A) | $Ca_2Mg_5(Si_8O_{22})(OH)_2$ | 3-3.3 | 5-6 | Mo |
| Tridymite | $SiO_2$ | 2.26 | 7 | Or |
| Triphylite | $Li(Fe,Mn)PO_4$ | 3.4-3.6 | 4.5-5 | Or |
| Troilite | Pyrrhotite | | | |
| Trona | $Na_2CO_3 \bullet NaHCO_3 \bullet 2H_2O$ | 2.13 | 3 | Mo |
| Troostite | Willemite, manganiferous | | | |
| Tungstite | $WO_3 \bullet nH_2O$ | | 2.5 | Or |
| Turgite | $2Fe_2O_3 \bullet nH_2O$ | 4.2-4.6 | 6.5 | ? |
| Turquoise | $CuAl_6(PO_4)_4(OH)_8 \bullet 5H_2O$ | 2.6-2.8 | 6 | Tr |
| Tyuyamunite | $Ca(UO_2)_2(VO_4)_2 \bullet 5H_2O$ | 3.7-4.3 | 2 | Or |

| Name | Composition | Density | Hard | Sys |
|---|---|---|---|---|
| **– U –** | | | | |
| Ulexite | $NaCaB_5O_9 \bullet 8H_2O$ | 1.96 | 1 | Tr |
| Uralite (H) | Hornblende after pyroxene | | | |
| Uraninite | $UO_2$ to $UO_3$ | 9-9.7 | 5.5 | Is |
| Uranophane | $Ca(UO_2)_2Si_2O_7 \bullet 6H_2O$ | 3.8-3.9 | 2-3 | Or |
| Uvarovite (G) | $Ca_3Cr_2(SiO_4)_3$ green | 3.45 | 7.5 | Is |
| **– V –** | | | | |
| Vanadinite | $Pb_5(VO_4)_3Cl$ | 6.7-7.1 | 3 | He |
| Variscite | $Al(PO_4) \bullet 2H_2O$ | 2.4-2.6 | 3.5-4.5 | Or |
| Vermiculite (M) | Biotite, altered | 2.4 | 1.5 | Mo |
| Vesuvianite | Idocrase | | | |
| Violarite | $Ni_2FeS_4$ | 4.8 | 4.5-5.5 | Is |
| Vivianite | $Fe_3(PO_4)_2 \bullet 8H_2O$ | 2.6-2.7 | 1.5-2 | Mo |
| **– W –** | | | | |
| Wad | Manganese oxides | | | |
| Wavellite | $Al_3(OH)_3(PO_4)_2 \bullet 5H_2O$ | 2.33 | 3.5-4 | Or |
| Wernerite (Sc) | Scapolite | | | |
| White pyrite | Marcasite | | | |
| White mica (M) | Muscovite | | | |
| Willemite | $Zn_2SiO_4$ | 3.9-4.2 | 5.5 | Rh |
| Witherite | $BaCO_3$ | 4.3 | 3.5 | Or |
| Wolframite | $(Fe,Mn)WO_4$ | 7-7.5 | 5-5.5 | Mo |
| Wollastonite | $Ca(SiO_3)$ | 2.8-2.9 | 5-5.5 | Tr |
| Wood tin | Cassiterite | | | |
| Wulfenite | $PbMoO_4$ | 6.5-7.5 | 3 | Te |
| Wurtzite | $(Zn,Fe)S$ | 4 | 4 | He |
| **– X –** | | | | |
| Xenotime | $YPO_4$ | 4.4-5.1 | 4-5 | Te |
| **– Z –** | | | | |
| Zeolite | A group of minerals | | | |
| Zincite | $ZnO$ | 5.68 | 4-4.5 | He |
| Zinc spinel | Gahnite | | | |
| Zinkenite (J) | $Pb_6Sb_{14}S_{27}$ | 5.3 | 3-3.5 | He |
| Zinnwaldite | Fe,Li mica | 3 | 2.5-3 | Mo |
| Zircon | $ZrSiO_4$ | 4.68 | 7.5 | Te |
| Zoisite | $Ca_2Al_3Si_3O_{12}(OH)$ | 3.3 | 6 | Or |

# ELEMENT–OXIDE CONVERSION

| Element | Multiply by | To get this Oxide |
|---|---|---|
| Al | 1.889 | $Al_2O_3$ |
| As | 1.320 | $As_2O_3$ |
| As | 1.534 | $As_2O_5$ |
| B | 3.220 | $B_2O_3$ |
| Ba | 1.117 | $BaO$ |
| Be | 2.775 | $BeO$ |
| Bi | 1.115 | $Bi_2O_3$ |
| Ca | 1.399 | $CaO$ |
| Ce | 1.171 | $Ce_2O_3$ |
| Co | 1.271 | $CoO$ |
| Cr | 1.462 | $Cr_2O_3$ |
| Cs | 1.060 | $Cs_2O$ |
| Cu | 1.252 | $CuO$ |
| F | 2.055 | $CaF_2$ |
| Fe | 1.430 | $Fe_2O_3$ |
| Fe | 1.382 | $Fe_3O_4$ |
| Fe | 1.286 | $FeO$ |
| K | 1.205 | $K_2O$ |
| La | 1.173 | $La_2O_3$ |
| Mg | 1.658 | $MgO$ |
| Mn | 1.291 | $MnO$ |
| Mn | 1.582 | $MnO_2$ |
| Mo | 1.500 | $MoO_3$ |
| Na | 1.348 | $Na_2O$ |
| Nb | 1.431 | $Nb_2O_5$ |
| Ni | 1.273 | $NiO$ |
| P | 2.291 | $P_2O_5$ |
| Pb | 1.077 | $PbO$ |
| Rb | 1.094 | $Rb_2O$ |
| Sb | 1.197 | $Sb_2O_3$ |
| Si | 2.139 | $SiO_2$ |
| Sn | 1.270 | $SnO_2$ |
| Sr | 1.183 | $SrO$ |
| Ta | 1.221 | $Ta_2O_5$ |
| Th | 1.138 | $ThO_2$ |
| Ti | 1.668 | $TiO_2$ |
| U | 1.179 | $U_3O_8$ |
| U | 1.202 | $UO_3$ |
| U | 1.134 | $UO_2$ |
| V | 1.785 | $V_2O_5$ |
| W | 1.261 | $WO_3$ |
| Y | 1.270 | $Y_2O_3$ |
| Zn | 1.245 | $ZnO$ |
| Zr | 1.351 | $ZrO_2$ |

# MINERALS SORTED BY DENSITY

| Name | Density | Name | Density |
|---|---|---|---|
| Diatomite | 0.4-0.6 | Hauynite (Fp) | 2.4-2.5 |
| Ice | 0.917 | Lazurite | 2.4-2.5 |
| Carnallite | 1.6 | Leucite (Fp) | 2.4-2.5 |
| Borax | 1.7 | Petalite (Fp) | 2.4 |
| Epsomite | 1.75 | Variscite | 2.4-2.6 |
| Potash Alum | 1.75 | Vermiculite (M) | 2.4 |
| Allophane (C) | 1.8-1.9 | Colemanite | 2.42 |
| Melanterite | 1.90 | Cancrinite(Fp) | 2.45 |
| Opal | 1.9-2.2 | Harmotome (Z) | 2.45 |
| Kernite | 1.95 | Hectorite (C) | 2.5 |
| Ulexite | 1.96 | Microcline (F) | 2.5-2.6 |
| Goslarite | 1.98 | Montmorillonite | 2.5 |
| Gaylussite | 1.99 | Nepheline (Fp) | 2.5-2.7 |
| Sylvite | 1.99 | Nontronite (C) | 2.5 |
| Chabazite (Z) | 2.0-2.2 | Saponite (C) | 2.5 |
| Chrysocolla | 2.0-2.4 | Orthoclase (F) | 2.57 |
| Gmelinite (Z) | 2.0-2.2 | Anorthoclase (O) | 2.58 |
| Halloysite (C) | 2.0-2.2 | Alunite | 2.6-2.8 |
| Nitre | 2-2.1 | Beidellite (C) | 2.6 |
| Sepiolite | 2.0 | Chalcedony | 2.6-2.7 |
| Sulphur | 2-2.1 | Chlorite | 2.6-2.9 |
| Chalcanthite | 2.1-2.3 | Cordierite | 2.6-2.7 |
| Kainite | 2.1 | Dickite (C) | 2.6 |
| Stilbite (Z) | 2.1-2.2 | Kaolinite | 2.6-2.7 |
| Trona | 2.13 | Nacrite (C) | 2.6 |
| Halite | 2.16 | Scapolite | 2.6-2.7 |
| Garnierite | 2.2-2.8 | Turquoise | 2.6-2.8 |
| Heulandite | 2.2 | Vivianite | 2.6-2.7 |
| Lechatelierite | 2.2 | Kaliophilite | 2.61 |
| Noselite (Fp) | 2.2-2.4 | Albite (P) | 2.62 |
| Phillipsite (Z) | 2.2 | Chert | 2.65 |
| Scolecite (Z) | 2.2-2.4 | Flint | 2.65 |
| Serpentine | 2.2 | Oligoclase (P) | 2.65 |
| Sodalite (Fp) | 2.2-2.3 | Quartz | 2.65 |
| Natrolite (Z) | 2.25 | Eucryptite | 2.67 |
| Tridymite | 2.26 | Thenardite | 2.68 |
| Analcime (Fp) | 2.27 | Andesine (P) | 2.69 |
| Laumontite (Z) | 2.28 | Beryl | 2.7-2.8 |
| Soda nitre | 2.29 | Glauberite | 2.7-2.8 |
| Apophyllite | 2.3-2.4 | Marialite (Sc) | 2.7 |
| Cristobalite | 2.3 | Meionite (Sc) | 2.7 |
| Gibbsite | 2.3-2.4 | Muscovite (M) | 2.7-3.1 |
| Glauconite (M) | 2.3 | Pectolite | 2.7-2.8 |
| Graphite | 2.3 | Talc | 2.7-2.8 |
| Thomsonite (Z) | 2.3 | Labradorite (P) | 2.71 |
| Gypsum | 2.32 | Calcite | 2.72 |
| Wavellite | 2.33 | Bytownite (P) | 2.74 |
| Brucite | 2.39 | Anorthite (P) | 2.76 |

# MINERALS SORTED BY DENSITY

| Name | Density |
|---|---|
| Polyhalite | 2.78 |
| Anhydrite | 2.8-3.0 |
| Anthophyllite(A) | 2.8-3.2 |
| Biotite (M) | 2.8-3.2 |
| Datolite | 2.8-3 |
| Hyalophane (O) | 2.8 |
| Lepidolite (M) | 2.8-3 |
| Prehnite | 2.8-2.9 |
| Pyrophyllite | 2.8-2.9 |
| Wollastonite | 2.8-2.9 |
| Langbeinite | 2.83 |
| Dolomite | 2.85 |
| Paragonite (M) | 2.85 |
| Phlogopite (M) | 2.86 |
| Ankerite | 2.9-3 |
| Boracite | 2.9-3 |
| Cryolite | 2.9-3 |
| Danburite | 2.9-3.0 |
| Jarosite | 2.9-3.3 |
| Melilite | 2.9-3.1 |
| Phenacite | 2.9-3.0 |
| Pollucite | 2.9 |
| Aragonite | 2.95 |
| Erythrite | 2.95 |
| Roscoelite (M) | 2.97 |
| Actinolite (A) | 3.0-3.2 |
| Amblygonite | 3.0-3.1 |
| Annabergite | 3.0 |
| Boehmite (B) | 3.0-3.1 |
| Edenite (H) | 3 |
| Ferrimolybdite | 3 |
| Glaucophane (A) | 3.0-3.2 |
| Lazulite | 3.0-3.1 |
| Magnesite | 3.0-3.2 |
| Margarite (M) | 3.0-3.1 |
| Pargasite (H) | 3-3.5 |
| Tourmaline | 3-3.2 |
| Tremolite (A) | 3-3.3 |
| Zinnwaldite | 3 |
| Lawsonite | 3.09 |
| Andalusite | 3.1-3.2 |
| Apatite | 3.1-3.2 |
| Autunite | 3.1-3.2 |
| Bronzite (E) | 3.1-3.3 |
| Chondrodite | 3.1-3.2 |
| Clinohumite | 3.1-3.2 |
| Cummingtonite | 3.1-3.6 |
| Euclase | 3.1 |
| Humite | 3.1-3.2 |
| Norbergite | 3.1-3.2 |
| Scorodite | 3.1-3.3 |
| Spodumene (P) | 3.1-3.2 |
| Fluorite | 3.18 |
| Clinoenstatite(E) | 3.19 |
| Augite (P) | 3.2-3.4 |
| Aurichalcite | 3.2-3.7 |
| Axinite | 3.2-3.4 |
| Clinozoisite | 3.2-3.4 |
| Crocidolite | 3.2-3.3 |
| Diopside (P) | 3.2-3.3 |
| Dumortierite | 3.2-3.4 |
| Enstatite (P) | 3.2-3.5 |
| Forsterite (Ov) | 3.2 |
| Hastingsite (H) | 3.2 |
| Helvite | 3.2-3.4 |
| Hornblende | 3.2 |
| Monticellite | 3.2 |
| Pigeonite (P) | 3.2-3.4 |
| Sillimanite | 3.2 |
| Torbernite | 3.22 |
| Mullite | 3.23 |
| Diaspore | 3.3-3.4 |
| Dioptase | 3.3 |
| Epidote | 3.3-3.5 |
| Idocrase | 3.3-3.4 |
| Jadeite (P) | 3.3-3.5 |
| Olivine (Ov) | 3.3-4.4 |
| Zoisite | 3.3 |
| Scorzalite | 3.35 |
| Celsian (F) | 3.37 |
| Acmite (P) | 3.4-3.6 |
| Hemimorphite | 3.4-3.5 |
| Hypersthene (P) | 3.4-3.5 |
| Piedmontite | 3.4 |
| Sphene | 3.4-3.5 |
| Topaz | 3.4-3.6 |
| Triphylite | 3.4-3.6 |
| Riebeckite (A) | 3.44 |
| Arfvedsonite(A) | 3.45 |
| Uvarovite (G) | 3.45 |
| Realgar | 3.48 |
| Orpiment | 3.49 |
| Allanite | 3.5-4.2 |
| Chloritoid (M) | 3.5 |
| Diamond | 3.5 |
| Garnet (G) | 3.5-4.3 |

# MINERALS SORTED BY DENSITY

| Name | Density |
|---|---|
| Iddingsite | 3.5-3.8 |
| Lithiophilite | 3.5 |
| Ottrelite (M) | 3.5 |
| Rhodochrosite | 3.5-3.6 |
| Pyrope (G) | 3.51 |
| Grossularite (G) | 3.53 |
| Sklodowskite | 3.54 |
| Hedenbergite (P) | 3.55 |
| Chrysoberyl | 3.6-3.8 |
| Clinoferrosilite | 3.6 |
| Ferrosilite (P) | 3.6 |
| Hydrozincite | 3.6-3.8 |
| Kyanite | 3.6-3.7 |
| Limonite | 3.6-4 |
| Rhodonite | 3.6-3.7 |
| Spinel group | 3.6-4 |
| Staurolite | 3.6-3.8 |
| Atacamite | 3.7-3.8 |
| Strontianite | 3.7 |
| Tyuyamunite | 3.7-4.3 |
| Andradite (G) | 3.75 |
| Azurite | 3.77 |
| Siderite | 3.8-3.9 |
| Uranophane | 3.8-3.9 |
| Rhodolite (G) | 3.84 |
| Anatase | 3.9 |
| Antlerite | 3.9 |
| Brochantite | 3.9 |
| Brookite | 3.9-4.1 |
| Celestite | 3.9-4.0 |
| Gummite | 3.9-6.4 |
| Malachite | 3.9-4.0 |
| Sphalerite | 3.9-4.1 |
| Willemite | 3.9-4.2 |
| Alabandite | 4.0 |
| Cervanite | 4.0-5.0 |
| Cubanite | 4.0-4.2 |
| Gadolinite (R) | 4-4.5 |
| Ilvaite | 4.0 |
| Libethenite | 4 |
| Wurtzite | 4 |
| Corundum | 4.02 |
| Galaxite (S) | 4.03 |
| Perovskite | 4.03 |
| Lepidocrocite | 4.09 |
| Carnotite | 4.1 |
| Chalcopyrite | 4.1-4.3 |
| Samarskite | 4.1-6.2 |
| Sternbergite | 4.1-4.2 |
| Tephroite (Ov) | 4.1 |
| Fayalite (Ov) | 4.14 |
| Spessartite (G) | 4.18 |
| Fergusonite (R) | 4.2-5.8 |
| Pyrochlore | 4.2-4.5 |
| Rutile | 4.2-4.3 |
| Turgite | 4.2-4.6 |
| Powellite | 4.23 |
| Almandite (G) | 4.25 |
| Chromite | 4.3-4.6 |
| Manganite | 4.3 |
| Smithsonite | 4.3-4.4 |
| Witherite | 4.3 |
| Goethite | 4.37 |
| Clinoclase | 4.38 |
| Hercynite (S) | 4.39 |
| Enargite | 4.4-4.5 |
| Stannite | 4.4 |
| Xenotime | 4.4-5.1 |
| Barite | 4.5 |
| Brannerite | 4.5-5.4 |
| Stibnite | 4.5-4.6 |
| Gahnite (S) | 4.55 |
| Stillwellite | 4.57 |
| Covellite | 4.6-4.7 |
| Molybdenite | 4.6-4.7 |
| Pentlandite | 4.6-5 |
| Pyrrhotite | 4.6 |
| Tennantite | 4.6-5.1 |
| Tetrahedrite | 4.6-5.1 |
| Bravoite | 4.66 |
| Zircon | 4.68 |
| Ilmenite | 4.7 |
| Polycrase (R) | 4.7-5.9 |
| Pyrolusite | 4.75 |
| Braunite | 4.8 |
| Linnaeite | 4.8 |
| Siegenite | 4.8 |
| Violarite | 4.8 |
| Hausmannite | 4.84 |
| Marcasite | 4.89 |
| Bastnaesite(R) | 4.9-5.2 |
| Greenockite | 4.9 |
| Bornite | 5.0-5.1 |
| Euxenite | 5-5.9 |
| Manganosite | 5.0-5.4 |
| Monazite | 5.0-5.3 |

# MINERALS SORTED BY DENSITY

| Name | Density |
|---|---|
| Polianite | 5.0 |
| Pyrite | 5.02 |
| Jacobsite (S) | 5.1 |
| Franklinite | 5.15 |
| Magnetite (S) | 5.18 |
| Columbite | 5.2-6.7 |
| Miargyrite | 5.2-5.3 |
| Hematite | 5.26 |
| Linarite | 5.3 |
| Millerite | 5.3-5.7 |
| Thorite | 5.3 |
| Zinkenite (J) | 5.3 |
| Cerargyrite | 5.5-6 |
| Chalcocite | 5.5-5.8 |
| Jamesonite | 5.5-6 |
| Proustite | 5.55 |
| Plagionite (J) | 5.56 |
| Digenite | 5.6 |
| Embolite | 5.6 |
| Zincite | 5.68 |
| Arsenic | 5.7 |
| Iodobromite | 5.7 |
| Iodyrite | 5.7 |
| Allemontite | 5.8-6.2 |
| Bournonite | 5.8-5.9 |
| Semseyite | 5.8 |
| Tenorite | 5.8-6.4 |
| Pyrargyrite | 5.85 |
| Arsenopyrite | 5.9-6.2 |
| Crocoite | 5.9-6.1 |
| Danaite | 5.9-6.2 |
| Gersdorffite | 5.9 |
| Larsenite (Ov) | 5.9 |
| Scheelite | 5.9-6.1 |
| Boulangerite | 6-6.3 |
| Bromyrite | 6-6.5 |
| Cuprite | 6 |
| Phosgenite | 6.0-6.3 |
| Polybasite | 6.0-6.2 |
| Ni skutterudite | 6.1-6.9 |
| Skutterudite | 6.1-6.9 |
| Pearceite | 6.15 |
| Anglesite | 6.2-6.4 |
| Stephanite | 6.2-6.3 |
| Stromeyerite | 6.2-6.3 |
| Tantalite | 6.2-8 |
| Geocronite | 6.3-6.5 |
| Cobaltite | 6.33 |
| Microlite | 6.33 |
| Menaghinite (J) | 6.36 |
| Pyromorphite | 6.5-7.1 |
| Wulfenite | 6.5-7.5 |
| Cerussite | 6.55 |
| Vanadinite | 6.7-7.1 |
| Cassiterite | 6.8-7.1 |
| Huebnerite (W) | 7.0 |
| Mimetite | 7.0-7.2 |
| Wolframite | 7-7.5 |
| Rammelsbergite | 7.1 |
| Acanthite | 7.2-7.3 |
| Calomel | 7.2 |
| Argentite | 7.3 |
| Tin | 7.3 |
| Galena | 7.4-7.6 |
| Loellingite | 7.4-7.5 |
| Nagyagite | 7.4 |
| Ferberite (W) | 7.5 |
| Awaruite | 7.7-8.1 |
| Niccolite | 7.78 |
| Nickel iron | 7.8-8.2 |
| Bismite | 8 |
| Sylvanite | 8-8.2 |
| Cinnabar | 8.10 |
| Altaite | 8.16 |
| Stolzite | 8.3-8.4 |
| Hessite | 8.4 |
| Krennerite | 8.62 |
| Petzite | 8.7-9 |
| Minium | 8.9-9.2 |
| Uraninite | 9-9.7 |
| Calaverite | 9.35 |
| Thorianite | 9.7 |
| Bismuth | 9.8 |
| Silver | 10.5 |
| Sperrylite | 10.5 |
| Palladium | 11.9 |
| Electrum | 13.5-17 |
| Mercury | 13.6 |
| Platinum | 14-19 |
| Gold | 15-19.3 |
| Iridosmine | 19.3-21 |
| Iridium | 22.7 |

# MINERALS SORTED BY HARDNESS

| Name | Hardness | Name | Hardness |
|---|---|---|---|
| Boehmite (B) | 0 soft | Gypsum | 2 |
| Carnotite | 0 soft | Hydrozincite | 2-2.5 |
| Mercury | 0 | Jamesonite | 2-3 |
| Carnallite | 1 | Kaolinite | 2-2.5 |
| Embolite | 1-1.5 | Krennerite | 2-3 |
| Graphite | 1-2 | Melanterite | 2 |
| Halloysite (C) | 1-2 | Muscovite (M) | 2-2.5 |
| Hectorite (C) | 1-1.5 | Nacrite (C) | 2-2.5 |
| Iodobromite | 1-1.5 | Nitre | 2 |
| Iodyrite | 1-1.5 | Paragonite (M) | 2 |
| Molybdenite | 1-1.5 | Polybasite | 2-3 |
| Montmorillonite | 1-1.5 | Potash Alum | 2-2.5 |
| Nagyagite | 1-1.5 | Proustite | 2-2.5 |
| Nontronite (C) | 1-1.5 | Sepiolite | 2-2.5 |
| Pyrolusite | 1-2 | Serpentine | 2-5 |
| Pyrophyllite | 1-2 | Stephanite | 2-2.5 |
| Saponite (C) | 1-1.5 | Stibnite | 2 |
| Soda nitre | 1-2 | Sylvite | 2 |
| Sternbergite | 1-1.5 | Tin | 2 |
| Talc | 1 | Torbernite | 2-2.5 |
| Ulexite | 1 | Tyuyamunite | 2 |
| Beidellite (C) | 1.5 | Uranophane | 2-3 |
| Calomel | 1.5 | Biotite (M) | 2.5-3 |
| Covellite | 1.5-2 | Boulangerite | 2.5-3 |
| Erythrite | 1.5-2.5 | Bournonite | 2.5-3 |
| Ferrimolybdite | 1.5 | Bromyrite | 2.5 |
| Ice | 1.5 | Brucite | 2.5 |
| Orpiment | 1.5-2 | Calaverite | 2.5 |
| Realgar | 1.5-2 | Cerargyrite | 2.5 |
| Sulphur | 1.5-2.5 | Chalcanthite | 2.5 |
| Sylvanite | 1.5-2 | Chalcocite | 2.5-3 |
| Vermiculite (M) | 1.5 | Cinnabar | 2.5 |
| Vivianite | 1.5-2 | Clinoclase | 2.5-3 |
| Acanthite | 2-2.5 | Crocoite | 2.5-3 |
| Argentite | 2-2.5 | Cryolite | 2.5 |
| Aurichalcite | 2 | Digenite | 2.5-3 |
| Autunite | 2-2.5 | Galena | 2.5 |
| Bismuth | 2-2.5 | Geocronite | 2.5 |
| Borax | 2-2.5 | Gibbsite | 2.5-3.5 |
| Chlorite | 2-2.5 | Glauberite | 2.5-3 |
| Chrysocolla | 2-4 | Gold | 2.5-3 |
| Diatomite | 2 | Gummite | 2.5-5 |
| Dickite (C) | 2-2.5 | Halite | 2.5 |
| Epsomite | 2-2.5 | Hessite | 2.5-3 |
| Garnierite | 2-3 | Langbeinite | 2.5-3.5 |
| Gaylussite | 2-3 | Lepidolite (M) | 2.5-4 |
| Glauconite (M) | 2 | Linarite | 2.5 |
| Goslarite | 2-2.5 | Menaghinite (J) | 2.5 |

# MINERALS SORTED BY HARDNESS

| Name | Hardness |
|---|---|
| Miargyrite | 2.5 |
| Minium | 2.5 |
| Petzite | 2.5-3 |
| Phlogopite (M) | 2.5-3 |
| Phosphuranylite | 2.5 |
| Plagionite (J) | 2.5 |
| Polyhalite | 2.5-3 |
| Pyrargyrite | 2.5 |
| Roscoelite (M) | 2.5 |
| Semseyite | 2.5 |
| Silver | 2.5-3 |
| Stolzite | 2.5-3 |
| Stromeyerite | 2.5-3 |
| Thenardite | 2.5 |
| Tungstite | 2.5 |
| Zinnwaldite | 2.5-3 |
| Allemontite | 3-4 |
| Allophane (C) | 3 |
| Altaite | 3 |
| Anglesite | 3 |
| Anhydrite | 3-3.5 |
| Atacamite | 3-3.5 |
| Barite | 3-3.5 |
| Bornite | 3 |
| Calcite | 3 |
| Celestite | 3-3.5 |
| Cerussite | 3-3.5 |
| Electrum | 3 |
| Enargite | 3 |
| Eucryptite | ? |
| Greenockite | 3-3.5 |
| Iddingsite | 3 |
| Jarosite | 3 |
| Kainite | 3 |
| Kernite | 3 |
| Larsenite (Ov) | 3 |
| Millerite | 3-3.5 |
| Pearceite | 3 |
| Phosgenite | 3 |
| Tennantite | 3-4.5 |
| Tenorite | 3-4 |
| Tetrahedrite | 3-4.5 |
| Trona | 3 |
| Vanadinite | 3 |
| Wulfenite | 3 |
| Zinkenite (J) | 3-3.5 |
| Alabandite | 3.4-4 |
| Ankerite | 3.5 |
| Annabergite | 3.5-3 |
| Antlerite | 3.5-4 |
| Aragonite | 3.5-4 |
| Arsenic | 3.5 |
| Azurite | 3.5-4 |
| Brochantite | 3.5-4 |
| Chalcopyrite | 3.5-4 |
| Cubanite | 3.5 |
| Cuprite | 3.5-4 |
| Dolomite | 3.5-4 |
| Heulandite | 3.5-4 |
| Magnesite | 3.5-5 |
| Malachite | 3.5-4 |
| Margarite (M) | 3.5-5 |
| Mimetite | 3.5 |
| Pentlandite | 3.5-4 |
| Powellite | 3.5-4 |
| Pyromorphite | 3.5-4 |
| Rhodochrosite | 3.5-4.5 |
| Scorodite | 3.5-4 |
| Siderite | 3.5-4 |
| Sphalerite | 3.5-4 |
| Stilbite (Z) | 3.5-4 |
| Strontianite | 3.5-4 |
| Variscite | 3.5-4.5 |
| Wavellite | 3.5-4 |
| Witherite | 3.5 |
| Alunite | 4 |
| Bastnaesite (R) | 4-4.5 |
| Cervanite | 4-5 |
| Chabazite (Z) | 4-5 |
| Colemanite | 4-4.5 |
| Crocidolite | 4 |
| Fluorite | 4 |
| Laumontite (Z) | 4 |
| Libethenite | 4 |
| Manganite | 4 |
| Platinum | 4-4.5 |
| Pyrrhotite | 4 |
| Riebeckite (A) | 4 |
| Stannite | 4 |
| Wurtzite | 4 |
| Xenotime | 4-5 |
| Zincite | 4-4.5 |
| Apophyllite | 4.5-5 |
| Bismite | 4.5 |
| Brannerite | 4.5 |
| Gmelinite (Z) | 4.5 |

# MINERALS SORTED BY HARDNESS

| Name | Hardness |
|---|---|
| Harmotome (Z) | 4.5 |
| Hemimorphite | 4.5-5 |
| Linnaeite | 4.5-5.5 |
| Palladium | 4.5-5 |
| Phillipsite (Z) | 4.5-5 |
| Scheelite | 4.5-5 |
| Siegenite | 4.5-5.5 |
| Triphylite | 4.5-5 |
| Violarite | 4.5-5.5 |
| Actinolite (A) | 5-6 |
| Analcime (Fp) | 5-5.5 |
| Apatite | 5 |
| Augite (P) | 5-6 |
| Awaruite | 5 |
| Cancrinite(Fp) | 5-6 |
| Datolite | 5-5.5 |
| Diopside (P) | 5-6 |
| Dioptase | 5 |
| Ferberite (W) | 5 |
| Goethite | 5-5.5 |
| Hedenbergite (P) | 5-6 |
| Hornblende | 5-6 |
| Huebnerite (W) | 5 |
| Hypersthene (P) | 5-6 |
| Kyanite | 5-7 |
| Lazulite | 5-5.5 |
| Lazurite | 5-5.5 |
| Lepidocrocite | 5 |
| Limonite | 5-5.5 |
| Lithiophilite | 5 |
| Loellingite | 5-5.5 |
| Melilite | 5 |
| Monazite | 5-5.5 |
| Monticellite | 5 |
| Natrolite (Z) | 5-5.5 |
| Niccolite | 5-5.5 |
| Nickel iron | 5 |
| Opal | 5-6 |
| Pectolite | 5 |
| Pigeonite (P) | 5-6 |
| Pyrochlore | 5 |
| Samarskite | 5-6 |
| Scapolite | 5-6 |
| Scolecite (Z) | 5-5.5 |
| Skutterudite | 5 |
| Smithsonite | 5 |
| Sphene | 5-5.5 |
| Thomsonite (Z) | 5 |
| Thorite | 5 |
| Tremolite (A) | 5-6 |
| Wolframite | 5-5.5 |
| Wollastonite | 5-5.5 |
| Allanite | 5.5-6 |
| Anatase | 5.5-6 |
| Anthophyllite(A) | 5.5-6 |
| Arsenopyrite | 5.5-6 |
| Bravoite | 5.5-6 |
| Bronzite (E) | 5.5 |
| Brookite | 5.5-6 |
| Chromite | 5.5 |
| Cobaltite | 5.5 |
| Danaite | 5.5-6 |
| Enstatite (P) | 5.5 |
| Euxenite | 5.5-6.5 |
| Fergusonite (R) | 5.5-6.5 |
| Gersdorffite | 5.5 |
| Hausmannite | 5.5 |
| Hauynite (Fp) | 5.5-6 |
| Hematite | 5.5-6.5 |
| Ilmenite | 5.5-6 |
| Ilvaite | 5.5-6 |
| Jacobsite (S) | 5.5-6.5 |
| Leucite (Fp) | 5.5-6 |
| Manganosite | 5.5 |
| Marialite (Sc) | 5.5-6 |
| Meionite (Sc) | 5.5-6 |
| Microlite | 5.5 |
| Nepheline (Fp) | 5.5-6 |
| Ni skutterudite | 5.5-6 |
| Pargasite (H) | 5.5 |
| Perovskite | 5.5 |
| Polycrase (R) | 5.5-6.5 |
| Rammelsbergite | 5.5-6 |
| Rhodonite | 5.5-6 |
| Scorzalite | 5.5-6 |
| Sodalite (Fp) | 5.5-6 |
| Uraninite | 5.5 |
| Willemite | 5.5 |
| Acmite (P) | 6-6.5 |
| Albite (P) | 6 |
| Amblygonite | 6 |
| Andesine (P) | 6 |
| Anorthite (P) | 6 |
| Anorthoclase (O) | 6 |
| Arfvedsonite(A) | 6 |
| Braunite | 6-6.5 |

# MINERALS SORTED BY HARDNESS

| Name | Hardness |
|---|---|
| Bytownite (P) | 6 |
| Cassiterite | 6-7 |
| Celsian (F) | 6 |
| Chloritoid (M) | 6-7 |
| Chondrodite | 6-6.5 |
| Clinoenstatite(E) | 6 |
| Clinoferrosilite | 6 |
| Clinohumite | 6 |
| Clinozoisite | 6-6.5 |
| Columbite | 6 |
| Cummingtonite | 6 |
| Edenite (H) | 6 |
| Epidote | 6-7 |
| Ferrosilite (P) | 6 |
| Franklinite | 6 |
| Glaucophane (A) | 6-6.5 |
| Hastingsite (H) | 6.0 |
| Helvite | 6-6.5 |
| Humite | 6 |
| Hyalophane (O) | 6 |
| Iridium | 6-7 |
| Iridosmine | 6-7 |
| Kaliophilite | 6 |
| Labradorite (P) | 6 |
| Lechatelierite | 6-7 |
| Magnetite (S) | 6 |
| Marcasite | 6-6.5 |
| Microcline (F) | 6 |
| Mullite | 6-7 |
| Norbergite | 6 |
| Noselite (Fp) | 6 |
| Oligoclase (P) | 6 |
| Orthoclase (F) | 6 |
| Ottrelite (M) | 6-7 |
| Petalite (Fp) | 6-6.5 |
| Polianite | 6-6.5 |
| Prehnite | 6-6.5 |
| Pyrite | 6-6.5 |
| Rutile | 6-6.5 |
| Sillimanite | 6-7 |
| Sperrylite | 6-7 |
| Tantalite | 6-6.5 |
| Tephroite (Ov) | 6 |
| Turquoise | 6 |
| Zoisite | 6 |
| Axinite | 6.5-7 |
| Diaspore | 6.5-7 |
| Fayalite (Ov) | 6.5 |
| Forsterite (Ov) | 6.5 |
| Gadolinite (R) | 6.5-7 |
| Garnet (G) | 6.5-7.5 |
| Grossularite (G) | 6.5 |
| Idocrase | 6.5 |
| Jadeite (P) | 6.5-7 |
| Olivine (Ov) | 6.5-7 |
| Piedmontite | 6.5 |
| Pollucite | 6.5 |
| Spodumene (P) | 6.5-7 |
| Thorianite | 6.5 |
| Turgite | 6.5 |
| Almandite (G) | 7 |
| Andradite (G) | 7 |
| Boracite | 7 |
| Chalcedony | 7 |
| Chert | 7 |
| Cordierite | 7-7.5 |
| Cristobalite | 7 |
| Danburite | 7 |
| Dumortierite | 7 |
| Flint | 7 |
| Pyrope (G) | 7 |
| Quartz | 7 |
| Rhodolite (G) | 7 |
| Spessartite (G) | 7 |
| Staurolite | 7-7.5 |
| Tourmaline | 7-7.5 |
| Tridymite | 7 |
| Andalusite | 7.5 |
| Beryl | 7.5-8 |
| Euclase | 7.5 |
| Gahnite (S) | 7.5-8 |
| Galaxite (S) | 7.5-8 |
| Hercynite (S) | 7.5-8 |
| Phenacite | 7.5-8 |
| Uvarovite (G) | 7.5 |
| Zircon | 7.5 |
| Lawsonite | 8 |
| Spinel group | 8 |
| Topaz | 8 |
| Chrysoberyl | 8.5 |
| Corundum | 9 |
| Diamond | 10 |

# METAL CONTENT OF MINERALS

| Name | % Metal |
|---|---|
| **Aluminum:** | |
| Bauxite | 74 |
| Corundum | 53 |
| **Antimony:** | |
| Jamesonite | 29 |
| Stibnite | 71 |
| **Arsenic:** | |
| Arsenopyrite | 31 |
| Orpiment | 61 |
| Realgar | 70 |
| **Barium:** Barite | 59 |
| **Beryllium:** Beryl | 5 |
| **Bismuth:** | |
| Bismuthinite | 81 |
| **Chromium:**Chromite | 46 |
| **Cobalt:** | |
| Cobaltite | 36 |
| Linnaeite | 48 |
| Smaltite | 28 |
| **Copper:** | |
| Azurite | 55 |
| Bornite | 63 |
| Chalcocite | 80 |
| Chalcopyrite | 35 |
| Chrysocolla | 36 |
| Covellite | 66 |
| Cuprite | 89 |
| Malachite | 57 |
| Tetrahedrite | 52 |
| **Iron:** | |
| Hematite | 70 |
| Limonite | 60 |
| Magnetite | 72 |
| Marcasite | 46 |
| Pyrite | 46 |
| Pyrrhotite | 61 |
| Siderite | 48 |
| **Lead:** | |
| Anglesite | 68 |
| Cerussite | 77 |
| Galena | 87 |
| **Magnesium:** | |
| Magnesite | 29 |
| Periclase | 60 |
| **Manganese:** | |
| Pyrolusite | 63 |
| Rhodochrosite | 62 |
| Rhodonite | 48 |
| **Mercury:** | |
| Calomel | 85 |
| Cinnabar | 86 |
| Metacinnabarite | 86 |
| **Molybdenum:** | |
| Molybdenite | 60 |
| Wulfenite | 26 |
| **Nickel:** | |
| Chloanthite | 28 |
| Millerite | 65 |
| Niccolite | 44 |
| Pentlandite | 22 |
| **Niobium Pentoxide:** | |
| Columbite-Tantal. | 83 |
| **Silver:** | |
| Argentite | 87 |
| Cerargyrite | 75 |
| Polybasite | 76 |
| Proustite | 65 |
| Pyrargyrite | 60 |
| Stephanite | 68 |
| **Tantalum Pentoxide:** | |
| Columbite-Tantal. | 86 |
| **Tin:** | |
| Cassiterite | 79 |
| Stannite | 28 |
| **Titanium:** | |
| Illmenite | 32 |
| Rutile | 60 |
| **Tungsten:** | |
| Ferberite | 64 |
| Huebnerite | 61 |
| Scheelite | 64 |
| Wolframite | 51 |
| **Uranium:** % in $U_3O_8$ | 85 |
| **Vanadium:** | |
| Vanadinite-$V_2O_5$% | 19 |
| % in $V_2O_5$ | 56 |
| **Zinc:** | |
| Calamine | 54 |
| Franklinite | 16 |
| Smithsonite | 52 |
| Sphalerite | 67 |
| Willemite | 59 |
| Zincite | 80 |
| **Zirconium:** | |
| Zircon | 50 |

# DISTINCT COLOR MINERALS

| Color | Mineral | Composition |
|---|---|---|
| Blue–Gray | Chalcocite | $Cu_2S$ |
| Brass–Yellow | Chalcopyrite | $CuFeS_2$ |
| Brass–Yellow(pale) | Electrum | Au,Ag |
| | Marcasite | $FeS_2$ |
| | Millerite | NiS |
| | Pentlandite | (Fe,Ni)S |
| | Pyrite | $FeS_2$ |
| Copper–Pink | Copper | Cu |
| | Niccolite | NiAs |
| | Breithauptite | NiSb |
| Copper–Pink (pale) | Maucherite | $Ni_3As_2$ |
| | Melonite | $NiTe_2$ |
| Cream | Emplecite | $Cu_2S \bullet Bi_2S_3$ |
| | Calaverite / Krennerite | $(Au,Ag)Te_2$ |
| Gold–Yellow | Gold | Au |
| Indigo–Blue | Covellite | CuS |
| Orange–Red | Crocoite | $PbCrO_4$ |
| | Wulfenite | $PbMoO_4$ |
| Pink | Erythrite | $Co_3(AsO_4)_2 \bullet 8H_2O$ |
| | Kunzite | $LiAl(Si_2O_6)$ |
| | Rhodochrosite | $MnCO_3$ |
| | Rhodonite | $Mn(SiO_3)$ |
| Pink to lilac | Lepidolite | $K_2Li_3Al_3(AlSi_3O_{10})_2$ |
| Pink–Buffish | Bornite | $Cu_5FeS_4$ |
| Pink–Cream | Cobaltite | CoAsS |
| | Bismuth | Bi |
| | Pyrrhotite | FeS |
| | Cubanite | $Cu_2S.Fe_4S_5$ |
| Pink–Gray | Enargite | $Cu_2S \bullet CuS \bullet As_2S_3$ |
| | Famatinite | $Cu_2S \bullet CuS \bullet Sb_2S_3$ |
| | Coloradote | HgTe |
| Purple | Bornite | $Cu_5FeS_4$ |
| | Rickardite | $Cu_3Te_2$ |
| | Umangite | $Cu_3Se_3$ |
| | Germanite | $Cu_3(Fe,Ge)S_4$ |
| Red | Cinnabar | HgS |
| | Lepidocrocite | FeO(OH) |
| | Realgar | AsS |
| | Zoisite (Thulite) | $Ca_2Al_3(SiO_4)_3(OH)$ |
| Violet | Violarite | $(Ni,Fe)_3S_4$ |
| | Bravoite | $(Ni, Fe)S_2$ |
| Yellow | Carnotite | $K_2(UO_2)_2(VO_4)_2 \bullet nH_2O$ |
| | Orpiment | $As_2S_3$ |
| | Perovskite | $CaTiO_3$ |
| | Serpentine | $Mg_6(Si_4O_{10})(OH)_8$ |
| | Tyuyamunite | $Ca(UO_2)_2(VO_4)_2$ |
| Yellow–Brown | Jarosite | $KFe_3(OH)_6(SO_4)_2$ |
| Yellow–Green | Autunite | $Ca(UO_2)_2(PO_4)_2 \bullet$ 10 to 12 $H_2O$ |
| Yellow–Orange | Greenockite | CdS |
| Yellow–Red | Chondrodite | $Mg_5(SiO_4)_2(F,OH)_2$ |

# MINERAL CRYSTAL SYSTEM

| Xtal System | Xtal Class | Xtal Symmetry |
|---|---|---|
| Isometric | Hexoctahedral * | C,3A4,4A3,6A2,9P |
| | Gyroidal | 3A4,4A3,6A2 |
| | Hextetrahedral * | 3A2,4A3,6P |
| | Diploidal * | C,3A2,4A3,3P |
| | Tetartoidal | 3A2,4A3 |
| Hexagonal: | | |
| Hexagonal | Dihexagonal-dipyramidal * | C,1A6,6A2,7P |
| | Hexagonal-trapezohedral | 1A6,6A2 |
| | Dihexagonal-pyramidal * | 1A6,6P |
| | Ditrigonal-dipyramidal | 1A3,3A2,4P |
| | Hexagonal-dipyramidal * | C,1A6,1P |
| | Hexagonal-pyramidal | 1A6 |
| | Trigonal-dipyramidal | 1A3,1P |
| Rhombohedral | Hexagonal-scalenohedral * | C,1A3,3A2,2P |
| | Trigonal-trapezohedral * | 1A3,3P |
| | Ditrigonal-pyramidal * | 1A3,3P |
| | Rhombohedral | C,1A3 |
| | Trigonal-pyramidal | 1A3 |
| Tetragonal | Ditetragonal-dipyramidal * | C,1A4,4A2,5P |
| | Tetragonal-trapezohedral | 1A4,4A2 |
| | Ditetragonal-pyramidal | 1A4,4P |
| | Tetragonal-scalenohedral * | 3A2,2P |
| | Tetragonal-dipyramidal | C,1A4,1P |
| | Tetragonal-pyramidal | 1A4 |
| | Tetragonal-disphenoidal | 1A2 |
| Orthorhombic | Rhombic-dipyramidal * | C,3A2,3P |
| | Rhombic-disphenoidal | 3A2 |
| | Rhombic-pyramidal * | 1A2,2P |
| Monoclinic | Prismatic * | C,1A2,1P |
| | Sphenoidal | 1A2 |
| | Domatic | 1P |
| Triclinic | Pinacoidal * | C |
| | Pedial | None |

**Symmetry values** are coded as in the following example:
"C,1A4,4A2,3P" is **C**enter of symmetry, **1 A**xis of **4** fold symmetry, **4 A**xes of **2** fold symmetry, and **3 P**lanes of symmetry.

There are a total of 32 crystal classes, however most minerals crystallize in only 15 of those classes. The 15 common classes are marked with an " * " after their name.
The above data was obtained from *Danas Manual of Mineralogy, by James D. Dana, 1959.*

# MINOR ELEMENTS IN SED ROCK

Average Concentration in parts per million (ppm)

| Element | Earth Crust | Soil | Sedimentary Rocks Types: Calcareous | Arenaceous | Argillaceous |
|---|---|---|---|---|---|
| Antimony | 0.2 | 2 | ... | 1 | 3 |
| Arsenic | 1.8 | 7.5 | 0.5 | 0.5 | 20 |
| Barium | 425 | 300 | 60 | 250 | 450 |
| Beryllium | 2.8 | 0.5-4 | ... | ... | 3 |
| Bismuth | 0.17 | 0.8 | ... | 0.3 | 1 |
| Boron | 10 | 29 | 18 | 90 | 220 |
| Cadmium | 0.2 | 0.3 | 0.05 | ... | 0.2 |
| Chlorine | 130 | ... | 200 | ... | ... |
| Chromium | 100 | 43 | 4 | 70 | 450 |
| Cobalt | 25 | 10 | 1 | 1 | 18 |
| Copper | 55 | 15 | 5 | 10 | 140 |
| Fluorine | 625 | 300 | 250 | ... | 550 |
| Gold | 0.004 | 0.002 | 0.003 | 0.004 | 0.004 |
| Iodine | 0.5 | ... | 4 | 0.4 | 1.7 |
| Iron | | 21000 | 3800 | 9700 | 46000 |
| Lead | 0.004 | 17 | 7 | 10 | 22 |
| Lithium | 20 | 22 | 10 | 15 | 51 |
| Manganese | 950 | 320 | 400 | 152 | 750 |
| Mercury | 0.08 | 0.056 | 0.03 | 0.4 | 0.8 |
| Molybdenum | 1.5 | 2.5 | 0.4 | 0.6 | 3 |
| Nickel | 75 | 17 | 3 | 5 | 44 |
| Niobium | 20 | ... | 0.3 | ... | 20 |
| Phosphorus | | 300 | 200 | 300 | 740 |
| Platinum Gp | 0.006 | ... | ... | ... | ... |
| Potassium | | 11000 | 2700 | 10600 | 12000 |
| Rare Earths | | | 22 | 17 | 100 |
| Rhenium | 0.0004 | 0.005 | ... | 0.0003 | 0.0005 |
| Rubidium | 90 | 35 | 56 | 40 | 143 |
| Selenium | 0.05 | 0.31 | 0.07 | 1 | 0.7 |
| Silver | 0.07 | 0.09 | 0.20 | 0.4 | 250 |
| Strontium | 375 | 67 | 600 | 20 | 260 |
| Sulphur | | 100-2000 | 2000 | 2400 | 1850 |
| Tantalum | 2 | ... | ... | ... | 0.8 |
| Tellerium | 0.001 | ... | ... | ... | ... |
| Thallium | 0.45 | 0.1 | ... | 2 | 2 |
| Thorium | 10 | 13 | 2 | 2 | 12 |
| Tin | 2 | 10 | ... | ... | 40 |
| Titanium | 5700 | 5000 | 400 | 1200 | 4500 |
| Tungsten | 1.5 | 1 | 0.6 | ... | 3 |
| Uranium | 2.7 | 1 | 2 | 0.4 | 4 |
| Vanadium | 135 | 57 | 10 | 20 | 130 |
| Zinc | 70 | 36 | 12 | 15 | 200 |
| Zirconium | 165 | 270 | 25 | 260 | 160 |

# MINOR ELEMENTS IN IGN ROCK

Average Concentration in parts per million (ppm)

| | Igneous Rocks Types | | | | |
|---|---|---|---|---|---|
| Element | Granite/ Rhyolite | Syenite/ Trachyte | Diorite/ Andesite | Gabbro/ Basalt | Ultra Mafic |
| Antimony | 0.2 | 0.5 | 0.2 | 0.15 | 0.1 |
| Arsenic | 1.8 | 3 | 2.5 | 1.7 | 1.1 |
| Barium | 600 | 1100 | 230 | 220 | 1.0 |
| Beryllium | 3.5 | 3.5 | 2.0 | 1 | 0.1 |
| Bismuth | 0.02 | ... | 0.008 | 0.4 | 1.0 |
| Boron | 13 | ... | 16 | 8 | 10 |
| Cadmium | 0.1 | ... | 0.01 | 0.15 | 0.1 |
| Chlorine | 300 | 400 | 550 | 230 | ... |
| Chromium | 10 | 1.3 | 55 | 225 | 2700 |
| Cobalt | 2.5 | 1 | 7 | 46 | 140 |
| Copper | 13 | 6 | 40 | 90 | 40 |
| Fluorine | 805 | ... | ... | 375 | 60 |
| Gold | 0.002 | 0.003 | 0.004 | 0.003 | 0.006 |
| Iodine | 0.17 | 0.15 | 0.2 | 0.11 | 0.12 |
| Iron | 14200 | 27000 | 32000 | 86500 | 94300 |
| Lead | 19 | 13 | 15 | 6 | 1.2 |
| Lithium | 40 | 30 | 25 | 16 | 0.2 |
| Manganese | 425 | 750 | 900 | 1550 | 1200 |
| Mercury | 0.06 | ... | 0.06 | 0.05 | 0.008 |
| Molybdenum | 1.3 | 0.6 | 1.0 | 1.6 | 0.3 |
| Nickel | 5.0 | 4 | 35 | 135 | 2000 |
| Niobium | 20 | 35 | 15 | 19 | 11 |
| Phosphorus | 630 | 1300 | 1300 | 2000 | 1000 |
| Platinum Gp | 0.009 | ... | ... | 0.15 | 0.4 |
| Potassium | 42000 | 51000 | 21000 | 8300 | 34 |
| Rare Earths | 200 | ... | 40 | 50 | 20 |
| Rhenium | 0.0006 | ... | 0.0005 | 0.0006 | ... |
| Rubidium | 276 | ... | 10 | 32 | 0.14 |
| Selenium | 0.1 | 0.07 | 0.07 | 0.1 | 0.1 |
| Silver | 0.4 | ... | 0.07 | 0.1 | 0.06 |
| Strontium | 125 | ... | 350 | 460 | 15 |
| Sulphur | 350 | ... | ... | 1000 | 2000 |
| Tantalum | 4.0 | 2 | 1.7 | 0.7 | 0.2 |
| Tellerium | 0.007 | ... | ... | ... | ... |
| Thallium | 2.3 | 1.5 | 0.4 | 0.3 | 0.02 |
| Thorium | 18 | 13 | 8.5 | 3 | 0.004 |
| Tin | 3.1 | ... | 1.4 | 1.5 | 0.5 |
| Titanium | 1100 | ... | 7400 | 12300 | 2700 |
| Tungsten | 1.7 | 1.2 | 1.5 | 0.9 | 0.3 |
| Uranium | 3.7 | 4.0 | 2.8 | 0.6 | 0.02 |
| Vanadium | 40 | 30 | 106 | 240 | 30 |
| Zinc | 50 | 80 | 55 | 100 | 56 |
| Zirconium | 180 | 500 | 180 | 130 | 43 |

# IGNEOUS ROCK CLASSIFICATION[1]

**Potash (K) Feldspar > 2/3 of Total Feldspar**

Accessory Minerals: biotite, hornblende, pyroxene, muscovite

| Coarse Grain | Fine Grain | Components |
|---|---|---|
| Granite | Rhyolite | Quartz > 10% |
| Syenite | Trachyte | Quartz & Feldspathoid < 10% |
| XXX Syenite | Phonolite | XXX Feldspathoid > 10% |

**Potash (K) Feldspar 1/3 to 2/3 of Total Feldspar**

Accessory Minerals: biotite, hornblende, pyroxene

| Coarse Grain | Fine Grain | Components |
|---|---|---|
| Quartz monzonite | Quartz Latite | Quartz > 10% |
| Monzonite | Latite | Quartz & Feldspathoid > 10% |
| XXX Monzonite | XXX Latite | XXX Feldspathoid > 10% |

**Plagioclase Feldspar > 2/3 of Total Feldspar**
**Potash (K) Feldspar > 10% of Total Feldspar**

Accessory Minerals: hornblende, biotite, pyroxene

| Coarse Grain | Fine Grain | Components |
|---|---|---|
| Granodiorite | Dacite | Quartz > 10% |

**Soda Plagioclase, Potash Feldspar<10% of Total Feldspar**

Accessory Minerals: hornblende, biotite, pyroxene

| Coarse Grain | Fine Grain | Components |
|---|---|---|
| Quartz Diorite | Dacite | Quartz > 10% |
| Diorite | Andesite | Quartz & Feldspathoid<10% |

**Calcic Plagioclase, Potash Feldspar<10% of Total Feldspar**

Accessory Minerals: pyroxene, olivine, uralite

| Coarse Grain | Fine Grain | Components |
|---|---|---|
| Gabbro,anorthosite | Basalt | Quartz & Feldspathoid < 10% |
| Diabase | | Quartz & Feldspathoid < 10% |
| Theralite | Tephrite | Feldspathoid & Pyroxene > 10% |

**Minor or No Feldspar – Mainly Pyroxene and/or Olivine**

Accessory Minerals: Serpentine, iron ore

| Coarse Grain | Fine Grain | Components |
|---|---|---|
| Peridotite,dunite | Limburgite | Pyroxene & olivine |

**Minor or No Feldspar – Mainly FerroMags & Feldspathoids**

Accessory Minerals: hornblende, biotite, iron ore

| Coarse Grain | Fine Grain | Components |
|---|---|---|
| Fergusite,Missourite | Leucite | FerroMags & Feldspathoids |

Trap = Dark aphanitic rock, Felsite = Light aphanitic rock
Porphyry is >50% phenocrysts, porphyritic is <50% phenocrysts.
XXX is a descriptor, such as "biotite latite" for "XXX latite"

(1) *Classification of Rocks, 1955, Russell Travis, Colorado School Mines, Golden, Colorado.* See this book for more detail.

## IGNEOUS ROCK CLASS BY COLOR

There is no standard for the classification of igneous rocks by the percentage of dark minerals, however, the following three classes are most common:

**S.J. Shand, 1947, Eruptive Rocks, John Wiley, New York**

| % Dark Minerals | Class Name |
|---|---|
| 0 to 30 | Leucocratic |
| 30 to 60 | Mesocratic |
| 60 to 90 | Melanocratic |
| > 90 | Hypermelanic |

**S.J. Ellis, 1948, Minerology Magazine, Vol 28, p447-469**

| % Dark Minerals | Class Name |
|---|---|
| 0 to 10 | Holofelsic |
| 10 to 40 | Felsic |
| 40 to 70 | Mafelsic |
| > 70 | Mafic |

**I.U.G.S, Anon, 1973, Geotimes, October 1973, p26-30**

| % Dark Minerals | Class Name |
|---|---|
| 0 to 35 | Leucocratic |
| 35 to 65 | Mesocratic |
| 65 to 90 | Melanocratic |
| > 90 | Ultramafic |

# SEDIMENTARY ROCK CLASSES[1]

**Grain Size < 1/256 mm - Clastic and Crystalline**

Mudstone: Includes claystone and siltstone
Shale: Clay based unit, finely fissile
Argillite: Indurated shale, recrystallized
Bentonite: Clay that swells when wet
Chert: Cryptocrystalline varieties of silica, flint is a variety
Diatomite: Rock made of silica frustules of diatom plants
Limestone: > 80% Calcium carbonate, crystalline
Dolomite: >80% Magnesium carbonate, crystalline
Chalk: Soft lime unit made of microorganism tests, calcite matrix
Caliche: Lime unit formed near surface, calcium carbonate cap
Marlstone: 25 to 75% clay and calcium carbonate
Siderite: Iron carbonate
Coal: Indurated, dense, carbon rock, made from lignite. Types range from bituminous to anthracite to graphite, 8400 btu to 16000+ btu
Phosphorite: Phosphate (Collophane) Rock, massive
Halite (rock salt), Gypsum, & Anhydrite: Massive evaporites

**Grain Size 1/256 – 2 mm - Clastic and Crystalline**

Oolite: Spherical with concentric or radial structure parts
Limestone: > 80% Calcium carbonate, crystalline
Dolomite: >80% Magnesium carbonate, crystalline
Sandstone: Compacted clastic sediment, usually quartz grains
Arkose: Sandstone with > 25% feldspar grains
Graywacke: Sandstone w/ large quartz and feldspar fragments in a clay matrix, angular frags, well indurated
Subgraywacke: Graywacke w/ less feldspar and more quartz
Peat: Residual of partially decomposed vegetation in a bog
Lignite: Consolidated peat, between peat & coal, <8400 btu

**Grain Size > 2 mm - Clastic**

Conglomerate: Consolidated & rounded parts
Breccia: Consolidated & angular fragments
Gravel: Unconsolidated & rounded fragments
Rubble: Unconsolidated & angular fragments
Till: Unsorted glacial debris, mix of clay, sand, gravel, boulders

**Grain Size 1/256 to >2 mm - Clastic Volcanics**

Agglomerate: Consolidated & rounded fragments > 32 mm
Volcanic Breccia: Consolidated & angular fragments > 4mm
Tuff: Consolidated volcanic ash
Ash: Unconsolidated particles < 4 mm

(1) Based in part on classes set up by *Classification of Rocks, 1955, Russell Travis, Colorado School Mines, Golden, CO.*

# METAMORPHIC ROCK CLASSES[1]

## Primary Minerals

Quartz, feldspar, calcite, dolomite, talc, muscovite, sericite, chlorite, hornblende, serpentine, biotite, pyroxene, actinolite, epidote, olivine, magnetite

## Accessory Minerals

Muscovite, sericite, sillimanite, kyanite, cordierite, tremolite, wollastonite, albite, andalusite, garnet, phlogopite, diopside, enstatite, staurolite, glaucophane, anthophyllite, pyrophyllite, chloritoid, actinolite, tourmaline, epidote, chiastolite, olivine, serpentine, chlorite, biotite, graphite, chondrodite, scapolite

## Massive / Granular Structure (Contact Metamorphism) Nondirectional – Fine, to Medium, to Coarse Grained:

Hornfels (catch–all term for nondirectional metamorphic unit)
Metaquartzite – primarily quartz, with silica cement
Marble – metamorphosed calcite or dolomite
Serpentine – metamorphosed olivine & pyroxene forming antigorite or chrysotile (asbestos)
Soapstone – massive talc

## Lineate / Foliate Structure (Mechanical Metamorphism) Cataclastic:

Mylonite - foliated, fine ground
Augen Gneiss – Augen (eye) structures in a gneissic rock of alternating bands of coarse granular minerals & schist minerals
Flaser Unite – Small lenses of granular material separated by wavy ribbons & streaks of fine crystalline, foliated material

## Lineate / Foliate Structure (Regional Metamorphism) Slaty, Phyllitic, Schistose, and Gneissose Structure:

Slate – metamorphosed shale, fissile, slatey cleavage
Phyllite – Argillic unit between slate and schist, silky sheen
Schist – med to coarse grained, mica minerals parallel orientation
Amphibolite – amphibole (hornblende) & plagioclase schist
Gneiss – coarse grained unit of alternating bands of coarse granular minerals & schistose minerals
Granulite – alternating coarse & fine bands of hornblende and mica, very planar schistosity, high temperature unit

## Lineate / Foliate Structure (Plutonic Metamorphism) Migmatitic Structure:

Migmatites – mixed unit of metamorphic material, this alternating layers or lenses of granitic and schistose materlal

(1) See previous page footnote for reference.

# GEOCHEM DETECTION LIMITS [1]

| Element | Lower Detection Limit | Standard Analysis Method |
|---|---|---|
| Aluminum | 0.1% | Total acid digestion |
| Antimony | 1 ppm | GC, Fusion, organic extraction |
| Arsenic | 1 ppm | GC, Perchloric/nitric, hydride |
| Barium | 0.005% | Lithium meta-borate fusion |
| Beryllium | 0.2 / 0.02 ppm | Perchloric-nitric / 4 acid digestion |
| Bismuth | 0.2 ppm | GC, HCl (soils), Cl2 (rock) |
| Cadmium | 0.2 / 0.02 ppm | Perchloric-nitric / 4 acid digestion |
| Calcium | 0.005% | Lithium meta-borate fusion |
| Cobalt | 1 / 0.1 ppm | Perchloric-nitric / 4 acid digestion |
| Copper | 1 / 0.1 ppm | Perchloric-nitric / 4 acid digestion |
| Fluorine | 0.01% | Fusion, specific ion electrode |
| Gold | 0.002 ppm | GC,Roast, aqua regia, organic extr |
| Gold | 0.005 oz/ton | Fire assay |
| Gold | 0.001 ppm | Fire assay – atomic absorption |
| Iron | 0.005% | Lithium meta-borate fusion |
| Lead | 1 / 0.1 ppm | Perchloric-nitric / 4 acid digestion |
| Lithium | 0.005% | Total acid digestion |
| Lithium | 1 / 0.1 ppm | Perchloric-nitric / 4 acid digestion |
| Magnesium | 0.005% | Lithium meta-borate fusion |
| Manganese | 0.005% | Total acid digestion |
| Manganese | 1 / 0.1 ppm | Perchloric-nitric / 4 acid digestion |
| Mercury | 0.01 ppm | GC, Perchloric/nitric, cold vapor |
| Molybdenum | 1 / 0.1 ppm | Perchloric-nitric / 4 acid digestion |
| Potassium | 0.005% | Lithium meta-borate fusion |
| Rubidium | 0.005% | Lithium meta-borate fusion |
| Rubidium | 1 / 0.1 ppm | Perchloric-nitric / 4 acid digestion |
| Silicon | 0.1% | Lithium meta-borate fusion, color |
| Silver | 0.05 oz/ton | Fire assay |
| Silver | 0.2 / 0.02 ppm | Perchloric-nitric / 4 acid digestion |
| Sodium | 0.005% | Lithium meta-borate fusion |
| Strontium | 0.005% | Lithium meta-borate fusion |
| Strontium | 1 / 0.1 ppm | Perchloric-nitric / 4 acid digestion |
| Tellurium | 0.2 ppm | GC, HBr/Br, organic extraction |
| Thallium | 0.02 ppm | GC, Total digestion, organic extract |
| Tin | 1 ppm | GC, Fusion, organic extraction |
| Tungsten | 1 ppm | GC, Fusion, colorimetric |
| Uranium | 0.2 ppm | GC, Total, extraction, fluorimetric |
| Uranium | 0.2 ppm | GC, 1N nitric, extraction, fluorimetric |
| Vanadium | 1 / 0.1 ppm | Perchloric-nitric / 4 acid digestion |
| Zinc | 1 / 0.1 ppm | Perchloric-nitric / 4 acid digestion |

GC stands for Standard Geochemical Analysis. There are many more analysis methods available, however, these are common.

(1) Data courtesy of *Cone Geochemical Inc, Denver, Colorado.*

# MOHS SCALE OF HARDNESS

| Mohs Index | Rock |
|---|---|
| 1 | Talc |
| 2 | Gypsum |
| 3 | Calcite |
| 4 | Fluorite |
| 5 | Apatite |
| 6 | Orthoclase |
| 7 | Quartz |
| 8 | Topaz |
| 9 | Corundum |
| 10 | Diamond |

# PARTICLE SIZE DESCRIPTIONS

| Size Term | Particle Diameter |
|---|---|
| **Sedimentary Units:** | |
| Boulder | >256 mm |
| Cobble | 64 to 256 mm |
| Pebble | 4 to 64 mm |
| Granule | 2 to 4 mm |
| Very Coarse Sand | 1 to 2 mm |
| Coarse Sand | 1/2 to 1 mm |
| Medium Sand | 1/4 to 1/2 mm |
| Fine Sand | 1/8 to 1/4 mm |
| Very Fine Sand | 1/16 to 1/8 mm |
| Silt | 1/256 to 1/16 mm |
| Clay | < 1/256 mm |
| **Pyroclastic Units:** | |
| Bomb or block | >32 mm |
| Lapilli | 4 to 32 mm |
| Coarse Ash | 1/4 to 4 mm |
| Fine Ash | < 1/4 mm |
| **Igneous Rocks:** | |
| Pegmatitic | > 30 mm |
| Coarse Grained | 5 to 30 mm |
| Medium Grained | 1 to 5 mm |
| Fine Grained | < 1 mm |

# RICHTER EARTHQUAKE SCALE

| Richter Magnitude | Mercalli Intensity | Description |
|---|---|---|
| 2 | I | Usually not felt, detected by instruments. |
| 2 | II | Felt by few, especially on upper floors of buildings, detected by instruments. |
| 3 | III | Felt noticeably indoors, vibration like a passing vehicle, cars may rock. |
| | IV | Felt indoors by many, outdoors by few, dishes & doors disturbed, like heavy truck nearby, walls–cracking sound. |
| 4 | V | Felt by most people, slight damage; some dishes & windows broken, some cracked plaster, trees disturbed. |
| 5 | VI | Felt by all, many frightened and run outdoors, damage minor to moderate. |
| 5 to 6 | VII | Everyone runs outdoors, much damage to poor design buildings, minor damage to good design buildings, some chimneys broken, noticed by people driving cars. |
| 6 | VIII | Everyone runs outdoors, damage is moderate to major. Damage minor in well designed structures, major in poor designs; chimneys, columns, & walls fall, heavy furniture turned, well water changes; sand & mud ejected. |
| 7 | IX | Major damage in all structures, ground cracked, pipes broken, shift foundation. |
| 7 & 8 | X | Major damage, most masonry & frame structures destroyed, ground badly cracked, landslides, water sloshed over river banks, rails bent. |
| 8 | XI | Almost all masonry structures destroyed bridges fall, big fissures in ground, land slumps, rails bent greatly. |
| 8 | XII | Total destruction. ground surface waves seen, objects thrown up into the air. All construction destroyed. |

Richter Magnitudes increase energy logarithmically, 10 times for each number jump, so 8 is not twice as large as 4, it is 10,000 times as large! Richter Magnitudes are measured on instruments.

Mercalli Intensity is based on actual observations of the resulting damage, and therefore can not be measured on instruments.

# CORE DRILL SPECS

| Core Size | Core Diameter Inch | Core Diameter mm | Core Volume cu inch/foot | Hole Diameter Inch | Hole Diameter mm |
|---|---|---|---|---|---|
| **CONVENTIONAL:** | | | | | |
| EX or EWM | 0.845 | 21.5 | 6.7 | 1.485 | 37.7 |
| EXT | 0.905 | 23.0 | 7.7 | 1.485 | 37.7 |
| E17 | 0.968 | 24.6 | 8.8 | 1.485 | 37.7 |
| AX or AWM | 1.185 | 30.1 | 13.2 | 1.890 | 48.0 |
| AXT | 1.280 | 32.5 | 15.4 | 1.890 | 48.0 |
| A17 | 1.310 | 33.3 | 16.2 | 1.890 | 48.0 |
| BX,BXM,BWM | 1.655 | 42.0 | 25.8 | 2.360 | 59.9 |
| NX, NXM, NXMS, NWM | 2.155 | 54.7 | 43.8 | 2.980 | 75.7 |
| BM | 1.281 | 32.5 | 15.5 | 2.360 | 59.9 |
| BMLC | 1.386 | 35.2 | 18.1 | 2.360 | 59.9 |
| NMLC | 2.045 | 51.9 | 39.4 | 2.980 | 75.7 |
| A19DT | 1.156 | 29.4 | 12.6 | 1.890 | 48.0 |
| A19TT | 1.062 | 27.0 | 10.6 | 1.890 | 48.0 |
| B19DT | 1.565 | 39.8 | 23.1 | 2.360 | 59.9 |
| B19TT | 1.500 | 38.1 | 21.2 | 2.360 | 59.9 |
| N19DT | 2.095 | 53.2 | 41.4 | 2.980 | 75.7 |
| N19TT | 2.045 | 51.9 | 39.4 | 2.980 | 75.7 |
| H19DT | 2.500 | 63.5 | 58.9 | 3.783 | 96.1 |
| H19TT | 2.406 | 61.1 | 54.6 | 3.783 | 96.1 |
| **WIRELINE:** | | | | | |
| AQ | 1.062 | 27.0 | 10.6 | 1.890 | 48.0 |
| BQ | 1.433 | 36.4 | 19.3 | 2.360 | 59.9 |
| NQ | 1.875 | 47.6 | 33.1 | 2.980 | 75.7 |
| HQ | 2.500 | 63.5 | 58.9 | 3.783 | 96.1 |
| BQ3 | 1.320 | 33.5 | 16.4 | 2.360 | 59.9 |
| NQ3 | 1.775 | 45.1 | 29.7 | 2.980 | 75.7 |
| HQ3 | 2.406 | 61.1 | 54.6 | 3.783 | 96.1 |
| PQ3 | 3.270 | 83.1 | 100.8 | 4.828 | 122.6 |
| B18DT | 1.565 | 39.8 | 23.1 | 2.360 | 59.9 |
| B18TT | 1.500 | 38.1 | 21.2 | 2.360 | 59.9 |
| N18DT | 2.095 | 53.2 | 41.4 | 2.980 | 75.7 |
| N18TT | 2.045 | 51.9 | 39.4 | 2.980 | 75.7 |
| H18DT | 2.500 | 63.5 | 58.9 | 3.783 | 96.1 |
| H18TT | 2.406 | 61.1 | 54.6 | 3.783 | 96.1 |

# GEOLOGIC TIME SCALE

| Era | Period or System | | Epoch, Age or Series | Approximate[1] Million Years Before Present |
|---|---|---|---|---|
| **Phanerozoic Eon:** | | | | 0 to 600 |
| Cainozoic or Cenozoic: | | | | 0 to 65 |
| | Quaternary | | | 0 to 2 |
| | | Holocene | | 0 to 0.011 |
| | | | Bronze Age | |
| | | | Iron Age | |
| | | | Neolithic | |
| | | | Mesolithic | |
| | | Pleistocene | | 0.011 to 2 |
| | | | Paleolithic: | 0.01 to start |
| | | | Young Pal. | 0.01 to 0.033 |
| | | | Perigordian | 0.023 to 0.033 |
| | | | Middle Pal. | 0.033 to 0.07 |
| | | | Old Pal. | 0.07 to start |
| | Tertiary | | | 2 to 65 |
| | | Neogene | | 2 to 22.5 |
| | | | Pliocene | 2 to 6 |
| | | | Miocene | 6 to 22.5 |
| | | Paleogene | | 22.5 to 65 |
| | | | Oligocene | 22.5 to 36 |
| | | | Eocene | 36 to 58 |
| | | | Paleocene | 58 to 65 |
| Mesozoic | | | | 65 to 230 |
| | Cretaceous | | | 65 to 141 |
| | Jurassic | | | 141 to 195 |
| | Triassic | | | 195 to 230 |
| Paleozoic | | | | 230 to 600 |
| | Permian | | | 230 to 280 |
| | Carboniferous | | | 280 to 345 |
| | | Pennsylvanian | | 280 to 310 |
| | | Mississippian | | 310 to 345 |
| | Devonian | | | 345 to 395 |
| | Silurian | | | 395 to 435 |
| | Ordovician | | | 435 to 500 |
| | Cambrian | | | 500 to 600 |
| **Proterozoic & Archaen:** | | | | 600 to start |
| Precambrian Z | | | | 600 to 800 |
| Precambrian Y | | | | 800 to 1600 |
| Precambrian X | | | | 1600 to 2500 |
| Precambrian W or Archaean | | | | 2500 to start |

(1) Accepted dates vary greatly, these represent an average set.

# Glues, Solvents, Paints & Finishes

# GLUE TYPES & APPLICATIONS

| Glue Type | Characteristics (1) |
|---|---|
| Acrylic Resin | Bonds to anything porous & nonporous, waterproof, very strong, fast setting (3 to 30 min), oil and gas resistant, good gap & hole filling.<br>2 part – liquid and powder, expensive, tan.<br>Brands: "3 Ton Adhesive" & "P.A.C.", both by Tridox Labs, Philadelphia, PA, F88 Adhesive. |
| Acrylonitrile | Bonds to anything porous & nonporous, not recommended for wood, flexible, waterproof, similar to rubber cement.<br>1 part liquid, flammable, brown.<br>Brands: "Pliobond" (Goodyear) by W.J.Ruscoe Co, Akron, Ohio. |
| Aliphatic Resin (yellow glue) | Use mainly for wood, moisture resistant (not waterproof), very strong, dries translucent, instant sticky but dries 45 min to 24 hrs, high resistance to solvents and heat, sandable, will set at temps from 45°F to 110°F, dries hard, glue can be colored with water soluble dyes.<br>1 part liquid, non-toxic,non-flammable,no stain.<br>Brands: "Titebond Glue" by Franklin Glue Co, Columbus, Ohio (many others, common glue). |
| Casein ( protein glue ) | Use mainly for wood, high water resistance (not waterproof), resistant to oil, grease, gas, good gap filling, set at temps above 32°F, clamps recommended, dries in 8 hours.<br>1 part powder, mix with water, inexpensive.<br>Brands: "No 30 Casein Glue" by National Casein Co, Chicago, IL. |
| Cellulose Nitrate | Bonds to many porous & nonporous, good water resistance (not waterproof), fast setting (2 hrs to 24 hrs), moderately high strength (up to 3500 psi), shrinks some on drying.<br>1 part liquid, flammable, clear to amber.<br>Brands: "Ever Fast Liquid Cement" by Ambroid Co, Taunton, MA and "Duco Cement" by E.I. du Pont de Nemours & Co, Wilmington, DE. |
| Contact Cement | Bonds to many porous & nonporous but mainly for laminates to wood, water resistant (not waterproof), requires application to both pieces, let dry 40 minutes then put pieces together for an instant bond, moderate to high strength.<br>1 part liquid, flammable (solvent).<br>Brand: "Weldwood Contact Cement" by Roberts Consolidated Industries, City of Industry, CA and "Veneer Glue" by Albert Constantine & Son Bronx, NY. |

# GLUE TYPES & APPLICATIONS

| Glue Type | Characteristics (1) |
|---|---|
| Cyanoacrylate Ester | Bonds many materials including metals, rubber, & most plastics, non-porous, oil, water, and chemical resistant, very fast setting (<5 secs), not a gap filler, poor shock and peel resistance, no clamping needed. Show extreme care when using this glue, IT BONDS SKIN ! 1 part liquid, non-flammable. Brand: "Zip Grip" by Devcon Corp, Danvers, MA and "Permabond" by Edmund Scientific, Barrington, New Jersey. |
| Epoxy | Bonds to many materials, porous & non-porous, waterproof, resistant to most solvents and acid, setting ranges from very fast (5 min) to very long (weeks), very high strength, usually dries transparent to honey colored, can be thinned & cleaned with acetone, non–shrinking if no thinner, hardens without evaporation (uses chemical reaction & heat ). 2 part liquid, non-flammable, not flexible, clear. Brand: Extremely common; for unusual purpose epoxies contact "Miller–Stephenson Chemical Co, Danbury, CT. |
| Hide Glue – flake | Bonds wood primarily (this is the main glue type used in cabinet work in the past; known as "Hot Glue"), non–staining, fast setting, high strength ( 2000 psi ), not waterproof. Flakes, water mixed, non-flammable, apply hot and be careful of "cold joints" on the work. Brand: "Constantine's Cabinet Flake Glue" by Albert Constantine & Son, Bronx, New York. |
| Hide Glue – liquid | Bonds to wood primarily, similar to flake Hide Glue but requires no mixing or heating, long setting time, heat resistant, resists most sealers, lacquers, water, mold & varnish, high strength, water resistant, not flexible but not brittle. 1 part liquid, non-flammable, honey color. Brand: "Franklin Liquid Hide Glue" by Franklin Glue Co, Columbus, Ohio. |
| Hot Melt Glue | Bonds most materials but is primarily suited for plastics, very fast setting time, moderate to no flexibility, not cured by evaporation, medium strength, waterproof. 1 part solid cartridges, non-flammable. Brand: "Hot–Grip" by Adhesive Products Corp, New York, New York. |

# GLUE TYPES & APPLICATIONS

| Glue Type | Characteristics (1) |
|---|---|
| Latex Combo | Sticks a variety of materials, porous and non-porous, especially good for fabric and paper items, water resistant (some are waterproof ), moderate to weak strength, very flexible ( becomes a synthetic rubber on set )<br>1 part liquid or paste, non-flammable.<br>Brand: "Flexible Patch–Stix" by Adhesive Products Corp, New York, New York. |
| Neoprene Base | Bonds a variety of materials, porous and non-porous but primarily used to bond paneling to walls, water resistant, moderate to high strength, setting time is two part – apply and separate for 10 minutes to increase tack, then join parts, final set 24 hours.<br>1 part viscous liquid, flammable.<br>Brand: "Weldwood Panel Adhesive" by Roberts Consolidated Industries, City of Industry, CA. |
| Polyester Resin | Bonds a variety of materials but used mainly with fiberglass cloth to bond to wood for boat hulls and car bodies, waterproof, not flexible, high strength, use at temps from $70^{\circ}$ to $80^{\circ}$F, setting time < 30 minutes, color is usually clear to amber but can be tinted, can be sanded and painted.<br>2 part liquid, flammable, catalyst amount is critical so measure precisely.<br>Brand: Numerous, Fiber-Glass-Evercoat Co Cincinnati, Ohio; Pettit Polyester Resin. |
| Polyethylene Hot Melt | Sticks to most materials, porous and non-porous, waterproof, moderately strong, very fast setting (< 1 minute), moderately flexible, usually cream colored, some are clear, applied with a special hot melt "gun".<br>1 part solid cartridges, non-flammable.<br>Brand: Several, USM Corp, Middleton, MA and Swingline Inc, Long Island, New York. |
| Polysulfide | Bonds to a variety of materials but is primarily for sealing seams, basically a caulking type adhesive that when it dries it is completely waterproof, setting time varies from several days to several weeks depending on humidity, medium strength, when cured it becomes a synthetic rubber, flexible.<br>1 or 2 part.<br>Brand: "Exide Polysulfide Caulk" by Atlas Minerals, Mertztown, Pennsylvania. |

# GLUE TYPES & APPLICATIONS

| Glue Type | Characteristics (1) |
|---|---|
| Polyvinyl Acetate Resin ( PVA ) (Elmers Glue) (White glue) | Bonds primarily wood and paper products, not waterproof, very strong if no moisture, setting times vary from several hours to several days, dries transparent, poor gap filling and do not use where glue must support load, corrodes metal, very common glue. 1 part liquid, non-flammable, white-dries clear. Brand: "Weldwood Presto–Set" by Roberts Consolidated Industries, City of Industry, CA. |
| Polyvinyl Chloride ( PVC ) | Bonds glass, china, porcelain, metal, marble, hard plastics, and other materials including some porous ( treat both sides for porous ), not generally for wood, waterproof, resistant to gas, oil, and alcohol, fast setting (minutes), clean up with lacquer thinner. 1 part liquid, flammable, clear. Brand: "Sheer Magic" by Miracle Adhesives Corp, Long Island, New York. |
| Resorcinol Resin | Bonds to a variety of materials but is used mainly as a boat building glue ( the main one), completely waterproof, thinning and cleanup before setting with alcohol and water, when cured it is resistant to gas, oil, acids, alkalis, and many solvents, setting time varies on temperature – 10 hrs @ 70°F to 3 hours @ 100°F, do not use at temperatures below 70°F, very strong, cures to a very dark color, can be sanded and painted, good gap filler. 2 part – liquid & powder, caustic powder, red. Brand: "Weldwood Resoricinol Waterproof Glue", by Roberts Consolidated Industries, City of Industry, California. |
| Rubber Base | Bonds to almost anything, porous or non-porous, waterproof, moderately flexible, good gap filling, setting time 24 hours, 1 part viscous liquid. Brand: "Black Magic Tough Glue" by Miracle Adhesives Corp, Long Island, New York. |
| Silicon Base | Although this group is primarily a sealer or caulking compound, it does have adhesive characteristics, porous or non-porous, mod. to weak strength, waterproof, setting time from 2 hours to 2 days, can withstand temps of 400 to 600°F, flexible, resists oil & some solvents. 1 part viscous liquid, non-flammable. Brand: "Hi Temp Silicone" by General Electric. |

# GLUE TYPES & APPLICATIONS

| Glue Type | Characteristics (1) |
|---|---|
| Urea–Resin Glue (Plastic-Resin) | Bonds wood primarily, resistant to water, oil, gas, and many solvents when cured, setting time ranges from 3 to 7 hours (less at high temperatures), very high strength ( stronger than wood usually), not a gap filler, non-staining, light tan to black color.<br>Powder, mix with water to form cream.<br>Brand: "Weldwood Plastic Resin Glue" by Roberts Consolidated Industries, City of Industry, California. |
| Water–Phase Epoxy | Bonds to many materials but is used primarily with fiberglass as a repair tool, water soluble when liquid and completely waterproof when hard, med to high strength, fast setting < 30 minutes, can be sanded and painted<br>2 part liquid.<br>Brand: "Dur–A–Poxy" by Dur–A–Flex, Hartford Connecticut. |

## Other Glues

| | |
|---|---|
| Albumin Glue | Made from blood and casein, used in plywood not as strong as animal glues, doesn't resist mold and fungi. |
| Bone Glue | Made from bones, used mostly in making of cartons and paper boxes, there are 15 grades of Bone Glue based on quality of raw material, the method of extraction and the blend. Green Bone Glue is used for gummed paper and tapes for cartons. |
| Cellulose Acetate | Typically used as the bonding cement for the soles of shoes. 10 psi strength. |
| Ceramic Adhesive | Made with porcelain enamel grit, iron oxide, & stainless steel powder. Heat resistant to 1500°F, shear strength of 1500 psi, must be heated to 1750°F in order to cure. |
| Fish Glue | Made from the jelly separated from fish oil or the solutions of the skins. It is used mainly for photo mounting, gummed paper, household use and in paints. The best Fish Glue is made from Russian isinglass. |
| Furan Cement | Made with furfural alcohol resins and is very strong and highly resistant to chemicals. Commonly used for bonding acid resistant brick and tile. |
| Latex Pastes | Rub-off latex, used mainly in photographic mounting, does not shrink. |

| | |
|---|---|
| Pyroxylin Cement | Solution of nitrocellulose in a solvent which is sometimes mixed with resin, gum or synthetic, poor tack but excellent adhesion to almost everything. Typically called household cement. |
| Soybean Glue | Made from soybean cake, used in plywoods, better water resistance than most vegetable pastes and better adhesive power. |
| Tapioca Paste | Typically known as vegetable glue, it is used in cheap plywoods, postage stamps, envelopes, and labels. Quick tack and cheap, but deteriorates. |
| Ultraviolet Glue | Glues that are liquid on application and do not cure until exposed to ultraviolet light, typically used for glass bonding. |

(1) In all of the above glue descriptions, the term "wood" also refers to "wood products" such as plywood, particle board, and aspen board.

Hints and general rules:

1. Apply glues and adhesives to clean, dry surfaces.
2. Drying times can usually be reduced by increasing the temperature. 70°F or higher is generally preferred.
3. Be careful of the solvents and catalysts used in many adhesives, most are toxic and can also hurt your eyes.
4. Hardwoods require less clamping time than softwoods.
5. The end grain of any wood is highly absorbent and will create a weak joint. To prevent this, apply a thin coat of glue to the end grain before the rest of the work and then give it a second coat when doing the normal gluing.
6. Precision alignment of parts glued with contact cement can be obtained by placing a thin sheet of paper between the pieces after the cement has been applied and is no longer tacky, align the pieces, press together and then pull out the piece of paper for final bonding.
7. Don't glue green or damp wood.
8. Clamp glue joints whenever possible to be safe.
9. Don't apply too much glue, this can actually weaken a joint in some cases. FOLLOW DIRECTIONS.

**An excellent book of common glues and adhesives is** *Home and Workshop Guide to Glues and Adhesives, 1979, by George Daniels, Popular Science, Harper & Row, New York.* It contains an abundance of info on glue types, glue techniques, and hints and is and absolute must for the good reference library!

# SOLVENTS

A solvent is a material, usually a liquid, that has the power to dissolve another material and form a homogeneous mixture known as a solution. There are literally thousands of solvents available commercially but most are not readily available to the average person. The following list of solvents are readily available in hardware stores and provide an excellent range of capabilities. Note that most of these are toxic and flammable, so exercise caution when using them and keep out of the reach of children. NOTE: These have been arranged in an approximate order of "strength", ie, top of the list dissolves a lot of plastics in particular.

| Solvent | Characteristics |
|---|---|
| Lacquer Thinner | A mixture of toluene, isopropanol, methyl isobutyl keytone, acetone, propylene glycol, monomethyl ether acetate and ethyl acetate. Photochemically reactive. Used to thin lacquers and epoxies but can be used as a general cleaner and degreaser also. Highly flammable. Dissolves or softens many plastics. |
| Acetone | 2–Propanone or Dimethyl ketone is the actual chemical, $CH_3COCH_3$, soluble in water and alcohol, non–photochemically reactive; used to clean and remove epoxy resins, polyester, ink, adhesives, contact cement, and fiberglass cleanup. Dissolves or softens many plastics. |
| Finger Nail Polish Remover | Mixture of acetone, cocamidopropyl dimethylamine propionate, and amp isostearic hydrolyzed animal protein. Good for various apps., dissolves plastics. |
| Weldwood Cleaner & Solvent | Mixture of 1,1,1 Trichloroethane ( methyl – chloroform $CH_3CCl_3$ ) and Dichloromethane ( $CH_2Cl_2$ ) A very powerful solvent that is typically used with adhesives, non–photochemically reactive, this solvent will clean up brushes and tools that have dried adhesives on them if you let them soak for an hour. Good for cleaning rubber platins, rollers and other parts in printers and type–writers. Dissolves or softens some plastics Avoid breathing fumes for long periods. |
| Plastic Cement | Methyl ethyl keytone or 2–Butanone, $CH_3CH_2COCH_3$, soluble in water and alcohols, actually a solvent that dissolves plastic and is typically used in making model airplanes. |

# SOLVENTS

| | |
|---|---|
| Methylene Chloride | Dichloromethane, $CH_2Cl_2$ , not a very common solvent but when mixed with xylene ( dimethylbenzene ) makes a strong solvent for things like crayon marks, lipstick, ink, magic marker, gum, latex, oil, and wax. Dissolves or softens some plastics. Marketed as "Klean-Clean" by Klean-Strip, W.W. Barr Inc, Memphis, TN. |
| Naptha | Naphthalene, slight smell but good for some applications, non-photochemically reactive. Very fast evaporation. |
| Turpentine ( Paint thinner ) | "Steam Distilled" or "Gum Spirits", made from pine trees; used as a thinner and cleaner for oil based paint, varnish, enamel, and stain. Photochemically reactive. |
| Solvent Alcohol | Methanol or Methyl Alcohol or wood alcohol, $CH_3OH$ , non-photochemically reactive, poisonous, used primarily as a thinner for shellac and shellac base primers. Do not use with oil or latex paints, stains, or varnishes. Also used in marine alcohol stoves, soluble in water and other alcohols. Can be mixed with gasoline in the gas tank to eliminate moisture problems ( 1/2 pint per 15 gallons ). Good cleaner for computer plastic parts. |
| Freon | Trichlortrifluoroethane, Freon TF, available in spray cans, non-flammable, non-conductive, low toxicity, odorless and does not attack plastic, rubber, paints or metal; low surface tension, evaporates fast. Although not commonly seen, freon is an excellent solvent and is typically used to clean electrical connectors and computer components. |
| Denatured Alcohol | Ethyl or grain alcohol made unfit for drinking by the addition of compounds. Soluble in water and other alcohols. Non-photochemically reactive, typically used to thin shellac, clean glass and metal, to clean ink from rubber rollers, and as a fuel in marine stoves. To clean glass, porcelain and piano keys, mix 1:1 with water. |
| Rubbing Alcohol ( Isopropyl ) | 2-Propanol is the actual chemical, $CH_3CHOHCH_3$, soluble in water and other alcohols, general cleaner and disinfectant, specifically used to clean tape recorder heads and computer disk drive heads. |

# PAINTS AND FINISHES

| Type | Characteristics |
|---|---|
| House Paints ...... | There are basically 5 groups of house paints. Oil Base, Alkyd, Emulsion, Water Thinned, and Catalytic. Each of these classes is subdivided into Exterior and Interior. |
| Oil Base ....... | Interior and Exterior, oil vehicle, thinned by solvents such as turpentine and mineral spirits, very slow drying, strong smell. Mainly used as Exterior paint. Use in well ventilated area. Good adhesion to chalky surfaces. |
| Alkyd .......... | Synthetic oil vehicle of a resin known as Alkyd. Interior and Exterior enamels, easy to apply, fast drying, odorless, and produce a tough coating. Easy cleanup and thinning with mineral spirits. Excellent interior paint, not resistant to chemicals, solvents, or corrosives. |
| Emulsion ...... | Water based paint mixture. Latex paints fall into this category and the most common are acrylic and vinyl (PVA). Available as interior and exterior, and as flat, gloss, and semigloss enamels. Very quick drying (sometimes less than 1 hour) but do not wash for 2 to 4 weeks, paints over damp surfaces, odorless, alkalis resistant, doesn't usually blister and peel. Excellent cover and blending characteristics, but poor adhesion to chalky surfaces, easy cleanup. Use special latex primers for painting bare wood. Paint at temperatures above 45°F. By far the most popular paint today. |
| Water–Thinned | Generally used to describe non–emulsion paint such as calcimine and casein and white wash. These paints are used primarily on masonry surfaces. The most common water thinned paint being Portland Cement Paints. |
| Catalytic ....... | This class of paints cures by a chemical heat process, not by evaporation of a solvent or water as in the other paints. Catalytic paints are usually two part paints which means that you have to mix two |

| Type | Characteristics |
|---|---|
| | parts to start the curing process. Included in this class are the epoxy and polyurethane resins. They are extremely tough and durable and are highly resistant to water, wear, acids, solvents, abrasion, salt water, and chemicals. Drying times are very fast (several hours). Good ventilation is necessary when working with these paints. Catalytic paints can not be applied over other paint types. Follow the manufacturers instructions very closely, these are not easy paints to use. |

House paints are further subgroupped into Exterior and Interior types as described below:

| | |
|---|---|
| Exterior Paint | These paints are designed to have long life spans, good adhesion and resistance to moisture, ultraviolet light, mildew, and sulfide and acid fumes. This class also includes the varnish and stain groups described later. Never use interior paints in place of exterior paints, they will not hold up under the weather. |
| Interior Paint | Interior paints are designed to maximize the hiding ability of the paint with only 1 or 2 coats. Flat paints contain more pigment than high sheen paints but are less durable. Good interior paints can be touched up easily without major changes in the sheen or color. |

| | |
|---|---|
| Varnish | Varnish is a solution of a hard resin, a drying oil, metallics for driers, and solvent. There are two types, natural and synthetic. Natural types are slow drying (24 to 48 hrs) and are subclassed as "long oil" (meaning high oil content) and "short oil" (meaning less oil content). Naturals are tough and used mainly for exterior and marine applications. Synthetics contain resins such as alkyd (the most common), polyurethane, vinyl, and phenolic and are more durable and faster drying than naturals. Apply with natural bristle brushes; apply 3 to 4 coats |

| Type | Characteristics |
|---|---|
| | total, let dry between coats and sand with 240 grit sandpaper. Varnishes are usually transparent and are excellent sealers. |
| Shellac .......... | Shellac is one of the oldest wood finishes. It is made from a mixture of the dry resinous secretions of the lac bug (SE Asia) and alcohol. Once mixed, shellac has a very short shelf life, so store it in flake form. Shellac is mixed in what is called a "cut". A "3 pound cut" is 3 pounds of shellac in 1 gallon of alcohol. Initial coats are typically 1 or 2 pound cuts. Shellac is applied with a brush and the better finishes use 6 to 8 coats. Each coat should be sanded with 220 to 240 grit sand paper after it has dried (1 to 2 hours). The final coat is typically rubbed out with a fine 3/0 steel wool. |
| Automotive Paints: | |
| Urethane Enamel | The best of the car finishes, lasts over 10 years, has the best look, and is the most expensive. Paint jobs can run over $1000 and paint cost alone ranges from $50 to $100 per gallon. |
| "Clear Coat" .... | This is the top coat of a two part paint. The Clear Coat is applied over a base coat of acrylic enamel or acrylic lacquer and produces a beautiful "wet look" finish just like a factory paint job. This type of finish is very difficult to apply and should be done by an expert. Has a life of 8 to 10 years and costs between $400 & $600. |
| Acrylic Lacquer . | Mid range auto paint, very fast drying, much higher gloss and better durability than the alkyd enamels. Must be machine polished after drying so it is more expensive than the Acrylic Enamel paints. Acrylic lacquer must not be painted over acrylic enamel. Expect to pay $300 to $500 for this paint job. Life span is 5 to 7 years. |

## PAINTS AND FINISHES

| Type | Characteristics |
|---|---|
| Acrylic Enamel | Mid range auto paint, very slow drying, much higher gloss and better durability than the alkyd enamels and acrylic lacquer. Acrylic enamel should not be painted over acrylic lacquer. Usually requires a heat booth to aid drying. Expect to pay $200 to $300 for a paint job. |
| Alkyd Enamel | Cheap paint with low durability (will some–times loose its gloss in less than 2 months). Paint life will only be 1 to 3 years. The paint job will probably only cost $100 to $200 and is commonly referred to as the "baked enamel" job since the vehicle is baked at 150°F in a heat booth to set the paint. |
| Lacquer | Lacquer is a fast–drying, high gloss varnish used by most furniture manufacturers as the top–coat finish. It is very hard, dries crystal clear and is highly resistant to alcohols, water, heat, and mild acids. Although the original lacquers came from insects and the sumac tree, almost all produced today are synthetic and are mixed with some combination of resins (better adhesion), nitrocellulose, linseed or castor oil (improves flexibility), vinyls, acrylics or synthetic polymers. The main problem with lacquers is that they dry so fast that it is sometimes difficult to get a good finish. Use a spray gun if possible. Multiple coats are usually necessary |
| Primers | Primers are paints intended to produce a good foundation for the overlying coats of paint. Exterior wood primers penetrate deeply into the surface, adhere tightly to the surface, and seal off the wood. Primers typically have an abundance of pigment to allow sanding if necessary. Metal primers are specifically designed to adhere to the metal and stop any oxidation (rusting). Automotive primers usually have a lot of resin included also. |

## PAINTS AND FINISHES

| Type | Characteristics |
|---|---|
| Oils ............. | Penetrating oils such as linseed oil, tung oil, and Danish rubbing oil make up a class of finishes that protect wood while leaving the grain and natural texture visible. Oils won't crack, chip, or scale off and provide a beautiful surface. The addition of resins such as polyurethane greatly increase the toughness of the surface and still maintain the clear finish. The oils are applied with a soft rag, allowed to sit for 30 minutes to allow the oil to soak in, then buffed with a soft clean rag. Buffing with fine 4/0 steel wood will improve the sheen. |

Miscellaneous:

| | |
|---|---|
| Fire–Retardants . | Paints that decompose by melting into a thick mass of cellular charred material that insulates the material it is painted on. The decomposition begins at a temperature below the combustion point of the substrate; ratings are based on the ability to suppress combustion. |
| Floor Paint ..... | Specialized coatings that hard substances such as epoxy and phenolic modified alkyds, chlorinated rubber, & varnish. The coatings must also be water resistant. |
| Texture Paint ... | Interior coatings for ceilings and walls that produce a matte finish. They can contain sand, styrene fragments, nut shells, perlite, volcanic ash, or any other coarse material. |
| Two Part Paints . | This class of paint is generally expensive & includes the epoxies, polyesters, urethanes, and styrene–solubilized polyesters. They are all thermosetting, i.e. they cure by heat once a reactant has been added. These paints are extremely tough and durable and chemically resistant. |

# POCKET REF

## Hardware



(See also TOOLS, page 361, for drill, tap, & die info)

# BOLT TORQUE SPECIFICATIONS

| Bolt Size Inches | Coarse Thread / inch | SAE 0–1–2 74,000 psi Low Carbon Steel | SAE Grade 3 100,000 psi Med. Carbon Steel | SAE Grade 5 120,000 psi Med.Carbon Heat T. Steel |
|---|---|---|---|---|
| | | Standard Dry Torque in Foot–Pounds | | |
| 1/4 | 20 | 6 | 9 | 10 |
| 5/16 | 18 | 12 | 17 | 19 |
| 3/8 | 16 | 20 | 30 | 33 |
| 7/16 | 14 | 32 | 47 | 54 |
| 1/2 | 13 | 47 | 69 | 78 |
| 9/16 | 12 | 69 | 103 | 114 |
| 5/8 | 11 | 96 | 145 | 154 |
| 3/4 | 10 | 155 | 234 | 257 |
| 7/8 | 9 | 206 | 372 | 382 |
| 1 | 8 | 310 | 551 | 587 |
| 1–1/8 | 7 | 480 | 872 | 794 |
| 1–1/4 | 7 | 675 | 1211 | 1105 |
| 1–3/8 | 6 | 900 | 1624 | 1500 |
| 1–1/2 | 6 | 1100 | 1943 | 1775 |
| 1–5/8 | 5.5 | 1470 | 2660 | 2425 |
| 1–3/4 | 5 | 1900 | 3463 | 3150 |
| 1–7/8 | 5 | 2360 | 4695 | 4200 |
| 2 | 4.5 | 2750 | 5427 | 4550 |

In order to determine the torque for a fine thread bolt, increase the above coarse thread ratings by 9%.

# Effect of Lubrication on Torque

| Lubricant | Torque Rating in Foot–Pounds 5/16–18 thread/inch | 1/2–13 thread/inch |
|---|---|---|
| NO LUBE, steel | 29 | 121 |
| Plated & cleaned | 19 ( 66%) | 90 ( 26%) |
| SAE 20 oil | 18 ( 38%) | 87 ( 28%) |
| SAE 40 oil | 17 ( 41%) | 83 ( 31%) |
| Plated & SAE 30 | 16 ( 45%) | 79 ( 35%) |
| White grease | 16 ( 45%) | 79 ( 35%) |
| Dry Moly film | 14 ( 52%) | 66 ( 45%) |
| Graphite & oil | 13 ( 55%) | 62 ( 49%) |

Use the above lubrication percentages to calculate the approximate decrease in torque rating for other bolt sizes.

# BOLT TORQUE SPECIFICATIONS

| Bolt Size Inches | Coarse Thread / inch | SAE Grade 6 133,000 psi Med Carbon Temp. Steel | SAE Grade 7 133,000 psi Med. Carbon Alloy Steel | SAE Grade 8 150,000 psi Med.Carbon Alloy Steel |
|---|---|---|---|---|
| | | Standard Dry Torque in Foot–Pounds | | |
| 1/4 | 20 | 12.5 | 13 | 14 |
| 5/16 | 18 | 24 | 25 | 29 |
| 3/8 | 16 | 43 | 44 | 47 |
| 7/16 | 14 | 69 | 71 | 78 |
| 1/2 | 13 | 106 | 110 | 119 |
| 9/16 | 12 | 150 | 154 | 169 |
| 5/8 | 11 | 209 | 215 | 230 |
| 3/4 | 10 | 350 | 360 | 380 |
| 7/8 | 9 | 550 | 570 | 600 |
| 1 | 8 | 825 | 840 | 700 |
| 1–1/8 | 7 | 1304 | 1325 | 1430 |
| 1–1/4 | 7 | 1815 | 1825 | 1975 |
| 1–3/8 | 6 | 2434 | 2500 | 2650 |
| 1–1/2 | 6 | 2913 | 3000 | 3200 |
| 1–5/8 | 5.5 | 3985 | 4000 | 4400 |
| 1–3/4 | 5 | 5189 | 5300 | 5650 |
| 1–7/8 | 5 | 6980 | 7000 | 7600 |
| 2 | 4.5 | 7491 | 7500 | 8200 |

In order to determine the torque for a fine thread bolt increase the above coarse thread ratings by 9%.

Grades over Grade 8 are not common commercially, except in aircraft use. The following are a few of those types:

Supertanium, 160,000 psi, 8 points on head, quenched and tempered special alloy steel.

A354BD;A490, 150,000 psi, no markings, med. carbon quenched and tempered steel.

N.A.S. 144, MS2000, Military and Aircraft Std, 160,000 psi, high carbon alloy, quenched and tempered.

N.A.S. 623, National Aircraft Standard, 180,000 psi, high carbon alloy, quenched and tempered.

Aircraft Assigned Steel, no number, 220,000 psi, high carbon alloy, quenched and tempered.

Torque ratings for the above special alloy bolts should be obtained from the manufacturer.

# BOLT TORQUE SPECIFICATIONS

| Bolt Size Inches or Number | Coarse Thread / inch | Allen Head 160,000 psi High Carbon CaseH Steel | Machine Scr. 60,000 psi Yellow Brass | Machine Scr. 70,000 psi Silicone Bronze |
|---|---|---|---|---|
| | | Standard Dry Torque in Foot–Pounds | | |
| #2 | 56 | ... | 2 in# | 2.3 in# |
| #3 | 48 | ... | 3.3 in# | 3.7 in# |
| #4 | 40 | ... | 4.4 in# | 4.9 in# |
| #5 | 40 | ... | 6.4 in# | 7.2 in# |
| #6 | 32 | 21 | 8 in# | 10 in# |
| #8 | 32 | 46 | 16 in# | 19 in# |
| #10 | 24 | 60 | 20 in# | 22 in# |
| 1/4 | 20 | 16 | 65 in# | 70 in# |
| 5/16 | 18 | 33 | 110 in# | 125 in# |
| 3/8 | 16 | 54 | 17 | 20 |
| 7/16 | 14 | 84 | 27 | 30 |
| 1/2 | 13 | 125 | 37 | 41 |
| 9/16 | 12 | 180 | 49 | 53 |
| 5/8 | 11 | 250 | 78 | 88 |
| 3/4 | 10 | 400 | 104 | 117 |
| 7/8 | 9 | 640 | 160 | 180 |
| 1 | 8 | 970 | 215 | 250 |
| 1–1/8 | 7 | 1520 | 325 | 365 |
| 1–1/4 | 7 | 2130 | 400 | 450 |
| 1–3/8 | 6 | 2850 | ... | ... |
| 1–1/2 | 6 | 3450 | 595 | 655 |
| 1–5/8 | | 4700 | ... | ... |
| 1–3/4 | 5 | 6100 | ... | ... |
| 1–7/8 | | 8200 | ... | ... |
| 2 | 4-1/2 | 8800 | ... | ... |

In order to determine the torque for a fine thread bolt increase the above coarse thread ratings by 9%.

Socket Set Screws (looks like an allen head without the head) are usually rated at 212,000 psi, and are high carbon, case hardened steel. Torque ratings are as follows: #6 = 9 in-lbs, #8 = 16 in-lbs, #10 = 30 in-lbs, 1/4 in = 70 in-lbs, 5/16 in = 140 in-lbs, 3/8 in = 18 ft-lbs, 7/16 in = 29 ft-lbs, 1/2 in = 43 ft-lbs, 9/16 in = 63 ft-lbs, 5/8 in = 100 ft-lbs, and 3/4 in = 146 ft-lbs.

# BOLT TORQUE SPECIFICATIONS

## METRIC

| Bolt Size Milli– meters | Coarse Thread Pitch | 5D<br>Standard 5D 71,160 psi Med Carbon Steel | 8G<br>Standard 8G 113,800 psi Med Carbon Steel | 10K<br>Standard 10K 142,000 psi Med Carbon Steel |
|---|---|---|---|---|
| | | Standard Dry Torque in Foot–Pounds | | |
| 6 mm | 1.00 | 5 | 6 | 8 |
| 8 mm | 1.00 | 10 | 16 | 22 |
| 10 mm | 1.25 | 19 | 31 | 40 |
| 12 mm | 1.25 | 34 | 54 | 70 |
| 14 mm | 1.25 | 55 | 89 | 117 |
| 16 mm | 2.00 | 83 | 132 | 175 |
| 18 mm | 2.00 | 111 | 182 | 236 |
| 22 mm | 2.50 | 182 | 284 | 394 |
| 24 mm | 3.00 | 261 | 419 | 570 |

## METRIC

| Bolt Size Milli– meters | Coarse Thread Pitch | 12K<br>Standard 12K 170,674 psi Med Carbon Steel |
|---|---|---|
| | | Standard Dry Torque in Foot–Pounds |
| 6 mm | 1.00 | 10 |
| 8 mm | 1.00 | 27 |
| 10 mm | 1.25 | 49 |
| 12 mm | 1.25 | 86 |
| 14 mm | 1.25 | 137 |
| 16 mm | 2.00 | 208 |
| 18 mm | 2.00 | 283 |
| 22 mm | 2.50 | 464 |
| 24 mm | 3.00 | 689 |

In order to determine the torque for a fine thread bolt increase the above coarse thread ratings by 9%. See first page of Bolt Torque Specifications for information on the Effects of Lubrication on Bolt Torque.

## Whitworth

| Bolt Size Inches | Coarse Thread / inch | Grades A & B 62,720 psi Med Carbon Steel | Grade S 112,000 psi Med Carbon Steel | Grade T 123,200 psi Med Carbon Steel |
|---|---|---|---|---|
| | | Standard Dry Torque in Foot–Pounds | | |
| 1/4 | 20 | 5 | 7 | 9 |
| 5/16 | 18 | 9 | 15 | 18 |
| 3/8 | 16 | 15 | 27 | 31 |
| 7/16 | 14 | 24 | 43 | 51 |
| 1/2 | 12 | 36 | 64 | 79 |
| 9/16 | 12 | 52 | 94 | 111 |
| 5/8 | 11 | 73 | 128 | 155 |
| 3/4 | 11 | 118 | 213 | 259 |
| 7/8 | 9 | 186 | 322 | 407 |
| 1 | 8 | 276 | 497 | 611 |

## Whitworth

| Bolt Size Inches | Coarse Thread / inch | Grade V 145,600 psi Med Carbon Steel |
|---|---|---|
| | | Standard Dry Torque in Foot–Pounds |
| 1/4 | 20 | 10 |
| 5/16 | 18 | 21 |
| 3/8 | 16 | 36 |
| 7/16 | 14 | 58 |
| 1/2 | 12 | 89 |
| 9/16 | 12 | 128 |
| 5/8 | 11 | 175 |
| 3/4 | 11 | 287 |
| 7/8 | 9 | 459 |
| 1 | 8 | 693 |

In order to determine the torque for a fine thread bolt increase the above coarse thread ratings by 9%. See first page of Bolt Torque Specifications for information on the Effects of Lubrication on Bolt Torque.

# WOOD SCREW SPECIFICATIONS

| Screw Number | Pilot Hole Sizes Hard Wood Drill Number | Soft Wood Drill Number | Shank Diameter Inches | Shank Hole Clearance Drill Number |
|---|---|---|---|---|
| 0 | 66 | 75 | 0.060 | 52 |
| 1 | 57 | 71 | 0.073 | 47 |
| 2 | 54 | 65 | 0.086 | 42 |
| 3 | 53 | 58 | 0.099 | 37 |
| 4 | 51 | 55 | 0.112 | 32 |
| 5 | 47 | 53 | 0.125 | 30 |
| 6 | 44 | 52 | 0.138 | 27 |
| 7 | 39 | 51 | 0.151 | 22 |
| 8 | 35 | 48 | 0.164 | 18 |
| 9 | 33 | 45 | 0.177 | 14 |
| 10 | 31 | 43 | 0.190 | 10 |
| 11 | 29 | 40 | 0.203 | 4 |
| 12 | 25 | 38 | 0.216 | 2 |
| 14 | 14 | 32 | 0.242 | D |
| 16 | 10 | 29 | 0.268 | I |
| 18 | 6 | 26 | 0.294 | N |
| 20 | 3 | 19 | 0.320 | P |
| 24 | D | 15 | 0.372 | V |

See the chapter on Tools for additional drill information and Drill Number to Inch conversions.

# Wood Screw Number vs Std Lengths

| Screw Number | Standard Lengths in Inches |
|---|---|
| 0 | 1/4 |
| 1 | 1/4, 3/8 |
| 2 | 1/4, 3/8, 1/2 |
| 3 | 1/4, 3/8, 1/2, 5/8 |
| 4 | 3/8, 1/2, 5/8, 3/4 |
| 5 | 3/8, 1/2, 5/8, 3/4 |
| 6 | 3/8, 1/2, 5/8, 3/4, 7/8, 1, 1–1/4, 1–1/2 |
| 7 | 3/8, 1/2, 5/8, 3/4, 7/8, 1, 1–1/4, 1–1/2 |
| 8 | 1/2, 5/8, 3/4, 7/8, 1, 1–1/4, 1–1/2, 1–3/4, 2 |
| 9 | 5/8, 3/4, 7/8, 1, 1–1/4, 1–1/2, 1–3/4, 2, 2–1/4 |
| 10 | 5/8, 3/4, 7/8, 1, 1–1/4, 1–1/2, 1–3/4, 2, 2–1/4 2-1/2 |
| 11 | 3/4, 7/8, 1, 1–1/4, 1–1/2, 1–3/4, 2, 2–1/4, 2–1/2, 2–3/4, 3 |
| 12 | 7/8, 1, 1–1/4, 1–1/2, 1–3/4, 2, 2–1/4, 2–1/2, 2–3/4, 3, 3–1/2 |
| 14 | 1, 1–1/4, 1–1/2, 1–3/4, 2, 2–1/4, 2–1/2, 2–3/4 3, 3–1/2, 4, 4-1/2 |
| 16 | 1–1/4, 1–1/2, 1–3/4, 2, 2–1/4, 2–1/2, 2–3/4, 3 3–1/2, 4, 4–1/2, 5, 5–1/2 |
| 18 | 1–1/2, 1–3/4, 2, 2–1/4, 2–1/2, 2–3/4, 3, 3-1/2,4 4–1/2, 5, 5–1/2, 6 |
| 20 | 1–3/4, 2, 2–1/4, 2–1/2, 2–3/4, 3, 3–1/2, 4, 4-1/2 5, 5–1/2, 6 |
| 24 | 3–1/2, 4, 4–1/2, 5, 5–1/2, 6 |

# SHEET METAL SCREW SPECS

| Screw Diameter # (inch) | Thickness of Metal Gauge # | Diameter of Pierced Hole (inch) | Drilled Hole Size Drill Number |
|---|---|---|---|
| #4 (0.112) | 28 | 0.086 | 44 |
| | 26 | 0.086 | 44 |
| | 24 | 0.093 | 42 |
| | 22 | 0.098 | 42 |
| | 20 | 0.100 | 40 |
| #6 (0.138) | 28 | 0.111 | 39 |
| | 26 | 0.111 | 39 |
| | 24 | 0.111 | 39 |
| | 22 | 0.111 | 38 |
| | 20 | 0.111 | 36 |
| #7 (0.155) | 28 | 0.121 | 37 |
| | 26 | 0.121 | 37 |
| | 24 | 0.121 | 35 |
| | 22 | 0.121 | 33 |
| | 20 | 0.121 | 32 |
| | 18 | .... | 31 |
| #8 (0.165) | 26 | 0.137 | 33 |
| | 24 | 0.137 | 33 |
| | 22 | 0.137 | 32 |
| | 20 | 0.137 | 31 |
| | 18 | .... | 30 |
| #10 (0.191) | 26 | 0.158 | 30 |
| | 24 | 0.158 | 30 |
| | 22 | 0.158 | 30 |
| | 20 | 0.158 | 29 |
| | 18 | 0.158 | 25 |
| #12 (0.218) | 24 | .... | 26 |
| | 22 | 0.185 | 25 |
| | 20 | 0.185 | 24 |
| | 18 | 0.185 | 22 |
| #14 (0.251) | 24 | .... | 15 |
| | 22 | 0.212 | 12 |
| | 20 | 0.212 | 11 |
| | 18 | 0.212 | 9 |

Note: The above values are recommended average values only. Variations in materials and local conditions may require significant deviations from the recommended values.

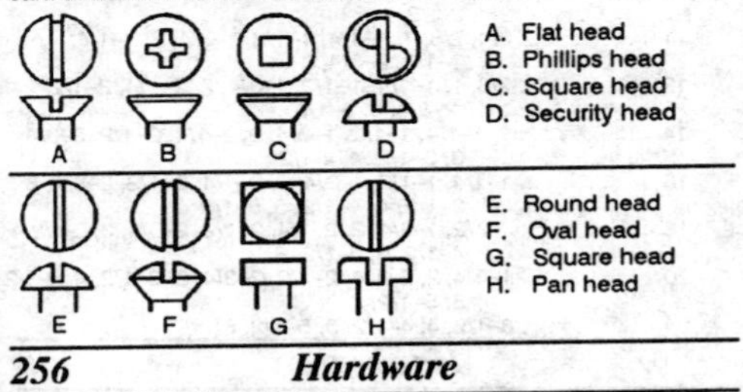


A. Flat head
B. Phillips head
C. Square head
D. Security head

E. Round head
F. Oval head
G. Square head
H. Pan head

# CABLE CLAMPS FOR WIRE ROPE

| Rope Diameter Inches | Number of Clamps Required | Clip Spacing Inches | Rope Turn-back Inches |
|---|---|---|---|
| 1/8 | 2 | 3 | 3–1/4 |
| 3/16 | 2 | 3 | 3–3/4 |
| 1/4 | 2 | 3–1/4 | 4–3/4 |
| 5/16 | 2 | 3–1/4 | 5–1/4 |
| 3/8 | 2 | 4 | 6–1/2 |
| 7/16 | 2 | 4–1/2 | 7 |
| 1/2 | 3 | 5 | 11–1/2 |
| 9/16 | 3 | 5–1/2 | 12 |
| 5/8 | 3 | 5–3/4 | 12 |
| 3/4 | 4 | 6–3/4 | 18 |
| 7/8 | 4 | 8 | 19 |
| 1 | 5 | 8–3/4 | 26 |
| 1–1/8 | 6 | 9–3/4 | 34 |
| 1–1/4 | 6 | 10–3/4 | 37 |
| 1–7/16 | 7 | 11–1/2 | 44 |
| 1–1/2 | 7 | 12–1/2 | 48 |
| 1–5/8 | 7 | 13–1/4 | 51 |
| 1–3/4 | 7 | 14–1/2 | 53 |
| 2 | 8 | 16–1/2 | 71 |
| 2–1/4 | 8 | 16–1/2 | 73 |
| 2–1/2 | 9 | 17–3/4 | 84 |
| 2–3/4 | 10 | 18 | 100 |
| 3 | 10 | 18 | 106 |

The above Number of Clamps and Spacing is based on data for Crosby–Laughlin clamps, Fort Wayne, Indiana, A Division of American Hoist. If more specific or detailed information is needed, contact Crosby–Laughlin. The data assumes approximately 80% efficiency and is based on the estimated breaking strength of a right regular or Lang lay wire rope in the 6 x 19 or 6 x 37 class.

When placing cable clamps on the wire, it is imperative that the U–bolt side of the clip is placed on the short, turn–back side and the saddle goes on the long side (the "live" end). Torque the nuts down to the specified torque for the particular U–bolt diameter, place a load on the wire, and then re–torque the clamps. If torque specs are not available from the manufacturer, refer to page number 249 in the HARDWARE chapter for approximate torque ratings.

# NAILS

The "d" listed after each Size Number stands for "penny", which was originally used in old England as a way of describing the number of pennies needed to buy 100 nails. Today, "penny" is used only to define the length of the nail.

## COMMON NAILS
### For General Construction

| Size Number | Length Inches | Shaft Diameter Gauge (inches) | Diameter of Head, inches | Number per Pound |
|---|---|---|---|---|
| 2d | 1 | 15 (0.072) | 11/64 | 840 |
| 3d | 1-1/4 | 14 (0.080) | 13/64 | 530 |
| 4d | 1-1/2 | 12-1/2 (0.095) | 1/4 | 300 |
| 5d | 1-3/4 | 12-1/2 (0.095) | 1/4 | 260 |
| 6d | 2 | 11-1/2 (0.113) | 17/64 | 170 |
| 7d | 2-1/4 | 11-1/2 (0.113) | 17/64 | 150 |
| 8d | 2-1/2 | 10-1/4 (0.131) | 9/32 | 105 |
| 9d | 2-3/4 | 10-1/4 (0.131) | 9/32 | 95 |
| 10d | 3 | 9 (0.148) | 5/16 | 65 |
| 12d | 3-1/4 | 9 (0.148) | 5/16 | 60 |
| 16d | 3-1/2 | 8 (0.162) | 11/32 | 44 |
| 20d | 4 | 6 (0.192) | 13/32 | 30 |
| 30d | 4-1/2 | 5 (0.207) | 7/16 | 22 |
| 40d | 5 | 4 (0.225) | 15/32 | 18 |
| 50d | 5-1/2 | 3 (0.244) | 1/2 | 14 |
| 60d | 6 | 2 (0.263) | 17/32 | 10 |

## BOX NAILS
### For Light Construction

| Size Number | Length Inches | Shaft Diameter Gauge (inches) | Diameter of Head, inches | Number per Pound |
|---|---|---|---|---|
| 2d | 1 | 15-1/2 (0.067) | 11/64 | 1010 |
| 3d | 1-1/4 | 14-1/2 (0.073) | 13/64 | 620 |
| 4d | 1-1/2 | 14 (0.080) | 1/4 | 450 |
| 5d | 1-3/4 | 14 (0.080) | 1/4 | 375 |
| 6d | 2 | 12-1/2 (0.095) | 17/64 | 230 |
| 7d | 2-1/4 | 12-1/2 (0.095) | 17/64 | 200 |
| 8d | 2-1/2 | 11-1/2 (0.109) | 9/32 | 130 |
| 10d | 3 | 10-1/2 (0.128) | 5/16 | 88 |
| 12d | 3-1/4 | 10-1/2 (0.128) | 5/16 | 80 |
| 16d | 3-1/2 | 10 (0.135) | 11/32 | 70 |
| 20d | 4 | 9 (0.148) | 13/32 | 52 |
| 30d | 4-1/2 | 9 (0.148) | 7/16 | 45 |
| 40d | 5 | 8 (0.162) | 15/32 | 35 |

## COMMON WIRE SPIKES
### For Heavy Construction

| Size Number | Length Inches | Shaft Diameter Gauge (inches) | Diameter of Head, Gauge | Number per Pound |
|---|---|---|---|---|
| 10d | 3 | 6 (0.192) | 3 | 43 |
| 12d | 3–1/4 | 6 (0.192) | 3 | 39 |
| 16d | 3–1/2 | 5 (0.207) | 2 | 31 |
| 20d | 4 | 4 (0.225) | 1 | 23 |
| 30d | 4–1/2 | 3 (0.244) | 0 | 18 |
| 40d | 5 | 2 (0.263) | 2/0 | 14 |
| 50d | 5–1/2 | 1 (0.283) | 3/0 | 11 |
| 60d | 6 | 1 (0.283) | 3/0 | 9 |
| 5/6 in | 7 | (0.312) | 0.370 in | 7 |
| 3/8 in | 8–1/2 | (0.375) | 0.433 in | 4 |

## CASING NAILS
### For Interior Trim

| Size Number | Length Inches | Shaft Diameter Gauge (inches) | Diameter of Head, Gauge | Number per Pound |
|---|---|---|---|---|
| 3d | 1–1/4 | 14–1/2 (0.073) | 11–1/2 | 625 |
| 4d | 1–1/2 | 14 (0.080) | 11 | 490 |
| 6d | 2 | 12–1/2 (0.095) | 9–1/2 | 250 |
| 8d | 2–1/2 | 11–1/2 (0.113) | 8–1/2 | 145 |
| 10d | 3 | 10–1/2 (0.128) | 7–1/2 | 95 |
| 16d | 3–1/2 | 10 (0.135) | 7 | 70 |
| 20d | 4 | 9 (0.148) | 6 | 52 |

## FINISHING NAILS
### For Cabinet Work and Interior Trim

| Size Number | Length Inches | Shaft Diameter Gauge (inches) | Diameter of Head, Gauge | Number per Pound |
|---|---|---|---|---|
| 2d | 1 | 16–1/2 (0.058) | 13–1/2 | 1350 |
| 3d | 1–1/4 | 15–1/2 (0.067) | 12–1/2 | 850 |
| 4d | 1–1/2 | 15 (0.072) | 12 | 550 |
| 5d | 1–3/4 | 15 (0.072) | 12 | 500 |
| 6d | 2 | 13 (0.091) | 10 | 300 |
| 8d | 2–1/2 | 12–1/2 (0.095) | 9–1/2 | 190 |
| 10d | 3 | 11–1/2 (0.113) | 8–1/2 | 125 |
| 16d | 3–1/2 | 11 (0.120) | 8 | 90 |
| 20d | 4 | 10 (0.135) | 7 | 60 |

## CONCRETE NAILS
### Round, Square, or Fluted
### For Fastening to Concrete

| Length Inches | Number per Pound |
|---|---|
| **5 Gauge (0.207 inch) Nail with 1/2 inch head:** | |
| 1/2 | 190 |
| 5/8 | 150 |
| 3/4 | 130 |
| 7/8 | 115 |
| 1 | 99 |
| 1–1/8 | 89 |
| 1–1/4 | 80 |
| 1–1/2 | 65 |
| 1–3/4 | 60 |
| 2 | 51 |
| 2–1/4 | 45 |
| 2–1/2 | 40 |
| 2–3/4 | 37 |
| 3 | 34 |
| **7 Gauge (0.177 inch) Nail with 3/8 inch head:** | |
| 1/2 | 330 |
| 5/8 | 260 |
| 3/4 | 210 |
| 7/8 | 175 |
| 1 | 155 |
| 1–1/8 | 135 |
| 1–1/4 | 120 |
| 1–1/2 | 100 |
| 1–3/4 | 85 |
| 2 | 75 |
| 2–1/4 | 65 |
| 2–1/2 | 60 |
| 2–3/4 | 55 |
| 3 | 50 |
| **9 Gauge (0.148 inch) Nail with 21/64 inch head:** | |
| 1/2 | 440 |
| 5/8 | 350 |
| 3/4 | 285 |
| 7/8 | 240 |
| 1 | 210 |
| 1–1/8 | 185 |
| 1–1/4 | 165 |
| 1–1/2 | 140 |
| 1–3/4 | 120 |
| 2 | 105 |
| 2–1/4 | 90 |
| 2–1/2 | 85 |
| 2–3/4 | 75 |
| 3 | 69 |
| 3–3/4 | 64 |

# NAILS

## ROOFING NAILS

| Size Number | Length Inches | Number per Pound |
|---|---|---|
| | 7/8 | 250 |
| 2d | 1 | 225 |
| 3d | 1-1/4 | 190 |
| 4d | 1-1/2 | 165 |
| 5d | 1-3/4 | 145 |

## SPIRAL FLOORING NAILS

| Size Number | Length Inches | Number per Pound |
|---|---|---|
| 6d | 2 | 177 |
| 7d | 2-1/4 | 158 |
| 8d | 2-1/2 | 142 |

## FENCE STAPLES

| Length Inches | Number per Pound |
|---|---|
| 7/8 | 125 |
| 1 | 105 |
| 1-1/4 | 88 |
| 1-1/2 | 72 |
| 1-3/4 | 60 |

## WIRE TACKS

| Size Oz | Length Inches | Shaft Diameter Gauge |
|---|---|---|
| 1 | 3/16 | 18 |
| 1-1/2 | 7/32 | 18 |
| 2 | 1/4 | 17 |
| 2-1/2 | 5/16 | 17 |
| 3 | 3/8 | 16 |
| 4 | 7/16 | 16 |
| 6 | 1/2 | 15 |
| 8 | 9/16 | 15 |
| 10 | 5/8 | 14-1/2 |
| 12 | 11/16 | 14-1/2 |
| 14 | 3/4 | 14 |
| 16 | 13/16 | 14 |
| 18 | 7/8 | 13-1/2 |
| 20 | 15/16 | 13-1/2 |
| 22 | 1 | 13-1/2 |
| 24 | 1-1/8 | 13 |

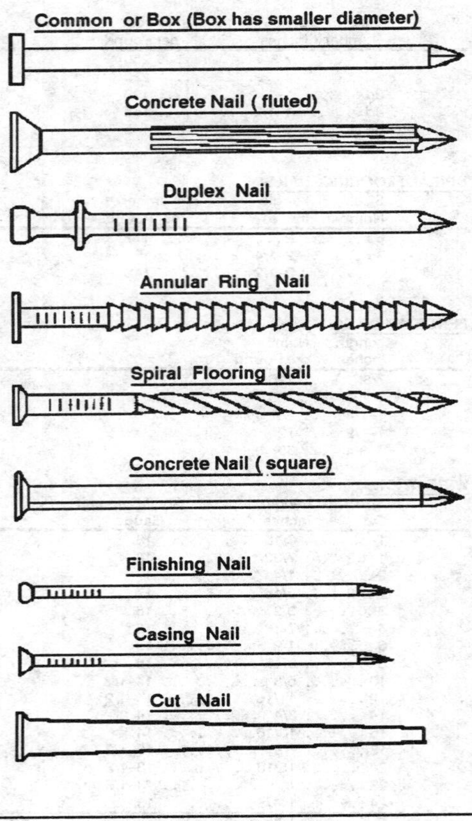
Common or Box (Box has smaller diameter)
Concrete Nail ( fluted)
Duplex Nail
Annular Ring Nail
Spiral Flooring Nail
Concrete Nail ( square)
Finishing Nail
Casing Nail
Cut Nail

# POCKET REF

# Math




(See also CONSTANTS, page 104 )

(See also SURVEYING, page 349)

(See also COMPUTER, page 61)

## NUMERIC PREFIXES

| Prefix | Abbreviation | Pronounce | Multiplier |
|---|---|---|---|
| atto | a | at–to | $10^{-18}$ |
| femto | f | fem–to | $10^{-15}$ |
| pico | p | pe–ko | $10^{-12}$ |
| nano | n | nan–o | $10^{-9}$ |
| micro | $\mu$ | mi–kro | $10^{-6}$ |
| milli | m | mil – I | $10^{-3}$ |
| centi | c | sent–ti | $10^{-2}$ |
| deci | d | des – I | $10^{-1}$ |
| deka | da | dek–a | $10^{1}$ |
| hecto | h | hek–to | $10^{2}$ |
| kilo | k | kil–o | $10^{3}$ |
| mega | M | meg–a | $10^{6}$ |
| giga | G | ji–ga | $10^{9}$ |
| tera | T | ter–a | $10^{12}$ |
| peta | P | pe–ta | $10^{15}$ |
| exa | E | ex–a | $10^{18}$ |
| | | sextillion | $10^{21}$ |
| | | septillion | $10^{24}$ |
| | | octillion | $10^{27}$ |
| | | nonillion | $10^{30}$ |

## ROMAN NUMERALS

| Roman | Arabic | Roman | Arabic |
|---|---|---|---|
| I | 1 | LXX | 70 |
| II | 2 | LXXX | 80 |
| III | 3 | XC | 90 |
| IV | 4 | C | 100 |
| V | 5 | CC | 200 |
| VI | 6 | CCC | 300 |
| VII | 7 | CD | 400 |
| VIII | 8 | D | 500 |
| IX | 9 | DC | 600 |
| X | 10 | CM | 900 |
| XI | 11 | M | 1000 |
| XII | 12 | MD | 1500 |
| XX | 20 | M$\overline{V}$ | 4000 |
| XXX | 30 | $\overline{V}$ | 5000 |
| XL | 40 | $\overline{X}$ | 10000 |
| L | 50 | $\overline{XX}$ | 20000 |
| LX | 60 | $\overline{C}$ | 100000 |

| Decimals of Inch | Fractions of Inch | Decimals of Foot | Millimeters | Drill Number |
|---|---|---|---|---|
| 0.001 | | 0.00008 | 0.0254 | |
| 0.002 | | 0.00017 | 0.0508 | |
| 0.003 | | 0.00025 | 0.0762 | |
| 0.004 | | 0.00033 | 0.1016 | |
| 0.005 | | 0.00042 | 0.1270 | |
| 0.006 | | 0.00050 | 0.1524 | |
| 0.007 | | 0.00058 | 0.1778 | |
| 0.0078 | 1/128 | 0.00065 | 0.1981 | |
| 0.008 | | 0.00067 | 0.2032 | |
| 0.009 | | 0.00075 | 0.2286 | |
| 0.010 | | 0.00083 | 0.2540 | |
| 0.011 | | 0.00092 | 0.2794 | |
| 0.012 | | 0.00100 | 0.3048 | |
| 0.013 | | 0.00108 | 0.3302 | |
| 0.0135 | | 0.00112 | 0.3429 | 80 |
| 0.014 | | 0.00117 | 0.3556 | |
| 0.0145 | | 0.00121 | 0.3683 | 79 |
| 0.015 | | 0.00125 | 0.3810 | |
| 0.0156 | 1/64 | 0.00130 | 0.3962 | |
| 0.016 | | 0.00133 | 0.4064 | 78 |
| 0.017 | | 0.00142 | 0.4318 | |
| 0.018 | | 0.00150 | 0.4572 | 77 |
| 0.019 | | 0.00158 | 0.4826 | |
| 0.020 | | 0.00167 | 0.5080 | 76 |
| 0.021 | | 0.00175 | 0.5334 | 75 |
| 0.022 | | 0.00183 | 0.5588 | |
| 0.0225 | | 0.00187 | 0.5715 | 74 |
| 0.023 | | 0.00192 | 0.5842 | |
| 0.0234 | 3/128 | 0.00195 | 0.5944 | |
| 0.024 | | 0.00200 | 0.6096 | 73 |
| 0.025 | | 0.00208 | 0.6350 | 72 |
| 0.026 | | 0.00217 | 0.6604 | 71 |
| 0.027 | | 0.00225 | 0.6858 | |
| 0.028 | | 0.00233 | 0.7112 | 70 |
| 0.029 | | 0.00242 | 0.7366 | |
| 0.0292 | | 0.00243 | 0.7417 | 69 |
| 0.030 | | 0.00250 | 0.7620 | |
| 0.031 | | 0.00258 | 0.7874 | 68 |
| 0.0312 | 1/32 | 0.00260 | 0.7925 | |
| 0.032 | | 0.00267 | 0.8128 | 67 |
| 0.033 | | 0.00275 | 0.8382 | 66 |
| 0.034 | | 0.00283 | 0.8636 | |
| 0.035 | | 0.00292 | 0.8890 | 65 |
| 0.036 | | 0.00300 | 0.9144 | 64 |
| 0.037 | | 0.00308 | 0.9398 | 63 |
| 0.038 | | 0.00317 | 0.9652 | 62 |
| 0.039 | 5/128 | 0.00325 | 0.9906 | 61 |
| 0.040 | | 0.00333 | 1.0160 | 60 |
| 0.041 | | 0.00342 | 1.0414 | 59 |
| 0.042 | | 0.00350 | 1.0668 | 58 |
| 0.043 | | 0.00358 | 1.0922 | 57 |
| 0.044 | | 0.00367 | 1.1176 | |
| 0.045 | | 0.00375 | 1.1430 | |

| Decimals of Inch | Fractions of Inch | Decimals of Foot | Millimeters | Drill Number |
|---|---|---|---|---|
| 0.046 | | 0.00383 | 1.1684 | |
| 0.0465 | | 0.00387 | 1.1582 | 56 |
| 0.0469 | 3/64 | 0.00390 | 1.1913 | |
| 0.047 | | 0.00392 | 1.1938 | |
| 0.048 | | 0.00400 | 1.2192 | |
| 0.049 | | 0.00408 | 1.2446 | |
| 0.050 | | 0.00417 | 1.2700 | |
| 0.051 | | 0.00425 | 1.2954 | |
| 0.052 | | 0.00433 | 1.3208 | 55 |
| 0.053 | | 0.00442 | 1.3462 | |
| 0.054 | | 0.00450 | 1.3716 | |
| 0.0547 | 7/128 | 0.00456 | 1.3894 | |
| 0.055 | | 0.00458 | 1.3970 | 54 |
| 0.056 | | 0.00467 | 1.4224 | |
| 0.057 | | 0.00475 | 1.4478 | |
| 0.058 | | 0.00483 | 1.4732 | |
| 0.059 | | 0.00492 | 1.4986 | |
| 0.0595 | | 0.00496 | 1.5113 | 53 |
| 0.060 | | 0.00500 | 1.5240 | |
| 0.061 | | 0.00508 | 1.5494 | |
| 0.062 | | 0.00517 | 1.5748 | |
| 0.0625 | 1/16 | 0.00521 | 1.5875 | |
| 0.063 | | 0.00525 | 1.6002 | |
| 0.0635 | | 0.00529 | 1.6129 | 52 |
| 0.064 | | 0.00533 | 1.6256 | |
| 0.065 | | 0.00542 | 1.6510 | |
| 0.066 | | 0.00550 | 1.6764 | |
| 0.067 | | 0.00558 | 1.7018 | 51 |
| 0.068 | | 0.00567 | 1.7272 | |
| 0.069 | | 0.00575 | 1.7526 | |
| 0.070 | 9/128 | 0.00583 | 1.7780 | 50 |
| 0.071 | | 0.00592 | 1.8034 | |
| 0.072 | | 0.00600 | 1.8288 | |
| 0.073 | | 0.00608 | 1.8542 | 49 |
| 0.074 | | 0.00617 | 1.8796 | |
| 0.075 | | 0.00625 | 1.9050 | |
| 0.076 | | 0.00633 | 1.9304 | 48 |
| 0.077 | | 0.00642 | 1.9558 | |
| 0.078 | | 0.00650 | 1.9812 | |
| 0.0781 | 5/64 | 0.00651 | 1.9837 | |
| 0.0785 | | 0.00654 | 1.9939 | 47 |
| 0.079 | | 0.00658 | 2.0066 | |
| 0.080 | | 0.00667 | 2.0320 | |
| 0.081 | | 0.00675 | 2.0574 | 46 |
| 0.082 | | 0.00683 | 2.0828 | 45 |
| 0.083 | | 0.00692 | 2.1082 | |
| 0.084 | | 0.00700 | 2.1336 | |
| 0.085 | | 0.00708 | 2.1590 | |
| 0.086 | 11/128 | 0.00717 | 2.1844 | 44 |
| 0.087 | | 0.00725 | 2.2098 | |
| 0.088 | | 0.00733 | 2.2352 | |
| 0.089 | | 0.00742 | 2.2606 | 43 |

# CONVERT INCH–FOOT–MM–DRILL #

| Decimals of Inch | Fractions of Inch | Decimals of Foot | Millimeters | Drill Number |
|---|---|---|---|---|
| 0.090 | | 0.00750 | 2.2860 | |
| 0.091 | | 0.00758 | 2.3114 | |
| 0.092 | | 0.00767 | 2.3368 | |
| 0.093 | | 0.00775 | 2.3622 | |
| 0.0935 | | 0.00779 | 2.3749 | 42 |
| 0.0937 | 3/32 | 0.00781 | 2.3800 | |
| 0.094 | | 0.00783 | 2.3876 | |
| 0.095 | | 0.00792 | 2.4130 | |
| 0.096 | | 0.00800 | 2.4384 | 41 |
| 0.097 | | 0.00808 | 2.4638 | |
| 0.098 | | 0.00817 | 2.4892 | 40 |
| 0.099 | | 0.00825 | 2.5146 | |
| 0.0995 | | 0.00829 | 2.5273 | 39 |
| 0.100 | | 0.00833 | 2.5400 | |
| 0.101 | | 0.00842 | 2.5654 | |
| 0.1015 | | 0.00846 | 2.5781 | 38 |
| 0.1016 | 13/128 | 0.00847 | 2.5806 | |
| 0.102 | | 0.00850 | 2.5908 | |
| 0.103 | | 0.00858 | 2.6162 | |
| 0.104 | | 0.00867 | 2.6416 | 37 |
| 0.105 | | 0.00875 | 2.6670 | |
| 0.106 | | 0.00883 | 2.6924 | |
| 0.1065 | | 0.00887 | 2.7051 | 36 |
| 0.107 | | 0.00892 | 2.7178 | |
| 0.108 | | 0.00900 | 2.7432 | |
| 0.109 | | 0.00908 | 2.7686 | |
| 0.1094 | 7/64 | 0.00912 | 2.7788 | |
| 0.110 | | 0.00917 | 2.7940 | 35 |
| 0.111 | | 0.00925 | 2.8194 | 34 |
| 0.112 | | 0.00933 | 2.8448 | |
| 0.113 | | 0.00942 | 2.8702 | 33 |
| 0.114 | | 0.00950 | 2.8956 | |
| 0.115 | | 0.00958 | 2.9210 | |
| 0.116 | | 0.00967 | 2.9464 | 32 |
| 0.117 | | 0.00975 | 2.9718 | |
| 0.1172 | 15/128 | 0.00977 | 2.9769 | |
| 0.118 | | 0.00983 | 2.9972 | |
| 0.119 | | 0.00992 | 3.0226 | |
| 0.120 | | 0.01000 | 3.0480 | 31 |
| 0.121 | | 0.01008 | 3.0734 | |
| 0.122 | | 0.01017 | 3.0988 | |
| 0.123 | | 0.01025 | 3.1242 | |
| 0.124 | | 0.01033 | 3.1496 | |
| 0.125 | 1/8 | 0.01042 | 3.1750 | |
| 0.126 | | 0.01050 | 3.2004 | |
| 0.127 | | 0.01058 | 3.2258 | |
| 0.128 | | 0.01067 | 3.2512 | |
| 0.1285 | | 0.01071 | 3.2639 | 30 |
| 0.129 | | 0.01075 | 3.2766 | |
| 0.130 | | 0.01083 | 3.3020 | |
| 0.131 | | 0.01092 | 3.3274 | |
| 0.132 | | 0.01100 | 3.3528 | |
| 0.1328 | 17/128 | 0.01107 | 3.3731 | |

# CONVERT INCH–FOOT–MM–DRILL #

| Decimals of Inch | Fractions of Inch | Decimals of Foot | Millimeters | Drill Number |
|---|---|---|---|---|
| 0.133 | | 0.01108 | 3.3782 | |
| 0.134 | | 0.01117 | 3.4036 | |
| 0.135 | | 0.01125 | 3.4290 | |
| 0.136 | | 0.01133 | 3.4544 | 29 |
| 0.137 | | 0.01142 | 3.4798 | |
| 0.138 | | 0.01150 | 3.5052 | |
| 0.139 | | 0.01158 | 3.5306 | |
| 0.140 | | 0.01167 | 3.5560 | |
| 0.1405 | | 0.01171 | 3.5687 | 28 |
| 0.1406 | 9/64 | 0.01172 | 3.5712 | |
| 0.141 | | 0.01175 | 3.5814 | |
| 0.142 | | 0.01183 | 3.6068 | |
| 0.143 | | 0.01192 | 3.6322 | |
| 0.144 | | 0.01200 | 3.6576 | 27 |
| 0.145 | | 0.01208 | 3.6830 | |
| 0.146 | | 0.01217 | 3.7084 | |
| 0.147 | | 0.01225 | 3.7338 | 26 |
| 0.148 | | 0.01233 | 3.7592 | |
| 0.1484 | 19/128 | 0.01237 | 3.7694 | |
| 0.149 | | 0.01242 | 3.7846 | |
| 0.1495 | | 0.01246 | 3.7973 | 25 |
| 0.150 | | 0.01250 | 3.8100 | |
| 0.151 | | 0.01258 | 3.8354 | |
| 0.152 | | 0.01267 | 3.8608 | 24 |
| 0.153 | | 0.01275 | 3.8862 | |
| 0.154 | | 0.01283 | 3.9116 | 23 |
| 0.155 | | 0.01292 | 3.9370 | |
| 0.156 | | 0.01300 | 3.9624 | |
| 0.1562 | 5/32 | 0.01302 | 3.9675 | |
| 0.157 | | 0.01308 | 3.9878 | 22 |
| 0.158 | | 0.01317 | 4.0132 | |
| 0.159 | | 0.01325 | 4.0386 | 21 |
| 0.160 | | 0.01333 | 4.0640 | |
| 0.161 | | 0.01342 | 4.0894 | 20 |
| 0.162 | | 0.01350 | 4.1148 | |
| 0.163 | | 0.01358 | 4.1402 | |
| 0.164 | | 0.01367 | 4.1656 | |
| 0.1641 | 21/128 | 0.01368 | 4.1681 | |
| 0.165 | | 0.01375 | 4.1910 | |
| 0.166 | | 0.01383 | 4.2164 | 19 |
| 0.167 | | 0.01392 | 4.2418 | |
| 0.168 | | 0.01400 | 4.2672 | |
| 0.169 | | 0.01408 | 4.2926 | |
| 0.1695 | | 0.01413 | 4.3053 | 18 |
| 0.170 | | 0.01417 | 4.3180 | |
| 0.171 | | 0.01425 | 4.3434 | |
| 0.1719 | 11/64 | 0.01433 | 4.3663 | |
| 0.172 | | 0.01433 | 4.3688 | |
| 0.173 | | 0.01442 | 4.3942 | 17 |
| 0.174 | | 0.01450 | 4.4196 | |
| 0.175 | | 0.01458 | 4.4450 | |
| 0.176 | | 0.01467 | 4.4704 | |
| 0.177 | | 0.01475 | 4.4958 | 16 |

| Decimals of Inch | Fractions of Inch | Decimals of Foot | Millimeters | Drill Number |
|---|---|---|---|---|
| 0.178 | | 0.01483 | 4.5212 | |
| 0.179 | | 0.01492 | 4.5466 | |
| 0.1797 | 23/128 | 0.01498 | 4.5644 | |
| 0.180 | | 0.01500 | 4.5720 | 15 |
| 0.181 | | 0.01508 | 4.5974 | |
| 0.182 | | 0.01517 | 4.6228 | 14 |
| 0.183 | | 0.01525 | 4.6482 | |
| 0.184 | | 0.01533 | 4.6736 | |
| 0.185 | | 0.01542 | 4.6990 | 13 |
| 0.186 | | 0.01550 | 4.7244 | |
| 0.187 | | 0.01558 | 4.7498 | |
| 0.1875 | 3/16 | 0.01563 | 4.7625 | |
| 0.188 | | 0.01567 | 4.7752 | |
| 0.189 | | 0.01575 | 4.8006 | 12 |
| 0.190 | | 0.01583 | 4.8260 | |
| 0.191 | | 0.01592 | 4.8514 | 11 |
| 0.192 | | 0.01600 | 4.8768 | |
| 0.193 | | 0.01608 | 4.9022 | |
| 0.1935 | | 0.01613 | 4.9149 | 10 |
| 0.194 | | 0.01617 | 4.9276 | |
| 0.195 | | 0.01625 | 4.9530 | |
| 0.1953 | 25/128 | 0.01628 | 4.9606 | |
| 0.196 | | 0.01633 | 4.9784 | 9 |
| 0.197 | | 0.01642 | 5.0038 | |
| 0.198 | | 0.01650 | 5.0292 | |
| 0.199 | | 0.01658 | 5.0546 | 8 |
| 0.200 | | 0.01667 | 5.0800 | |
| 0.201 | | 0.01675 | 5.1054 | 7 |
| 0.202 | | 0.01683 | 5.1308 | |
| 0.203 | | 0.01692 | 5.1562 | |
| 0.2031 | 13/64 | 0.01693 | 5.1587 | |
| 0.204 | | 0.01700 | 5.1816 | 6 |
| 0.205 | | 0.01708 | 5.2070 | |
| 0.2055 | | 0.01713 | 5.2197 | 5 |
| 0.206 | | 0.01717 | 5.2324 | |
| 0.207 | | 0.01725 | 5.2578 | |
| 0.208 | | 0.01733 | 5.2832 | |
| 0.209 | | 0.01742 | 5.3086 | 4 |
| 0.210 | | 0.01750 | 5.3340 | |
| 0.2109 | 27/128 | 0.01756 | 5.3569 | |
| 0.211 | | 0.01758 | 5.3594 | |
| 0.212 | | 0.01767 | 5.3848 | |
| 0.213 | | 0.01775 | 5.4102 | 3 |
| 0.214 | | 0.01783 | 5.4356 | |
| 0.215 | | 0.01792 | 5.4610 | |
| 0.216 | | 0.01800 | 5.4864 | |
| 0.217 | | 0.01808 | 5.5118 | |
| 0.218 | | 0.01817 | 5.5372 | |
| 0.2187 | 7/32 | 0.01823 | 5.5550 | |
| 0.219 | | 0.01825 | 5.5626 | |
| 0.220 | | 0.01833 | 5.5880 | |
| 0.221 | | 0.01842 | 5.6134 | 2 |

# CONVERT INCH–FOOT–MM–DRILL #

| Decimals of Inch | Fractions of Inch | Decimals of Foot | Millimeters | Drill Number |
|---|---|---|---|---|
| 0.222 | | 0.01850 | 5.6388 | |
| 0.223 | | 0.01858 | 5.6642 | |
| 0.224 | | 0.01867 | 5.6896 | |
| 0.225 | | 0.01875 | 5.7150 | |
| 0.226 | | 0.01883 | 5.7404 | |
| 0.2266 | 29/128 | 0.01888 | 5.7556 | |
| 0.227 | | 0.01892 | 5.7658 | |
| 0.228 | | 0.01900 | 5.7912 | 1 |
| 0.229 | | 0.01908 | 5.8166 | |
| 0.230 | | 0.01917 | 5.8420 | |
| 0.231 | | 0.01925 | 5.8674 | |
| 0.232 | | 0.01933 | 5.8928 | |
| 0.233 | | 0.01942 | 5.9182 | |
| 0.234 | | 0.01950 | 5.9436 | A |
| 0.2344 | 15/64 | 0.01953 | 5.9538 | |
| 0.235 | | 0.01958 | 5.9690 | |
| 0.236 | | 0.01967 | 5.9944 | |
| 0.237 | | 0.01975 | 6.0198 | |
| 0.238 | | 0.01983 | 6.0452 | B |
| 0.239 | | 0.01992 | 6.0706 | |
| 0.240 | | 0.02000 | 6.0960 | |
| 0.241 | | 0.02008 | 6.1214 | |
| 0.242 | | 0.02017 | 6.1468 | C |
| 0.2422 | 31/128 | 0.02018 | 6.1519 | |
| 0.243 | | 0.02025 | 6.1722 | |
| 0.244 | | 0.02033 | 6.1976 | |
| 0.245 | | 0.02042 | 6.2230 | |
| 0.246 | | 0.02050 | 6.2484 | D |
| 0.247 | | 0.02058 | 6.2738 | |
| 0.248 | | 0.02067 | 6.2992 | |
| 0.249 | | 0.02075 | 6.3246 | |
| 0.250 | 1/4 | 0.02083 | 6.3500 | E |
| 0.251 | | 0.02092 | 6.3754 | |
| 0.252 | | 0.02100 | 6.4008 | |
| 0.253 | | 0.02108 | 6.4262 | |
| 0.254 | | 0.02117 | 6.4516 | |
| 0.255 | | 0.02125 | 6.4770 | |
| 0.256 | | 0.02133 | 6.5024 | |
| 0.257 | | 0.02142 | 6.5278 | F |
| 0.2578 | 33/128 | 0.02148 | 6.5481 | |
| 0.258 | | 0.02150 | 6.5532 | |
| 0.259 | | 0.02158 | 6.5786 | |
| 0.260 | | 0.02167 | 6.6040 | |
| 0.261 | | 0.02175 | 6.6294 | G |
| 0.262 | | 0.02183 | 6.6548 | |
| 0.263 | | 0.02192 | 6.6802 | |
| 0.264 | | 0.02200 | 6.7056 | |
| 0.265 | | 0.02208 | 6.7310 | |
| 0.2656 | 17/64 | 0.02213 | 6.7462 | |
| 0.266 | | 0.02217 | 6.7564 | H |
| 0.267 | | 0.02225 | 6.7818 | |
| 0.268 | | 0.02233 | 6.8072 | |

| Decimals of Inch | Fractions of Inch | Decimals of Foot | Millimeters | Drill Number |
|---|---|---|---|---|
| 0.269 | | 0.02242 | 6.8326 | |
| 0.270 | | 0.02250 | 6.8580 | |
| 0.271 | | 0.02258 | 6.8834 | |
| 0.272 | | 0.02267 | 6.9088 | I |
| 0.273 | | 0.02275 | 6.9342 | |
| 0.2734 | 35/128 | 0.02278 | 6.9444 | |
| 0.274 | | 0.02283 | 6.9596 | |
| 0.275 | | 0.02292 | 6.9850 | |
| 0.276 | | 0.02300 | 7.0104 | |
| 0.277 | | 0.02308 | 7.0358 | J |
| 0.278 | | 0.02317 | 7.0612 | |
| 0.279 | | 0.02325 | 7.0866 | |
| 0.280 | | 0.02333 | 7.1120 | |
| 0.281 | | 0.02342 | 7.1374 | K |
| 0.2812 | 9/32 | 0.02343 | 7.1425 | |
| 0.282 | | 0.02350 | 7.1628 | |
| 0.283 | | 0.02358 | 7.1882 | |
| 0.284 | | 0.02367 | 7.2136 | |
| 0.285 | | 0.02375 | 7.2390 | |
| 0.286 | | 0.02383 | 7.2644 | |
| 0.287 | | 0.02392 | 7.2898 | |
| 0.288 | | 0.02400 | 7.3152 | |
| 0.289 | | 0.02408 | 7.3406 | |
| 0.2891 | 37/128 | 0.02409 | 7.34314 | |
| 0.290 | | 0.02417 | 7.3660 | L |
| 0.291 | | 0.02425 | 7.3914 | |
| 0.292 | | 0.02433 | 7.4168 | |
| 0.293 | | 0.02442 | 7.4422 | |
| 0.294 | | 0.02450 | 7.4676 | |
| 0.295 | | 0.02458 | 7.4930 | M |
| 0.296 | | 0.02467 | 7.5184 | |
| 0.29687 | 19/64 | 0.02474 | 7.5405 | |
| 0.297 | | 0.02475 | 7.5438 | |
| 0.298 | | 0.02483 | 7.5692 | |
| 0.299 | | 0.02492 | 7.5946 | |
| 0.300 | | 0.02500 | 7.6200 | |
| 0.301 | | 0.02508 | 7.6454 | |
| 0.302 | | 0.02517 | 7.6708 | N |
| 0.303 | | 0.02525 | 7.6962 | |
| 0.304 | | 0.02533 | 7.7216 | |
| 0.30469 | 39/128 | 0.02539 | 7.7391 | |
| 0.305 | | 0.02542 | 7.7470 | |
| 0.306 | | 0.02550 | 7.7724 | |
| 0.307 | | 0.02558 | 7.7978 | |
| 0.308 | | 0.02567 | 7.8232 | |
| 0.309 | | 0.02575 | 7.8486 | |
| 0.310 | | 0.02583 | 7.8740 | |
| 0.311 | | 0.02592 | 7.8994 | |
| 0.312 | | 0.02600 | 7.9248 | |
| 0.3125 | 5/16 | 0.02604 | 7.9375 | |
| 0.313 | | 0.02608 | 7.9502 | |
| 0.314 | | 0.02617 | 7.9756 | |
| 0.315 | | 0.02625 | 8.0010 | |

| Decimals of Inch | Fractions of Inch | Decimals of Foot | Millimeters | Drill Number |
|---|---|---|---|---|
| 0.316 | | 0.02633 | 8.0264 | O |
| 0.317 | | 0.02642 | 8.0518 | |
| 0.318 | | 0.02650 | 8.0772 | |
| 0.319 | | 0.02658 | 8.1026 | |
| 0.320 | | 0.02667 | 8.1280 | |
| 0.32031 | 41/128 | 0.02669 | 8.1359 | |
| 0.321 | | 0.02675 | 8.1534 | |
| 0.322 | | 0.02683 | 8.1788 | |
| 0.323 | | 0.02692 | 8.2042 | P |
| 0.324 | | 0.02700 | 8.2296 | |
| 0.325 | | 0.02708 | 8.2550 | |
| 0.326 | | 0.02717 | 8.2804 | |
| 0.327 | | 0.02725 | 8.3058 | |
| 0.328 | | 0.02733 | 8.3312 | |
| 0.32812 | 21/64 | 0.27344 | 8.3342 | |
| 0.329 | | 0.02742 | 8.3566 | |
| 0.330 | | 0.02750 | 8.3820 | |
| 0.331 | | 0.02758 | 8.4074 | |
| 0.332 | | 0.02767 | 8.4328 | Q |
| 0.333 | | 0.02775 | 8.4582 | |
| 0.334 | | 0.02783 | 8.4836 | |
| 0.335 | | 0.02792 | 8.5090 | |
| 0.33594 | 43/128 | 0.02799 | 8.5329 | |
| 0.336 | | 0.02800 | 8.5344 | |
| 0.337 | | 0.02808 | 8.5598 | |
| 0.338 | | 0.02817 | 8.5852 | |
| 0.339 | | 0.02825 | 8.6106 | R |
| 0.340 | | 0.02833 | 8.6360 | |
| 0.341 | | 0.02842 | 8.6614 | |
| 0.342 | | 0.02850 | 8.6868 | |
| 0.343 | | 0.02858 | 8.7122 | |
| 0.34375 | 11/32 | 0.02865 | 8.7312 | |
| 0.344 | | 0.02867 | 8.7376 | |
| 0.345 | | 0.02875 | 8.7630 | |
| 0.346 | | 0.02883 | 8.7884 | |
| 0.347 | | 0.02892 | 8.8138 | |
| 0.348 | | 0.02900 | 8.8392 | S |
| 0.349 | | 0.02908 | 8.8646 | |
| 0.350 | | 0.02917 | 8.8900 | |
| 0.351 | | 0.02925 | 8.9154 | |
| 0.35156 | 45/128 | 0.02930 | 8.9296 | |
| 0.352 | | 0.02933 | 8.9408 | |
| 0.353 | | 0.02942 | 8.9662 | |
| 0.354 | | 0.02950 | 8.9916 | |
| 0.355 | | 0.02958 | 9.0170 | |
| 0.356 | | 0.02967 | 9.0424 | |
| 0.357 | | 0.02975 | 9.0678 | |
| 0.358 | | 0.02983 | 9.0932 | T |
| 0.359 | | 0.02992 | 9.1186 | |
| 0.35937 | 23/64 | 0.02995 | 9.1280 | |
| 0.360 | | 0.03000 | 9.1440 | |
| 0.361 | | 0.03008 | 9.1694 | |
| 0.362 | | 0.03017 | 9.1948 | |

| Decimals of Inch | Fractions of Inch | Decimals of Foot | Millimeters | Drill Number |
|---|---|---|---|---|
| 0.363 | | 0.03025 | 9.2202 | |
| 0.364 | | 0.03033 | 9.2456 | |
| 0.365 | | 0.03042 | 9.2710 | |
| 0.366 | | 0.03050 | 9.2964 | |
| 0.367 | | 0.03058 | 9.3218 | |
| 0.36719 | 47/128 | 0.03060 | 9.3266 | |
| 0.368 | | 0.03067 | 9.3472 | U |
| 0.369 | | 0.03075 | 9.3726 | |
| 0.370 | | 0.03083 | 9.3980 | |
| 0.371 | | 0.03092 | 9.4234 | |
| 0.372 | | 0.03100 | 9.4488 | |
| 0.373 | | 0.03108 | 9.4742 | |
| 0.374 | | 0.03117 | 9.4996 | |
| 0.375 | 3/8 | 0.03125 | 9.5250 | |
| 0.376 | | 0.03133 | 9.5504 | |
| 0.377 | | 0.03142 | 9.5758 | V |
| 0.378 | | 0.03150 | 9.6012 | |
| 0.379 | | 0.03158 | 9.6266 | |
| 0.380 | | 0.03167 | 9.6520 | |
| 0.381 | | 0.03175 | 9.6774 | |
| 0.382 | | 0.03183 | 9.7028 | |
| 0.38281 | 49/128 | 0.03190 | 9.7234 | |
| 0.383 | | 0.03192 | 9.7282 | |
| 0.384 | | 0.03200 | 9.7536 | |
| 0.385 | | 0.03208 | 9.7790 | |
| 0.386 | | 0.03217 | 9.8044 | W |
| 0.387 | | 0.03225 | 9.8298 | |
| 0.388 | | 0.03233 | 9.8552 | |
| 0.389 | | 0.03242 | 9.8806 | |
| 0.390 | | 0.03250 | 9.9060 | |
| 0.39062 | 25/64 | 0.03255 | 9.9217 | |
| 0.391 | | 0.03258 | 9.9314 | |
| 0.392 | | 0.03267 | 9.9568 | |
| 0.393 | | 0.03275 | 9.9822 | |
| 0.394 | | 0.03283 | 10.0076 | |
| 0.395 | | 0.03292 | 10.0330 | |
| 0.396 | | 0.03300 | 10.0584 | |
| 0.397 | | 0.03308 | 10.0838 | X |
| 0.398 | | 0.03317 | 10.1092 | |
| 0.39844 | 51/128 | 0.03320 | 10.1204 | |
| 0.399 | | 0.03325 | 10.1346 | |
| 0.400 | | 0.03333 | 10.1600 | |
| 0.401 | | 0.03342 | 10.1854 | |
| 0.402 | | 0.03350 | 10.2108 | |
| 0.403 | | 0.03358 | 10.2362 | |
| 0.404 | | 0.03367 | 10.2616 | Y |
| 0.405 | | 0.03375 | 10.2870 | |
| 0.406 | | 0.03383 | 10.3124 | |
| 0.40625 | 13/32 | 0.03385 | 10.3187 | |
| 0.407 | | 0.03392 | 10.3378 | |
| 0.408 | | 0.03400 | 10.3632 | |
| 0.409 | | 0.03408 | 10.3886 | |

| Decimals of Inch | Fractions of Inch | Decimals of Foot | Millimeters | Drill Number |
|---|---|---|---|---|
| 0.410 | | 0.03417 | 10.4140 | |
| 0.411 | | 0.03425 | 10.4394 | |
| 0.412 | | 0.03433 | 10.4648 | |
| 0.413 | | 0.03442 | 10.4902 | Z |
| 0.414 | | 0.03450 | 10.5156 | |
| 0.41406 | 53/128 | 0.03451 | 10.5171 | |
| 0.415 | | 0.03458 | 10.5410 | |
| 0.416 | | 0.03467 | 10.5664 | |
| 0.417 | | 0.03475 | 10.5918 | |
| 0.418 | | 0.03483 | 10.6172 | |
| 0.419 | | 0.03492 | 10.6426 | |
| 0.420 | | 0.03500 | 10.6680 | |
| 0.421 | | 0.03508 | 10.6934 | |
| 0.42187 | 27/64 | 0.03516 | 10.7155 | |
| 0.422 | | 0.03517 | 10.7188 | |
| 0.423 | | 0.03525 | 10.7442 | |
| 0.424 | | 0.03533 | 10.7696 | |
| 0.425 | | 0.03542 | 10.7950 | |
| 0.426 | | 0.03550 | 10.8204 | |
| 0.427 | | 0.03558 | 10.8458 | |
| 0.428 | | 0.03567 | 10.8712 | |
| 0.429 | | 0.03575 | 10.8966 | |
| 0.42968 | 55/128 | 0.03581 | 10.9139 | |
| 0.430 | | 0.03583 | 10.9220 | |
| 0.431 | | 0.03592 | 10.9474 | |
| 0.432 | | 0.03600 | 10.9728 | |
| 0.433 | | 0.03608 | 10.9982 | |
| 0.434 | | 0.03617 | 11.0236 | |
| 0.435 | | 0.03625 | 11.0490 | |
| 0.436 | | 0.03633 | 11.0744 | |
| 0.437 | | 0.03642 | 11.0998 | |
| 0.4375 | 7/16 | 0.03646 | 11.1125 | |
| 0.438 | | 0.03650 | 11.1252 | |
| 0.439 | | 0.03658 | 11.1506 | |
| 0.440 | | 0.03667 | 11.1760 | |
| 0.441 | | 0.03675 | 11.2014 | |
| 0.442 | | 0.03683 | 11.2268 | |
| 0.443 | | 0.03692 | 11.2522 | |
| 0.444 | | 0.03700 | 11.2776 | |
| 0.445 | | 0.03708 | 11.3030 | |
| 0.44531 | 57/128 | 0.03711 | 11.3109 | |
| 0.446 | | 0.03717 | 11.3284 | |
| 0.447 | | 0.03725 | 11.3538 | |
| 0.448 | | 0.03733 | 11.3792 | |
| 0.449 | | 0.03742 | 11.4046 | |
| 0.450 | | 0.03750 | 11.4300 | |
| 0.451 | | 0.03758 | 11.4554 | |
| 0.452 | | 0.03767 | 11.4808 | |
| 0.453 | | 0.03775 | 11.5062 | |
| 0.45312 | 29/64 | 0.03776 | 11.5092 | |
| 0.454 | | 0.03783 | 11.5316 | |
| 0.455 | | 0.03792 | 11.5570 | |

# CONVERT INCH–FOOT–MM–DRILL #

| Decimals of Inch | Fractions of Inch | Decimals of Foot | Millimeters | Drill Number |
|---|---|---|---|---|
| 0.456 | | 0.03800 | 11.5824 | |
| 0.457 | | 0.03808 | 11.6078 | |
| 0.458 | | 0.03817 | 11.6332 | |
| 0.459 | | 0.03825 | 11.6586 | |
| 0.460 | | 0.03833 | 11.6840 | |
| 0.46094 | 59/128 | 0.03841 | 11.7079 | |
| 0.461 | | 0.03842 | 11.7094 | |
| 0.462 | | 0.03850 | 11.7348 | |
| 0.463 | | 0.03858 | 11.7602 | |
| 0.464 | | 0.03867 | 11.7856 | |
| 0.465 | | 0.03875 | 11.8110 | |
| 0.466 | | 0.03883 | 11.8364 | |
| 0.467 | | 0.03892 | 11.8618 | |
| 0.468 | | 0.03900 | 11.8872 | |
| 0.46875 | 15/32 | 0.03906 | 11.9062 | |
| 0.469 | | 0.03908 | 11.9126 | |
| 0.470 | | 0.03917 | 11.9380 | |
| 0.471 | | 0.03925 | 11.9634 | |
| 0.472 | | 0.03933 | 11.9888 | |
| 0.473 | | 0.03942 | 12.0142 | |
| 0.474 | | 0.03950 | 12.0396 | |
| 0.475 | | 0.03958 | 12.0650 | |
| 0.476 | | 0.03967 | 12.0904 | |
| 0.47656 | 61/128 | 0.03971 | 12.1046 | |
| 0.477 | | 0.03975 | 12.1158 | |
| 0.478 | | 0.03983 | 12.1412 | |
| 0.479 | | 0.03992 | 12.1666 | |
| 0.480 | | 0.04000 | 12.1920 | |
| 0.481 | | 0.04008 | 12.2174 | |
| 0.482 | | 0.04017 | 12.2428 | |
| 0.483 | | 0.04025 | 12.2682 | |
| 0.484 | | 0.04033 | 12.2936 | |
| 0.48437 | 31/64 | 0.04036 | 12.3030 | |
| 0.485 | | 0.04042 | 12.3190 | |
| 0.486 | | 0.04050 | 12.3444 | |
| 0.487 | | 0.04058 | 12.3698 | |
| 0.488 | | 0.04067 | 12.3952 | |
| 0.489 | | 0.04075 | 12.4206 | |
| 0.490 | | 0.04083 | 12.4460 | |
| 0.491 | | 0.04092 | 12.4714 | |
| 0.492 | | 0.04100 | 12.4968 | |
| 0.49219 | 63/128 | 0.04102 | 12.5016 | |
| 0.493 | | 0.04108 | 12.5222 | |
| 0.494 | | 0.04117 | 12.5476 | |
| 0.495 | | 0.04125 | 12.5730 | |
| 0.496 | | 0.04133 | 12.5984 | |
| 0.497 | | 0.04142 | 12.6238 | |
| 0.498 | | 0.04150 | 12.6492 | |
| 0.499 | | 0.04158 | 12.6746 | |
| 0.500 | 1/2 | 0.04167 | 12.7000 | |

# SQUARES, CUBES, AND ROOTS

| n | Square | Cube | Square Root | Cube Root |
|---|---|---|---|---|
| 1 | 1 | 1 | 1.00000 | 1.00000 |
| 2 | 4 | 8 | 1.41421 | 1.25992 |
| 3 | 9 | 27 | 1.73205 | 1.44225 |
| 4 | 16 | 64 | 2.00000 | 1.58740 |
| 5 | 25 | 125 | 2.23607 | 1.70998 |
| 6 | 36 | 216 | 2.44949 | 1.81712 |
| 7 | 49 | 343 | 2.64575 | 1.91293 |
| 8 | 64 | 512 | 2.82843 | 2.00000 |
| 9 | 81 | 729 | 3.00000 | 2.08008 |
| 10 | 100 | 1000 | 3.16228 | 2.15443 |
| 11 | 121 | 1331 | 3.31662 | 2.22398 |
| 12 | 144 | 1728 | 3.46410 | 2.28943 |
| 13 | 169 | 2197 | 3.60555 | 2.35133 |
| 14 | 196 | 2744 | 3.74166 | 2.41014 |
| 15 | 225 | 3375 | 3.87298 | 2.46621 |
| 16 | 256 | 4096 | 4.00000 | 2.51984 |
| 17 | 289 | 4913 | 4.12311 | 2.57128 |
| 18 | 324 | 5832 | 4.24264 | 2.62074 |
| 19 | 361 | 6859 | 4.35890 | 2.66840 |
| 20 | 400 | 8000 | 4.47214 | 2.71442 |
| 21 | 441 | 9261 | 4.58258 | 2.75892 |
| 22 | 484 | 10648 | 4.69042 | 2.80204 |
| 23 | 529 | 12167 | 4.79583 | 2.84387 |
| 24 | 576 | 13824 | 4.89898 | 2.88450 |
| 25 | 625 | 15625 | 5.00000 | 2.92402 |
| 26 | 676 | 17576 | 5.09902 | 2.96250 |
| 27 | 729 | 19683 | 5.19615 | 3.00000 |
| 28 | 784 | 21952 | 5.29150 | 3.03659 |
| 29 | 841 | 24389 | 5.38516 | 3.07232 |
| 30 | 900 | 27000 | 5.47723 | 3.10723 |
| 31 | 961 | 29791 | 5.56776 | 3.14138 |
| 32 | 1024 | 32768 | 5.65685 | 3.17480 |
| 33 | 1089 | 35937 | 5.74456 | 3.20753 |
| 34 | 1156 | 39304 | 5.83095 | 3.23961 |
| 35 | 1225 | 42875 | 5.91608 | 3.27107 |
| 36 | 1296 | 46656 | 6.00000 | 3.30193 |
| 37 | 1369 | 50653 | 6.08276 | 3.33222 |
| 38 | 1444 | 54872 | 6.16441 | 3.36198 |
| 39 | 1521 | 59319 | 6.24500 | 3.39121 |
| 40 | 1600 | 64000 | 6.32456 | 3.41995 |
| 41 | 1681 | 68921 | 6.40312 | 3.44822 |
| 42 | 1764 | 74088 | 6.48074 | 3.47603 |
| 43 | 1849 | 79507 | 6.55744 | 3.50340 |
| 44 | 1936 | 85184 | 6.63325 | 3.53035 |
| 45 | 2025 | 91125 | 6.70820 | 3.55689 |
| 46 | 2116 | 97336 | 6.78233 | 3.58305 |
| 47 | 2209 | 103823 | 6.85565 | 3.60883 |
| 48 | 2304 | 110592 | 6.92820 | 3.63424 |
| 49 | 2401 | 117649 | 7.00000 | 3.65931 |
| 50 | 2500 | 125000 | 7.07107 | 3.68403 |
| 51 | 2601 | 132651 | 7.14143 | 3.70843 |
| 52 | 2704 | 140608 | 7.21110 | 3.73251 |
| 53 | 2809 | 148877 | 7.28011 | 3.75629 |
| 54 | 2916 | 157464 | 7.34847 | 3.77976 |

| n | Square | Cube | Square Root | Cube Root |
|---|---|---|---|---|
| 55 | 3025 | 166375 | 7.41620 | 3.80295 |
| 56 | 3136 | 175616 | 7.48331 | 3.82586 |
| 57 | 3249 | 185193 | 7.54983 | 3.84850 |
| 58 | 3364 | 195112 | 7.61577 | 3.87088 |
| 59 | 3481 | 205379 | 7.68115 | 3.89300 |
| 60 | 3600 | 216000 | 7.74597 | 3.91487 |
| 61 | 3721 | 226981 | 7.81025 | 3.93650 |
| 62 | 3844 | 238328 | 7.87401 | 3.95789 |
| 63 | 3969 | 250047 | 7.93725 | 3.97906 |
| 64 | 4096 | 262144 | 8.00000 | 4.00000 |
| 65 | 4225 | 274625 | 8.06226 | 4.02073 |
| 66 | 4356 | 287496 | 8.12404 | 4.04124 |
| 67 | 4489 | 300763 | 8.18535 | 4.06155 |
| 68 | 4624 | 314432 | 8.24621 | 4.08166 |
| 69 | 4761 | 328509 | 8.30662 | 4.10157 |
| 70 | 4900 | 343000 | 8.36660 | 4.12129 |
| 71 | 5041 | 357911 | 8.42615 | 4.14082 |
| 72 | 5184 | 373248 | 8.48528 | 4.16017 |
| 73 | 5329 | 389017 | 8.54400 | 4.17934 |
| 74 | 5476 | 405224 | 8.60233 | 4.19834 |
| 75 | 5625 | 421875 | 8.66025 | 4.21716 |
| 76 | 5776 | 438976 | 8.71780 | 4.23582 |
| 77 | 5929 | 456533 | 8.77496 | 4.25432 |
| 78 | 6084 | 474552 | 8.83176 | 4.27266 |
| 79 | 6241 | 493039 | 8.88819 | 4.29084 |
| 80 | 6400 | 512000 | 8.94427 | 4.30887 |
| 81 | 6561 | 531441 | 9.00000 | 4.32675 |
| 82 | 6724 | 551368 | 9.05539 | 4.34448 |
| 83 | 6889 | 571787 | 9.11043 | 4.36207 |
| 84 | 7056 | 592704 | 9.16515 | 4.37952 |
| 85 | 7225 | 614125 | 9.21954 | 4.39683 |
| 86 | 7396 | 636056 | 9.27362 | 4.41400 |
| 87 | 7569 | 658503 | 9.32738 | 4.43105 |
| 88 | 7744 | 681472 | 9.38083 | 4.44796 |
| 89 | 7921 | 704969 | 9.43398 | 4.46475 |
| 90 | 8100 | 729000 | 9.48683 | 4.48140 |
| 91 | 8281 | 753571 | 9.53939 | 4.49794 |
| 92 | 8464 | 778688 | 9.59166 | 4.51436 |
| 93 | 8649 | 804357 | 9.64365 | 4.53065 |
| 94 | 8836 | 830584 | 9.69536 | 4.54684 |
| 95 | 9025 | 857375 | 9.74679 | 4.56290 |
| 96 | 9216 | 884736 | 9.79796 | 4.57886 |
| 97 | 9409 | 912673 | 9.84886 | 4.59470 |
| 98 | 9604 | 941192 | 9.89949 | 4.61044 |
| 99 | 9801 | 970299 | 9.94987 | 4.62607 |
| 100 | 10000 | 1000000 | 10.00000 | 4.64159 |
| 110 | 12100 | 1331000 | 10.48809 | 4.79142 |
| 120 | 14400 | 1728000 | 10.95445 | 4.93242 |
| 130 | 16900 | 2197000 | 11.40175 | 5.06580 |
| 140 | 19600 | 2744000 | 11.83216 | 5.19249 |
| 150 | 22500 | 3375000 | 12.24745 | 5.31329 |
| 160 | 25600 | 4096000 | 12.64911 | 5.42884 |
| 170 | 28900 | 4913000 | 13.03840 | 5.53966 |
| 180 | 32400 | 5832000 | 13.41641 | 5.64622 |

# SQUARES, CUBES, AND ROOTS

| n | Square | Cube | Square Root | Cube Root |
|---|---|---|---|---|
| 190 | 36100 | 6859000 | 13.78405 | 5.74890 |
| 200 | 40000 | 8000000 | 14.14214 | 5.84804 |
| 210 | 44100 | 9261000 | 14.49138 | 5.94392 |
| 220 | 48400 | 10648000 | 14.83240 | 6.03681 |
| 230 | 52900 | 12167000 | 15.16575 | 6.12693 |
| 240 | 57600 | 13824000 | 15.49193 | 6.21447 |
| 250 | 62500 | 15625000 | 15.81139 | 6.29961 |
| 260 | 67600 | 17576000 | 16.12452 | 6.38250 |
| 270 | 72900 | 19683000 | 16.43168 | 6.46330 |
| 280 | 78400 | 21952000 | 16.73320 | 6.54213 |
| 290 | 84100 | 24389000 | 17.02939 | 6.61911 |
| 300 | 90000 | 27000000 | 17.32051 | 6.69433 |
| 310 | 96100 | 29791000 | 17.60682 | 6.76790 |
| 320 | 102400 | 32768000 | 17.88854 | 6.83990 |
| 330 | 108900 | 35937000 | 18.16590 | 6.91042 |
| 340 | 115600 | 39304000 | 18.43909 | 6.97953 |
| 350 | 122500 | 42875000 | 18.70829 | 7.04730 |
| 360 | 129600 | 46656000 | 18.97367 | 7.11379 |
| 370 | 136900 | 50653000 | 19.23538 | 7.17905 |
| 380 | 144400 | 54872000 | 19.49359 | 7.24316 |
| 390 | 152100 | 59319000 | 19.74842 | 7.30614 |
| 400 | 160000 | 64000000 | 20.00000 | 7.36806 |
| 410 | 168100 | 68921000 | 20.24846 | 7.42896 |
| 420 | 176400 | 74088000 | 20.49390 | 7.48887 |
| 430 | 184900 | 79507000 | 20.73644 | 7.54784 |
| 440 | 193600 | 85184000 | 20.97618 | 7.60590 |
| 450 | 202500 | 91125000 | 21.21320 | 7.66309 |
| 460 | 211600 | 97336000 | 21.44761 | 7.71944 |
| 470 | 220900 | 103823000 | 21.67948 | 7.77498 |
| 480 | 230400 | 110592000 | 21.90890 | 7.82974 |
| 490 | 240100 | 117649000 | 22.13594 | 7.88374 |
| 500 | 250000 | 125000000 | 22.36068 | 7.93701 |
| 510 | 260100 | 132651000 | 22.58318 | 7.98957 |
| 520 | 270400 | 140608000 | 22.80351 | 8.04145 |
| 530 | 280900 | 148877000 | 23.02173 | 8.09267 |
| 540 | 291600 | 157464000 | 23.23790 | 8.14325 |
| 550 | 302500 | 166375000 | 23.45208 | 8.19321 |
| 560 | 313600 | 175616000 | 23.66432 | 8.24257 |
| 570 | 324900 | 185193000 | 23.87467 | 8.29134 |
| 580 | 336400 | 195112000 | 24.08319 | 8.33955 |
| 590 | 348100 | 205379000 | 24.28992 | 8.38721 |
| 600 | 360000 | 216000000 | 24.49490 | 8.43433 |
| 610 | 372100 | 226981000 | 24.69818 | 8.48093 |
| 620 | 384400 | 238328000 | 24.89980 | 8.52702 |
| 630 | 396900 | 250047000 | 25.09980 | 8.57262 |
| 640 | 409600 | 262144000 | 25.29822 | 8.61774 |
| 650 | 422500 | 274625000 | 25.49510 | 8.66239 |
| 660 | 435600 | 287496000 | 25.69047 | 8.70659 |
| 670 | 448900 | 300763000 | 25.88436 | 8.75034 |
| 680 | 462400 | 314432000 | 26.07681 | 8.79366 |
| 690 | 476100 | 328509000 | 26.26785 | 8.83656 |
| 700 | 490000 | 343000000 | 26.45751 | 8.87904 |
| 710 | 504100 | 357911000 | 26.64583 | 8.92112 |
| 720 | 518400 | 373248000 | 26.83282 | 8.96281 |

## SQUARES, CUBES, AND ROOTS

| n | Square | Cube | Square Root | Cube Root |
|---|---|---|---|---|
| 730 | 532900 | 389017000 | 27.01851 | 9.00411 |
| 740 | 547600 | 405224000 | 27.20294 | 9.04504 |
| 750 | 562500 | 421875000 | 27.38613 | 9.08560 |
| 760 | 577600 | 438976000 | 27.56810 | 9.12581 |
| 770 | 592900 | 456533000 | 27.74887 | 9.16566 |
| 780 | 608400 | 474552000 | 27.92848 | 9.20516 |
| 790 | 624100 | 493039000 | 28.10694 | 9.24434 |
| 800 | 640000 | 512000000 | 28.28427 | 9.28318 |
| 810 | 656100 | 531441000 | 28.46050 | 9.32170 |
| 820 | 672400 | 551368000 | 28.63564 | 9.35990 |
| 830 | 688900 | 571787000 | 28.80972 | 9.39780 |
| 840 | 705600 | 592704000 | 28.98275 | 9.43539 |
| 850 | 722500 | 614125000 | 29.15476 | 9.47268 |
| 860 | 739600 | 636056000 | 29.32576 | 9.50969 |
| 870 | 756900 | 658503000 | 29.49576 | 9.54640 |
| 880 | 774400 | 681472000 | 29.66479 | 9.58284 |
| 890 | 792100 | 704969000 | 29.83287 | 9.61900 |
| 900 | 810000 | 729000000 | 30.00000 | 9.65489 |
| 910 | 828100 | 753571000 | 30.16621 | 9.69052 |
| 920 | 846400 | 778688000 | 30.33150 | 9.72589 |
| 930 | 864900 | 804357000 | 30.49590 | 9.76100 |
| 940 | 883600 | 830584000 | 30.65942 | 9.79586 |
| 950 | 902500 | 857375000 | 30.82207 | 9.83048 |
| 960 | 921600 | 884736000 | 30.98387 | 9.86485 |
| 970 | 940900 | 912673000 | 31.14482 | 9.89898 |
| 980 | 960400 | 941192000 | 31.30495 | 9.93288 |
| 990 | 980100 | 970299000 | 31.46427 | 9.96655 |
| 1000 | 1000000 | 1000000000 | 31.62278 | 10.00000 |

## DEGREES & TRIG FUNCTIONS

| n | n Radians | Sine | Cosine | Tangent |
|---|---|---|---|---|
| 1 | 0.01745 | 0.01745 | 0.99985 | 0.01746 |
| 2 | 0.03491 | 0.03490 | 0.99939 | 0.03492 |
| 3 | 0.05236 | 0.05234 | 0.99863 | 0.05241 |
| 4 | 0.06981 | 0.06976 | 0.99756 | 0.06993 |
| 5 | 0.08727 | 0.08716 | 0.99619 | 0.08749 |
| 6 | 0.10472 | 0.10453 | 0.99452 | 0.10510 |
| 7 | 0.12217 | 0.12187 | 0.99255 | 0.12278 |
| 8 | 0.13963 | 0.13917 | 0.99027 | 0.14054 |
| 9 | 0.15708 | 0.15643 | 0.98769 | 0.15838 |
| 10 | 0.17453 | 0.17365 | 0.98481 | 0.17633 |
| 11 | 0.19199 | 0.19081 | 0.98163 | 0.19438 |
| 12 | 0.20944 | 0.20791 | 0.97815 | 0.21256 |
| 13 | 0.22689 | 0.22495 | 0.97437 | 0.23087 |
| 14 | 0.24435 | 0.24192 | 0.97030 | 0.24933 |
| 15 | 0.26180 | 0.25882 | 0.96593 | 0.26795 |
| 16 | 0.27925 | 0.27564 | 0.96126 | 0.28675 |
| 17 | 0.29671 | 0.29237 | 0.95630 | 0.30573 |
| 18 | 0.31416 | 0.30902 | 0.95106 | 0.32492 |
| 19 | 0.33161 | 0.32557 | 0.94552 | 0.34433 |

# DEGREES & TRIG FUNCTIONS

| n | n Radians | Sine | Cosine | Tangent |
|---|---|---|---|---|
| 20 | 0.34907 | 0.34202 | 0.93969 | 0.36397 |
| 21 | 0.36652 | 0.35837 | 0.93358 | 0.38386 |
| 22 | 0.38397 | 0.37461 | 0.92718 | 0.40403 |
| 23 | 0.40143 | 0.39073 | 0.92050 | 0.42447 |
| 24 | 0.41888 | 0.40674 | 0.91355 | 0.44523 |
| 25 | 0.43633 | 0.42262 | 0.90631 | 0.46631 |
| 26 | 0.45379 | 0.43837 | 0.89879 | 0.48773 |
| 27 | 0.47124 | 0.45399 | 0.89101 | 0.50953 |
| 28 | 0.48869 | 0.46947 | 0.88295 | 0.53171 |
| 29 | 0.50615 | 0.48481 | 0.87462 | 0.55431 |
| 30 | 0.52360 | 0.50000 | 0.86603 | 0.57735 |
| 31 | 0.54105 | 0.51504 | 0.85717 | 0.60086 |
| 32 | 0.55851 | 0.52992 | 0.84805 | 0.62487 |
| 33 | 0.57596 | 0.54464 | 0.83867 | 0.64941 |
| 34 | 0.59341 | 0.55919 | 0.82904 | 0.67451 |
| 35 | 0.61087 | 0.57358 | 0.81915 | 0.70021 |
| 36 | 0.62832 | 0.58779 | 0.80902 | 0.72654 |
| 37 | 0.64577 | 0.60182 | 0.79864 | 0.75355 |
| 38 | 0.66323 | 0.61566 | 0.78801 | 0.78129 |
| 39 | 0.68068 | 0.62932 | 0.77715 | 0.80978 |
| 40 | 0.69813 | 0.64279 | 0.76604 | 0.83910 |
| 41 | 0.71558 | 0.65606 | 0.75471 | 0.86929 |
| 42 | 0.73304 | 0.66913 | 0.74314 | 0.90040 |
| 43 | 0.75049 | 0.68200 | 0.73135 | 0.93252 |
| 44 | 0.76794 | 0.69466 | 0.71934 | 0.96569 |
| 45 | 0.78540 | 0.70711 | 0.70711 | 1.00000 |
| 46 | 0.80285 | 0.71934 | 0.69466 | 1.03553 |
| 47 | 0.82030 | 0.73135 | 0.68200 | 1.07237 |
| 48 | 0.83776 | 0.74314 | 0.66913 | 1.11061 |
| 49 | 0.85521 | 0.75471 | 0.65606 | 1.15037 |
| 50 | 0.87266 | 0.76604 | 0.64279 | 1.19175 |
| 51 | 0.89012 | 0.77715 | 0.62932 | 1.23490 |
| 52 | 0.90757 | 0.78801 | 0.61566 | 1.27994 |
| 53 | 0.92502 | 0.79864 | 0.60182 | 1.32704 |
| 54 | 0.94248 | 0.80902 | 0.58779 | 1.37638 |
| 55 | 0.95993 | 0.81915 | 0.57358 | 1.42815 |
| 56 | 0.97738 | 0.82904 | 0.55919 | 1.48256 |
| 57 | 0.99484 | 0.83867 | 0.54464 | 1.53986 |
| 58 | 1.01229 | 0.84805 | 0.52992 | 1.60033 |
| 59 | 1.02974 | 0.85717 | 0.51504 | 1.66428 |
| 60 | 1.04720 | 0.86603 | 0.50000 | 1.73205 |
| 61 | 1.06465 | 0.87462 | 0.48481 | 1.80405 |
| 62 | 1.08210 | 0.88295 | 0.46947 | 1.88073 |
| 63 | 1.09956 | 0.89101 | 0.45399 | 1.96261 |
| 64 | 1.11701 | 0.89879 | 0.43837 | 2.05030 |
| 65 | 1.13446 | 0.90631 | 0.42262 | 2.14451 |
| 66 | 1.15192 | 0.91355 | 0.40674 | 2.24604 |
| 67 | 1.16937 | 0.92050 | 0.39073 | 2.35585 |
| 68 | 1.18682 | 0.92718 | 0.37461 | 2.47509 |
| 69 | 1.20428 | 0.93358 | 0.35837 | 2.60509 |
| 70 | 1.22173 | 0.93969 | 0.34202 | 2.74748 |
| 71 | 1.23918 | 0.94552 | 0.32557 | 2.90421 |
| 72 | 1.25664 | 0.95106 | 0.30902 | 3.07768 |
| 73 | 1.27409 | 0.95630 | 0.29237 | 3.27085 |

| n | n Radians | Sine | Cosine | Tangent |
|---|---|---|---|---|
| 74 | 1.29154 | 0.96126 | 0.27564 | 3.48741 |
| 75 | 1.30900 | 0.96593 | 0.25882 | 3.73205 |
| 76 | 1.32645 | 0.97030 | 0.24192 | 4.01078 |
| 77 | 1.34390 | 0.97437 | 0.22495 | 4.33148 |
| 78 | 1.36136 | 0.97815 | 0.20791 | 4.70463 |
| 79 | 1.37881 | 0.98163 | 0.19081 | 5.14455 |
| 80 | 1.39626 | 0.98481 | 0.17365 | 5.67128 |
| 81 | 1.41372 | 0.98769 | 0.15643 | 6.31375 |
| 82 | 1.43117 | 0.99027 | 0.13917 | 7.11537 |
| 83 | 1.44862 | 0.99255 | 0.12187 | 8.14435 |
| 84 | 1.46608 | 0.99452 | 0.10453 | 9.51436 |
| 85 | 1.48353 | 0.99619 | 0.08716 | 11.43005 |
| 86 | 1.50098 | 0.99756 | 0.06976 | 14.30067 |
| 87 | 1.51844 | 0.99863 | 0.05234 | 19.08114 |
| 88 | 1.53589 | 0.99939 | 0.03490 | 28.63625 |
| 89 | 1.55334 | 0.99985 | 0.01745 | 57.28996 |
| 90 | 1.57080 | 1.00000 | 0.00000 | infinity |
| 91 | 1.58825 | 0.99985 | –0.01745 | –57.28996 |
| 92 | 1.60570 | 0.99939 | –0.03490 | –28.63625 |
| 93 | 1.62316 | 0.99863 | –0.05234 | –19.08114 |
| 94 | 1.64061 | 0.99756 | –0.06976 | –14.30067 |
| 95 | 1.65806 | 0.99619 | –0.08716 | –11.43005 |
| 96 | 1.67552 | 0.99452 | –0.10453 | –9.51436 |
| 97 | 1.69297 | 0.99255 | –0.12187 | –8.14435 |
| 98 | 1.71042 | 0.99027 | –0.13917 | –7.11537 |
| 99 | 1.72788 | 0.98769 | –0.15643 | –6.31375 |
| 100 | 1.74533 | 0.98481 | –0.17365 | –5.67128 |
| 101 | 1.76278 | 0.98163 | –0.19081 | –5.14455 |
| 102 | 1.78024 | 0.97815 | –0.20791 | –4.70463 |
| 103 | 1.79769 | 0.97437 | –0.22495 | –4.33148 |
| 104 | 1.81514 | 0.97030 | –0.24192 | –4.01078 |
| 105 | 1.83260 | 0.96593 | –0.25882 | –3.73205 |
| 106 | 1.85005 | 0.96126 | –0.27564 | –3.48741 |
| 107 | 1.86750 | 0.95630 | –0.29237 | –3.27085 |
| 108 | 1.88496 | 0.95106 | –0.30902 | –3.07768 |
| 109 | 1.90241 | 0.94552 | –0.32557 | –2.90421 |
| 110 | 1.91986 | 0.93969 | –0.34202 | –2.74748 |
| 111 | 1.93732 | 0.93358 | –0.35837 | –2.60509 |
| 112 | 1.95477 | 0.92718 | –0.37461 | –2.47509 |
| 113 | 1.97222 | 0.92050 | –0.39073 | –2.35585 |
| 114 | 1.98968 | 0.91355 | –0.40674 | –2.24604 |
| 115 | 2.00713 | 0.90631 | –0.42262 | –2.14451 |
| 116 | 2.02458 | 0.89879 | –0.43837 | –2.05030 |
| 117 | 2.04204 | 0.89101 | –0.45399 | –1.96261 |
| 118 | 2.05949 | 0.88295 | –0.46947 | –1.88073 |
| 119 | 2.07694 | 0.87462 | –0.48481 | –1.80405 |
| 120 | 2.09440 | 0.86603 | –0.50000 | –1.73205 |
| 121 | 2.11185 | 0.85717 | –0.51504 | –1.66428 |
| 122 | 2.12930 | 0.84805 | –0.52992 | –1.60033 |
| 123 | 2.14675 | 0.83867 | –0.54464 | –1.53986 |
| 124 | 2.16421 | 0.82904 | –0.55919 | –1.48256 |
| 125 | 2.18166 | 0.81915 | –0.57358 | –1.42815 |
| 126 | 2.19911 | 0.80902 | –0.58779 | –1.37638 |
| 127 | 2.21657 | 0.79864 | –0.60182 | –1.32704 |

| n | n Radians | Sine | Cosine | Tangent |
|---|---|---|---|---|
| 128 | 2.23402 | 0.78801 | −0.61566 | −1.27994 |
| 129 | 2.25147 | 0.77715 | −0.62932 | −1.23490 |
| 130 | 2.26893 | 0.76604 | −0.64279 | −1.19175 |
| 131 | 2.28638 | 0.75471 | −0.65606 | −1.15037 |
| 132 | 2.30383 | 0.74314 | −0.66913 | −1.11061 |
| 133 | 2.32129 | 0.73135 | −0.68200 | −1.07237 |
| 134 | 2.33874 | 0.71934 | −0.69466 | −1.03553 |
| 135 | 2.35619 | 0.70711 | −0.70711 | −1.00000 |
| 136 | 2.37365 | 0.69466 | −0.71934 | −0.96569 |
| 137 | 2.39110 | 0.68200 | −0.73135 | −0.93252 |
| 138 | 2.40855 | 0.66913 | −0.74314 | −0.90040 |
| 139 | 2.42601 | 0.65606 | −0.75471 | −0.86929 |
| 140 | 2.44346 | 0.64279 | −0.76604 | −0.83910 |
| 141 | 2.46091 | 0.62932 | −0.77715 | −0.80978 |
| 142 | 2.47837 | 0.61566 | −0.78801 | −0.78129 |
| 143 | 2.49582 | 0.60182 | −0.79864 | −0.75355 |
| 144 | 2.51327 | 0.58779 | −0.80902 | −0.72654 |
| 145 | 2.53073 | 0.57358 | −0.81915 | −0.70021 |
| 146 | 2.54818 | 0.55919 | −0.82904 | −0.67451 |
| 147 | 2.56563 | 0.54464 | −0.83867 | −0.64941 |
| 148 | 2.58309 | 0.52992 | −0.84805 | −0.62487 |
| 149 | 2.60054 | 0.51504 | −0.85717 | −0.60086 |
| 150 | 2.61799 | 0.50000 | −0.86603 | −0.57735 |
| 151 | 2.63545 | 0.48481 | −0.87462 | −0.55431 |
| 152 | 2.65290 | 0.46947 | −0.88295 | −0.53171 |
| 153 | 2.67035 | 0.45399 | −0.89101 | −0.50953 |
| 154 | 2.68781 | 0.43837 | −0.89879 | −0.48773 |
| 155 | 2.70526 | 0.42262 | −0.90631 | −0.46631 |
| 156 | 2.72271 | 0.40674 | −0.91355 | −0.44523 |
| 157 | 2.74017 | 0.39073 | −0.92050 | −0.42447 |
| 158 | 2.75762 | 0.37461 | −0.92718 | −0.40403 |
| 159 | 2.77507 | 0.35837 | −0.93358 | −0.38386 |
| 160 | 2.79253 | 0.34202 | −0.93969 | −0.36397 |
| 161 | 2.80998 | 0.32557 | −0.94552 | −0.34433 |
| 162 | 2.82743 | 0.30902 | −0.95106 | −0.32492 |
| 163 | 2.84489 | 0.29237 | −0.95630 | −0.30573 |
| 164 | 2.86234 | 0.27564 | −0.96126 | −0.28675 |
| 165 | 2.87979 | 0.25882 | −0.96593 | −0.26795 |
| 166 | 2.89725 | 0.24192 | −0.97030 | −0.24933 |
| 167 | 2.91470 | 0.22495 | −0.97437 | −0.23087 |
| 168 | 2.93215 | 0.20791 | −0.97815 | −0.21256 |
| 169 | 2.94961 | 0.19081 | −0.98163 | −0.19438 |
| 170 | 2.96706 | 0.17365 | −0.98481 | −0.17633 |
| 171 | 2.98451 | 0.15643 | −0.98769 | −0.15838 |
| 172 | 3.00197 | 0.13917 | −0.99027 | −0.14054 |
| 173 | 3.01942 | 0.12187 | −0.99255 | −0.12278 |
| 174 | 3.03687 | 0.10453 | −0.99452 | −0.10510 |
| 175 | 3.05433 | 0.08716 | −0.99619 | −0.08749 |
| 176 | 3.07178 | 0.06976 | −0.99756 | −0.06993 |
| 177 | 3.08923 | 0.05234 | −0.99863 | −0.05241 |
| 178 | 3.10669 | 0.03490 | −0.99939 | −0.03492 |
| 179 | 3.12414 | 0.01745 | −0.99985 | −0.01746 |
| 180 | 3.14159 | 0.00000 | −1.00000 | 0.00000 |
| 181 | 3.15905 | −0.01745 | −0.99985 | 0.01746 |

# DEGREES & TRIG FUNCTIONS

| n | n Radians | Sine | Cosine | Tangent |
|---|---|---|---|---|
| 182 | 3.17650 | –0.03490 | –0.99939 | 0.03492 |
| 183 | 3.19395 | –0.05234 | –0.99863 | 0.05241 |
| 184 | 3.21141 | –0.06976 | –0.99756 | 0.06993 |
| 185 | 3.22886 | –0.08716 | –0.99619 | 0.08749 |
| 186 | 3.24631 | –0.10453 | –0.99452 | 0.10510 |
| 187 | 3.26377 | –0.12187 | –0.99255 | 0.12278 |
| 188 | 3.28122 | –0.13917 | –0.99027 | 0.14054 |
| 189 | 3.29867 | –0.15643 | –0.98769 | 0.15838 |
| 190 | 3.31613 | –0.17365 | –0.98481 | 0.17633 |
| 191 | 3.33358 | –0.19081 | –0.98163 | 0.19438 |
| 192 | 3.35103 | –0.20791 | –0.97815 | 0.21256 |
| 193 | 3.36849 | –0.22495 | –0.97437 | 0.23087 |
| 194 | 3.38594 | –0.24192 | –0.97030 | 0.24933 |
| 195 | 3.40339 | –0.25882 | –0.96593 | 0.26795 |
| 196 | 3.42085 | –0.27564 | –0.96126 | 0.28675 |
| 197 | 3.43830 | –0.29237 | –0.95630 | 0.30573 |
| 198 | 3.45575 | –0.30902 | –0.95106 | 0.32492 |
| 199 | 3.47321 | –0.32557 | –0.94552 | 0.34433 |
| 200 | 3.49066 | –0.34202 | –0.93969 | 0.36397 |
| 201 | 3.50811 | –0.35837 | –0.93358 | 0.38386 |
| 202 | 3.52557 | –0.37461 | –0.92718 | 0.40403 |
| 203 | 3.54302 | –0.39073 | –0.92050 | 0.42447 |
| 204 | 3.56047 | –0.40674 | –0.91355 | 0.44523 |
| 205 | 3.57792 | –0.42262 | –0.90631 | 0.46631 |
| 206 | 3.59538 | –0.43837 | –0.89879 | 0.48773 |
| 207 | 3.61283 | –0.45399 | –0.89101 | 0.50953 |
| 208 | 3.63028 | –0.46947 | –0.88295 | 0.53171 |
| 209 | 3.64774 | –0.48481 | –0.87462 | 0.55431 |
| 210 | 3.66519 | –0.50000 | –0.86603 | 0.57735 |
| 211 | 3.68264 | –0.51504 | –0.85717 | 0.60086 |
| 212 | 3.70010 | –0.52992 | –0.84805 | 0.62487 |
| 213 | 3.71755 | –0.54464 | –0.83867 | 0.64941 |
| 214 | 3.73500 | –0.55919 | –0.82904 | 0.67451 |
| 215 | 3.75246 | –0.57358 | –0.81915 | 0.70021 |
| 216 | 3.76991 | –0.58779 | –0.80902 | 0.72654 |
| 217 | 3.78736 | –0.60182 | –0.79864 | 0.75355 |
| 218 | 3.80482 | –0.61566 | –0.78801 | 0.78129 |
| 219 | 3.82227 | –0.62932 | –0.77715 | 0.80978 |
| 220 | 3.83972 | –0.64279 | –0.76604 | 0.83910 |
| 221 | 3.85718 | –0.65606 | –0.75471 | 0.86929 |
| 222 | 3.87463 | –0.66913 | –0.74314 | 0.90040 |
| 223 | 3.89208 | –0.68200 | –0.73135 | 0.93252 |
| 224 | 3.90954 | –0.69466 | –0.71934 | 0.96569 |
| 225 | 3.92699 | –0.70711 | –0.70711 | 1.00000 |
| 226 | 3.94444 | –0.71934 | –0.69466 | 1.03553 |
| 227 | 3.96190 | –0.73135 | –0.68200 | 1.07237 |
| 228 | 3.97935 | –0.74314 | –0.66913 | 1.11061 |
| 229 | 3.99680 | –0.75471 | –0.65606 | 1.15037 |
| 230 | 4.01426 | –0.76604 | –0.64279 | 1.19175 |
| 231 | 4.03171 | –0.77715 | –0.62932 | 1.23490 |
| 232 | 4.04916 | –0.78801 | –0.61566 | 1.27994 |
| 233 | 4.06662 | –0.79864 | –0.60182 | 1.32704 |
| 234 | 4.08407 | –0.80902 | –0.58779 | 1.37638 |
| 235 | 4.10152 | –0.81915 | –0.57358 | 1.42815 |

| n | n Radians | Sine | Cosine | Tangent |
|---|---|---|---|---|
| 236 | 4.11898 | –0.82904 | –0.55919 | 1.48256 |
| 237 | 4.13643 | –0.83867 | –0.54464 | 1.53986 |
| 238 | 4.15388 | –0.84805 | –0.52992 | 1.60033 |
| 239 | 4.17134 | –0.85717 | –0.51504 | 1.66428 |
| 240 | 4.18879 | –0.86603 | –0.50000 | 1.73205 |
| 241 | 4.20624 | –0.87462 | –0.48481 | 1.80405 |
| 242 | 4.22370 | –0.88295 | –0.46947 | 1.88073 |
| 243 | 4.24115 | –0.89101 | –0.45399 | 1.96261 |
| 244 | 4.25860 | –0.89879 | –0.43837 | 2.05030 |
| 245 | 4.27606 | –0.90631 | –0.42262 | 2.14451 |
| 246 | 4.29351 | –0.91355 | –0.40674 | 2.24604 |
| 247 | 4.31096 | –0.92050 | –0.39073 | 2.35585 |
| 248 | 4.32842 | –0.92718 | –0.37461 | 2.47509 |
| 249 | 4.34587 | –0.93358 | –0.35837 | 2.60509 |
| 250 | 4.36332 | –0.93969 | –0.34202 | 2.74748 |
| 251 | 4.38078 | –0.94552 | –0.32557 | 2.90421 |
| 252 | 4.39823 | –0.95106 | –0.30902 | 3.07768 |
| 253 | 4.41568 | –0.95630 | –0.29237 | 3.27085 |
| 254 | 4.43314 | –0.96126 | –0.27564 | 3.48741 |
| 255 | 4.45059 | –0.96593 | –0.25882 | 3.73205 |
| 256 | 4.46804 | –0.97030 | –0.24192 | 4.01078 |
| 257 | 4.48550 | –0.97437 | –0.22495 | 4.33148 |
| 258 | 4.50295 | –0.97815 | –0.20791 | 4.70463 |
| 259 | 4.52040 | –0.98163 | –0.19081 | 5.14455 |
| 260 | 4.53786 | –0.98481 | –0.17365 | 5.67128 |
| 261 | 4.55531 | –0.98769 | –0.15643 | 6.31375 |
| 262 | 4.57276 | –0.99027 | –0.13917 | 7.11537 |
| 263 | 4.59022 | –0.99255 | –0.12187 | 8.14435 |
| 264 | 4.60767 | –0.99452 | –0.10453 | 9.51436 |
| 265 | 4.62512 | –0.99619 | –0.08716 | 11.43005 |
| 266 | 4.64258 | –0.99756 | –0.06976 | 14.30067 |
| 267 | 4.66003 | –0.99863 | –0.05234 | 19.08114 |
| 268 | 4.67748 | –0.99939 | –0.03490 | 28.63625 |
| 269 | 4.69494 | –0.99985 | –0.01745 | 57.28996 |
| 270 | 4.71239 | –1.00000 | 0.00000 | infinity |
| 271 | 4.72984 | –0.99985 | 0.01745 | –57.28996 |
| 272 | 4.74730 | –0.99939 | 0.03490 | –28.63625 |
| 273 | 4.76475 | –0.99863 | 0.05234 | –19.08114 |
| 274 | 4.78220 | –0.99756 | 0.06976 | –14.30067 |
| 275 | 4.79966 | –0.99619 | 0.08716 | –11.43005 |
| 276 | 4.81711 | –0.99452 | 0.10453 | –9.51436 |
| 277 | 4.83456 | –0.99255 | 0.12187 | –8.14435 |
| 278 | 4.85202 | –0.99027 | 0.13917 | –7.11537 |
| 279 | 4.86947 | –0.98769 | 0.15643 | –6.31375 |
| 280 | 4.88692 | –0.98481 | 0.17365 | –5.67128 |
| 281 | 4.90438 | –0.98163 | 0.19081 | –5.14455 |
| 282 | 4.92183 | –0.97815 | 0.20791 | –4.70463 |
| 283 | 4.93928 | –0.97437 | 0.22495 | –4.33148 |
| 284 | 4.95674 | –0.97030 | 0.24192 | –4.01078 |
| 285 | 4.97419 | –0.96593 | 0.25882 | –3.73205 |
| 286 | 4.99164 | –0.96126 | 0.27564 | –3.48741 |
| 287 | 5.00909 | –0.95630 | 0.29237 | –3.27085 |
| 288 | 5.02655 | –0.95106 | 0.30902 | –3.07768 |
| 289 | 5.04400 | –0.94552 | 0.32557 | –2.90421 |

| n | n Radians | Sine | Cosine | Tangent |
|---|---|---|---|---|
| 290 | 5.06145 | −0.93969 | 0.34202 | −2.74748 |
| 291 | 5.07891 | −0.93358 | 0.35837 | −2.60509 |
| 292 | 5.09636 | −0.92718 | 0.37461 | −2.47509 |
| 293 | 5.11381 | −0.92050 | 0.39073 | −2.35585 |
| 294 | 5.13127 | −0.91355 | 0.40674 | −2.24604 |
| 295 | 5.14872 | −0.90631 | 0.42262 | −2.14451 |
| 296 | 5.16617 | −0.89879 | 0.43837 | −2.05030 |
| 297 | 5.18363 | −0.89101 | 0.45399 | −1.96261 |
| 298 | 5.20108 | −0.88295 | 0.46947 | −1.88073 |
| 299 | 5.21853 | −0.87462 | 0.48481 | −1.80405 |
| 300 | 5.23599 | −0.86603 | 0.50000 | −1.73205 |
| 301 | 5.25344 | −0.85717 | 0.51504 | −1.66428 |
| 302 | 5.27089 | −0.84805 | 0.52992 | −1.60033 |
| 303 | 5.28835 | −0.83867 | 0.54464 | −1.53986 |
| 304 | 5.30580 | −0.82904 | 0.55919 | −1.48256 |
| 305 | 5.32325 | −0.81915 | 0.57358 | −1.42815 |
| 306 | 5.34071 | −0.80902 | 0.58779 | −1.37638 |
| 307 | 5.35816 | −0.79864 | 0.60182 | −1.32704 |
| 308 | 5.37561 | −0.78801 | 0.61566 | −1.27994 |
| 309 | 5.39307 | −0.77715 | 0.62932 | −1.23490 |
| 310 | 5.41052 | −0.76604 | 0.64279 | −1.19175 |
| 311 | 5.42797 | −0.75471 | 0.65606 | −1.15037 |
| 312 | 5.44543 | −0.74314 | 0.66913 | −1.11061 |
| 313 | 5.46288 | −0.73135 | 0.68200 | −1.07237 |
| 314 | 5.48033 | −0.71934 | 0.69466 | −1.03553 |
| 315 | 5.49779 | −0.70711 | 0.70711 | −1.00000 |
| 316 | 5.51524 | −0.69466 | 0.71934 | −0.96569 |
| 317 | 5.53269 | −0.68200 | 0.73135 | −0.93252 |
| 318 | 5.55015 | −0.66913 | 0.74314 | −0.90040 |
| 319 | 5.56760 | −0.65606 | 0.75471 | −0.86929 |
| 320 | 5.58505 | −0.64279 | 0.76604 | −0.83910 |
| 321 | 5.60251 | −0.62932 | 0.77715 | −0.80978 |
| 322 | 5.61996 | −0.61566 | 0.78801 | −0.78129 |
| 323 | 5.63741 | −0.60182 | 0.79864 | −0.75355 |
| 324 | 5.65487 | −0.58779 | 0.80902 | −0.72654 |
| 325 | 5.67232 | −0.57358 | 0.81915 | −0.70021 |
| 326 | 5.68977 | −0.55919 | 0.82904 | −0.67451 |
| 327 | 5.70723 | −0.54464 | 0.83867 | −0.64941 |
| 328 | 5.72468 | −0.52992 | 0.84805 | −0.62487 |
| 329 | 5.74213 | −0.51504 | 0.85717 | −0.60086 |
| 330 | 5.75959 | −0.50000 | 0.86603 | −0.57735 |
| 331 | 5.77704 | −0.48481 | 0.87462 | −0.55431 |
| 332 | 5.79449 | −0.46947 | 0.88295 | −0.53171 |
| 333 | 5.81195 | −0.45399 | 0.89101 | −0.50953 |
| 334 | 5.82940 | −0.43837 | 0.89879 | −0.48773 |
| 335 | 5.84685 | −0.42262 | 0.90631 | −0.46631 |
| 336 | 5.86431 | −0.40674 | 0.91355 | −0.44523 |
| 337 | 5.88176 | −0.39073 | 0.92050 | −0.42447 |
| 338 | 5.89921 | −0.37461 | 0.92718 | −0.40403 |
| 339 | 5.91667 | −0.35837 | 0.93358 | −0.38386 |
| 340 | 5.93412 | −0.34202 | 0.93969 | −0.36397 |
| 341 | 5.95157 | −0.32557 | 0.94552 | −0.34433 |
| 342 | 5.96903 | −0.30902 | 0.95106 | −0.32492 |
| 343 | 5.98648 | −0.29237 | 0.95630 | −0.30573 |

## DEGREES & TRIG FUNCTIONS

| n | n Radians | Sine | Cosine | Tangent |
|---|---|---|---|---|
| 344 | 6.00393 | −0.27564 | 0.96126 | −0.28675 |
| 345 | 6.02139 | −0.25882 | 0.96593 | −0.26795 |
| 346 | 6.03884 | −0.24192 | 0.97030 | −0.24933 |
| 347 | 6.05629 | −0.22495 | 0.97437 | −0.23087 |
| 348 | 6.07375 | −0.20791 | 0.97815 | −0.21256 |
| 349 | 6.09120 | −0.19081 | 0.98163 | −0.19438 |
| 350 | 6.10865 | −0.17365 | 0.98481 | −0.17633 |
| 351 | 6.12611 | −0.15643 | 0.98769 | −0.15838 |
| 352 | 6.14356 | −0.13917 | 0.99027 | −0.14054 |
| 353 | 6.16101 | −0.12187 | 0.99255 | −0.12278 |
| 354 | 6.17847 | −0.10453 | 0.99452 | −0.10510 |
| 355 | 6.19592 | −0.08716 | 0.99619 | −0.08749 |
| 356 | 6.21337 | −0.06976 | 0.99756 | −0.06993 |
| 357 | 6.23083 | −0.05234 | 0.99863 | −0.05241 |
| 358 | 6.24828 | −0.03490 | 0.99939 | −0.03492 |
| 359 | 6.26573 | −0.01745 | 0.99985 | −0.01746 |
| 360 | 6.28319 | 0.00000 | 1.00000 | 0.00000 |

## LOG, LOG e, CIRCUMFRENCE & AREA

| n | Log 10 | Log e | Circumference @ Diameter n | Circle Area @ Radius n |
|---|---|---|---|---|
| 1 | 0.00000 | 0.00000 | 3.1416 | 0.7854 |
| 2 | 0.30103 | 0.69315 | 6.2832 | 3.1416 |
| 3 | 0.47712 | 1.09861 | 9.4248 | 7.0686 |
| 4 | 0.60206 | 1.38629 | 12.5664 | 12.5664 |
| 5 | 0.69897 | 1.60944 | 15.7080 | 19.6350 |
| 6 | 0.77815 | 1.79176 | 18.8496 | 28.2743 |
| 7 | 0.84510 | 1.94591 | 21.9911 | 38.4845 |
| 8 | 0.90309 | 2.07944 | 25.1327 | 50.2655 |
| 9 | 0.95424 | 2.19722 | 28.2743 | 63.6173 |
| 10 | 1.00000 | 2.30259 | 31.4159 | 78.5398 |
| 11 | 1.04139 | 2.39790 | 34.5575 | 95.0332 |
| 12 | 1.07918 | 2.48491 | 37.6991 | 113.0973 |
| 13 | 1.11394 | 2.56495 | 40.8407 | 132.7323 |
| 14 | 1.14613 | 2.63906 | 43.9823 | 153.9380 |
| 15 | 1.17609 | 2.70805 | 47.1239 | 176.7146 |
| 16 | 1.20412 | 2.77259 | 50.2655 | 201.0619 |
| 17 | 1.23045 | 2.83321 | 53.4071 | 226.9801 |
| 18 | 1.25527 | 2.89037 | 56.5487 | 254.4690 |
| 19 | 1.27875 | 2.94444 | 59.6903 | 283.5287 |
| 20 | 1.30103 | 2.99573 | 62.8319 | 314.1593 |
| 21 | 1.32222 | 3.04452 | 65.9734 | 346.3606 |
| 22 | 1.34242 | 3.09104 | 69.1150 | 380.1327 |
| 23 | 1.36173 | 3.13549 | 72.2566 | 415.4756 |
| 24 | 1.38021 | 3.17805 | 75.3982 | 452.3893 |
| 25 | 1.39794 | 3.21888 | 78.5398 | 490.8739 |
| 26 | 1.41497 | 3.25810 | 81.6814 | 530.9292 |
| 27 | 1.43136 | 3.29584 | 84.8230 | 572.5553 |
| 28 | 1.44716 | 3.33220 | 87.9646 | 615.7522 |
| 29 | 1.46240 | 3.36730 | 91.1062 | 660.5199 |
| 30 | 1.47712 | 3.40120 | 94.2478 | 706.8583 |

# LOG–LOG e–CIRCUMFERENCE–AREA

| n | Log 10 | Log e | Circumference @ Diameter n | Circle Area @ Radius n |
|---|---|---|---|---|
| 31 | 1.49136 | 3.43399 | 97.3894 | 754.7676 |
| 32 | 1.50515 | 3.46574 | 100.5310 | 804.2477 |
| 33 | 1.51851 | 3.49651 | 103.6726 | 855.2986 |
| 34 | 1.53148 | 3.52636 | 106.8142 | 907.9203 |
| 35 | 1.54407 | 3.55535 | 109.9557 | 962.1128 |
| 36 | 1.55630 | 3.58352 | 113.0973 | 1017.8760 |
| 37 | 1.56820 | 3.61092 | 116.2389 | 1075.2101 |
| 38 | 1.57978 | 3.63759 | 119.3805 | 1134.1149 |
| 39 | 1.59106 | 3.66356 | 122.5221 | 1194.5906 |
| 40 | 1.60206 | 3.68888 | 125.6637 | 1256.6371 |
| 41 | 1.61278 | 3.71357 | 128.8053 | 1320.2543 |
| 42 | 1.62325 | 3.73767 | 131.9469 | 1385.4424 |
| 43 | 1.63347 | 3.76120 | 135.0885 | 1452.2012 |
| 44 | 1.64345 | 3.78419 | 138.2301 | 1520.5308 |
| 45 | 1.65321 | 3.80666 | 141.3717 | 1590.4313 |
| 46 | 1.66276 | 3.82864 | 144.5133 | 1661.9025 |
| 47 | 1.67210 | 3.85015 | 147.6549 | 1734.9445 |
| 48 | 1.68124 | 3.87120 | 150.7964 | 1809.5574 |
| 49 | 1.69020 | 3.89182 | 153.9380 | 1885.7410 |
| 50 | 1.69897 | 3.91202 | 157.0796 | 1963.4954 |
| 51 | 1.70757 | 3.93183 | 160.2212 | 2042.8206 |
| 52 | 1.71600 | 3.95124 | 163.3628 | 2123.7166 |
| 53 | 1.72428 | 3.97029 | 166.5044 | 2206.1834 |
| 54 | 1.73239 | 3.98898 | 169.6460 | 2290.2210 |
| 55 | 1.74036 | 4.00733 | 172.7876 | 2375.8294 |
| 56 | 1.74819 | 4.02535 | 175.9292 | 2463.0086 |
| 57 | 1.75587 | 4.04305 | 179.0708 | 2551.7586 |
| 58 | 1.76343 | 4.06044 | 182.2124 | 2642.0794 |
| 59 | 1.77085 | 4.07754 | 185.3540 | 2733.9710 |
| 60 | 1.77815 | 4.09434 | 188.4956 | 2827.4334 |
| 61 | 1.78533 | 4.11087 | 191.6372 | 2922.4666 |
| 62 | 1.79239 | 4.12713 | 194.7787 | 3019.0705 |
| 63 | 1.79934 | 4.14313 | 197.9203 | 3117.2453 |
| 64 | 1.80618 | 4.15888 | 201.0619 | 3216.9909 |
| 65 | 1.81291 | 4.17439 | 204.2035 | 3318.3072 |
| 66 | 1.81954 | 4.18965 | 207.3451 | 3421.1944 |
| 67 | 1.82607 | 4.20469 | 210.4867 | 3525.6524 |
| 68 | 1.83251 | 4.21951 | 213.6283 | 3631.6811 |
| 69 | 1.83885 | 4.23411 | 216.7699 | 3739.2807 |
| 70 | 1.84510 | 4.24850 | 219.9115 | 3848.4510 |
| 71 | 1.85126 | 4.26268 | 223.0531 | 3959.1921 |
| 72 | 1.85733 | 4.27667 | 226.1947 | 4071.5041 |
| 73 | 1.86332 | 4.29046 | 229.3363 | 4185.3868 |
| 74 | 1.86923 | 4.30407 | 232.4779 | 4300.8403 |
| 75 | 1.87506 | 4.31749 | 235.6194 | 4417.8647 |
| 76 | 1.88081 | 4.33073 | 238.7610 | 4536.4598 |
| 77 | 1.88649 | 4.34381 | 241.9026 | 4656.6257 |
| 78 | 1.89209 | 4.35671 | 245.0442 | 4778.3624 |
| 79 | 1.89763 | 4.36945 | 248.1858 | 4901.6699 |
| 80 | 1.90309 | 4.38203 | 251.3274 | 5026.5482 |
| 81 | 1.90849 | 4.39445 | 254.4690 | 5152.9973 |
| 82 | 1.91381 | 4.40672 | 257.6106 | 5281.0173 |
| 83 | 1.91908 | 4.41884 | 260.7522 | 5410.6079 |

# LOG–LOG e–CIRCUMFERENCE–AREA

| n | Log 10 | Log e | Circumference @ Diameter n | Circle Area @ Radius n |
|---|---|---|---|---|
| 84 | 1.92428 | 4.43082 | 263.8938 | 5541.7694 |
| 85 | 1.92942 | 4.44265 | 267.0354 | 5674.5017 |
| 86 | 1.93450 | 4.45435 | 270.1770 | 5808.8048 |
| 87 | 1.93952 | 4.46591 | 273.3186 | 5944.6787 |
| 88 | 1.94448 | 4.47734 | 276.4602 | 6082.1234 |
| 89 | 1.94939 | 4.48864 | 279.6017 | 6221.1389 |
| 90 | 1.95424 | 4.49981 | 282.7433 | 6361.7251 |
| 91 | 1.95904 | 4.51086 | 285.8849 | 6503.8822 |
| 92 | 1.96379 | 4.52179 | 289.0265 | 6647.6101 |
| 93 | 1.96848 | 4.53260 | 292.1681 | 6792.9087 |
| 94 | 1.97313 | 4.54329 | 295.3097 | 6939.7782 |
| 95 | 1.97772 | 4.55388 | 298.4513 | 7088.2184 |
| 96 | 1.98227 | 4.56435 | 301.5929 | 7238.2295 |
| 97 | 1.98677 | 4.57471 | 304.7345 | 7389.8113 |
| 98 | 1.99123 | 4.58497 | 307.8761 | 7542.9640 |
| 99 | 1.99564 | 4.59512 | 311.0177 | 7697.6874 |
| 100 | 2.00000 | 4.60517 | 314.1593 | 7853.9816 |
| 110 | 2.04139 | 4.70048 | 345.5752 | 9503.3178 |
| 120 | 2.07918 | 4.78749 | 376.9911 | 11309.7336 |
| 130 | 2.11394 | 4.86753 | 408.4070 | 13273.2290 |
| 140 | 2.14613 | 4.94164 | 439.8230 | 15393.8040 |
| 150 | 2.17609 | 5.01064 | 471.2389 | 17671.4587 |
| 160 | 2.20412 | 5.07517 | 502.6548 | 20106.1930 |
| 170 | 2.23045 | 5.13580 | 534.0708 | 22698.0069 |
| 180 | 2.25527 | 5.19296 | 565.4867 | 25446.9005 |
| 190 | 2.27875 | 5.24702 | 596.9026 | 28352.8737 |
| 200 | 2.30103 | 5.29832 | 628.3185 | 31415.9265 |
| 210 | 2.32222 | 5.34711 | 659.7345 | 34636.0590 |
| 220 | 2.34242 | 5.39363 | 691.1504 | 38013.2711 |
| 230 | 2.36173 | 5.43808 | 722.5663 | 41547.5628 |
| 240 | 2.38021 | 5.48064 | 753.9822 | 45238.9342 |
| 250 | 2.39794 | 5.52146 | 785.3982 | 49087.3852 |
| 260 | 2.41497 | 5.56068 | 816.8141 | 53092.9158 |
| 270 | 2.43136 | 5.59842 | 848.2300 | 57255.5261 |
| 280 | 2.44716 | 5.63479 | 879.6459 | 61575.2160 |
| 290 | 2.46240 | 5.66988 | 911.0619 | 66051.9855 |
| 300 | 2.47712 | 5.70378 | 942.4778 | 70685.8347 |
| 310 | 2.49136 | 5.73657 | 973.8937 | 75476.7635 |
| 320 | 2.50515 | 5.76832 | 1005.3096 | 80424.7719 |
| 330 | 2.51851 | 5.79909 | 1036.7256 | 85529.8600 |
| 340 | 2.53148 | 5.82895 | 1068.1415 | 90792.0277 |
| 350 | 2.54407 | 5.85793 | 1099.5574 | 96211.2750 |
| 360 | 2.55630 | 5.88610 | 1130.9734 | 101787.6020 |
| 370 | 2.56820 | 5.91350 | 1162.3893 | 107521.0086 |
| 380 | 2.57978 | 5.94017 | 1193.8052 | 113411.4948 |
| 390 | 2.59106 | 5.96615 | 1225.2211 | 119459.0606 |
| 400 | 2.60206 | 5.99146 | 1256.6371 | 125663.7061 |
| 410 | 2.61278 | 6.01616 | 1288.0530 | 132025.4313 |
| 420 | 2.62325 | 6.04025 | 1319.4689 | 138544.2360 |
| 430 | 2.63347 | 6.06379 | 1350.8848 | 145220.1204 |
| 440 | 2.64345 | 6.08677 | 1382.3008 | 152053.0844 |
| 450 | 2.65321 | 6.10925 | 1413.7167 | 159043.1281 |
| 460 | 2.66276 | 6.13123 | 1445.1326 | 166190.2514 |

| n | Log 10 | Log e | Circumference @ Diameter n | Circle Area @ Radius n |
|---|---|---|---|---|
| 470 | 2.67210 | 6.15273 | 1476.5485 | 173494.4543 |
| 480 | 2.68124 | 6.17379 | 1507.9645 | 180955.7368 |
| 490 | 2.69020 | 6.19441 | 1539.3804 | 188574.0990 |
| 500 | 2.69897 | 6.21461 | 1570.7963 | 196349.5408 |
| 510 | 2.70757 | 6.23441 | 1602.2123 | 204282.0623 |
| 520 | 2.71600 | 6.25383 | 1633.6282 | 212371.6634 |
| 530 | 2.72428 | 6.27288 | 1665.0441 | 220618.3441 |
| 540 | 2.73239 | 6.29157 | 1696.4600 | 229022.1044 |
| 550 | 2.74036 | 6.30992 | 1727.8760 | 237582.9444 |
| 560 | 2.74819 | 6.32794 | 1759.2919 | 246300.8640 |
| 570 | 2.75587 | 6.34564 | 1790.7078 | 255175.8633 |
| 580 | 2.76343 | 6.36303 | 1822.1237 | 264207.9422 |
| 590 | 2.77085 | 6.38012 | 1853.5397 | 273397.1007 |
| 600 | 2.77815 | 6.39693 | 1884.9556 | 282743.3388 |
| 610 | 2.78533 | 6.41346 | 1916.3715 | 292246.6566 |
| 620 | 2.79239 | 6.42972 | 1947.7874 | 301907.0540 |
| 630 | 2.79934 | 6.44572 | 1979.2034 | 311724.5310 |
| 640 | 2.80618 | 6.46147 | 2010.6193 | 321699.0877 |
| 650 | 2.81291 | 6.47697 | 2042.0352 | 331830.7240 |
| 660 | 2.81954 | 6.49224 | 2073.4512 | 342119.4400 |
| 670 | 2.82607 | 6.50728 | 2104.8671 | 352565.2355 |
| 680 | 2.83251 | 6.52209 | 2136.2830 | 363168.1107 |
| 690 | 2.83885 | 6.53669 | 2167.6989 | 373928.0656 |
| 700 | 2.84510 | 6.55108 | 2199.1149 | 384845.1001 |
| 710 | 2.85126 | 6.56526 | 2230.5308 | 395919.2142 |
| 720 | 2.85733 | 6.57925 | 2261.9467 | 407150.4079 |
| 730 | 2.86332 | 6.59304 | 2293.3626 | 418538.6813 |
| 740 | 2.86923 | 6.60665 | 2324.7786 | 430084.0343 |
| 750 | 2.87506 | 6.62007 | 2356.1945 | 441786.4669 |
| 760 | 2.88081 | 6.63332 | 2387.6104 | 453645.9792 |
| 770 | 2.88649 | 6.64639 | 2419.0263 | 465662.5711 |
| 780 | 2.89209 | 6.65929 | 2450.4423 | 477836.2426 |
| 790 | 2.89763 | 6.67203 | 2481.8582 | 490166.9938 |
| 800 | 2.90309 | 6.68461 | 2513.2741 | 502654.8246 |
| 810 | 2.90849 | 6.69703 | 2544.6900 | 515299.7350 |
| 820 | 2.91381 | 6.70930 | 2576.1060 | 528101.7251 |
| 830 | 2.91908 | 6.72143 | 2607.5219 | 541060.7947 |
| 840 | 2.92428 | 6.73340 | 2638.9378 | 554176.9441 |
| 850 | 2.92942 | 6.74524 | 2670.3538 | 567450.1730 |
| 860 | 2.93450 | 6.75693 | 2701.7697 | 580880.4816 |
| 880 | 2.94448 | 6.77992 | 2764.6015 | 608212.3377 |
| 890 | 2.94939 | 6.79122 | 2796.0175 | 622113.8852 |
| 900 | 2.95424 | 6.80239 | 2827.4334 | 636172.5123 |
| 910 | 2.95904 | 6.81344 | 2858.8493 | 650388.2191 |
| 920 | 2.96379 | 6.82437 | 2890.2652 | 664761.0055 |
| 930 | 2.96848 | 6.83518 | 2921.6812 | 679290.8715 |
| 940 | 2.97313 | 6.84588 | 2953.0971 | 693977.8172 |
| 950 | 2.97772 | 6.85646 | 2984.5130 | 708821.8424 |
| 960 | 2.98227 | 6.86693 | 3015.9289 | 723822.9474 |
| 970 | 2.98677 | 6.87730 | 3047.3449 | 738981.1319 |
| 980 | 2.99123 | 6.88755 | 3078.7608 | 754296.3961 |
| 990 | 2.99564 | 6.89770 | 3110.1767 | 769768.7399 |
| 1000 | 3.00000 | 6.90776 | 3141.5927 | 785398.1634 |

## RIGHT TRIANGLE TRIG FORMULAS

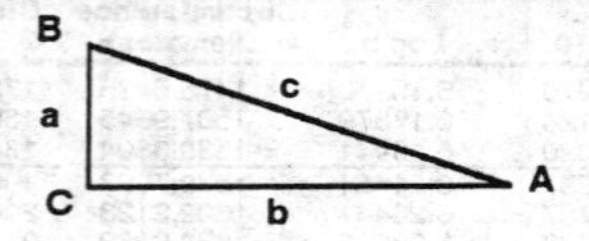


A, B, C = Angles a, b, c = Distances

$$\sin A = \frac{a}{c}, \quad \cos A = \frac{b}{c}, \quad \tan A = \frac{a}{b}$$

$$\cot A = \frac{b}{a}, \quad sec\ A = \frac{c}{b}, \quad cosec\ A = \frac{c}{a}$$

**Given a and b, Find A, B, and c**

$$\tan A = \frac{a}{b} = \cot B, \quad c = \sqrt{a^2+b^2} = a\sqrt{1+\frac{b^2}{a^2}}$$

**Given a and c, Find A, B, b**

$$\sin A = \frac{a}{c} = \cos B, \quad b = \sqrt{(c+a)(c-a)} = c\sqrt{1-\frac{a^2}{c^2}}$$

**Given A and a, Find B, b, c**

$$B = 90^o - A, \quad b = a \cot A, \quad c = \frac{a}{\sin A}$$

**Given A and b, Find B, a, c**

$$B = 90^o - A, \quad a = b \tan A, \quad c = \frac{b}{\cos A}$$

**Given A and c, Find B, a, b**

$$B = 90^o - A, \quad a = c \sin A, \quad b = c \cos A$$

## OBLIQUE TRIANGLE FORMULAS

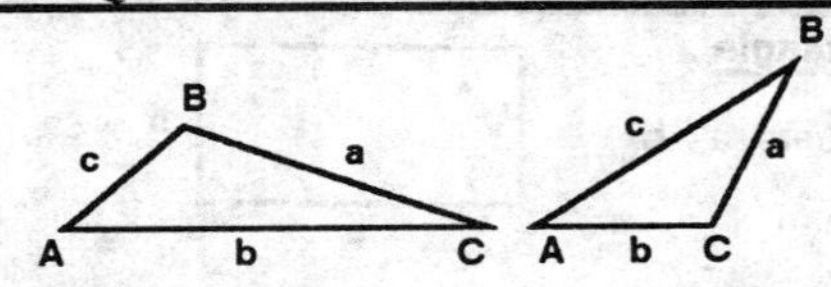


**Given A, B and a, Find b, C, and c**

$$b = \frac{a \sin B}{\sin A} \;,\; C = 180^o - (A+B) \;,\; c = \frac{a \sin C}{\sin A}$$

**Given A, a and b, Find B, C, and c**

$$\sin B = \frac{b \sin A}{a} \;,\; C = 180^o - (A+B) \;,\; c = \frac{a \sin C}{\sin A}$$

**Given a, b and C, Find A, B, and c**

$$A+B = 180^o - C \;,\; c = \frac{a \sin C}{\sin A}$$

$$\tan \tfrac{1}{2}(A-B) = \frac{(a-b) \tan \tfrac{1}{2}(A+B)}{a+b}$$

**Given a, b and c, Find A, B, and C**

$$s = \frac{a+b+c}{2} \;,\; \sin \tfrac{1}{2} A = \sqrt{\frac{(s-b)(s-c)}{bc}}$$

$$\sin \tfrac{1}{2} B = \sqrt{\frac{(s-a)(s-c)}{ac}} \;,\; C = 180^o - (A+B)$$

**Given a, b and c, Find Area**

$$s = \frac{a+b+c}{2} \;,\; Area = \sqrt{s(s-a)(s-b)(s-c)}$$

$$Area = bc \sin \frac{A}{2} \;,\; Area = \frac{a^2 \sin B \sin C}{2 \sin A}$$

# PLANE FIGURE FORMULAS

## Rectangle

If square, a = b

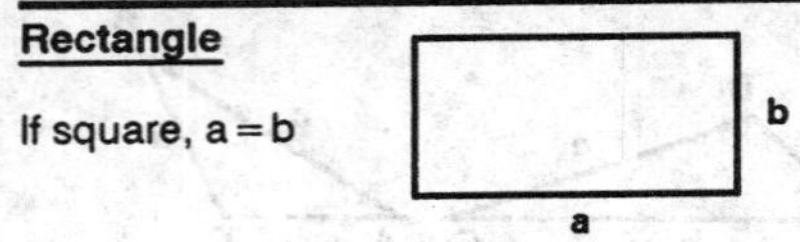


$$Area = ab$$
$$Perimeter = 2\,(a + b)\ ,\ Diagonal = \sqrt{a^2 + b^2}$$

## Parallelogram

All sides are parallel
$\theta$ = degrees

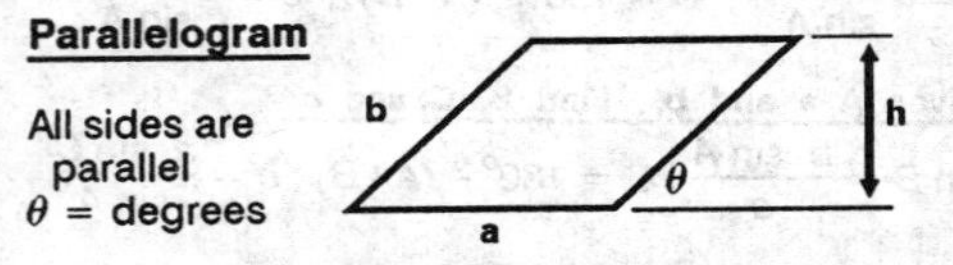


$$Area = ah = ab \sin\theta\ ,\ Perimeter = 2\,(a + b)$$

## Trapezoid

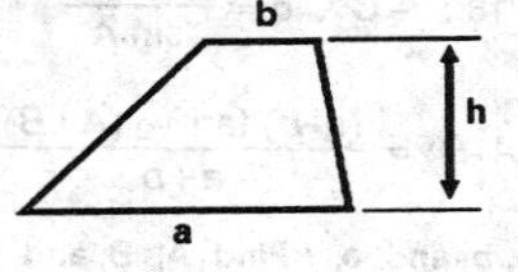


$$Area = \frac{(a + b)}{2}h$$
Perimeter = Sum of lengths of sides

## Quadrilateral

$\theta$ = degrees

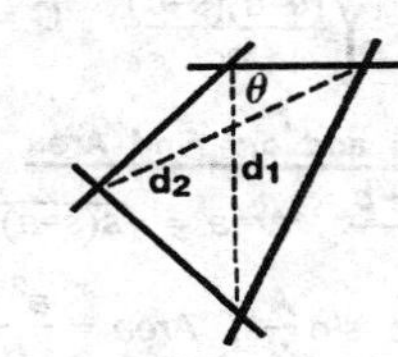


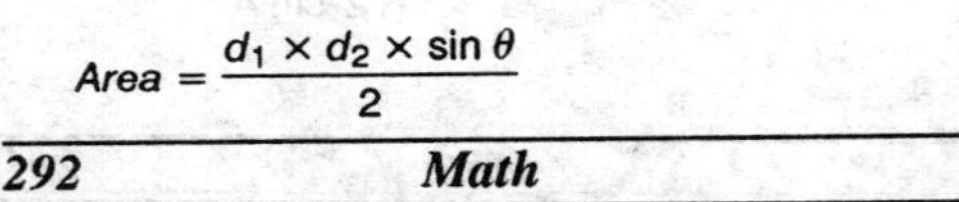

$$Area = \frac{d_1 \times d_2 \times \sin\theta}{2}$$

## Trapezium

a to g = lengths

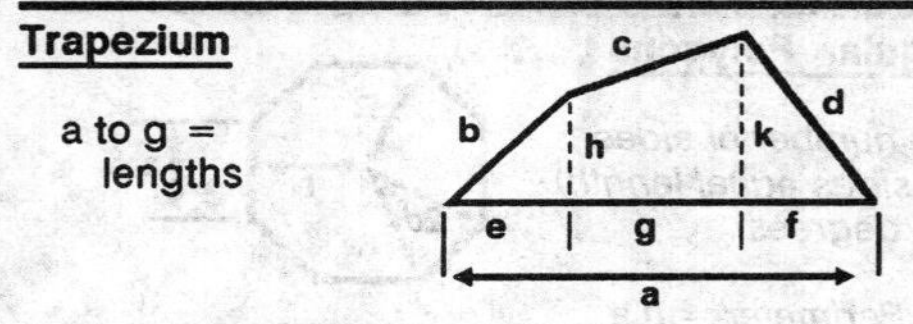


$$Perimeter = a + b + c + d$$

$$Area = \frac{(h + k)\,g + e\,h + f\,k}{2}$$

## Equilateral Triangle

a = all sides equal

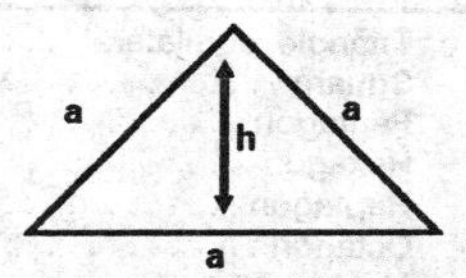


$$Perimeter = 3\,a\ ,\quad h = \frac{a}{2}\sqrt{3} = 0.866\,a$$

$$Area = a^2\,\frac{\sqrt{3}}{4} = 0.433\,a^2$$

## Annulus

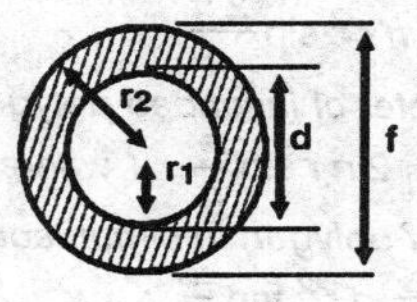


$$Area = 0.7854\,(d^2 - f^2)$$

$$Area = \pi\,(r_1 + r_2)\,(r_2 - r_1)$$

# PLANE FIGURE FORMULAS

## Regular Polygons

*n* = number of sides
(all sides equal length)
$\theta$ = degrees

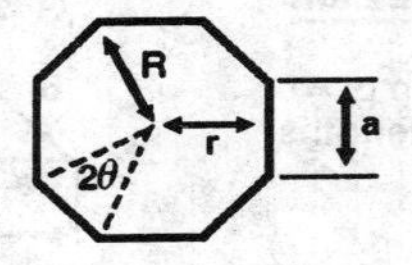


Perimeter = $n\,a$

$$Area = \frac{n\,a\,r}{2} = n\,r^2 \tan\theta = \frac{n\,R^2}{2}\sin 2\theta$$

| Polygon | Number of Sides | Area |
|---|---|---|
| Triangle, equilateral | 3 | $0.4330\ a^2$ |
| Square | 4 | $1.0000\ a^2$ |
| Pentagon | 5 | $1.7205\ a^2$ |
| Hexagon | 6 | $2.5981\ a^2$ |
| Heptagon | 7 | $3.6339\ a^2$ |
| Octagon | 8 | $4.8284\ a^2$ |
| Nonagon | 9 | $6.1818\ a^2$ |
| Decagon | 10 | $7.6942\ a^2$ |
| Undecagon | 11 | $9.3656\ a^2$ |
| Dodecagon | 12 | $11.1961\ a^2$ |

*Area of inscribed polygon in a circle of radius r =*

$$\tfrac{1}{2}\,n\,r^2 \sin\frac{2\pi}{n}$$

*Perimeter of inscribed polygon in circle of radius*

$$r = 2\,n\,r\sin\frac{\pi}{n} \quad (\text{where } \pi \text{ radians} = 180^\circ)$$

*Area of polygon in circumscribed circle , radius*

$$R = n\,r^2\tan\frac{\pi}{n}$$

*Perimeter of circumscribed polygon in circle ,*

*with radius r =* $2\,n\,r\tan\frac{\pi}{n}$

# PLANE FIGURE FORMULAS

## Circle

Z = point
X = point
θ = degrees
c,d,r,m = lengths
π = 3.14159
c = cord
r = radius

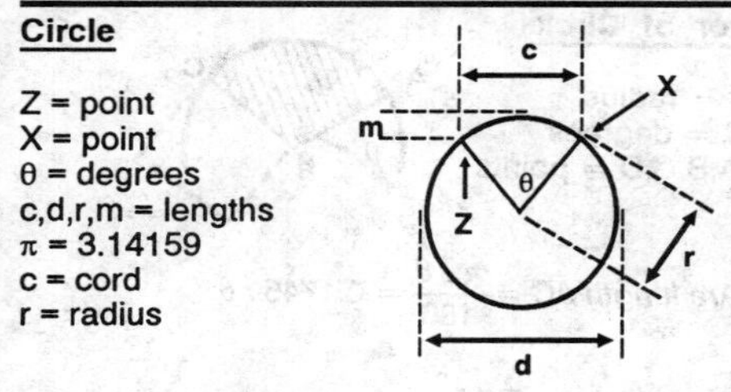


$$Circumference = 2\,\pi\,r = \pi\,d = 3.14159\,d$$

$$Circumference\ or\ Perimeter = 2\,\pi\,r = \pi\,d$$

$$Area = \pi\,r^2 = \pi\,\frac{d^2}{4} = 0.78539\,d^2$$

$$Area = \frac{Perimeter^2}{4\,\pi} = 0.07958\,Perimeter^2$$

$$Length\ of\ \text{arc}\ XZ = \theta\,\frac{\pi}{180}\,r = 0.017453\,\theta\,r$$

$$r = \frac{m^2 + \frac{1}{4}c^2}{2m} = \frac{\frac{1}{2}c}{\sin \frac{1}{2}\theta}$$

$$c = 2\sqrt{2\,mr - m^2} = 2\,r\,\sin \tfrac{1}{2}\,\theta$$

$$m = r \pm \sqrt{r^2 - \frac{c^2}{4}} \quad (use + \text{ if arc} \geq 180^o,$$

$$- \text{ if arc} < 180^o)$$

$$m = \tfrac{1}{2}\,c\,\tan \tfrac{1}{4}\,\theta = 2\,r\,\sin^2 \tfrac{1}{4}\,\theta$$

# PLANE FIGURE FORMULAS

## Sector of Circle

$r$ = radius
$\theta$ = degrees
A,B,C,D = points

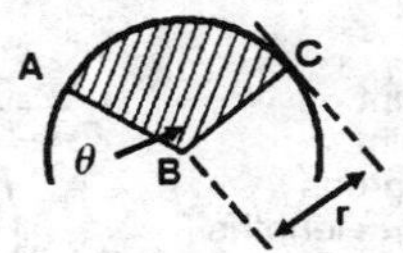


$$\text{Arc length } AC = \frac{\pi\, r\, \theta}{180} = 0.1745\, r\, \theta$$

$$\text{Area } ABCA = \frac{\pi\, \theta\, r^2}{360} = 0.008727\, \theta\, r^2$$

$$\text{Area } ABCA = \frac{\text{Arc length } AC \times r}{2}$$

## Segment of Circle

$r$ = radius
$\theta$ = degrees
A,B,C,D = points

For $\theta < 90°$

$$\text{Area } ACDA = \frac{r^2}{2}\left(\frac{\pi\, \theta}{180} - \sin\theta\right)$$

For $\theta > 90°$

$$\text{Area } ACDA = \frac{r^2}{2}\left(\frac{\pi\, \theta}{180} - \sin(180 - \theta)\right)$$

## Circular Zone

*Area ACDFA =*
*Circle Area –*
*Segment Area ABCA –*
*Segment Area FDEF*

## Hollow Circle Sector

$\theta$ = degrees
A,B,C,D = points
r = radius

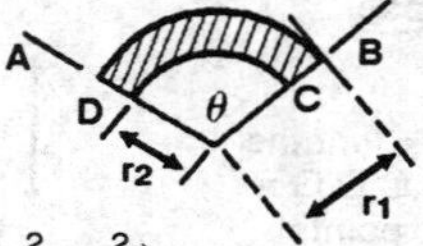


$$\text{Area } ABCDA = \frac{\pi\,\theta\,(r_2^2 - r_1^2)}{360}$$

$$\text{Area } ABCDA = \frac{r_1 - r_2}{2}\,(\text{Arc length } AB + \text{Arc length } CD)$$

## Fillet

r = radius

*Area of fillet* = $0.215\,r^2$

## Parabola

A,B,C = points
a,b = lengths

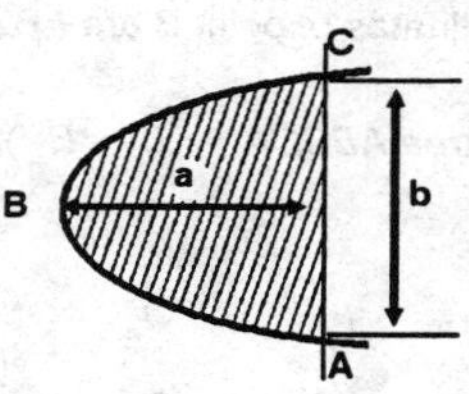


$$\text{Area } ABCA = \frac{2}{3}\,a\,b$$

$$\text{Arc Length } ABC = b\left\{\frac{1}{2}\left(1 + 16\left(\frac{a}{b}\right)^2\right)^{\frac{1}{2}} + \frac{1}{8\left(\frac{a}{b}\right)}\log_e \times \left[4n + \left(1 + 16\left(\frac{a}{b}\right)^2\right)^{\frac{1}{2}}\right]\right\}$$

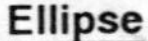

## Ellipse

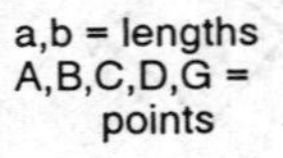

a,b = lengths
A,B,C,D,G = points

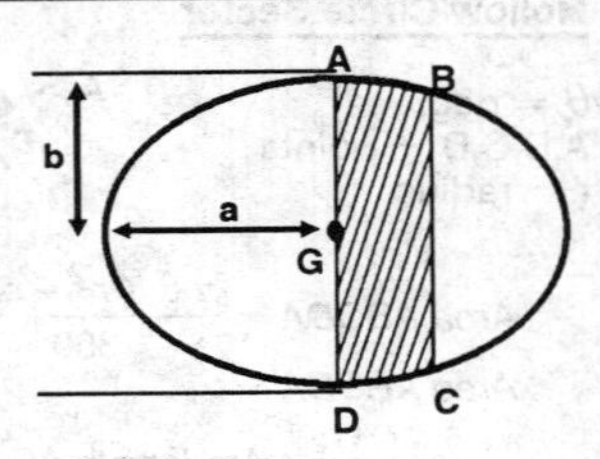


*Area of ellipse* = $\pi\, a\, b$

*Perimeter of ellipse* = $\pi\, [\, 1.5\, (\, a + b\, ) - \sqrt{ab}\, ]$
(approximate)

Assuming point G is the center of the ellipse, which has ( x, y ) coordinates of ( 0, 0 ), and the coordinates of point B are ( $B_x$, $B_y$ ) :

$$\text{Area } ABCDA = (\, B_x \times B_y\, ) + ab \sin^{-1} \left( \frac{B_x}{a} \right)$$

# SOLID FIGURE FORMULAS

**Parallelopiped and Cube**

a,b,c = lengths

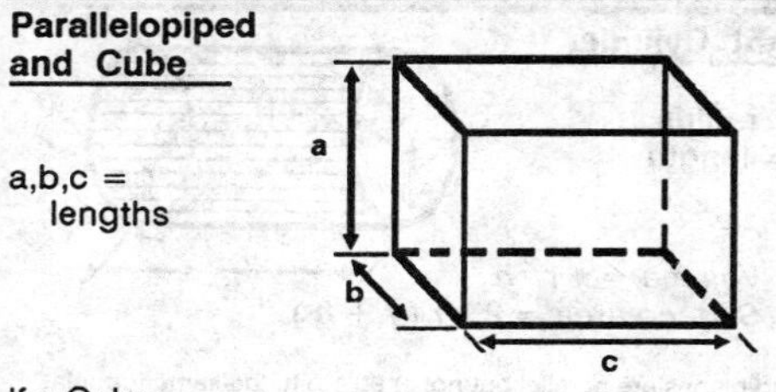


If a Cube:

*Volume* $= a^3$ *Surface area* $= 6\,a^2$

If a Parallelopiped (a, b, and c can be different):

*Volume* $= a\,b\,c$ *Area* $= 2\,(\,ab + bc + ac\,)$

**Prism – Right, or oblique, regular or irregular**

A = area
h = length
a,b,c = length

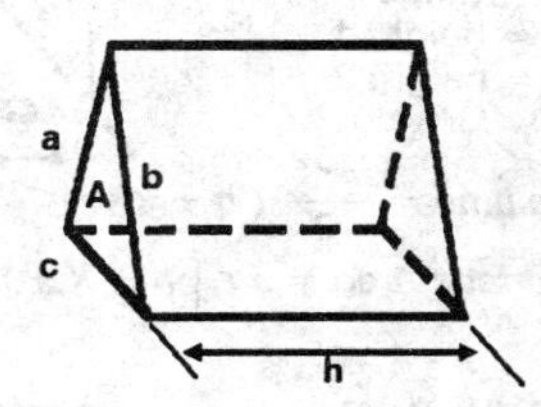


*Volume* $= A\,h$ where A is the area of the end plate abca. If the end plate has 3 or more sides, see page 294 for rules of calculating areas of polygons.

*Convex Surface area* $= h\,(\,a+b+c+....n\text{ sides}\,)$

If end planes are parallel but not at 90° to h, the same formulas apply but a slice at 90° through the prism must be used to determine a, b, and c.

## SOLID FIGURE FORMULAS

### Right Cylinder

r = radius
h = length

$$Volume = \pi r^2 h$$

$$Surface\ Area = 2 \pi r ( r + h )$$

If end planes are parallel but not at 90° to h, the same formulas apply but a slice at 90° through the prism must be used to determine r.

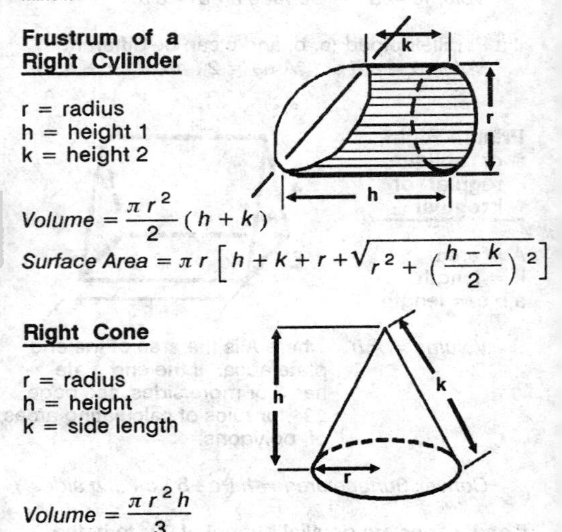


### Frustrum of a Right Cylinder

r = radius
h = height 1
k = height 2

$$Volume = \frac{\pi r^2}{2} ( h + k )$$

$$Surface\ Area = \pi r \left[ h + k + r + \sqrt{r^2 + \left( \frac{h - k}{2} \right)^2} \right]$$

### Right Cone

r = radius
h = height
k = side length

$$Volume = \frac{\pi r^2 h}{3}$$

$$Surface\ Area = \pi r ( r + k )$$

# SOLID FIGURE FORMULAS

## Right Pyramid

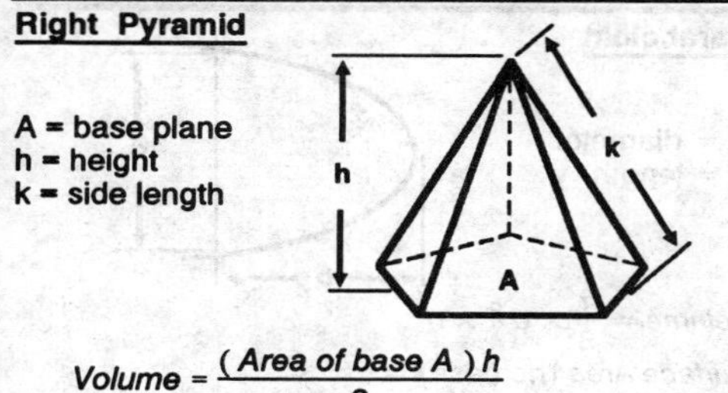


A = base plane
h = height
k = side length

$$Volume = \frac{(Area\ of\ base\ A)\,h}{3}$$

$$Surface\ Area\ (no\ base) = Perimeter\ of\ base\ A \times \frac{k}{2}$$

Use polygon areas on page 294 , if you want to include the base area

## Sphere

r = radius

$$Volume = \frac{4\pi r^3}{3}$$

$$Surface\ Area = 4\pi r^2$$

## Circular Ring

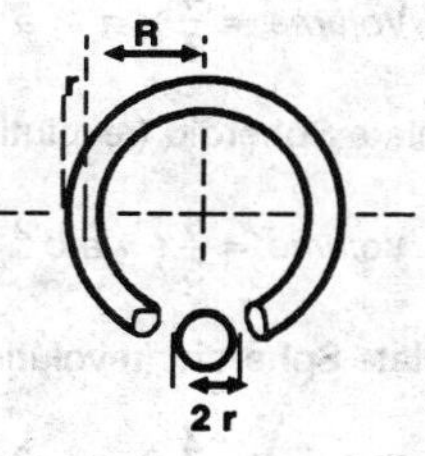


r = cross section radius
R = ring radius

$$Volume = 2\pi^2 R r$$

$$Surface\ Area = 2\pi^2 R r^2$$

**Paraboloid**

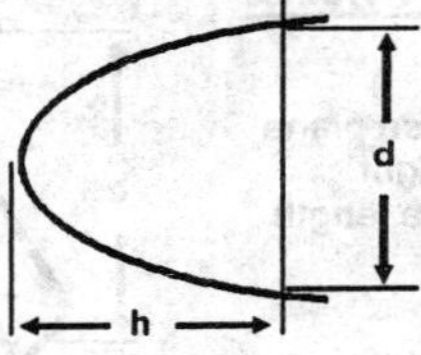


d = diameter
h = length

$$Volume = \frac{\pi}{8} \times d^2 \times h$$

*Surface Area (no base)* =

$$\frac{2}{3} \times \pi \times \frac{d}{h^2}\left[\left(\frac{d^2}{16} + h^2\right)^{\frac{3}{2}} - \left(\frac{d}{4}\right)^3\right]$$

**Ellipsoid and Spheroid**

a, b, c = axis radius

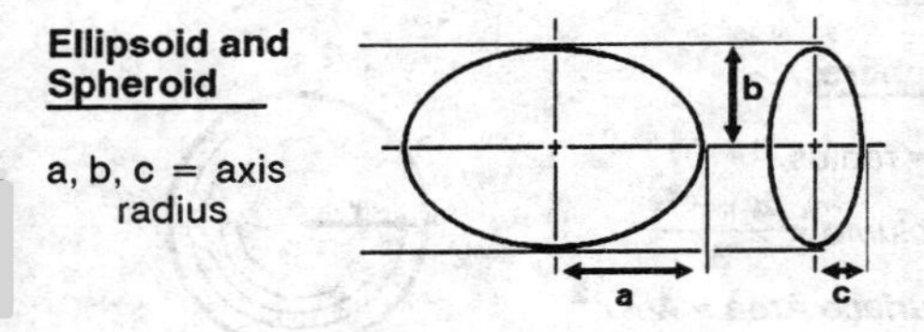


$$Volume = \frac{4}{3} \times \pi \times a \times b \times c$$

Prolate Spheroid (revolution about major axis b)

$$Volume = \frac{4}{3}(\pi\, a\, b^2)$$

Oblate Spheroid (revolution about minor axis a)

$$Volume = \frac{4}{3}(\pi\, b\, a^2)$$

# POCKET REF

## Mine and Mill [1]



(See also WEIGHTS OF MATERIALS, p. 389, for Angle of Repose, Rock Densities, etc)

[1] Two pocket sized handbooks are available that deal with roadway and milling equipment specifications and other general data: (a) *Pioneer Facts and Figures, Portec Pioneer Division, Minneapolis, MN* and (b) *Cedarapids Reference Book, Iowa Mfg Co. Cedar Rapids, Iowa*.

# STANDARD SIEVE SERIES

| Tyler Inch/Mesh # | US Standard Inch/Sieve # | Sieve Opening Inches | Sieve Opening Millimeters |
|---|---|---|---|
| | 4.24 inch | 4.24 | 107.6 |
| | 4 inch | 4.00 | 101.6 |
| | 2.12 inch | 2.12 | 53.8 |
| | 2 inch | 2.00 | 50.8 |
| | 1-1/2 inch | 1.50 | 38.1 |
| | 1-1/4 inch | 1.25 | 31.5 |
| 1.05 inch | 1.06 inch | 1.06 | 26.5 |
| | 1.00 inch | 1.00 | 25.0 |
| 0.883 inch | 7/8 inch | 0.875 | 22.4 |
| 0.742 inch | 3/4 inch | 0.750 | 19.0 |
| 0.624 inch | 5/8 inch | 0.625 | 16.0 |
| 0.525 inch | 0.530 inch | 0.530 | 13.2 |
| | 1/2 inch | 0.500 | 12.5 |
| 0.441 inch | 7/16 inch | 0.4375 | 11.2 |
| 0.371 inch | 3/8 inch | 0.375 | 9.5 |
| 2 1/2 | 5/16 inch | 0.3125 | 8.0 |
| 3 | 0.265 inch | 0.265 | 6.7 |
| | 1/4 inch | 0.250 | 6.3 |
| 3 1/2 | 3-1/2 | 0.223 | 5.66 |
| 4 | 4 | 0.187 | 4.76 |
| 5 | 5 | 0.157 | 4.00 |
| 6 | 6 | 0.132 | 3.66 |
| 7 | 7 | 0.111 | 2.83 |
| 8 | 8 | 0.0937 | 2.38 |
| 9 | 10 | 0.0787 | 2.00 |
| 10 | 12 | 0.0661 | 1.68 |
| 12 | 14 | 0.0555 | 1.41 |
| 14 | 16 | 0.0469 | 1.19 |
| 16 | 18 | 0.0394 | 1.00 |
| 20 | 20 | 0.0331 | 0.84 |
| 24 | 25 | 0.0280 | 0.71 |
| 28 | 30 | 0.0232 | 0.59 |
| 32 | 35 | 0.0197 | 0.50 |
| 35 | 40 | 0.0165 | 0.42 |
| 42 | 45 | 0.0138 | 0.35 |
| 48 | 50 | 0.0117 | 0.297 |
| 60 | 60 | 0.0098 | 0.250 |
| 65 | 70 | 0.0083 | 0.210 |
| 80 | 80 | 0.0070 | 0.177 |
| 100 | 100 | 0.0059 | 0.149 |
| 115 | 120 | 0.0049 | 0.125 |
| 150 | 140 | 0.0041 | 0.105 |
| 170 | 170 | 0.0035 | 0.088 |
| 200 | 200 | 0.0029 | 0.074 |
| 250 | 230 | 0.0024 | 0.062 |
| 270 | 270 | 0.0021 | 0.053 |
| 325 | 325 | 0.0017 | 0.044 |
| 400 | 400 | 0.0015 | 0.037 |

Note: 1 millimeter = 1000 microns

# MINERAL DRESSING SIZING SCALE

| Size | Mineral Dressing Method |
|---|---|
| +4 inch to 400 mesh | Screening |
| +4 inch to 65 mesh | Magnetic Separator (dry) |
| +4 inch to 325 mesh | Magnetic Separator (wet) |
| +4 inch to 4 mesh | Sink – Float |
| +4 inch to 65 mesh | Hammer Mill – Jaw Crusher |
| +4 inch to 65 mesh | Gyratory Crusher |
| +4 inch to 28 mesh | Rolls |
| 4 inch to 20 mesh | Jigging |
| 2 inch to 3 mesh | Rod Mill |
| 1 inch to 325 mesh | Ball Mill |
| 0.5 inch to 26 micron | Pulverizer |
| 3 mesh to 48 mesh | Weinig Jig |
| 4 mesh to 100 mesh | Humphreys Spiral |
| 8 mesh to 200 mesh | Shaking Table |
| 10 mesh to 18.5 micron | Isodynamic Separator |
| 35 mesh to 4.6 micron | Classification |
| 35 mesh to 6.5 micron | Flotation |
| 48 mesh to 3.25 micron | Turbidimetry |
| 65 mesh to 9.25 micron | Superpanner |
| 100 mesh to 6.5 micron | Infrasizer |
| 0.81 micron to 0.25 micron | Centrifuge |
| | |
| 400 mesh to 0.2 micron | Normal Microscope Range |
| ± 0.5 micron | Brownian Movement and the wavelength of visible light |
| 0.41 micron to 0.001 micron | Normal Electron Microscope |
| ± 0.025 micron | Thinnest files visible by light interference |
| 0.004 micron | Large Molecules |
| 0.0007 micron | Average Crystal Unit |

NOTE: The above size ranges are approximations only and the actual size range can vary considerably depending on the material and current technology.

# STOCKPILE VOLUME & WEIGHT

The following formula is used to calculate the volume of a stockpile if the diameter and height are known:

$$\text{Volume in cubic feet} = 0.2618 \times D^2 \times h$$

D = Diameter of the base of the cone in feet
h = Height of the cone in feet

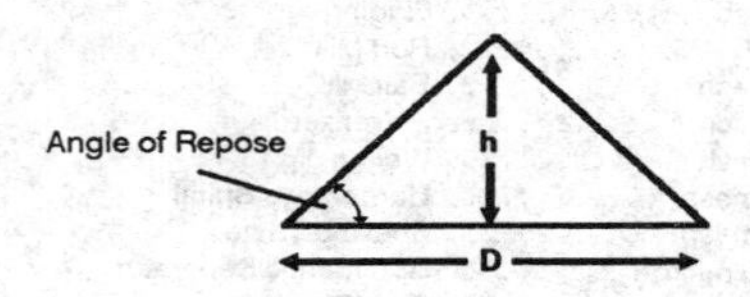


In order to calculate the actual weight of material in the stockpile, determine the density or weight/cubic foot (or look up an approximation of the density in the WEIGHTS OF MATERIALS chapter.)

$$\text{Weight in tons} = \text{Volume in cubic feet} \times \frac{\text{Density in lbs/cu. feet}}{2000}$$

**CONICAL STOCKPILE VOLUMES (37° Angle of Repose)**

| Diameter in feet | Height in feet | Volume in Cu Yds | Weight at 100 lbs/cu foot |
|---|---|---|---|
| 26.50 | 10 | 68 | 92 |
| 39.83 | 15 | 230 | 310 |
| 53.00 | 20 | 545 | 735 |
| 66.30 | 25 | 1065 | 1440 |
| 79.50 | 30 | 1845 | 2490 |
| 92.83 | 35 | 2930 | 3955 |
| 106.00 | 40 | 4370 | 5900 |
| 132.66 | 50 | 8540 | 11525 |
| 159.16 | 60 | 14755 | 19915 |
| 186.00 | 70 | 23375 | 31555 |
| 212.16 | 80 | 34970 | 47210 |
| 238.83 | 90 | 49795 | 67225 |
| 265.30 | 100 | 68300 | 92210 |

# MATERIAL DUMPING ANGLES

| Material | Dumping Angle in Degrees |
|---|---|
| Ashes, dry | 33 |
| Ashes, moist | 36 |
| Ashes, wet | 30 |
| Asphalt | 45 |
| Cinders, dry | 33 |
| Cinders, moist | 34 |
| Cinders, wet | 31 |
| Cinders and Clay | 30 |
| Clay | 45 |
| Coal, hard | 24 |
| Coal, soft | 30 |
| Coke | 23 |
| Concrete | 30 |
| Earth, loose | 28 |
| Earth, compact | 50 |
| Garbage | 30 |
| Gravel | 40 |
| Ore, dry | 30 |
| Ore, damp | 37 |
| Rubble | 45 |
| Sand, dry | 35 |
| Sand, damp | 40 |
| Sand, with crushed stone | 27 |
| Stone | 30 |
| Stone, broken | 27 |
| Stone, crushed | 30 |

# ROCK BULKING FACTORS

| Material | Density In Place | Density After Mined | Percent Expansion |
|---|---|---|---|
| Basalt | 3.00 | 1.72 | 75 to 80% |
| Clay | 1.86 | 1.49 | 20 to 30% |
| Dolomite | 2.56 | 1.73 | 50 to 60% |
| Gneiss | 2.69 | 1.54 | 75 to 80% |
| Granite | 2.72 | 1.55 | 75 to 80% |
| Gravel, dry | 1.80 | 1.40 | 20 to 30% |
| Gravel, wet | 2.00 | 1.60 | 20 to 30% |
| Gravel, wet w/clay | 1.92 | 1.28 | 50 to 60% |
| Limestone | 2.69 | 1.54 | 75 to 80% |
| Quartz | 2.64 | 1.51 | 75 to 80% |
| Sand, dry | 1.60 | 1.28 | 20 to 30% |
| Sand, wet | 1.95 | 1.56 | 20 to 30% |
| Sandstone | 2.42 | 1.38 | 75 to 80% |
| Slate | 2.80 | 1.52 | 85 to 90% |
| Soil, w/clay | 1.76 | 1.41 | 20 to 30% |

## LENGTH OF BELT IN A ROLL

In order to calculate the number of feet of conveyor belt in a tightly coiled roll, use the following equations:

A = Diameter of coil in inches + Diameter of coil hole in inches

Belt length in feet = A x Number of coils x 0.131

## CONVEYOR SLOPE MAXIMUMS

| Material | Maximum Slope Degrees |
|---|---|
| Cement, loose | 22 |
| Coke, screened | 18 |
| Coke, breeze | 20 |
| Concrete, 6 inch slump | 12 |
| Concrete, 4 inch slump | 20 |
| Concrete, 2 inch slump | 24 to 26 |
| Coal, +4 inch lump, soft | 15 |
| Coal, – 4 inch lump, soft | 16 |
| Coal, anthracite | 16 |
| Coal, unsized | 18 |
| Coal, soft, fine | 20 to 22 |
| Earth, loose | 20 |
| Earth, sluggish | 22 |
| Glass batch | 21 |
| Gravel, sized, washed | 12 |
| Gravel, sized, unwashed | 15 |
| Gravel, unsized | 18 to 20 |
| Grain, whole | 15 |
| Gypsum, powdered | 23 |
| Lime, powdered | 23 |
| Logs, no bark | 10 |
| Ore, +4 inch | 18 |
| Ore, – 4 inch | 20 |
| Ore, sized | 16 |
| Packages, paper wrapped, smooth belt | 16 |
| Packages, paper wrapped, ribflex belt | 25 to 45 |
| Salt | 20 |
| Sand, dry | 16 |
| Sand, moist, bank run | 20 |
| Sand, foundry | 24 |
| Sulphur, powdered | 21 |
| Stone, sized, +4 inch | 15 |
| Stone, sized, – 4 inch | 16 |
| Stone, unsized, +4 inch | 16 |
| Stone, unsized, – 4 inch | 18 |
| Stone, – 3/8 inch | 20 |
| Wood chips | 25 |

# CONVEYOR CAPACITIES

| Belt Width Inches | Material Size Inches | Belt Speed Feet/min | Tons/hour capacity @ lbs/cu ft material weight with 22° idlers. |
|---|---|---|---|
| 12 | 2 to 3 | 200 | 14 @ 30, 44 @ 100, 70 @ 150 |
| 16 | 3 to 5 | 300 | 36 @ 30, 123 @ 100, 183 @ 150 |
| 18 | 4 to 6 | 300 | 39 @ 30, 156 @ 100, 231 @ 150 |
| 24 | 6 to 8 | 300 | 84 @ 30, 276 @ 100, 414 @ 150 |
| 30 | 7 to 12 | 350 | 150 @ 30, 504 @ 100, 756 @ 150 |
| 36 | 8 to 16 | 350 | 228 @ 30, 728 @ 100, 1088 @ 150 |
| 42 | 10 to 20 | 400 | 340 @ 30, 1128 @ 100, 1692 @ 150 |
| 48 | 12 to 24 | 400 | 452 @ 30, 1512 @ 100, 2248 @ 150 |

Note: For capacities with 35° idlers, multiply 22° capacity by 1.15

# CONVEYOR HORSEPOWER vs LOAD

| Conv Length Feet | Horsepower Required for Transporting Material on Level Ground at the given Tons/Hour Capacity 100 | 200 | 400 | 600 | 800 | 1000 |
|---|---|---|---|---|---|---|
| 25 | 2.0 | 2.5 | 3.5 | 4.5 | 5.5 | 6.5 |
| 50 | 2.4 | 3.0 | 4.2 | 5.4 | 6.6 | 7.8 |
| 100 | 3.0 | 3.8 | 5.3 | 6.8 | 8.3 | 9.8 |
| 200 | 4.3 | 5.3 | 7.5 | 9.7 | 11.9 | 14.1 |
| 300 | 5.6 | 7.0 | 9.8 | 12.6 | 15.4 | 18.2 |
| 400 | 6.8 | 8.5 | 11.9 | 15.3 | 18.7 | 22.1 |
| 500 | 8.0 | 10.1 | 14.3 | 18.5 | 22.7 | 26.9 |

| Conv Lift Feet | Extra Horsepower Required in addition to above HP For Lifting at the given Tons/Hour Capacity 100 | 200 | 400 | 600 | 800 | 1000 |
|---|---|---|---|---|---|---|
| 10 | 1 | 2 | 4 | 6 | 8 | 10 |
| 20 | 2 | 4 | 8 | 12 | 16 | 20 |
| 30 | 3 | 6 | 12 | 18 | 24 | 30 |
| 40 | 4 | 8 | 16 | 24 | 32 | 40 |
| 50 | 5 | 10 | 20 | 30 | 40 | 50 |
| 60 | 6 | 12 | 24 | 36 | 48 | 60 |
| 80 | 8 | 16 | 32 | 48 | 64 | 80 |
| 100 | 10 | 20 | 40 | 60 | 80 | 100 |

The above data is for equipment manufactured by *Portec Pioneer Division, Minneapolis, MN 55414.*

# JAW CRUSHER HP vs TONS/HOUR

The following data is for *Pioneer, Portec Division Jaw Crushers. Contact Pioneer in Minneapolis, MN* for current, exact specs.

| Model Size(1) | Horsepower Elec/Diesel | Tons/Hr Capacity @ given Feed Size(in) 3/4 | 1 | 1-1/4 | 1-1/2 | 2 | 2-1/2 | 3 |
|---|---|---|---|---|---|---|---|---|
| 1016 | 15/25 | 7 | 10 | 12 | 14 | 19 | 24 | 28 |
| 1020 | 20/30 | 8 | 12 | 15 | 18 | 24 | 30 | 36 |
| 1024 | 25/40 | 10 | 15 | 18 | 22 | 29 | 36 | 44 |
| 1036 | 40/60 | ... | 22 | 27 | 33 | 44 | 55 | 67 |
| 1524 | 40/60 | ... | ... | ... | ... | 36 | 45 | 54 |
| 1536 | 75/110 | ... | ... | ... | ... | 54 | 68 | 81 |
| 1830 | 60/90 | ... | ... | ... | ... | ... | 61 | 74 |
| 2036 | 100/140 | ... | ... | ... | ... | ... | ... | 93 |
| 2148 | 125/170 | ... | ... | ... | ... | ... | ... | 124 |
| 2854 | 150/190 | 3-1/2 inch feed = 178 TPH | | | | | | |

| Model Size(1) | Horsepower Elec/Diesel | Tons/Hr Capacity @ given Feed Size(in) 4 | 5 | 6 | 7 | 8 | 9 | 10 |
|---|---|---|---|---|---|---|---|---|
| 1524 | 40/60 | 72 | ... | ... | ... | ... | ... | ... |
| 1536 | 75/110 | 109 | 136 | ... | ... | ... | ... | ... |
| 1830 | 60/90 | 98 | 123 | ... | ... | ... | ... | ... |
| 2036 | 100/140 | 124 | 156 | 187 | ... | ... | ... | ... |
| 2436 | 100/150 | 136 | 171 | 205 | 239 | 273 | ... | ... |
| 2148 | 125/170 | 165 | 207 | 248 | ... | ... | ... | ... |
| 2854 | 150/190 | 204 | 256 | 308 | 360 | 410 | ... | ... |
| 3042 | 150/190 | 178 | 223 | 268 | 313 | 357 | ... | ... |
| 3546 | 200/250 | 210 | 275 | 318 | 370 | 423 | 475 | ... |
| 4248 | 250/310 | ... | 315 | 365 | 425 | 485 | 546 | 607 |
| 4248 | 250/310 | 11 inch feed = 668 TPH | | | | | | |
| 4248 | 250/310 | 12 inch feed = 730 TPH | | | | | | |

(1) Model Size values in column 1 describe the dimensions at the top of the jaw opening. The first two digits, e.g. "15" in Model 1524, are the number of inches between the jaw plates. The second two digits, e.g. "24" in Model 1524, are the number of inches between the side plates.

Capacities in the above tables are based on material that weighs 2700 lbs/cu. yard (100 lbs/cu. foot) and the Jaw Crusher has closed side plates.

# Money

# CURRENCY EXCHANGE RATES 1-92

"Value" below is what 1 unit of Currency is worth in US Dollars

| Country | Currency | Value |
|---|---|---|
| Afghanistan | Afghani | 0.00086 |
| Algeria | Dinar | 0.047 |
| Argentina | Peso | 1.0099 |
| Australia | Aus. Dollar | 0.746 |
| Austria | Schilling | 0.0882 |
| Bahamas | Dollar | 1.000 |
| Bahrain | Dinar | 2.652 |
| Barbados | Dollar | 0.497 |
| Belgium | Franc | 0.030 |
| Belize | Dollar | 0.500 |
| Bermuda | Dollar | 1.000 |
| Brazil | Cruzeiro | 0.00078 |
| Bulgaria | Leva | 0.0418 |
| Burma | Kyat | 0.163 |
| Cameroun | CFA Franc | 0.00364 |
| Canada | Dollar | 0.8547 |
| Cayman Is. | Dollar | 1.2048 |
| Chad | CFA Franc | 0.00364 |
| Chile | Peso | 0.00266 |
| China - Renminbi Yuan | | 0.1833 |
| Columbia | Peso | 0.0014 |
| Costa Rica | Colon | 0.0072 |
| Cuba | Peso | 0.7576 |
| Cyprus | Pound | 2.195 |
| Czechoslovakia | Koruna | 0.0349 |
| Denmark | Krone | 0.16 |
| Dominican R | Peso | 0.0797 |
| Ecuador | Sucre | 0.00078 |
| Egypt | Pound | 0.300 |
| El Salvador | Colon | 0.1231 |
| Ethiopia | Birr | 0.4831 |
| Fiji | Dollar | 0.6711 |
| Finland | Markka | 0.2278 |
| France | Franc | 0.1818 |
| Gambia | Dalasi | 0.1125 |
| Germany | Mark | 0.6211 |
| Ghana | Cedi | 0.00256 |
| Greece | Drachma | 0.0054 |
| Guam/Panama - Dollar | | 1.000 |
| Guatemala | Quetzal | 0.1976 |
| Haiti | Gourde | 0.200 |
| Hong Kong | Dollar | 0.5682 |
| Hungary | Forint | 0.01286 |
| Iceland | Krona | 0.017 |
| India | Rupee | 0.0386 |
| Indonesia | Rupiah | 0.0005 |
| Iran | Rial | 0.0007 |
| Iraq | Dinar | 3.215 |
| Ireland (Eire) | Punt | 1.655 |
| Israel - New Shekel | | 0.435 |
| Italy | Lira | 0.00083 |
| Jamaica | Dollar | 0.0515 |
| Japan | Yen | 0.0079 |
| Kenya | Shilling | 0.0353 |
| Korea, North | Won | 1.0309 |
| Korea, South | Won | 0.0013 |
| Kuwait | Dinar | 3.448 |
| Laos | Kip | 0.0014 |
| Lebanon | Pound | 0.0011 |
| Libya | Dinar | 3.623 |
| Malta | Lira | 3.14 |
| Mexico | Peso | 0.00032 |
| Morocco | Dirham | 0.1191 |
| Nepal | Rupee | 0.0234 |
| Netherlands | Guilder | 0.5525 |
| New Zealand | Dollar | 0.5405 |
| Norway | Krone | 0.1582 |
| Oman | Rial | 2.597 |
| Pakistan | Rupee | 0.0406 |
| Papua NG | Kina | 1.047 |
| Paraguay | Guarani | 0.0007 |
| Peru | New Sol | 1.0101 |
| Philippines | Peso | 0.0382 |
| Poland | Zloty | 0.00009 |
| Portugal | Escudo | 0.0072 |
| Puerto Rico | Dollar | 1.00 |
| Romania | Leu | 0.0051 |
| Saudi Arabia | Riyal | 0.2667 |
| Singapore | Dollar | 0.6098 |
| South Africa | Rand | 0.3125 |
| Spain | Peseta | 0.0098 |
| Sweden | Krona | 0.1707 |
| Switzerland | Franc | 0.6993 |
| Syria | Pound | 0.0483 |
| Tahiti | CFP Franc | 0.00364 |
| Taiwan | Dollar | 0.04 |
| Thailand | Baht | 0.0394 |
| Tunisia | Dinar | 1.105 |
| Turkey | Lira | 0.00018 |
| Uganda | Shilling | 0.001 |
| United Kingdom - Pound | | 1.78 |
| Uruguay | Peso | 0.00039 |
| USSR | Rouble | 0.0091 |
| Venezuela | Bolivar | 0.016 |
| Vietnam | Dong | 0.00008 |
| Yemen,N | Dinar | 2.15 |
| Yugoslavia | Dinar | 0.0096 |
| Zaire | Zaire | 0.00002 |
| Zambia | Kwacha | 0.0105 |
| Zimbabwe | Dollar | 0.1961 |

Use these exchange rates only as a general guide. If you need current rates, call a foreign exchange company such as *Thomas Cook, 1580 Court Place, Denver, CO 80202, (303) 571–0808.*

# DISCOUNT FACTORS / PRESENT VAL

| Year | 5 | 6 | 7 | 8 | 9 | 10 | 11 |
|---|---|---|---|---|---|---|---|
| | Rate of Interest per Year in Percent | | | | | | |
| 1 | 0.952 | 0.943 | 0.935 | 0.926 | 0.917 | 0.909 | 0.901 |
| 2 | 0.907 | 0.890 | 0.873 | 0.857 | 0.842 | 0.826 | 0.812 |
| 3 | 0.864 | 0.840 | 0.816 | 0.794 | 0.772 | 0.751 | 0.731 |
| 4 | 0.823 | 0.792 | 0.763 | 0.735 | 0.708 | 0.683 | 0.659 |
| 5 | 0.784 | 0.747 | 0.713 | 0.681 | 0.650 | 0.621 | 0.593 |
| 6 | 0.746 | 0.705 | 0.666 | 0.630 | 0.596 | 0.564 | 0.535 |
| 7 | 0.711 | 0.665 | 0.623 | 0.583 | 0.547 | 0.513 | 0.482 |
| 8 | 0.677 | 0.627 | 0.582 | 0.540 | 0.502 | 0.467 | 0.434 |
| 9 | 0.645 | 0.592 | 0.544 | 0.500 | 0.460 | 0.424 | 0.391 |
| 10 | 0.614 | 0.558 | 0.508 | 0.463 | 0.422 | 0.386 | 0.352 |
| 11 | 0.585 | 0.527 | 0.475 | 0.429 | 0.388 | 0.350 | 0.317 |
| 12 | 0.557 | 0.497 | 0.444 | 0.397 | 0.356 | 0.319 | 0.286 |
| 13 | 0.530 | 0.469 | 0.415 | 0.368 | 0.326 | 0.290 | 0.258 |
| 14 | 0.505 | 0.442 | 0.388 | 0.340 | 0.299 | 0.263 | 0.232 |
| 15 | 0.481 | 0.417 | 0.362 | 0.315 | 0.275 | 0.239 | 0.209 |
| 20 | 0.377 | 0.312 | 0.258 | 0.215 | 0.178 | 0.149 | 0.124 |

| Year | 12 | 13 | 14 | 15 | 16 | 17 | 18 |
|---|---|---|---|---|---|---|---|
| | Rate of Interest per Year in Percent | | | | | | |
| 1 | 0.893 | 0.885 | 0.877 | 0.870 | 0.862 | 0.855 | 0.847 |
| 2 | 0.797 | 0.783 | 0.769 | 0.756 | 0.743 | 0.731 | 0.718 |
| 3 | 0.712 | 0.693 | 0.675 | 0.658 | 0.641 | 0.624 | 0.609 |
| 4 | 0.636 | 0.613 | 0.592 | 0.572 | 0.552 | 0.534 | 0.516 |
| 5 | 0.567 | 0.543 | 0.519 | 0.497 | 0.476 | 0.456 | 0.437 |
| 6 | 0.507 | 0.480 | 0.456 | 0.432 | 0.410 | 0.390 | 0.370 |
| 7 | 0.452 | 0.425 | 0.400 | 0.376 | 0.354 | 0.333 | 0.314 |
| 8 | 0.404 | 0.376 | 0.351 | 0.327 | 0.305 | 0.285 | 0.266 |
| 9 | 0.361 | 0.333 | 0.308 | 0.284 | 0.263 | 0.243 | 0.225 |
| 10 | 0.322 | 0.295 | 0.270 | 0.247 | 0.227 | 0.208 | 0.191 |
| 11 | 0.287 | 0.261 | 0.237 | 0.215 | 0.195 | 0.178 | 0.162 |
| 12 | 0.257 | 0.231 | 0.208 | 0.187 | 0.168 | 0.152 | 0.137 |
| 13 | 0.229 | 0.204 | 0.182 | 0.163 | 0.145 | 0.130 | 0.116 |
| 14 | 0.205 | 0.181 | 0.160 | 0.141 | 0.125 | 0.111 | 0.099 |
| 15 | 0.183 | 0.160 | 0.140 | 0.123 | 0.108 | 0.095 | 0.084 |
| 20 | 0.104 | 0.087 | 0.073 | 0.061 | 0.051 | 0.043 | 0.037 |

EXAMPLE:

What is the present value of $100 in 12 years at a 14% discount?

Net Present Value = $100 x 0.208 = $20.80

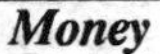

# SIMPLE INTEREST ON $100

| Days | 5 | 6 | 7 | 8 | 9 | 10 | 11 |
|---|---|---|---|---|---|---|---|
| | Rate of Interest in Percent | | | | | | |
| 1 | 0.014 | 0.016 | 0.019 | 0.022 | 0.025 | 0.027 | 0.030 |
| 2 | 0.027 | 0.033 | 0.038 | 0.044 | 0.049 | 0.055 | 0.060 |
| 3 | 0.041 | 0.049 | 0.058 | 0.066 | 0.074 | 0.082 | 0.090 |
| 4 | 0.055 | 0.066 | 0.077 | 0.088 | 0.099 | 0.110 | 0.121 |
| 5 | 0.069 | 0.082 | 0.096 | 0.110 | 0.123 | 0.137 | 0.151 |
| 6 | 0.082 | 0.099 | 0.115 | 0.132 | 0.148 | 0.164 | 0.181 |
| 7 | 0.096 | 0.115 | 0.134 | 0.153 | 0.173 | 0.192 | 0.211 |
| 8 | 0.110 | 0.132 | 0.153 | 0.175 | 0.197 | 0.219 | 0.241 |
| 9 | 0.123 | 0.148 | 0.173 | 0.197 | 0.222 | 0.247 | 0.271 |
| 10 | 0.137 | 0.164 | 0.192 | 0.219 | 0.247 | 0.274 | 0.301 |
| 20 | 0.274 | 0.329 | 0.384 | 0.438 | 0.493 | 0.548 | 0.603 |
| 30 | 0.411 | 0.493 | 0.575 | 0.658 | 0.740 | 0.822 | 0.904 |
| 40 | 0.548 | 0.658 | 0.767 | 0.877 | 0.986 | 1.10 | 1.21 |
| 50 | 0.685 | 0.822 | 0.959 | 1.10 | 1.23 | 1.37 | 1.51 |
| 60 | 0.822 | 0.986 | 1.15 | 1.32 | 1.48 | 1.64 | 1.81 |
| 70 | 0.959 | 1.15 | 1.34 | 1.53 | 1.73 | 1.92 | 2.11 |
| 80 | 1.10 | 1.32 | 1.53 | 1.75 | 1.97 | 2.19 | 2.41 |
| 90 | 1.23 | 1.48 | 1.73 | 1.97 | 2.22 | 2.47 | 2.71 |
| 100 | 1.37 | 1.64 | 1.92 | 2.19 | 2.47 | 2.74 | 3.01 |
| 200 | 2.74 | 3.29 | 3.84 | 4.38 | 4.93 | 5.48 | 6.03 |

| Days | 12 | 13 | 14 | 15 | 16 | 17 | 18 |
|---|---|---|---|---|---|---|---|
| | Rate of Interest in Percent | | | | | | |
| 1 | 0.033 | 0.036 | 0.038 | 0.041 | 0.044 | 0.047 | 0.049 |
| 2 | 0.066 | 0.071 | 0.077 | 0.082 | 0.088 | 0.093 | 0.099 |
| 3 | 0.099 | 0.107 | 0.115 | 0.123 | 0.132 | 0.140 | 0.148 |
| 4 | 0.132 | 0.142 | 0.153 | 0.164 | 0.175 | 0.186 | 0.197 |
| 5 | 0.164 | 0.178 | 0.192 | 0.205 | 0.219 | 0.233 | 0.247 |
| 6 | 0.197 | 0.214 | 0.230 | 0.247 | 0.263 | 0.279 | 0.296 |
| 7 | 0.230 | 0.249 | 0.268 | 0.288 | 0.307 | 0.326 | 0.345 |
| 8 | 0.263 | 0.285 | 0.307 | 0.329 | 0.351 | 0.373 | 0.395 |
| 9 | 0.296 | 0.321 | 0.345 | 0.370 | 0.395 | 0.419 | 0.444 |
| 10 | 0.329 | 0.356 | 0.384 | 0.411 | 0.438 | 0.466 | 0.493 |
| 20 | 0.658 | 0.712 | 0.767 | 0.822 | 0.877 | 0.932 | 0.986 |
| 30 | 0.986 | 1.07 | 1.15 | 1.23 | 1.32 | 1.40 | 1.48 |
| 40 | 1.32 | 1.42 | 1.53 | 1.64 | 1.75 | 1.86 | 1.97 |
| 50 | 1.64 | 1.78 | 1.92 | 2.05 | 2.19 | 2.33 | 2.47 |
| 60 | 1.97 | 2.14 | 2.30 | 2.47 | 2.63 | 2.79 | 2.96 |
| 70 | 2.30 | 2.49 | 2.68 | 2.88 | 3.07 | 3.26 | 3.45 |
| 80 | 2.63 | 2.85 | 3.07 | 3.29 | 3.51 | 3.73 | 3.95 |
| 90 | 2.96 | 3.21 | 3.45 | 3.70 | 3.95 | 4.19 | 4.44 |
| 100 | 3.29 | 3.56 | 3.84 | 4.11 | 4.38 | 4.66 | 4.93 |
| 200 | 6.58 | 7.12 | 7.67 | 8.22 | 8.77 | 9.32 | 9.86 |

EXAMPLE: If you put $100 in savings for 30 days at 12% simple interest, how much interest do you earn during that period?

$0.986 ( 98.6¢)    130 days = 3.29 + 0.986 = $4.27

# COMPOUND INTEREST

| Year | Rate of Interest per Year in Percent 5 | 6 | 7 | 8 | 9 | 10 | 11 |
|---|---|---|---|---|---|---|---|
| 1 | 1.05 | 1.06 | 1.07 | 1.08 | 1.09 | 1.10 | 1.11 |
| 2 | 1.10 | 1.12 | 1.14 | 1.17 | 1.19 | 1.21 | 1.23 |
| 3 | 1.16 | 1.19 | 1.22 | 1.26 | 1.30 | 1.33 | 1.37 |
| 4 | 1.22 | 1.26 | 1.31 | 1.36 | 1.41 | 1.46 | 1.52 |
| 5 | 1.28 | 1.34 | 1.40 | 1.47 | 1.54 | 1.61 | 1.68 |
| 6 | 1.34 | 1.42 | 1.50 | 1.59 | 1.68 | 1.77 | 1.86 |
| 7 | 1.41 | 1.50 | 1.60 | 1.71 | 1.83 | 1.95 | 2.07 |
| 8 | 1.48 | 1.59 | 1.72 | 1.86 | 1.99 | 2.14 | 2.30 |
| 9 | 1.55 | 1.69 | 1.84 | 2.00 | 2.17 | 2.36 | 2.55 |
| 10 | 1.63 | 1.79 | 1.97 | 2.16 | 2.37 | 2.59 | 2.83 |
| 11 | 1.71 | 1.90 | 2.10 | 2.33 | 2.58 | 2.85 | 3.14 |
| 12 | 1.80 | 2.01 | 2.25 | 2.52 | 2.81 | 3.14 | 3.49 |
| 13 | 1.89 | 2.13 | 2.41 | 2.72 | 3.07 | 3.45 | 3.87 |
| 14 | 1.98 | 2.26 | 2.58 | 2.94 | 3.34 | 3.79 | 4.29 |
| 15 | 2.08 | 2.40 | 2.76 | 3.17 | 3.64 | 4.17 | 4.77 |
| 20 | 2.65 | 3.21 | 3.86 | 4.66 | 5.60 | 6.72 | 8.03 |

| Year | Rate of Interest per Year in Percent 12 | 13 | 14 | 15 | 16 | 17 | 18 |
|---|---|---|---|---|---|---|---|
| 1 | 1.12 | 1.13 | 1.14 | 1.15 | 1.16 | 1.17 | 1.18 |
| 2 | 1.25 | 1.28 | 1.30 | 1.32 | 1.34 | 1.37 | 1.39 |
| 3 | 1.40 | 1.44 | 1.48 | 1.52 | 1.56 | 1.60 | 1.64 |
| 4 | 1.57 | 1.63 | 1.69 | 1.75 | 1.81 | 1.87 | 1.94 |
| 5 | 1.76 | 1.84 | 1.93 | 2.01 | 2.10 | 2.19 | 2.29 |
| 6 | 1.97 | 2.08 | 2.20 | 2.31 | 2.44 | 2.56 | 2.70 |
| 7 | 2.21 | 2.35 | 2.51 | 2.66 | 2.83 | 3.00 | 3.19 |
| 8 | 2.47 | 2.65 | 2.86 | 3.05 | 3.28 | 3.51 | 3.76 |
| 9 | 2.77 | 3.00 | 3.26 | 3.51 | 3.80 | 4.10 | 4.44 |
| 10 | 3.10 | 3.39 | 3.71 | 4.04 | 4.41 | 4.80 | 5.24 |
| 11 | 3.47 | 3.83 | 4.24 | 4.65 | 5.12 | 5.62 | 6.18 |
| 12 | 3.89 | 4.33 | 4.83 | 5.35 | 5.94 | 6.57 | 7.29 |
| 13 | 4.35 | 4.89 | 5.50 | 6.15 | 6.89 | 7.69 | 8.60 |
| 14 | 4.87 | 5.53 | 6.27 | 7.07 | 7.99 | 9.00 | 10.15 |
| 15 | 5.46 | 6.25 | 7.15 | 8.13 | 9.26 | 10.52 | 11.98 |
| 20 | 9.63 | 11.51 | 13.77 | 16.35 | 19.45 | 23.07 | 27.39 |

EXAMPLE (compounded annually):

If you put $100 in savings for 8 years at 12% compound interest, how much interest do you earn during that period?

[ $100 x 2.47 ] – $100 = $147 interest

# NUMBERED DAYS OF THE YEAR

| Jan 1 | 1 | Mar 15 | 74 | May 27 | 147 | Aug 8 | 220 | Oct 20 | 293 |
|---|---|---|---|---|---|---|---|---|---|
| Jan 2 | 2 | Mar 16 | 75 | May 28 | 148 | Aug 9 | 221 | Oct 21 | 294 |
| Jan 3 | 3 | Mar 17 | 76 | May 29 | 149 | Aug 10 | 222 | Oct 22 | 295 |
| Jan 4 | 4 | Mar 18 | 77 | May 30 | 150 | Aug 11 | 223 | Oct 23 | 296 |
| Jan 5 | 5 | Mar 19 | 78 | May 31 | 151 | Aug 12 | 224 | Oct 24 | 297 |
| Jan 6 | 6 | Mar 20 | 79 | Jun 1 | 152 | Aug 13 | 225 | Oct 25 | 298 |
| Jan 7 | 7 | Mar 21 | 80 | Jun 2 | 153 | Aug 14 | 226 | Oct 26 | 299 |
| Jan 8 | 8 | Mar 22 | 81 | Jun 3 | 154 | Aug 15 | 227 | Oct 27 | 300 |
| Jan 9 | 9 | Mar 23 | 82 | Jun 4 | 155 | Aug 16 | 228 | Oct 28 | 301 |
| Jan 10 | 10 | Mar 24 | 83 | Jun 5 | 156 | Aug 17 | 229 | Oct 29 | 302 |
| Jan 11 | 11 | Mar 25 | 84 | Jun 6 | 157 | Aug 18 | 230 | Oct 30 | 303 |
| Jan 12 | 12 | Mar 26 | 85 | Jun 7 | 158 | Aug 19 | 231 | Oct 31 | 304 |
| Jan 13 | 13 | Mar 27 | 86 | Jun 8 | 159 | Aug 20 | 232 | Nov 1 | 305 |
| Jan 14 | 14 | Mar 28 | 87 | Jun 9 | 160 | Aug 21 | 233 | Nov 2 | 306 |
| Jan 15 | 15 | Mar 29 | 88 | Jun 10 | 161 | Aug 22 | 234 | Nov 3 | 307 |
| Jan 16 | 16 | Mar 30 | 89 | Jun 11 | 162 | Aug 23 | 235 | Nov 4 | 308 |
| Jan 17 | 17 | Mar 31 | 90 | Jun 12 | 163 | Aug 24 | 236 | Nov 5 | 309 |
| Jan 18 | 18 | Apr 1 | 91 | Jun 13 | 164 | Aug 25 | 237 | Nov 6 | 310 |
| Jan 19 | 19 | Apr 2 | 92 | Jun 14 | 165 | Aug 26 | 238 | Nov 7 | 311 |
| Jan 20 | 20 | Apr 3 | 93 | Jun 15 | 166 | Aug 27 | 239 | Nov 8 | 312 |
| Jan 21 | 21 | Apr 4 | 94 | Jun 16 | 167 | Aug 28 | 240 | Nov 9 | 313 |
| Jan 22 | 22 | Apr 5 | 95 | Jun 17 | 168 | Aug 29 | 241 | Nov 10 | 314 |
| Jan 23 | 23 | Apr 6 | 96 | Jun 18 | 169 | Aug 30 | 242 | Nov 11 | 315 |
| Jan 24 | 24 | Apr 7 | 97 | Jun 19 | 170 | Aug 31 | 243 | Nov 12 | 316 |
| Jan 25 | 25 | Apr 8 | 98 | Jun 20 | 171 | Sep 1 | 244 | Nov 13 | 317 |
| Jan 26 | 26 | Apr 9 | 99 | Jun 21 | 172 | Sep 2 | 245 | Nov 14 | 318 |
| Jan 27 | 27 | Apr 10 | 100 | Jun 22 | 173 | Sep 3 | 246 | Nov 15 | 319 |
| Jan 28 | 28 | Apr 11 | 101 | Jun 23 | 174 | Sep 4 | 247 | Nov 16 | 320 |
| Jan 29 | 29 | Apr 12 | 102 | Jun 24 | 175 | Sep 5 | 248 | Nov 17 | 321 |
| Jan 30 | 30 | Apr 13 | 103 | Jun 25 | 176 | Sep 6 | 249 | Nov 18 | 322 |
| Jan 31 | 31 | Apr 14 | 104 | Jun 26 | 177 | Sep 7 | 250 | Nov 19 | 323 |
| Feb 1 | 32 | Apr 15 | 105 | Jun 27 | 178 | Sep 8 | 251 | Nov 20 | 324 |
| Feb 2 | 33 | Apr 16 | 106 | Jun 28 | 179 | Sep 9 | 252 | Nov 21 | 325 |
| Feb 3 | 34 | Apr 17 | 107 | Jun 29 | 180 | Sep 10 | 253 | Nov 22 | 326 |
| Feb 4 | 35 | Apr 18 | 108 | Jun 30 | 181 | Sep 11 | 254 | Nov 23 | 327 |
| Feb 5 | 36 | Apr 19 | 109 | Jul 1 | 182 | Sep 12 | 255 | Nov 24 | 328 |
| Feb 6 | 37 | Apr 20 | 110 | Jul 2 | 183 | Sep 13 | 256 | Nov 25 | 329 |
| Feb 7 | 38 | Apr 21 | 111 | Jul 3 | 184 | Sep 14 | 257 | Nov 26 | 330 |
| Feb 8 | 39 | Apr 22 | 112 | Jul 4 | 185 | Sep 15 | 258 | Nov 27 | 331 |
| Feb 9 | 40 | Apr 23 | 113 | Jul 5 | 186 | Sep 16 | 259 | Nov 28 | 332 |
| Feb 10 | 41 | Apr 24 | 114 | Jul 6 | 187 | Sep 17 | 260 | Nov 29 | 333 |
| Feb 11 | 42 | Apr 25 | 115 | Jul 7 | 188 | Sep 18 | 261 | Nov 30 | 334 |
| Feb 12 | 43 | Apr 26 | 116 | Jul 8 | 189 | Sep 19 | 262 | Dec 1 | 335 |
| Feb 13 | 44 | Apr 27 | 117 | Jul 9 | 190 | Sep 20 | 263 | Dec 2 | 336 |
| Feb 14 | 45 | Apr 28 | 118 | Jul 10 | 191 | Sep 21 | 264 | Dec 3 | 337 |
| Feb 15 | 46 | Apr 29 | 119 | Jul 11 | 192 | Sep 22 | 265 | Dec 4 | 338 |
| Feb 16 | 47 | Apr 30 | 120 | Jul 12 | 193 | Sep 23 | 266 | Dec 5 | 339 |
| Feb 17 | 48 | May 1 | 121 | Jul 13 | 194 | Sep 24 | 267 | Dec 6 | 340 |
| Feb 18 | 49 | May 2 | 122 | Jul 14 | 195 | Sep 25 | 268 | Dec 7 | 341 |
| Feb 19 | 50 | May 3 | 123 | Jul 15 | 196 | Sep 26 | 269 | Dec 8 | 342 |
| Feb 20 | 51 | May 4 | 124 | Jul 16 | 197 | Sep 27 | 270 | Dec 9 | 343 |
| Feb 21 | 52 | May 5 | 125 | Jul 17 | 198 | Sep 28 | 271 | Dec 10 | 344 |
| Feb 22 | 53 | May 6 | 126 | Jul 18 | 199 | Sep 29 | 272 | Dec 11 | 345 |
| Feb 23 | 54 | May 7 | 127 | Jul 19 | 200 | Sep 30 | 273 | Dec 12 | 346 |
| Feb 24 | 55 | May 8 | 128 | Jul 20 | 201 | Oct 1 | 274 | Dec 13 | 347 |
| Feb 25 | 56 | May 9 | 129 | Jul 21 | 202 | Oct 2 | 275 | Dec 14 | 348 |
| Feb 26 | 57 | May 10 | 130 | Jul 22 | 203 | Oct 3 | 276 | Dec 15 | 349 |
| Feb 27 | 58 | May 11 | 131 | Jul 23 | 204 | Oct 4 | 277 | Dec 16 | 350 |
| Feb 28 | 59 | May 12 | 132 | Jul 24 | 205 | Oct 5 | 278 | Dec 17 | 351 |
| Mar 1 | 60 | May 13 | 133 | Jul 25 | 206 | Oct 6 | 279 | Dec 18 | 352 |
| Mar 2 | 61 | May 14 | 134 | Jul 26 | 207 | Oct 7 | 280 | Dec 19 | 353 |
| Mar 3 | 62 | May 15 | 135 | Jul 27 | 208 | Oct 8 | 281 | Dec 20 | 354 |
| Mar 4 | 63 | May 16 | 136 | Jul 28 | 209 | Oct 9 | 282 | Dec 21 | 355 |
| Mar 5 | 64 | May 17 | 137 | Jul 29 | 210 | Oct 10 | 283 | Dec 22 | 356 |
| Mar 6 | 65 | May 18 | 138 | Jul 30 | 211 | Oct 11 | 284 | Dec 23 | 357 |
| Mar 7 | 66 | May 19 | 139 | Jul 31 | 212 | Oct 12 | 285 | Dec 24 | 358 |
| Mar 8 | 67 | May 20 | 140 | Aug 1 | 213 | Oct 13 | 286 | Dec 25 | 359 |
| Mar 9 | 68 | May 21 | 141 | Aug 2 | 214 | Oct 14 | 287 | Dec 26 | 360 |
| Mar 10 | 69 | May 22 | 142 | Aug 3 | 215 | Oct 15 | 288 | Dec 27 | 361 |
| Mar 11 | 70 | May 23 | 143 | Aug 4 | 216 | Oct 16 | 289 | Dec 28 | 362 |
| Mar 12 | 71 | May 24 | 144 | Aug 5 | 217 | Oct 17 | 290 | Dec 29 | 363 |
| Mar 13 | 72 | May 25 | 145 | Aug 6 | 218 | Oct 18 | 291 | Dec 30 | 364 |
| Mar 14 | 73 | May 26 | 146 | Aug 7 | 219 | Oct 19 | 292 | Dec 31 | 365 |

# POCKET REF

# Plumbing & Pipe



(See also TOOLS, p. 361 for pipe thread data)

(See also WATER, p. 377 for Friction Loss Values)

# COPPER PIPE & TUBING

When measuring copper pipe, sweat fittings are measured by their inside diameter ( ID ) and compression fittings are measured by their outside diameter ( OD ). Hard temper comes in 20 foot straight lengths and soft temper comes in 20 foot straight lengths or 60 foot coils. Copper tubing is normally designed to conform with ASTM Designation B88. See the code for specific information on each type.

Use 50/50 solid core solder ( NOT RESIN CORE ) and a high quality flux when soldering sweat fittings.

## TYPES OF COPPER PIPE

| Type | Characteristics |
|---|---|
| DWV | DWV stands for "Drain, Waste and Vent" and is recommended for above ground use only and no pressure applications. Sweat fittings only. Available only in hard type and in sizes from 1–1/4 inch to 6 inch. |
| K | A thick walled, flexible copper tubing. Much thicker wall than Type L and M and is required for all underground installations. Typical uses include water services, plumbing, heating, steam, gas, oil, oxygen, and other applications where thick walled tubing is required. Can be used with sweat, flared, and compression fittings. Available in hard and soft types. |
| L | Standard tubing used for interior, above ground plumbing. Uses include heating, air–conditioning, steam, gas and oil and for underground drainage lines. This is a flexible tubing but be very careful not to crimp the line when bending it. Special tools (inexpensive) are readily available to make bending much easier and safer. Although sweat, compression and flare fittings are available, only compression fittings are legal for gas lines. Available in hard and soft types. |
| M | Typically used with interior heating and pressure line applications. Wall thickness is slightly less than types K and L. Normally used with sweat fittings. Available in hard and soft types. |

# COPPER PIPE & TUBING

| Nominal Size Inches | Actual OD Inches | Type K Wall Th. Inch | Type K Weight Lbs/foot | Type L Wall Th. Inch | Type L Weight Lbs/foot |
|---|---|---|---|---|---|
| 1/4 | 0.375 | 0.035 | 0.145 | 0.030 | 0.126 |
| 3/8 | 0.500 | 0.049 | 0.269 | 0.035 | 0.198 |
| 1/2 | 0.625 | 0.049 | 0.344 | 0.040 | 0.285 |
| 5/8 | 0.750 | 0.049 | 0.418 | 0.042 | 0.362 |
| 3/4 | 0.875 | 0.065 | 0.641 | 0.045 | 0.455 |
| 1 | 1.125 | 0.065 | 0.839 | 0.050 | 0.655 |
| 1–1/4 | 1.375 | 0.065 | 1.040 | 0.055 | 0.884 |
| 1–1/2 | 1.625 | 0.072 | 1.360 | 0.060 | 1.140 |
| 2 | 2.125 | 0.083 | 2.060 | 0.070 | 1.750 |
| 2–1/2 | 2.625 | 0.095 | 2.930 | 0.080 | 2.480 |
| 3 | 3.125 | 0.109 | 4.000 | 0.090 | 3.330 |
| 3–1/2 | 3.625 | 0.120 | 5.120 | 0.100 | 4.290 |
| 4 | 4.125 | 0.134 | 6.510 | 0.110 | 5.380 |
| 5 | 5.125 | 0.160 | 9.670 | 0.125 | 7.610 |
| 6 | 6.125 | 0.192 | 13.90 | 0.140 | 10.20 |
| 8 | 8.125 | 0.271 | 25.90 | 0.200 | 19.30 |
| 10 | 10.125 | 0.338 | 40.30 | 0.250 | 30.10 |
| 12 | 12.125 | 0.405 | 57.80 | 0.280 | 40.40 |

| Nominal Size Inches | Actual OD Inches | Type M Wall Th. Inch | Type M Weight Lbs/foot | Type DWV Wall Th. Inch | Type DWV Weight Lbs/foot |
|---|---|---|---|---|---|
| 1–1/4 | 1.375 | 0.042 | 0.682 | 0.040 | 0.65 |
| 1–1/2 | 1.625 | 0.049 | 0.940 | 0.042 | 0.81 |
| 2 | 2.125 | 0.058 | 1.460 | 0.042 | 1.07 |
| 2–1/2 | 2.625 | 0.065 | 2.030 | ... | ... |
| 3 | 3.125 | 0.072 | 2.680 | 0.045 | 1.69 |
| 3–1/2 | 3.625 | 0.083 | 3.580 | ... | ... |
| 4 | 4.125 | 0.095 | 4.660 | 0.058 | 2.87 |
| 5 | 5.125 | 0.109 | 6.660 | 0.072 | 4.43 |
| 6 | 6.125 | 0.122 | 8.920 | 0.083 | 6.10 |
| 8 | 8.125 | 0.170 | 16.50 | ... | ... |
| 10 | 10.125 | 0.212 | 25.60 | ... | ... |
| 12 | 12.125 | 0.254 | 36.70 | ... | ... |

"Wall Th." stands for Wall Thickness
"OD" stands for Outside Diameter

Data included in this table is courtesy of *ITT–Grinnell Corporation, Providence, Rhode Island.*

# PLASTIC PIPE

Although there are many plastic pipe types listed below, PVC and ABS are by far the most common types. It is imperative that the correct primers and solvents be used on each type of pipe or the joints will not seal properly and the overall strength will be weakened.

## TYPES OF PLASTIC PIPE

| Type | Characteristics |
|---|---|
| PVC | Polyvinyl Chloride, Type 1, Grade 1. This pipe is strong, rigid and resistant to a variety of acids and bases. Some solvents and chlorinated hydrocarbons may damage the pipe. PVC is very common, easy to work with and readily available at most hardware stores. Maximum useable temperature is 140°F (60°C) and pressure ratings start at a minimum of 125 to 200 psi (check for specific ratings on the pipe or ask the seller). PVC can be used with water, gas, and drainage systems but NOT with hot water systems. |
| ABS | Acrylonitrile Butadiene Styrene, Type 1. This pipe is strong and rigid and resistant to a variety of acids and bases. Some solvents and chlorinated hydrocarbons may damage the pipe. ABS is very common, easy to work with and readily available at most hardware stores. Maximum useable temperature is 160°F (71°C) at low pressures. It is most common as a DWV pipe. |
| CPVC | Chlorinated polyvinyl chloride. Similar to PVC but designed specifically for piping water at up to 180°F (82°C) (can actually withstand 200°F for a limited time). Pressure rating is 100 psi. |
| PE | Polyethylene. A flexible pipe for pressurized water systems such as sprinklers. Not for hot water. |
| PB | Polybutylene. A flexible pipe for pressurized water systems both hot and cold. ONLY compression type joints can be used. |
| Polypropylene | Low pressure, lightweight material that is good up to 180°F (82°C). Highly resistant to acids, bases, and many solvents. Good for laboratory plumbing. |
| PVDF | Polyvinylidene fluoride. Strong, very tough, and resistant to abrasion, acids, bases, solvents, and much more. Good to 280°F (138°C). Good in lab. |
| FRP Epoxy | A thermosetting plastic over fiberglass. Very high strength and excellent chemical resistance. Good to 220°F (105°C). Excellent for labs. |

# PLASTIC PIPE

| Nominal Size Inches | Actual OD Inches | PVC Sched. 40 Wall Th. Inch | PVC Sched. 40 Weight Lbs/foot | PVC Sched. 80 Wall Th. Inch | PVC Sched. 80 Weight Lbs/foot |
|---|---|---|---|---|---|
| 1/4 | 0.540 | ... | ... | 0.119 | 0.10 |
| 1/2 | 0.840 | 0.109 | 0.16 | 0.147 | 0.21 |
| 3/4 | 1.050 | 0.113 | 0.22 | 0.154 | 0.28 |
| 1 | 1.315 | 0.133 | 0.32 | 0.179 | 0.40 |
| 1-1/4 | 1.660 | 0.140 | 0.43 | 0.191 | 0.57 |
| 1-1/2 | 1.990 | 0.145 | 0.52 | 0.200 | 0.69 |
| 2 | 2.375 | 0.154 | 0.70 | 0.218 | 0.95 |
| 2-1/2 | 2.875 | 0.203 | 1.10 | 0.276 | 1.45 |
| 3 | 3.500 | 0.216 | 1.44 | 0.300 | 1.94 |
| 4 | 4.500 | 0.237 | 2.05 | 0.337 | 2.83 |
| 6 | 6.625 | 0.280 | 3.61 | 0.432 | 5.41 |
| 8 | 8.625 | 0.322 | 5.45 | 0.500 | 8.22 |
| 10 | 10.750 | 0.365 | 7.91 | 0.593 | 12.28 |
| 12 | 12.750 | 0.406 | 10.35 | 0.687 | 17.10 |

| Nominal Size Inches | Actual OD Inches | CPVC Sched. 40 Wall Th. Inch | CPVC Sched. 40 Weight Lbs/foot | CPVC Sched. 80 Wall Th. Inch | CPVC Sched. 80 Weight Lbs/foot |
|---|---|---|---|---|---|
| 1/4 | 0.540 | ... | ... | 0.119 | 0.12 |
| 1/2 | 0.840 | 0.147 | 0.19 | 0.147 | 0.24 |
| 3/4 | 1.050 | 0.154 | 0.25 | 0.154 | 0.33 |
| 1 | 1.315 | 0.179 | 0.38 | 0.179 | 0.49 |
| 1-1/4 | 1.660 | 0.191 | 0.51 | 0.191 | 0.67 |
| 1-1/2 | 1.990 | 0.200 | 0.61 | 0.200 | 0.81 |
| 2 | 2.375 | 0.218 | 0.82 | 0.218 | 1.09 |
| 2-1/2 | 2.875 | 0.276 | 1.29 | 0.276 | 1.65 |
| 3 | 3.500 | 0.300 | 1.69 | 0.300 | 2.21 |
| 4 | 4.500 | 0.337 | 2.33 | 0.337 | 3.23 |
| 6 | 6.625 | 0.432 | 4.10 | 0.432 | 6.17 |
| 8 | 8.625 | ... | ... | 0.500 | 9.06 |

| Nominal Size Inches | Actual OD Inches | PVDF Sched. 80 Wall Th. Inch | PVDF Sched. 80 Weight Lbs/foot | Polypropylene 80 Wall Th. Inch | Polypropylene 80 Weight Lbs/foot |
|---|---|---|---|---|---|
| 1/2 | 0.840 | 0.147 | 0.24 | 0.147 | 0.14 |
| 3/4 | 1.050 | 0.154 | 0.33 | 0.154 | 0.19 |
| 1 | 1.315 | 0.179 | 0.49 | 0.179 | 0.27 |
| 1-1/4 | 1.660 | 0.191 | ... | 0.191 | 0.38 |
| 1-1/2 | 1.990 | 0.200 | 0.81 | 0.200 | 0.45 |
| 2 | 2.375 | 0.218 | 1.13 | 0.218 | 0.62 |

$$\text{Pipe Schedule Number} = 1000 \times \frac{\text{psi internal pressure}}{\text{psi allowable fiber stress}}$$

# STEEL PIPE

| Nominal Size & OD Inches | Schedule Numbers (1) a – b – c | Wall Thick Inches | Inside Diameter Inches | Pipe Weight Lbs/foot |
|---|---|---|---|---|
| | ...–...–10S | 0.049 | 0.307 | 0.18 |
| 1/8 | 40–Std–40S | 0.068 | 0.269 | 0.24 |
| 0.405 | 80–XS–80S | 0.095 | 0.215 | 0.31 |
| | ...–...–10S | 0.065 | 0.410 | 0.33 |
| 1/4 | 40–Std–40S | 0.088 | 0.364 | 0.42 |
| 0.540 | 80–XS–80S | 0.119 | 0.302 | 0.53 |
| | ...–...–5S | 0.065 | 0.710 | 0.53 |
| 3/8 | ...–...–10S | 0.065 | 0.545 | 0.42 |
| 0.675 | 40–Std–40S | 0.091 | 0.493 | 0.56 |
| | 80–XS–80S | 0.126 | 0.423 | 0.73 |
| | ...–...–5S | 0.065 | 0.710 | 0.53 |
| | ...–...–10S | 0.083 | 0.674 | 0.67 |
| 1/2 | 40–Std–40S | 0.109 | 0.622 | 0.85 |
| 0.840 | 80–XS–80S | 0.147 | 0.546 | 1.08 |
| | 160–...–... | 0.187 | 0.466 | 1.30 |
| | ...–XXS–... | 0.294 | 0.252 | 1.71 |
| | ...–...–5S | 0.065 | 0.920 | 0.68 |
| | ...–...–10S | 0.083 | 0.884 | 0.85 |
| 3/4 | 40–Std–40S | 0.113 | 0.824 | 1.13 |
| 1.050 | 80–XS–80S | 0.154 | 0.742 | 1.47 |
| | 160–...–... | 0.218 | 0.614 | 1.93 |
| | ...–XXS–... | 0.308 | 0.434 | 2.44 |
| | ...–...–5S | 0.065 | 1.185 | 0.86 |
| | ...–...–10S | 0.109 | 1.097 | 1.40 |
| 1 | 40–Std–40S | 0.133 | 1.049 | 1.67 |
| 1.315 | 80–XS–80S | 0.179 | 0.957 | 2.17 |
| | 160–...–... | 0.250 | 0.815 | 2.84 |
| | ...–XXS–... | 0.358 | 0.599 | 3.65 |
| | ...–...–5S | 0.065 | 1.530 | 1.10 |
| | ...–...–10S | 0.109 | 1.442 | 1.80 |
| 1–1/4 | 40–Std–40S | 0.140 | 1.380 | 2.27 |
| 1.660 | 80–XS–80S | 0.191 | 1.278 | 2.99 |
| | 160–...–... | 0.250 | 1.160 | 3.76 |
| | ...–XXS–... | 0.382 | 0.896 | 5.21 |
| | ...–...–5S | 0.065 | 1.770 | 1.27 |
| | ...–...–10S | 0.109 | 1.682 | 2.08 |
| | 40–Std–40S | 0.145 | 1.610 | 2.71 |
| 1–1/2 | 80–XS–80S | 0.200 | 1.500 | 3.63 |
| 1.900 | 160–...–... | 0.281 | 1.338 | 4.85 |
| | ...–XXS–... | 0.400 | 1.100 | 6.40 |
| | ...–...–... | 0.525 | 0.850 | 7.71 |
| | ...–...–... | 0.650 | 0.600 | 8.67 |
| | ...–...–5S | 0.065 | 2.245 | 1.60 |
| | ...–...–10S | 0.109 | 2.157 | 2.63 |
| | 40–Std–40S | 0.154 | 2.067 | 3.65 |
| 2 | 80–XS–80S | 0.218 | 1.939 | 5.02 |
| 2.375 | 160–...–... | 0.343 | 1.689 | 7.44 |
| | ...–XXS–... | 0.436 | 1.503 | 9.02 |
| | ...–...–... | 0.562 | 1.251 | 11 |
| | ...–...–... | 0.687 | 1.001 | 12 |

# STEEL PIPE

| Nominal Size & OD Inches | Schedule Numbers (1) a – b – c | Wall Thick Inches | Inside Diameter Inches | Pipe Weight Lbs/foot |
|---|---|---|---|---|
| | ...–...–5S | 0.083 | 2.709 | 2.0 |
| | ...–...–10S | 0.120 | 2.635 | 3.5 |
| | 40–Std–40S | 0.203 | 2.469 | 5.8 |
| 2-1/2 | 80–XS–80S | 0.276 | 2.323 | 7.7 |
| 2.875 | 160–...–... | 0.375 | 2.125 | 10 |
| | ...–XXS–... | 0.552 | 1.771 | 14 |
| | ...–...–... | 0.675 | 1.525 | 16 |
| | ...–...–... | 0.800 | 1.275 | 18 |
| | ...–...–5S | 0.083 | 3.334 | 3.0 |
| | ...–...–10S | 0.120 | 3.260 | 4.3 |
| | 40–Std–40S | 0.216 | 3.068 | 7.6 |
| 3 | 80–XS–80S | 0.300 | 2.900 | 10.2 |
| 3.500 | 160–...–... | 0.437 | 2.626 | 14.3 |
| | ...–XXS–... | 0.600 | 2.300 | 19 |
| | ...–...–... | 0.725 | 2.050 | 21 |
| | ...–...–... | 0.850 | 1.800 | 24 |
| | ...–...–5S | 0.083 | 3.834 | 3.5 |
| 3-1/2 | ...–...–10S | 0.120 | 3.760 | 5.0 |
| 4.000 | 40–Std–40S | 0.226 | 3.548 | 9.1 |
| | 80–XS–80S | 0.318 | 3.364 | 12 |
| | ...–XXS–... | 0.636 | 2.728 | 23 |
| | ...–...–5S | 0.083 | 4.334 | 3.9 |
| | ...–...–10S | 0.120 | 4.260 | 5.6 |
| | ...–...–... | 0.188 | 4.124 | 8.6 |
| | 40–Std–40S | 0.237 | 4.026 | 11 |
| 4 | 80–XS–80S | 0.337 | 3.826 | 15 |
| 4.500 | 120–...–... | 0.437 | 3.626 | 19 |
| | ...–...–... | 0.500 | 3.500 | 21 |
| | 160–...–... | 0.531 | 3.438 | 23 |
| | ...–XXS–... | 0.674 | 3.152 | 28 |
| | ...–...–... | 0.800 | 2.900 | 32 |
| | ...–...–... | 0.925 | 2.650 | 35 |
| | ...–...–5S | 0.109 | 5.345 | 6.3 |
| | ...–...–10S | 0.134 | 5.295 | 7.8 |
| | 40–Std–40S | 0.258 | 5.047 | 15 |
| 5 | 80–XS–80S | 0.375 | 4.813 | 21 |
| 5.563 | 120–...–... | 0.500 | 4.563 | 27 |
| | 160–...–... | 0.625 | 4.313 | 33 |
| | ...–XXS–... | 0.750 | 4.063 | 38 |
| | ...–...–... | 0.875 | 3.813 | 44 |
| | ...–...–... | 1.000 | 3.563 | 48 |
| | ...–...–5S | 0.109 | 6.407 | 5.4 |
| | ...–...–10S | 0.134 | 6.357 | 9.3 |
| | ...–...–... | 0.219 | 6.187 | 15 |
| | 40–Std–40S | 0.280 | 6.065 | 19 |
| 6 | 80–XS–80S | 0.432 | 5.761 | 28 |
| 6.625 | 120–...–... | 0.562 | 5.501 | 36 |
| | 160–...–... | 0.718 | 5.189 | 45 |
| | ...–XXS–... | 0.864 | 4.897 | 53 |
| | ...–...–... | 1.000 | 4.625 | 60 |

| Nominal Size & OD Inches | Schedule Numbers (1) a – b – c | Wall Thick Inches | Inside Diameter Inches | Pipe Weight Lbs/foot |
|---|---|---|---|---|
| | ...–...–... | 1.125 | 4.375 | 66 |
| | ...–...–5S | 0.109 | 8.407 | 9.9 |
| | ...–...–10S | 0.148 | 8.329 | 13 |
| | ...–...–... | 0.219 | 8.187 | 20 |
| | 20–...–... | 0.250 | 8.125 | 22 |
| | 30–...–... | 0.277 | 8.071 | 25 |
| | 40–Std–40S | 0.322 | 7.981 | 29 |
| 8 | 60–...–... | 0.406 | 7.813 | 36 |
| 8.625 | 80–XS–80S | 0.500 | 7.625 | 43 |
| | 100–...–... | 0.593 | 7.439 | 51 |
| | 120–...–... | 0.718 | 7.189 | 61 |
| | 140–...–... | 0.812 | 7.001 | 68 |
| | 160–...–... | 0.906 | 6.813 | 75 |
| | ...–...–... | 1.000 | 6.625 | 81 |
| | ...–...–... | 1.125 | 6.375 | 90 |
| | ...–...–5S | 0.134 | 10.482 | 15 |
| | ...–...–10S | 0.165 | 10.420 | 19 |
| | ...–...–... | 0.219 | 10.312 | 25 |
| | 20–...–... | 0.250 | 10.250 | 28 |
| | 30–...–... | 0.307 | 10.136 | 34 |
| | 40–Std–40S | 0.365 | 10.020 | 40 |
| | 60–XS–80S | 0.500 | 9.750 | 55 |
| 10 | 80–...–... | 0.593 | 9.564 | 64 |
| 10.750 | 100–...–... | 0.718 | 9.314 | 77 |
| | 120–...–... | 0.843 | 9.064 | 89 |
| | ...–...–... | 0.875 | 9.000 | 92 |
| | 140–...–... | 1.000 | 8.750 | 104 |
| | 160–...–... | 1.125 | 8.500 | 116 |
| | ...–...–... | 1.250 | 8.250 | 127 |
| | ...–...–... | 1.500 | 7.750 | 148 |
| | ...–...–5S | 0.156 | 12.438 | 21 |
| | ...–...–10S | 0.180 | 12.390 | 24 |
| | 20–...–... | 0.250 | 12.250 | 33 |
| | 30–...–... | 0.330 | 12.090 | 44 |
| | ...–Std–40S | 0.375 | 12.000 | 50 |
| | 40–...–... | 0.406 | 11.938 | 54 |
| | ...–XS...80S | 0.500 | 11.750 | 65 |
| 12 | 60–...–... | 0.562 | 11.626 | 73 |
| 12.750 | 80–...–... | 0.687 | 11.376 | 89 |
| | ...–...–... | 0.750 | 11.250 | 96 |
| | 100–...–... | 0.843 | 11.064 | 107 |
| | ...–...–... | 0.875 | 11.000 | 111 |
| | 120–...–... | 1.000 | 10.750 | 125 |
| | 140–...–... | 1.125 | 10.500 | 140 |
| | ...–...–... | 1.250 | 10.250 | 154 |
| | 160–...–... | 1.312 | 10.126 | 160 |
| | ...–...–5S | 0.156 | 13.688 | 23 |
| | ...–...–10S | 0.188 | 13.624 | 28 |
| | ...–...–... | 0.210 | 13.580 | 31 |
| | ...–...–... | 0.219 | 13.562 | 32 |

# STEEL PIPE

| Nominal Size & OD Inches | Schedule Numbers (1) a – b – c | Wall Thick Inches | Inside Diameter Inches | Pipe Weight Lbs/foot |
|---|---|---|---|---|
| | 10–...–... | 0.250 | 13.500 | 37 |
| | ...–...–... | 0.281 | 13.438 | 41 |
| | 20–...–... | 0.312 | 13.376 | 46 |
| | ...–...–... | 0.344 | 13.312 | 50 |
| | 30–Std–... | 0.375 | 13.250 | 55 |
| | 40–...–... | 0.437 | 13.126 | 63 |
| 14 | ...–...–... | 0.469 | 13.062 | 68 |
| 14.000 | ...–XS–... | 0.500 | 13.000 | 72 |
| | 60–...–... | 0.593 | 12.814 | 85 |
| | ...–...–... | 0.625 | 12.750 | 89 |
| | 80–...–... | 0.750 | 12.500 | 106 |
| | 100–...–... | 0.937 | 12.126 | 131 |
| | 120–...–... | 1.093 | 11.814 | 151 |
| | 140–...–... | 1.250 | 11.500 | 170 |
| | 160–...–... | 1.406 | 11.188 | 189 |
| | ...–...–5S | 0.165 | 15.670 | 28 |
| | ...–...–10S | 0.188 | 15.624 | 32 |
| | 10–...–... | 0.250 | 15.500 | 42 |
| | 20–...–... | 0.312 | 15.376 | 52 |
| | 30–Std–... | 0.375 | 15.250 | 63 |
| 16 | 40–XS–... | 0.500 | 15.000 | 83 |
| 16.000 | 60–...–... | 0.656 | 14.688 | 107 |
| | 80–...–... | 0.843 | 14.314 | 136 |
| | 100–...–... | 1.031 | 13.938 | 165 |
| | 120–...–... | 1.218 | 13.564 | 192 |
| | 140–...–... | 1.437 | 13.126 | 224 |
| | 160–...–... | 1.593 | 12.814 | 245 |
| | ...–...–5S | 0.165 | 17.670 | 31 |
| | ...–...–10S | 0.188 | 17.624 | 36 |
| | 10–...–... | 0.250 | 17.500 | 47 |
| | 20–...–... | 0.312 | 17.376 | 59 |
| | ...–Std–... | 0.375 | 17.250 | 71 |
| | 30–...–... | 0.437 | 17.126 | 82 |
| 18 | ...–XS–... | 0.500 | 17.000 | 93 |
| 18.000 | 40–...–... | 0.562 | 16.876 | 105 |
| | 60–...–... | 0.750 | 16.500 | 138 |
| | 80–...–... | 0.937 | 16.126 | 171 |
| | 100–...–... | 1.156 | 15.688 | 208 |
| | 120–...–... | 1.375 | 15.250 | 244 |
| | 140–...–... | 1.562 | 14.876 | 274 |
| | 160–...–... | 1.781 | 14.438 | 308 |
| | ...–...–5S | 0.188 | 19.634 | 40 |
| | ...–...–10S | 0.218 | 19.564 | 46 |
| | 10–...–... | 0.250 | 19.500 | 53 |
| 20 | 20–Std–... | 0.375 | 19.250 | 79 |
| 20.000 | 30–XS–... | 0.500 | 19.000 | 104 |
| | 40–...–... | 0.593 | 18.814 | 123 |
| | 60–...–... | 0.812 | 18.376 | 166 |
| | ...–...–... | 0.875 | 18.250 | 179 |
| | 80–...–... | 1.031 | 17.938 | 209 |

## STEEL PIPE

| Nominal Size & OD Inches | Schedule Numbers (1) a – b – c | Wall Thick Inches | Inside Diameter Inches | Pipe Weight Lbs/foot |
|---|---|---|---|---|
| | 100–...–... | 1.281 | 17.438 | 256 |
| | 120–...–... | 1.500 | 17.000 | 296 |
| | 140–...–... | 1.750 | 16.500 | 341 |
| | 160–...–... | 1.968 | 16.064 | 379 |
| 22 22.000 | ...–...–5S | 0.188 | 21.624 | 44 |
| | ...–...–10S | 0.218 | 21.564 | 51 |
| | 10–...–... | 0.250 | 21.500 | 58 |
| | 20–Std–... | 0.375 | 21.250 | 87 |
| | 30–XS–... | 0.500 | 21.000 | 115 |
| | ...–...–... | 0.625 | 20.750 | 143 |
| | ...–...–... | 0.750 | 20.500 | 170 |
| | 60–...–... | 0.875 | 20.250 | 197 |
| | 80–...–... | 1.125 | 19.750 | 251 |
| | 100–...–... | 1.375 | 19.250 | 303 |
| | 120–...–... | 1.625 | 18.750 | 354 |
| | 140–...–... | 1.875 | 18.250 | 403 |
| | 160–...–... | 2.125 | 17.750 | 451 |
| 24 24.000 | ...–...–5s | 0.218 | 23.564 | 55 |
| | 10–...–... | 0.250 | 23.500 | 63 |
| | 20–Std–... | 0.375 | 23.250 | 95 |
| | ...–XS–... | 0.500 | 22.000 | 125 |
| | 30–...–... | 0.562 | 22.876 | 141 |
| | ...–...–... | 0.625 | 22.750 | 156 |
| | 40–...–... | 0.687 | 22.626 | 171 |
| | ...–...–... | 0.750 | 22.500 | 186 |
| | ...–...–... | 0.875 | 22.250 | 216 |
| | 60–...–... | 0.968 | 22.064 | 238 |
| | 80–...–... | 1.218 | 21.564 | 296 |
| | 100–...–... | 1.531 | 20.938 | 367 |
| | 120–...–... | 1.812 | 20.376 | 429 |
| | 140–...–... | 2.062 | 19.876 | 483 |
| | 160–...–... | 2.343 | 19.314 | 542 |
| 26 26.000 | ...–...–... | 0.250 | 25.500 | 67 |
| | 10–...–... | 0.312 | 25.376 | 86 |
| | ...–Std–... | 0.375 | 25.250 | 103 |
| | 20–XS–... | 0.500 | 25.000 | 136 |
| | ...–...–... | 0.625 | 24.750 | 169 |
| | ...–...–... | 0.750 | 24.500 | 202 |
| | ...–...–... | 0.875 | 24.250 | 235 |
| | ...–...–... | 1.000 | 24.000 | 267 |
| | ...–...–... | 1.125 | 23.750 | 299 |
| 28 28.000 | ...–...–... | 0.250 | 27.500 | 74 |
| | 10–...–... | 0.312 | 27.376 | 92 |
| | ...–Std–... | 0.375 | 27.250 | 111 |
| | 20–XS–... | 0.500 | 27.000 | 147 |
| | 30–...–... | 0.625 | 26.750 | 183 |
| | ...–...–... | 0.750 | 26.500 | 218 |
| | ...–...–... | 0.875 | 26.250 | 253 |
| | ...–...–... | 1.000 | 26.000 | 288 |
| | ...–...–... | 1.125 | 25.750 | 323 |

# STEEL PIPE

| Nominal Size & OD Inches | Schedule Numbers (1) a – b – c | Wall Thick Inches | Inside Diameter Inches | Pipe Weight Lbs/foot |
|---|---|---|---|---|
| | ...–...–5S | 0.250 | 29.500 | 79 |
| | 10–...–10S | 0.312 | 29.376 | 99 |
| | ...–Std–... | 0.375 | 29.250 | 119 |
| 30 | 20–XS–... | 0.500 | 29.000 | 158 |
| 30.000 | 30–...–... | 0.625 | 28.750 | 196 |
| | 40–...–... | 0.750 | 28.500 | 234 |
| | ...–...–... | 0.875 | 28.250 | 272 |
| | ...–...–... | 1.000 | 28.000 | 310 |
| | ...–...–... | 1.125 | 27.750 | 347 |
| | ...–...–... | 0.250 | 31.500 | 85 |
| | 10–...–... | 0.312 | 31.376 | 106 |
| | ...–Std–... | 0.375 | 31.250 | 127 |
| | 20–XS–... | 0.500 | 31.000 | 168 |
| 32 | 30–...–... | 0.625 | 30.750 | 209 |
| 32.000 | 40–...–... | 0.688 | 30.624 | 230 |
| | ...–...–... | 0.750 | 30.500 | 250 |
| | ...–...–... | 0.875 | 30.250 | 291 |
| | ...–...–... | 1.000 | 30.000 | 331 |
| | ...–...–... | 1.125 | 29.750 | 371 |
| | ...–...–... | 0.250 | 33.500 | 90 |
| | 10–...–... | 0.312 | 33.376 | 112 |
| | ...–Std–... | 0.375 | 33.250 | 135 |
| | 20–XS–... | 0.500 | 33.000 | 179 |
| 34 | 30–...–... | 0.625 | 32.750 | 223 |
| 34.000 | 40–...–... | 0.688 | 32.624 | 245 |
| | ...–...–... | 0.750 | 32.500 | 266 |
| | ...–...–... | 0.875 | 32.250 | 310 |
| | ...–...–... | 1.000 | 32.000 | 353 |
| | ...–...–... | 1.125 | 31.750 | 395 |
| | ...–...–... | 0.250 | 35.500 | 96 |
| | 10–...–... | 0.312 | 35.376 | 119 |
| | ...–Std–... | 0.375 | 35.250 | 143 |
| 36 | 20–XS–... | 0.500 | 35.000 | 190 |
| 36.000 | 30–...–... | 0.625 | 34.750 | 236 |
| | 40–...–... | 0.750 | 34.500 | 282 |
| | ...–...–... | 0.875 | 34.250 | 328 |
| | ...–...–... | 1.000 | 34.000 | 374 |
| | ...–...–... | 1.125 | 33.750 | 419 |
| | ...–...–... | 0.250 | 41.500 | 112 |
| | ...–Std–... | 0.375 | 41.250 | 167 |
| | 20–XS–... | 0.500 | 41.000 | 222 |
| 42 | 30–...–... | 0.625 | 40.750 | 276 |
| 42.000 | 40–...–... | 0.750 | 40.500 | 330 |
| | ...–...–... | 1.000 | 40.000 | 438 |
| | ...–...–... | 1.250 | 39.500 | 544 |
| | ...–...–... | 1.500 | 39.000 | 649 |

(1) In the preceding tables, column 2 contains information on the schedules of various types of pipe. Specifically, these types of pipe for the a–b–c spec are as follows:

a – ANSI B36.10, Steel Pipe Schedule numbers
b – ANSI B36.10, Steel Pipe nominal wall thickness
c – ANSI B36.19, Stainless Steel Schedule numbers

Additional values pertaining to each steel pipe size can be calculated with the following formulas:

A = Square inches of metal
d = Inside diameter of pipe in inches
D = Outside diameter of pipe in inches
T = Thickness of pipe wall in inches
R = Radius of gyration of the pipe in inches
x = multiply

Weight of pipe in pounds per foot = $10.6802 \times T \times (D - T)$

Outside surface area in sq feet per foot = $0.2618 \times D$

Inside surface area in sq feet per foot = $0.2618 \times d$

Inside area of pipe in square inches = $0.785 \times d^2$

Total area of metal in square inches = $0.785 \times (D^2 - d^2)$

Moment of Inertia in inches$^4$ = $0.0491 \times (D^2 - d^2)$

Radius of Gyration in inches = $0.25 \times \sqrt{D^2 + d^2}$

Section Modulus in inches$^3$ = $(0.0982 \times (D^4 - d^4)) / D$

Weight of water in a pipe in pounds = $0.3405 \times d^2$

**Pressure Ratings of Standard Schedule 40 Steel Pipe**

1/8 to 1 inch continuous weld or seamless = 700 psi
1–1/4 to 3 inch continuous weld = 800 psi
3–1/2 to 4 inch continuous weld = 1200 psi
2 to 12 inch electric weld = 1000 to 1300 psi
1–1/4 to 3 inch seamless = 1000 psi
3 to 12 inch seamless = 1000 to 1300 psi

The basic steel data and formulas listed above are courtesy of *ITT–Grinnell Corporation, Providence, Rhode Island.*

POCKET REF

# Rope–Cable–Chain



(See also HARDWARE, page 257, for Cable Clamps)

# ROPE

| Diameter Inches | Polypropylene Break Lbs | Lbs/100 feet | Nylon Break Lbs | Lbs/100 feet | Manila Break Lbs | Lbs/100 feet | Safe Load Ratio |
|---|---|---|---|---|---|---|---|
| 3/16 | 800 | 0.7 | 1000 | 1.0 | 406 | 1.5 | 10:1 |
| 1/4 | 1250 | 1.2 | 1650 | 1.5 | 540 | 2.0 | 10:1 |
| 5/16 | 1900 | 1.8 | 2550 | 2.5 | 900 | 2.9 | 10:1 |
| 3/8 | 2700 | 2.8 | 3700 | 3.5 | 1220 | 4.1 | 10:1 |
| 7/16 | 3500 | 3.8 | 5000 | 5.0 | 1580 | 5.3 | 10:1 |
| 1/2 | 4200 | 4.7 | 6400 | 6.5 | 2380 | 7.5 | 9:1 |
| 9/16 | 5100 | 6.1 | 8000 | 8.3 | 3100 | 10.4 | 8:1 |
| 5/8 | 6200 | 7.5 | 10400 | 10.5 | 3960 | 13.3 | 8:1 |
| 3/4 | 8500 | 10.7 | 14200 | 14.5 | 4860 | 16.7 | 7:1 |
| 13/16 | 9900 | 12.7 | 17000 | 17.0 | 5850 | 19.5 | 7:1 |
| 7/8 | 11500 | 15.0 | 20000 | 20.0 | 6950 | 22.4 | 7:1 |
| 1 | 14000 | 18.0 | 25000 | 26.4 | 8100 | 27.0 | 7:1 |
| 1–1/16 | 16000 | 20.4 | 28800 | 29.0 | 9450 | 31.2 | 7:1 |
| 1–1/8 | 18300 | 23.8 | 33000 | 34.0 | 10800 | 36.0 | 7:1 |
| 1–1/4 | 21000 | 27.0 | 37500 | 40.0 | 12200 | 41.6 | 7:1 |
| 1–5/16 | 23500 | 30.4 | 43000 | 45.0 | 13500 | 47.8 | 7:1 |
| 1–1/2 | 29700 | 38.4 | 53000 | 55.0 | 16700 | 60.0 | 7:1 |
| 1–5/8 | 36000 | 47.6 | 65000 | 66.5 | 20200 | 74.5 | 7:1 |
| 1–3/4 | 43000 | 59.0 | 78000 | 83.0 | 23800 | 89.5 | 7:1 |
| 2 | 52000 | 69.0 | 92000 | 95.0 | 28000 | 108 | 7:1 |
| 2–1/8 | 61000 | 80.0 | 106000 | 109 | ... | ... | 7:1 |
| 2–1/4 | 69000 | 92.0 | 125000 | 129 | ... | ... | 6:1 |
| 2–1/2 | 80000 | 107 | 140000 | 149 | ... | ... | 6:1 |
| 2–5/8 | 90000 | 120 | 162000 | 168 | ... | ... | 6:1 |
| 2–7/8 | 101000 | 137 | 180000 | 189 | ... | ... | 6:1 |
| 3 | 114000 | 153 | 200000 | 210 | ... | ... | 6:1 |
| 3–1/4 | 137000 | 190 | 250000 | 264 | ... | ... | 6:1 |
| 3–1/2 | 162000 | 232 | 300000 | 312 | ... | ... | 6:1 |
| 4 | 190000 | 276 | 360000 | 380 | ... | ... | 6:1 |

"Break Lbs" is the breaking or tensile strength in pounds and "Safe Load Ratio" is the ratio of breaking strength to safe load. Example: 1 inch Poly rope break strength = 14000 lbs, safe load = 14000 lbs / 7 = 2000 lbs. Base your working loads on the "Safe Load Ratio" not the "Break Lbs". Note also that the strength ratings are based on tests at room temperature; rope strength decreases with an increase of temperature. At 212°F strength is 30% less and decreases as the temperature rises.

"Lbs/foot" is the weight of the rope in pounds per linear foot.

The above data is courtesy of *Continental Western Corp, Denver, Colorado and Wall Industries, Granite Quarry, North Carolina.*

## WIRE ROPE (6 strand x 19 wire)

| Diameter Inches | Weight Lbs/foot | Normal Temp with Vertical Pull Safe Load Lbs | Breaking Point Lbs |
|---|---|---|---|
| 1/4 | 0.10 | 675 | 4800 |
| 5/16 | 0.16 | 1000 | 7400 |
| 3/8 | 0.23 | 1500 | 10600 |
| 7/16 | 0.31 | 2000 | 14400 |
| 1/2 | 0.40 | 2400 | 18700 |
| 9/16 | 0.51 | 3300 | 23600 |
| 5/8 | 0.63 | 4000 | 29000 |
| 3/4 | 0.90 | 6000 | 41400 |
| 7/8 | 1.23 | 8000 | 56000 |
| 1 | 1.60 | 10000 | 72800 |
| 1-1/8 | 2.03 | 13000 | 91400 |
| 1-1/4 | 2.50 | 16000 | 112400 |
| 1-3/8 | 3.03 | 19000 | 135000 |
| 1-1/2 | 3.60 | 22000 | 160000 |
| 1-3/4 | 4.90 | 30500 | 216000 |
| 2 | 6.40 | 40000 | 278000 |
| 2-1/2 | 10.00 | 60000 | 424000 |

## CHAIN

| Rod Diameter Inches | Weight Lbs/foot | Normal Temp with Vertical Pull Safe Load Lbs | Breaking Point Lbs |
|---|---|---|---|
| 1/4 | 0.75 | 1200 | 5000 |
| 5/16 | 1 | 1700 | 7000 |
| 3/8 | 1.5 | 2500 | 10000 |
| 7/16 | 2 | 3500 | 14000 |
| 1/2 | 2.5 | 4500 | 18000 |
| 9/16 | 3.25 | 5500 | 22000 |
| 5/8 | 4 | 6700 | 27000 |
| 11/16 | 5 | 8100 | 32500 |
| 3/4 | 6.25 | 10000 | 40000 |
| 13/16 | 7 | 10500 | 42000 |
| 7/8 | 8 | 12000 | 48000 |
| 15/16 | 9 | 13500 | 54000 |
| 1 | 10 | 15200 | 61000 |
| 1-1/16 | 12 | 17200 | 69000 |
| 1-1/8 | 13 | 19500 | 78000 |
| 1-3/16 | 14.5 | 22000 | 88000 |
| 1-1/4 | 16 | 23700 | 95000 |
| 1-5/16 | 17.5 | 26000 | 104000 |
| 1-3/8 | 19 | 28500 | 114000 |
| 1-7/16 | 21.5 | 30500 | 122000 |
| 1-1/2 | 23 | 33500 | 134000 |
| 1-9/16 | 25 | 35500 | 142000 |
| 1-5/8 | 28 | 38500 | 154000 |
| 1-11/16 | 30 | 39500 | 158000 |
| 1-3/4 | 31 | 41500 | 166000 |
| 1-13/16 | 33 | 44500 | 178000 |
| 1-7/8 | 35 | 47500 | 190000 |
| 1-15/16 | 38 | 50500 | 202000 |
| 2 | 40 | 54000 | 216000 |
| 2-1/4 | 53 | 68200 | 273000 |
| 2-1/2 | 65 | 84200 | 337000 |
| 2-3/4 | 73 | 96700 | 387000 |
| 3 | 86 | 109000 | 436000 |

# FEET OF CABLE ON A REEL

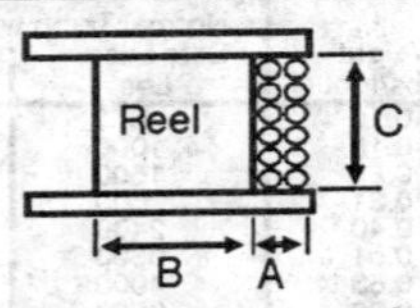


The following formula can be used to accurately determine the number of feet of rope or cable that is smoothly wound on a drum or reel ( A, B, and C are in inches ):

$$\textit{Length in Feet} = A \times [A + B] \times C \times K$$

Values of K for above equation

| Rope Diameter Inches | Value of K | Rope Diameter Inches | Value of K |
|---|---|---|---|
| 1/4 | 3.29 | 1–1/8 | 0.191 |
| 5/16 | 2.21 | 1–1/4 | 0.152 |
| 3/8 | 1.58 | 1–3/8 | 0.127 |
| 7/16 | 1.19 | 1–1/2 | 0.107 |
| 1/2 | 0.925 | 1–5/8 | 0.0886 |
| 9/16 | 0.741 | 1–3/4 | 0.0770 |
| 5/8 | 0.607 | 1–7/8 | 0.0675 |
| 3/4 | 0.428 | 2 | 0.0597 |
| 7/8 | 0.308 | 2–1/8 | 0.0532 |
| 1 | 0.239 | 2–1/4 | 0.0476 |

# PULL ANGLE vs STRENGTH LOSS

The load carrying capacity of a cable, rope, sling, etc decreases by the factor K as the angle $\theta$ increases.

| $\theta$ | K |
|---|---|
| 5 | 0.9962 |
| 10 | 0.9848 |
| 15 | 0.9659 |
| 20 | 0.9397 |
| 25 | 0.9063 |
| 30 | 0.8660 |
| 35 | 0.8792 |
| 40 | 0.7660 |
| 45 | 0.7071 |
| 50 | 0.6428 |
| 55 | 0.5736 |
| 60 | 0.5000 |
| 65 | 0.4226 |
| 70 | 0.3420 |
| 75 | 0.2588 |

# Steel



(See ALUMINUM, p. 60, for T Types)
(See also PLUMBING AND PIPE, p. 317)
(See also CHEMISTRY p.51 for melt and boil points and density of all metals)
(See also Rebar on page 44 )

# STEEL WIRE GAUGES

| Gauge Number | Diameter in Decimals of an Inch: American or Brown & Sharp | Birmingham or Stubs Iron | Washburn & Moen | W & M Music Wire | Imperial Wire |
|---|---|---|---|---|---|
| 7/0 | ... | ... | ... | 0.0087 | ... |
| 6/0 | ... | ... | ... | 0.0095 | 0.464 |
| 5/0 | ... | ... | ... | 0.010 | 0.432 |
| 4/0 | 0.460 | 0.454 | 0.3938 | 0.011 | 0.400 |
| 3/0 | 0.40964 | 0.425 | 0.3625 | 0.012 | 0.372 |
| 2/0 | 0.3648 | 0.380 | 0.3310 | 0.0133 | 0.348 |
| 0 | 0.32486 | 0.340 | 0.3065 | 0.0144 | 0.324 |
| 1 | 0.2893 | 0.300 | 0.2830 | 0.0156 | 0.300 |
| 2 | 0.25763 | 0.284 | 0.2625 | 0.0166 | 0.276 |
| 3 | 0.22942 | 0.259 | 0.2437 | 0.0178 | 0.252 |
| 4 | 0.20431 | 0.238 | 0.2253 | 0.0188 | 0.232 |
| 5 | 0.18194 | 0.220 | 0.2070 | 0.0202 | 0.212 |
| 6 | 0.16202 | 0.203 | 0.1920 | 0.0215 | 0.192 |
| 7 | 0.14428 | 0.180 | 0.1770 | 0.0230 | 0.176 |
| 8 | 0.12849 | 0.165 | 0.1620 | 0.0243 | 0.160 |
| 9 | 0.11443 | 0.148 | 0.1483 | 0.0256 | 0.144 |
| 10 | 0.10189 | 0.134 | 0.1350 | 0.0270 | 0.128 |
| 11 | 0.09074 | 0.120 | 0.1205 | 0.0284 | 0.116 |
| 12 | 0.08080 | 0.109 | 0.1055 | 0.0296 | 0.104 |
| 13 | 0.07196 | 0.095 | 0.0915 | 0.0314 | 0.092 |
| 14 | 0.06408 | 0.083 | 0.0800 | 0.0326 | 0.080 |
| 15 | 0.05706 | 0.072 | 0.0720 | 0.0345 | 0.072 |
| 16 | 0.05082 | 0.065 | 0.0625 | 0.0360 | 0.064 |
| 17 | 0.04525 | 0.058 | 0.0540 | 0.0377 | 0.056 |
| 18 | 0.04030 | 0.049 | 0.0475 | 0.0395 | 0.048 |
| 19 | 0.03589 | 0.042 | 0.0410 | 0.0414 | 0.040 |
| 20 | 0.03106 | 0.035 | 0.0348 | 0.0434 | 0.036 |
| 21 | 0.02846 | 0.032 | 0.0317 | 0.0460 | 0.032 |
| 22 | 0.02534 | 0.028 | 0.0286 | 0.0483 | 0.028 |
| 23 | 0.02257 | 0.025 | 0.0258 | 0.0510 | 0.024 |
| 24 | 0.0201 | 0.022 | 0.0230 | 0.0550 | 0.022 |
| 25 | 0.0179 | 0.020 | 0.0204 | 0.0586 | 0.020 |
| 26 | 0.01594 | 0.018 | 0.0181 | 0.0626 | 0.018 |
| 27 | 0.01419 | 0.016 | 0.0173 | 0.0658 | 0.0164 |
| 28 | 0.01264 | 0.014 | 0.0162 | 0.0720 | 0.0149 |
| 29 | 0.01125 | 0.013 | 0.0150 | 0.0760 | 0.0136 |
| 30 | 0.01002 | 0.012 | 0.0140 | 0.0800 | 0.0124 |
| 31 | 0.00892 | 0.010 | 0.0132 | ... | 0.0116 |
| 32 | 0.00795 | 0.009 | 0.0128 | ... | 0.0108 |
| 33 | 0.00708 | 0.008 | 0.0118 | ... | 0.0100 |
| 34 | 0.00630 | 0.007 | 0.0104 | ... | 0.0092 |
| 35 | 0.00561 | 0.005 | 0.0095 | ... | 0.0084 |
| 36 | 0.005 | 0.004 | 0.0090 | ... | 0.0076 |
| 37 | 0.00445 | ... | ... | ... | 0.0068 |
| 38 | 0.00396 | ... | ... | ... | 0.0060 |
| 39 | 0.00353 | ... | ... | ... | 0.0052 |
| 40 | 0.00314 | ... | ... | ... | 0.0048 |
| 41 | 0.00280 | ... | ... | ... | ... |
| 42 | 0.00249 | ... | ... | ... | ... |
| 43 | 0.00222 | ... | ... | ... | ... |
| 44 | 0.00198 | ... | ... | ... | ... |
| 45 | 0.00176 | ... | ... | ... | ... |
| 46 | 0.00157 | ... | ... | ... | ... |
| 47 | 0.00140 | ... | ... | ... | ... |
| 48 | 0.00124 | ... | ... | ... | ... |
| 49 | 0.00111 | ... | ... | ... | ... |
| 50 | 0.00099 | ... | ... | ... | ... |

## STEEL SHEET GAUGES

| Gauge Number | Steel Weight lbs per sq foot | Thickness Inches: US Standard Gauge | Thickness Inches: Manufacturers Standard | Weight lbs/sq ft: Galvanized Sheet | Weight lbs/sq ft: Stainless Steel |
|---|---|---|---|---|---|
| 7/0 | 20.00 | 0.5000 | ... | ... | ... |
| 6/0 | 18.75 | 0.4687 | ... | ... | ... |
| 5/0 | 17.50 | 0.4375 | ... | ... | ... |
| 4/0 | 16.25 | 0.4062 | ... | ... | ... |
| 3/0 | 15.00 | 0.3750 | ... | ... | ... |
| 2/0 | 13.75 | 0.3437 | ... | ... | ... |
| 0 | 12.50 | 0.3125 | ... | ... | ... |
| 1 | 11.25 | 0.2812 | ... | ... | ... |
| 2 | 10.62 | 0.2656 | ... | ... | ... |
| 3 | 10.00 | 0.2500 | 0.2391 | ... | ... |
| 4 | 9.37 | 0.2344 | 0.2242 | ... | ... |
| 5 | 8.75 | 0.2187 | 0.2092 | ... | ... |
| 6 | 8.12 | 0.2031 | 0.1943 | ... | ... |
| 7 | 7.50 | 0.1875 | 0.1793 | ... | ... |
| 8 | 6.87 | 0.1719 | 0.1644 | ... | ... |
| 9 | 6.25 | 0.1562 | 0.1495 | ... | ... |
| 10 | 5.62 | 0.1406 | 0.1345 | 5.7812 | 5.7937 |
| 11 | 5.00 | 0.1250 | 0.1196 | 5.1562 | 5.1500 |
| 12 | 4.37 | 0.1094 | 0.1046 | 4.5312 | 4.5063 |
| 13 | 3.75 | 0.0937 | 0.0897 | 3.9062 | 3.8625 |
| 14 | 3.12 | 0.0781 | 0.0747 | 3.2812 | 3.2187 |
| 15 | 2.81 | 0.0703 | 0.0673 | 2.9687 | 2.8968 |
| 16 | 2.50 | 0.0625 | 0.0598 | 2.6562 | 2.5750 |
| 17 | 2.25 | 0.0562 | 0.0538 | 2.4062 | 2.3175 |
| 18 | 2.00 | 0.0500 | 0.0478 | 2.1562 | 2.0600 |
| 19 | 1.75 | 0.0437 | 0.0418 | 1.9062 | 1.8025 |
| 20 | 1.50 | 0.0375 | 0.0359 | 1.6562 | 1.5450 |
| 21 | 1.37 | 0.0344 | 0.0329 | 1.5312 | 1.4160 |
| 22 | 1.25 | 0.0312 | 0.0299 | 1.4062 | 1.2875 |
| 23 | 1.12 | 0.0281 | 0.0269 | 1.2812 | 1.1587 |
| 24 | 1.00 | 0.0250 | 0.0239 | 1.1562 | 1.0300 |
| 25 | 0.875 | 0.0219 | 0.0209 | 1.0312 | 0.9013 |
| 26 | 0.750 | 0.0187 | 0.0179 | 0.9062 | 0.7725 |
| 27 | 0.687 | 0.0172 | 0.0164 | 0.8437 | 0.7081 |
| 28 | 0.625 | 0.0156 | 0.0149 | 0.7812 | 0.6438 |
| 29 | 0.562 | 0.0141 | 0.0135 | 0.7187 | 0.5794 |
| 30 | 0.500 | 0.0125 | 0.0120 | 0.6562 | 0.5150 |
| 31 | 0.437 | 0.0109 | 0.0105 | ... | ... |
| 32 | 0.406 | 0.0102 | 0.0097 | ... | ... |
| 33 | 0.375 | 0.0094 | 0.0090 | ... | ... |
| 34 | 0.344 | 0.0086 | 0.0082 | ... | ... |
| 35 | 0.312 | 0.0078 | 0.0075 | ... | ... |
| 36 | 0.281 | 0.0070 | 0.0067 | ... | ... |
| 37 | 0.266 | 0.0066 | 0.0064 | ... | ... |
| 38 | 0.250 | 0.0062 | 0.0060 | ... | ... |
| 39 | 0.234 | 0.0059 | ... | ... | ... |
| 40 | 0.219 | 0.0055 | ... | ... | ... |
| 41 | 0.211 | 0.0053 | ... | ... | ... |
| 42 | 0.203 | 0.0051 | ... | ... | ... |
| 43 | 0.195 | 0.0049 | ... | ... | ... |
| 44 | 0.187 | 0.0047 | ... | ... | ... |

# STEEL PLATE SIZES

| Thickness Inches | Weight Lbs/sq foot | Thickness Inches | Weight Lbs/sq foot |
|---|---|---|---|
| 3/16 | 7.65 | 2-1/8 | 86.70 |
| 1/4 | 10.20 | 2-1/4 | 91.80 |
| 5/16 | 12.75 | 2-1/2 | 102.00 |
| 3/8 | 15.30 | 2-3/4 | 112.20 |
| 7/16 | 17.85 | 3 | 122.40 |
| 1/2 | 20.40 | 3-1/4 | 132.60 |
| 9/16 | 22.95 | 3-1/2 | 142.80 |
| 5/8 | 25.50 | 3-3/4 | 153.00 |
| 11/16 | 28.05 | 4 | 163.20 |
| 3/4 | 30.60 | 4-1/4 | 173.40 |
| 13/16 | 33.15 | 4-1/2 | 183.60 |
| 7/8 | 35.70 | 5 | 204.00 |
| 1 | 40.80 | 5-1/2 | 224.40 |
| 1-1/8 | 45.90 | 6 | 244.80 |
| 1-1/4 | 51.00 | 6-1/2 | 265.20 |
| 1-3/8 | 56.10 | 7 | 285.60 |
| 1-1/2 | 61.20 | 7-1/2 | 306.00 |
| 1-5/8 | 66.30 | 8 | 326.40 |
| 1-3/4 | 71.40 | 9 | 367.20 |
| 1-7/8 | 76.50 | 10 | 408.00 |
| 2 | 81.60 | | |

# WIRE AND SHEET SPECIFICATIONS

Weights values listed on the previous three pages is based on a theoretical specific gravity of 7.7 for Iron (480 lbs/cubic foot) and 7.854 for Steel (489.6 lbs/cubic foot). B.W. gauge weights are based on a steel weight of 40.8 lbs/ square foot.

US Standard Gauge was established by Congress in 1893 and establishes that the weight determines the gauge, not the thickness. Galvanized Sheet Gauge is customarily assumed to be based on the US Standard Gauge except 2.5 ounces per square foot is added to the gauge weight of the same US Standard Gauge number.

# CHANNEL STEEL SPECS

| Size (Bar) Inches | Weight Lb/ ft |
|---|---|
| 3/4 x 5/16 x 1/8 | 0.50 |
| 3/4 x 3/8 x 1/8 | 0.56 |
| 7/8 x 3/8 x 1/8 | 0.61 |
| 7/8 x 7/16 x 1/8 | 0.69 |
| 1 x 3/8 x 1/8 | 0.68 |
| 1 x 1/2 x 1/8 | 0.84 |
| 1–1/8 x 9/16 x 3/16 | 1.16 |
| 1–1/4 x 1/2 x 1/8 | 1.01 |
| 1–1/2 x 1/2 x 1/8 | 1.12 |
| 1–1/2 x 9/16 x 3/16 | 1.44 |
| 1–1/2 x 3/4 x 1/8 | 1.17 |
| 1–1/2 x 1–1/2 x 3/16 | 2.65 |
| 1–3/4 x 1/2 x 3/16 | 1.55 |
| 2 x 1/2 x 1/8 | 1.33 |
| 2 x 9/16 x 3/16 | 1.86 |
| 2 x 5/8 x 1/4 | 2.28 |
| 2 x 1 x 1/8 | 1.78 |
| 2 x 1 x 3/16 | 2.57 |
| 2–1/2 x 5/8 x 3/16 | 2.27 |

**STRUCTURAL CHANNEL**

| Size Inches | Weight Lb/ ft |
|---|---|
| 3 x 1/2 x 0.170 | 4.1 |
| x 0.258 | 5.0 |
| x 0.356 | 6.0 |
| 3 x 1–7/8 x 0.313 | 7.1 |
| x 0.500 | 9.0 |
| 4 x 1–5/8 x 0.180 | 5.4 |
| x 0.320 | 7.25 |
| 4 x 2–1/2 x0.500 | 13.8 |
| 5 x 1–3/4 x 0.190 | 6.7 |
| x 0.325 | 9.0 |
| 6 x 2 x 0.0.200 | 8.2 |
| x 0.314 | 10.5 |
| x 0.437 | 13.0 |
| 6 x 2–1/2 x 0.313 | 12.0 |
| 6 x 3 x 0.313 | 15.1 |
| x 0.375 | 16.3 |
| 6 x 3–1/2 x 0.340 | 15.3 |
| x 0.375 | 18.0 |
| 7 x 2 x 0.210 | 9.8 |

**STRUCTURAL CHANNEL**

| Size Inches | Weight Lb/ ft |
|---|---|
| 7 x 2 x 0.314 | 12.25 |
| x 0.419 | 14.75 |
| 7 x 3 x 0.375 | 17.6 |
| 7 x 3–1/2 x 0.350 | 19.1 |
| 8 x 1–7/8 x 0.187 | 8.50 |
| 8 x 2–1/4 x 0.220 | 11.5 |
| x 0.303 | 13.75 |
| x 0.487 | 18.75 |
| 8 x 3 x 0.350 | 18.7 |
| x 0.400 | 20.0 |
| 8 x 3–1/2 x 0.375 | 21.4 |
| x 0.425 | 22.8 |
| 9 x 2–1/2 x 0.230 | 13.4 |
| x 0.285 | 15.0 |
| x 0.448 | 20.0 |
| 9 x 3–1/2 x 0.400 | 23.9 |
| x 0.450 | 25.4 |
| 10 x 2–5/8 x 0.240 | 15.3 |
| x 0.379 | 20.0 |
| x 0.526 | 25.0 |
| x 0.673 | 30.0 |
| 10 x 3–1/2 x 0.325 | 21.9 |
| x 0.375 | 24.9 |
| x 0.425 | 25.3 |
| x 0.475 | 28.3 |
| 10 x 4 x 0.425 | 28.5 |
| x 0.575 | 33.6 |
| x 0.794 | 41.1 |
| 12 x 3 x 0.280 | 20.7 |
| x 0.387 | 25.0 |
| x 0.510 | 30.0 |
| 13 x 4 x 0.375 | 31.8 |
| x 0.447 | 35.0 |
| x 0.560 | 40.0 |
| x 0.787 | 50.0 |
| 15 x 3–3/8 x 0.400 | 33.9 |
| x 0.520 | 40.0 |
| x 0.716 | 50.0 |
| 18 x 4 x 0.450 | 42.7 |
| x 0.500 | 45.8 |
| x 0.600 | 51.9 |
| x 0.700 | 58.0 |

# ANGLE STEEL SPECS

| Size Inches | Weight Lbs/ foot |
|---|---|
| 1/2 x 1/2 x 1/8 | 0.38 |
| 5/8 x 5/8 x 1/8 | 0.48 |
| 3/4 x 3/4 x 1/8 | 0.59 |
| x 3/32 | 0.463 |
| x 3/16 | 0.84 |
| 7/8 x 7/8 x 1/8 | 0.70 |
| 1 x 5/8 x 1/8 | 0.64 |
| 1 x 3/4 x 1/8 | 0.70 |
| 1 x 1 x 1/8 | 0.80 |
| x 3/16 | 1.16 |
| x 1/4 | 1.49 |
| 1–1/4 x 1–1/4 x 1/8 | 1.01 |
| x 3/16 | 1.48 |
| x 1/4 | 1.92 |
| 1–3/8 x 7/8 x 1/8 | 0.91 |
| x 3/16 | 1.32 |
| 1–1/2 x 1–1/4 x 3/16 | 1.64 |
| 1–1/2 x 1–1/2 x 1/8 | 1.23 |
| x 3/16 | 1.80 |
| x 1/4 | 2.34 |
| x 5/16 | 2.86 |
| x 3/8 | 3.35 |
| 1–3/4 x 1–1/4 x 1/8 | 1.23 |
| x 1/4 | 2.34 |
| 1–3/4 x 1–3/4 x 1/8 | 1.44 |
| x 3/16 | 2.12 |
| x 1/4 | 2.77 |
| x 5/16 | 3.39 |
| x 3/8 | 3.99 |
| 2 x 1–1/4 x 3/16 | 1.96 |
| x 1/4 | 2.55 |
| 2 x 1–1/2 x 1/8 | 1.44 |
| x 3/16 | 2.12 |
| x 1/4 | 2.77 |
| 2 x 2 x 1/8 | 1.65 |
| x 3/16 | 2.44 |
| x 1/4 | 3.19 |
| x 5/16 | 3.92 |
| x 3/8 | 4.70 |
| x 1/2 | 6.00 |
| 2–1/4 x 1–1/2 x 3/16 | 2.28 |
| x 1/4 | 2.98 |
| 2–1/4 x 2–1/4 x 3/16 | 2.75 |
| x 1/4 | 3.62 |
| x 5/16 | 4.50 |

| Size Inches | Weight Lbs/ foot |
|---|---|
| 2–1/4 x 2–1/4 x 3/8 | 5.30 |
| 2–1/2 x 1–1/2 x 3/16 | 2.44 |
| x 1/4 | 3.19 |
| x 5/16 | 3.92 |
| 2–1/2 x 2 x 1/8 | 1.86 |
| x 3/16 | 2.75 |
| x 1/4 | 3.62 |
| x 5/16 | 4.50 |
| x 3/8 | 5.30 |
| x 1/2 | 6.74 |
| 2–1/2 x 2–1/2 x 3/16 | 3.07 |
| x 1/4 | 4.10 |
| x 5/16 | 5.00 |
| x 3/8 | 5.90 |
| x 1/2 | 7.70 |
| 3 x 2 x 3/16 | 3.07 |
| x 1/4 | 4.1 |
| x 5/16 | 5.0 |
| x 3/8 | 5.9 |
| x 1/2 | 7.7 |
| 3 x 2–1/2 x 3/16 | 3.4 |
| x 1/4 | 4.5 |
| x 5/16 | 5.6 |
| x 3/8 | 6.6 |
| x 1/2 | 8.5 |
| 3 x 3 x 3/16 | 3.7 |
| x 1/4 | 4.9 |
| x 5/16 | 6.1 |
| x 3/8 | 7.2 |
| x 7/16 | 8.3 |
| x 1/2 | 9.4 |
| 3–1/2 x 2–1/2 x 1/4 | 4.9 |
| x 5/16 | 6.1 |
| x 3/8 | 7.2 |
| x 1/2 | 9.4 |
| 3–1/2 x 3 x 1/4 | 5.4 |
| x 5/16 | 6.6 |
| x 3/8 | 7.9 |
| x 1/2 | 10.2 |
| 3–1/2 x 3–1/2 x 1/4 | 5.8 |
| x 5/16 | 7.2 |
| x 3/8 | 8.5 |
| x 7/16 | 9.8 |
| x 1/2 | 11.1 |
| 4 x 3 x 1/4 | 5.8 |

# ANGLE STEEL SPECS

| Size Inches | Weight Lbs/ foot |
|---|---|
| x 5/16 | 7.2 |
| x 3/8 | 8.5 |
| x 7/16 | 9.8 |
| x 1/2 | 11.1 |
| x 5/8 | 13.6 |
| 4 x 3–1/2 x 1/4 | 6.2 |
| x 5/16 | 7.7 |
| x 3/8 | 9.1 |
| x 7/16 | 10.6 |
| x 1/2 | 11.9 |
| 4 x 4 x 1/4 | 6.6 |
| x 5/16 | 8.2 |
| x 3/8 | 9.8 |
| x 7/16 | 11.3 |
| x 1/2 | 12.8 |
| x 5/8 | 15.7 |
| x 3/4 | 18.5 |
| 5 x 3 x 1/4 | 6.6 |
| x 5/16 | 8.2 |
| x 3/8 | 9.8 |
| x 7/16 | 11.3 |
| x 1/2 | 12.8 |
| 5 x 3–1/2 x 1/4 | 7.2 |
| x 5/16 | 8.7 |
| x 3/8 | 10.4 |
| x 7/16 | 12.0 |
| x 1/2 | 13.6 |
| x 5/8 | 16.8 |
| x 3/4 | 19.8 |
| 5 x 5 x 5/16 | 10.3 |
| x 3/8 | 12.3 |
| x 7/16 | 14.3 |
| x 1/2 | 16.2 |
| x 5/8 | 20.0 |
| x 3/4 | 23.6 |
| 6 x 3–1/2 x 5/16 | 9.8 |
| x 3/8 | 11.7 |
| x 1/2 | 15.3 |
| 6 x 4 x 5/16 | 10.3 |
| x 3/8 | 12.3 |
| x 7/16 | 14.3 |
| x 1/2 | 16.2 |
| x 5/8 | 20.0 |
| x 3/4 | 23.6 |
| x 7/8 | 27.2 |
| 6 x 6 x 3/8 | 14.9 |
| x 7/16 | 17.2 |
| x 1/2 | 19.6 |
| x 5/8 | 24.2 |
| x 3/4 | 28.7 |
| x 7/8 | 33.1 |
| x 1 | 37.4 |
| 7 x 4 x 3/8 | 13.6 |
| x 7/16 | 15.8 |
| x 1/2 | 17.9 |
| x 5/8 | 22.1 |
| 8 x 4 x 1/2 | 19.6 |
| x 5/8 | 24.2 |
| x 3/4 | 28.7 |
| x 7/8 | 33.1 |
| 8 x 6 x 1/2 | 23.0 |
| x 5/8 | 28.5 |
| x 3/4 | 33.8 |
| x 1 | 44.2 |
| 8 x 8 x 1/2 | 26.4 |
| x 5/8 | 32.7 |
| x 3/4 | 38.9 |
| x 1 | 44.2 |
| 9 x 4 x 1/2 | 21.3 |

# TEE STEEL (BAR)

| Size Inches Flange x Stem | Weight Lbs/ foot |
|---|---|
| 3/4 x 3/4 x 1/8 | 0.61 |
| 1 x 1 x 1/8 | 0.85 |
| x 3/16 | 1.20 |
| 1–1/4 x 1–1/4 x 1/8 | 1.09 |
| x 3/16 | 1.55 |
| x 1/4 | 1.93 |
| 1–1/2 x 1–1/2 x 3/16 | 1.90 |
| x 1/4 | 2.43 |
| 1–3/4 x 1–3/4 x 3/16 | 2.26 |
| x 1/4 | 2.90 |
| 2 x 1–1/2 x 1/4 | 3.12 |
| 2 x 2 x 1/4 | 3.62 |
| x 5/16 | 4.30 |
| 2–1/4 x 2–1/4 x 1/4 | 4.10 |
| 2–1/2 x 2–1/2 x 1/4 | 4.60 |
| x 3/8 | 6.40 |

# ROUND STEEL BAR SPECS

| Size Inches | Weight Lbs/ foot |
|---|---|
| 1/8 | 0.042 |
| 3/16 | 0.094 |
| 1/4 | 0.167 |
| 5/16 | 0.261 |
| 3/8 | 0.376 |
| 7/16 | 0.511 |
| 1/2 | 0.668 |
| 9/16 | 0.845 |
| 5/8 | 1.040 |
| 11/16 | 1.260 |
| 3/4 | 1.500 |
| 13/16 | 1.760 |
| 7/8 | 2.040 |
| 15/16 | 2.350 |
| 1 | 2.670 |
| 1-1/16 | 3.010 |
| 1-1/8 | 3.380 |
| 1-3/16 | 3.770 |
| 1-1/4 | 4.170 |
| 1-5/16 | 4.600 |
| 1-3/8 | 5.050 |
| 1-7/16 | 5.517 |
| 1-1/2 | 6.010 |
| 1-9/16 | 6.519 |
| 1-5/8 | 7.050 |
| 1-11/16 | 7.604 |
| 1-3/4 | 8.180 |
| 1-13/16 | 8.773 |
| 1-7/8 | 9.390 |
| 1-15/16 | 10.024 |
| 2 | 10.700 |
| 2-1/16 | 11.360 |
| 2-1/8 | 12.058 |
| 2-3/16 | 12.778 |
| 2-1/4 | 13.519 |
| 2-5/16 | 14.280 |
| 2-3/8 | 15.063 |
| 2-7/16 | 15.866 |
| 2-1/2 | 16.690 |
| 2-9/16 | 17.535 |
| 2-5/8 | 18.400 |
| 2-11/16 | 19.287 |
| 2-3/4 | 20.195 |
| 2-13/16 | 21.123 |
| 2-7/8 | 22.042 |
| 2-15/16 | 23.042 |
| 3 | 24.033 |
| 3-1/16 | 25.045 |
| 3-1/8 | 26.078 |
| 3-3/16 | 27.142 |
| 3-1/4 | 28.206 |
| 3-5/16 | 28.301 |
| 3-3/8 | 30.417 |
| 3-7/16 | 31.554 |
| 3-1/2 | 32.712 |
| 3-9/16 | 33.891 |
| 3-5/8 | 35.090 |
| 3-11/16 | 36.311 |
| 3-3/4 | 37.552 |
| 3-13/16 | 38.814 |
| 3-7/8 | 40.097 |
| 3-15/16 | 41.401 |
| 4 | 42.726 |
| 4-1/16 | 44.071 |
| 4-1/8 | 45.438 |
| 4-3/16 | 46.825 |
| 4-1/4 | 48233 |
| 4-5/16 | 49.662 |
| 4-3/8 | 51.112 |
| 4-7/16 | 52.583 |
| 4-1/2 | 54.075 |
| 4-9/16 | 55.587 |
| 4-5/8 | 57.121 |
| 4-11/16 | 58.675 |
| 4-3/4 | 60.250 |
| 4-13/16 | 61.846 |
| 4-7/8 | 63.463 |
| 4-15/16 | 65.100 |
| 5 | 66.759 |
| 5-1/16 | 68.438 |
| 5-1/8 | 70.139 |
| 5-3/16 | 71.860 |
| 5-1/4 | 73.602 |
| 5-5/16 | 75.364 |
| 5-3/8 | 77.148 |
| 5-7/16 | 78.953 |
| 5-1/2 | 80.778 |
| 5-9/16 | 82.612 |
| 5-5/8 | 84.481 |
| 5-11/16 | 86.369 |
| 5-3/4 | 88.277 |
| 5-13/16 | 90.206 |
| 5-7/8 | 92.158 |
| 5-15/16 | 94.128 |
| 6 | 96.13 |
| 6-1/8 | 100.18 |
| 6-1/4 | 104.31 |
| 6-1/2 | 112.82 |
| 6-5/8 | 117.20 |
| 6-3/4 | 121.67 |
| 7 | 130.85 |
| 7-1/8 | 135.56 |
| 7-1/4 | 140.36 |
| 7-1/2 | 150.21 |
| 8 | 170.90 |

# SQUARE STEEL BAR SPECS

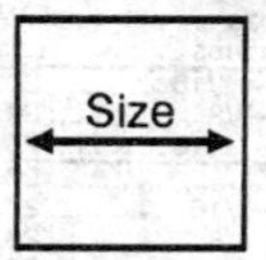


| Size Inches | Weight Lbs/ foot |
|---|---|
| 1/8 | 0.053 |
| 3/16 | 0.120 |
| 1/4 | 0.213 |
| 5/16 | 0.332 |
| 3/8 | 0.478 |
| 7/16 | 0.651 |
| 1/2 | 0.850 |
| 9/16 | 1.076 |
| 5/8 | 1.328 |
| 11/16 | 1.607 |
| 3/4 | 1.913 |
| 13/16 | 2.245 |
| 7/8 | 2.603 |
| 15/16 | 2.988 |
| 1 | 3.400 |
| 1–1/16 | 3.833 |
| 1–1/8 | 4.303 |
| 1–3/16 | 4.795 |
| 1–1/4 | 5.314 |
| 1–5/16 | 5.857 |
| 1–3/8 | 6.428 |
| 1–7/16 | 7.026 |
| 1–1/2 | 7.650 |
| 1–9/16 | 8.301 |
| 1–5/8 | 8.978 |
| 1–11/16 | 9.682 |
| 1–3/4 | 10.414 |
| 1–13/16 | 11.170 |
| 1–7/8 | 12.000 |
| 1–15/16 | 12.763 |
| 2 | 13.600 |
| 2–1/16 | 14.463 |
| 2–1/8 | 15.354 |
| 2–3/16 | 16.270 |
| 2–1/4 | 17.213 |
| 2–5/16 | 18.182 |
| 2–3/8 | 19.178 |
| 2–7/16 | 20.201 |
| 2–1/2 | 21.250 |
| 2–9/16 | 22.326 |
| 2–5/8 | 23.426 |
| 2–11/16 | 24.557 |
| 2–3/4 | 25.714 |

| Size Inches | Weight Lbs/ foot |
|---|---|
| 2–13/16 | 26.895 |
| 2–7/8 | 28.103 |
| 2–15/16 | 29.338 |
| 3 | 30.600 |
| 3–1/16 | 31.888 |
| 3–1/8 | 33.203 |
| 3–3/16 | 34.558 |
| 3–1/4 | 35.913 |
| 3–5/16 | 37.307 |
| 3–3/8 | 38.728 |
| 3–7/16 | 40.176 |
| 3–1/2 | 41.650 |
| 3–9/16 | 43.151 |
| 3–5/8 | 44.678 |
| 3–11/16 | 46.232 |
| 3–3/4 | 47.813 |
| 3–13/16 | 49.420 |
| 3–7/8 | 51.053 |
| 3–15/16 | 52.713 |
| 4 | 54.400 |
| 4–1/16 | 56.113 |
| 4–1/8 | 57.853 |
| 4–3/16 | 59.620 |
| 4–1/4 | 61.413 |
| 4–5/16 | 63.232 |
| 4–3/8 | 65.078 |
| 4–7/16 | 66.951 |
| 4–1/2 | 68.850 |
| 4–9/16 | 70.776 |
| 4–5/8 | 72.728 |
| 4–11/16 | 74.707 |
| 4–3/4 | 76.713 |
| 4–13/16 | 78.745 |
| 4–7/8 | 80.803 |
| 4–15/16 | 82.888 |
| 5 | 85.000 |
| 5–1/16 | 87.138 |
| 5–1/8 | 89.303 |
| 5–3/16 | 91.495 |
| 5–1/4 | 93.713 |
| 5–5/16 | 95.957 |
| 5–3/8 | 98.228 |
| 5–7/16 | 100.526 |
| 5–1/2 | 102.850 |
| 5–9/16 | 105.199 |
| 5–5/8 | 107.576 |
| 5–11/16 | 109.983 |
| 5–3/4 | 112.414 |
| 5–13/16 | 114.869 |
| 5–7/8 | 117.354 |
| 5–15/16 | 119.864 |
| 6 | 122.40 |

# HEXAGONAL STEEL BAR SPECS

Size

| Size Inches | Weight Lbs/ foot |
|---|---|
| 1/8 | 0.046 |
| 3/16 | 0.104 |
| 1/4 | 0.184 |
| 5/16 | 0.288 |
| 3/8 | 0.414 |
| 7/16 | 0.564 |
| 1/2 | 0.737 |
| 9/16 | 0.932 |
| 5/8 | 1.150 |
| 11/16 | 1.393 |
| 3/4 | 1.658 |
| 13/16 | 1.944 |
| 7/8 | 2.256 |
| 15/16 | 2.588 |
| 1 | 2.944 |
| 1-1/16 | 3.324 |
| 1-1/8 | 3.727 |
| 1-3/16 | 4.152 |
| 1-1/4 | 4.601 |
| 1-5/16 | 5.072 |
| 1-3/8 | 5.567 |
| 1-7/16 | 6.085 |
| 1-1/2 | 6.625 |
| 1-9/16 | 7.189 |
| 1-5/8 | 7.775 |
| 1-11/16 | 8.385 |
| 1-3/4 | 9.018 |
| 1-13/16 | 9.673 |
| 1-7/8 | 10.355 |
| 1-15/16 | 11.053 |
| 2 | 11.780 |
| 2-1/16 | 12.528 |
| 2-1/8 | 13.300 |
| 2-3/16 | 14.092 |
| 2-1/4 | 14.911 |
| 2-5/16 | 15.747 |
| 2-3/8 | 16.613 |
| 2-7/16 | 17.469 |
| 2-1/2 | 18.403 |
| 2-9/16 | 19.337 |
| 2-5/8 | 20.294 |
| 2-11/16 | 21.272 |
| 2-3/4 | 22.273 |

| Size Inches | Weight Lbs/ foot |
|---|---|
| 2-13/16 | 23.295 |
| 2-7/8 | 24.343 |
| 2-15/16 | 25.412 |
| 3 | 26.500 |
| 3-1/16 | 27.621 |
| 3-1/8 | 28.755 |
| 3-3/16 | 29.928 |
| 3-1/4 | 31.101 |
| 3-5/16 | 32.315 |
| 3-3/8 | 33.540 |
| 3-7/16 | 34.798 |
| 3-1/2 | 36.979 |
| 3-9/16 | 37.375 |
| 3-5/8 | 38.692 |
| 3-11/16 | 40.046 |
| 3-3/4 | 41.407 |
| 3-13/16 | 42.806 |
| 3-7/8 | 44.213 |
| 3-15/16 | 45.659 |
| 4 | 47.112 |
| 4-1/16 | 48.604 |
| 4-1/8 | 50.112 |
| 4-3/16 | 51.641 |
| 4-1/4 | 53.196 |
| 4-5/16 | 54.771 |
| 4-3/8 | 56.370 |
| 4-7/16 | 57.990 |
| 4-1/2 | 59.636 |
| 4-9/16 | 61.303 |
| 4-5/8 | 62.996 |
| 4-11/16 | 64.710 |
| 4-3/4 | 66.448 |
| 4-13/16 | 68.206 |
| 4-7/8 | 69.991 |
| 4-15/16 | 71.796 |
| 5 | 73.625 |
| 5-1/16 | 75.477 |
| 5-1/8 | 77.353 |
| 5-3/16 | 79.250 |
| 5-1/4 | 81.173 |
| 5-5/16 | 83.117 |
| 5-3/8 | 85.084 |
| 5-7/16 | 87.072 |
| 5-1/2 | 89.086 |
| 5-9/16 | 91.121 |
| 5-5/8 | 93.183 |
| 5-11/16 | 95.265 |
| 5-3/4 | 97.371 |
| 5-13/16 | 99.497 |
| 5-7/8 | 101.650 |
| 5-15/16 | 103.823 |
| 6 | 106.016 |

# OCTAGONAL STEEL BAR SPECS

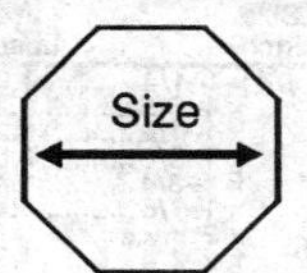


| Size Inches | Weight Lbs/ foot |
|---|---|
| 1/8 | 0.044 |
| 3/16 | 0.099 |
| 1/4 | 0.176 |
| 5/16 | 0.275 |
| 3/8 | 0.396 |
| 7/16 | 0.539 |
| 1/2 | 0.704 |
| 9/16 | 0.891 |
| 5/8 | 1.100 |
| 11/16 | 1.331 |
| 3/4 | 1.584 |
| 13/16 | 1.859 |
| 7/8 | 2.157 |
| 15/16 | 2.476 |
| 1 | 2.817 |
| 1-1/16 | 3.180 |
| 1-1/8 | 3.565 |
| 1-3/16 | 3.972 |
| 1-1/4 | 4.401 |
| 1-5/16 | 4.852 |
| 1-3/8 | 5.325 |
| 1-7/16 | 5.820 |
| 1-1/2 | 6.338 |
| 1-9/16 | 6.877 |
| 1-5/8 | 7.438 |
| 1-11/16 | 8.021 |
| 1-3/4 | 8.626 |
| 1-13/16 | 9.253 |
| 1-7/8 | 9.902 |
| 1-15/16 | 10.574 |
| 2 | 11.267 |
| 2-1/16 | 11.982 |
| 2-1/8 | 12.719 |
| 2-3/16 | 13.478 |
| 2-1/4 | 14.259 |
| 2-5/16 | 15.063 |
| 2-3/8 | 15.888 |
| 2-7/16 | 16.735 |
| 2-1/2 | 17.604 |
| 2-9/16 | 18.495 |
| 2-5/8 | 19.409 |
| 2-11/16 | 20.344 |
| 2-3/4 | 21.301 |
| 2-13/16 | 22.280 |
| 2-7/8 | 23.281 |
| 2-15/16 | 24.305 |
| 3 | 25.350 |
| 3-1/16 | 26.417 |
| 3-1/8 | 27.506 |
| 3-3/16 | 28.628 |
| 3-1/4 | 29.751 |
| 3-5/16 | 30.906 |
| 3-3/8 | 32.084 |
| 3-7/16 | 33.283 |
| 3-1/2 | 34.504 |
| 3-9/16 | 35.747 |
| 3-5/8 | 37.013 |
| 3-11/16 | 38.300 |
| 3-3/4 | 39.309 |
| 3-13/16 | 40.941 |
| 3-7/8 | 42.294 |
| 3-15/16 | 42.669 |
| 4 | 45.067 |
| 4-1/16 | 46.475 |
| 4-1/8 | 47.927 |
| 4-3/16 | 49.379 |
| 4-1/4 | 50.876 |
| 4-5/16 | 52.372 |
| 4-3/8 | 53.912 |
| 4-7/16 | 55.450 |
| 4-1/2 | 57.037 |
| 4-9/16 | 58.618 |
| 4-5/8 | 60.250 |
| 4-11/16 | 61.876 |
| 4-3/4 | 63.551 |
| 4-13/16 | 65.219 |
| 4-7/8 | 66.911 |
| 4-15/16 | 68.651 |
| 5 | 70.416 |
| 5-1/16 | 72.171 |
| 5-1/8 | 73.965 |
| 5-3/16 | 75.779 |
| 5-1/4 | 77.634 |
| 5-5/16 | 79.476 |
| 5-3/8 | 81.357 |
| 5-7/16 | 83.258 |
| 5-1/2 | 85.204 |
| 5-9/16 | 87.130 |
| 5-5/8 | 89.101 |
| 5-11/16 | 91.092 |
| 5-3/4 | 93.126 |
| 5-13/16 | 95.139 |
| 5-7/8 | 97.197 |
| 5-15/16 | 99.275 |
| 6 | 101.373 |

# FLAT STEEL SPECS

| Size Inches | Weight Lbs/ foot |
|---|---|
| 1/8 x 5/16 | 0.133 |
| x 3/8 | 0.159 |
| x 7/16 | 0.186 |
| x 1/2 | 0.213 |
| x 9/16 | 0.239 |
| x 5/8 | 0.266 |
| x 11/16 | 0.293 |
| x 3/4 | 0.319 |
| x 7/8 | 0.372 |
| x 1 | 0.425 |
| x 1–1/8 | 0.478 |
| x 1–1/4 | 0.531 |
| x 1–3/8 | 0.584 |
| x 1–1/2 | 0.638 |
| x 1–3/4 | 0.744 |
| x 1–7/8 | 0.797 |
| x 2 | 0.850 |
| x 2–1/4 | 0.956 |
| x 2–1/2 | 1.062 |
| x 2–3/4 | 1.169 |
| x 3 | 1.275 |
| x 3–1/2 | 1.488 |
| x 4 | 1.700 |
| x 4–1/2 | 1.913 |
| x 5 | 2.125 |
| x 6 | 2.550 |
| 3/16 x 5/16 | 0.199 |
| x 3/8 | 0.239 |
| x 7/16 | 0.279 |
| x 1/2 | 0.319 |
| x 9/16 | 0.359 |
| x 5/8 | 0.398 |
| x 11/16 | 0.438 |
| x 3/4 | 0.478 |
| x 7/8 | 0.558 |
| x 1 | 0.638 |
| x 1–1/8 | 0.717 |
| x 1–1/4 | 0.797 |
| x 1–3/8 | 0.877 |
| x 1–1/2 | 0.956 |
| x 1–3/4 | 1.116 |
| x 1–7/8 | 1.195 |
| x 2 | 1.275 |
| x 2–1/4 | 1.434 |
| x 2–1/2 | 1.594 |
| x 2–3/4 | 1.753 |
| x 3 | 1.912 |
| x 3–1/2 | 2.232 |
| x 4 | 2.550 |
| x 4–1/2 | 2.868 |
| x 5 | 3.188 |
| x 6 | 3.825 |
| 1/4 x 5/16 | 0.266 |
| x 3/8 | 0.319 |
| x 7/16 | 0.372 |
| x 1/2 | 0.425 |
| x 9/16 | 0.478 |
| x 5/8 | 0.531 |
| x 11/16 | 0.584 |
| x 3/4 | 0.638 |
| x 7/8 | 0.744 |
| x 1 | 0.850 |
| x 1–1/8 | 0.956 |
| 1/4 x 1–1/4 | 1.063 |
| x 1–3/8 | 1.169 |
| x 1–1/2 | 1.275 |
| x 1–5/8 | 1.531 |
| x 1–3/4 | 1.488 |
| x 1–7/8 | 1.594 |
| x 2 | 1.700 |
| x 2–1/4 | 1.913 |
| x 2–1/2 | 2.125 |
| x 2–3/4 | 2.338 |
| x 3 | 2.550 |
| x 3–1/2 | 2.975 |
| x 4 | 3.400 |
| x 4–1/2 | 3.825 |
| x 5 | 4.250 |
| x 6 | 5.100 |
| 5/16 x 3/8 | 0.398 |
| x 7/16 | 0.465 |
| x 1/2 | 0.531 |
| x 9/16 | 0.598 |
| x 5/8 | 0.664 |
| x 3/4 | 0.797 |
| x 7/8 | 0.930 |
| x 1 | 1.063 |
| x 1–1/8 | 1.195 |
| x 1–1/4 | 1.328 |
| x 1–3/8 | 1.461 |
| x 1–1/2 | 1.594 |
| x 1–3/4 | 1.859 |
| x 2 | 2.125 |
| x 2–1/4 | 2.391 |
| x 2–1/2 | 2.656 |
| x 2–3/4 | 2.922 |
| x 3 | 3.188 |
| x 3–1/2 | 3.719 |
| x 4 | 4.250 |
| x 4–1/2 | 4.781 |
| x 5 | 5.313 |
| x 6 | 6.375 |
| 3/8 x 7/16 | 0.558 |
| x 1/2 | 0.638 |
| x 9/16 | 0.717 |
| x 5/8 | 0.797 |
| x 11/16 | 0.877 |
| x 3/4 | 0.956 |
| x 7/8 | 1.116 |
| x 1 | 1.275 |
| x 1–1/8 | 1.434 |
| x 1–1/4 | 1.594 |
| x 1–3/8 | 1.753 |
| x 1–1/2 | 1.913 |
| x 1–3/4 | 2.231 |
| x 1–7/8 | 2.391 |
| x 2 | 2.550 |
| x 2–1/4 | 2.869 |
| x 2–1/2 | 3.188 |
| x 2–3/4 | 3.506 |
| x 3 | 3.825 |
| x 3–1/2 | 4.463 |
| x 4 | 5.100 |
| x 4–1/2 | 5.738 |
| x 5 | 6.375 |
| x 6 | 7.650 |

# FLAT STEEL SPECS

| Size Inches | Weight Lbs/ foot |
|---|---|
| 1/2 x 9/16 | 0.956 |
| x 5/8 | 1.063 |
| x 11/16 | 1.169 |
| x 3/4 | 1.275 |
| x 13/16 | 1.382 |
| x 7/8 | 1.488 |
| x 1 | 1.700 |
| x 1-1/8 | 1.913 |
| x 1-1/4 | 2.125 |
| x 1-3/8 | 2.338 |
| x 1-1/2 | 2.550 |
| x 1-3/4 | 2.975 |
| x 2 | 3.400 |
| x 2-1/4 | 3.825 |
| x 2-1/2 | 4.250 |
| x 2-3/4 | 4.675 |
| x 3 | 5.100 |
| x 3-1/2 | 5.950 |
| x 4 | 6.800 |
| x 5 | 8.500 |
| x 6 | 10.200 |
| 5/8 x 11/16 | 1.461 |
| x 3/4 | 1.594 |
| x 13/16 | 1.727 |
| x 7/8 | 1.859 |
| x 1 | 2.125 |
| x 1-1/8 | 2.391 |
| x 1-1/4 | 2.656 |
| x 1-3/8 | 2.922 |
| x 1-1/2 | 3.188 |
| x 1-3/4 | 3.719 |
| x 1-7/8 | 3.984 |
| x 2 | 4.250 |
| x 2-1/4 | 4.781 |
| x 2-1/2 | 5.313 |
| x 2-3/4 | 5.844 |
| x 3 | 6.375 |
| x 3-1/2 | 7.438 |
| x 4 | 8.500 |
| x 5 | 10.625 |
| x 6 | 12.750 |
| 3/4 x 7/8 | 2.231 |
| x 1 | 2.550 |
| x 1-1/8 | 2.869 |
| x 1-1/4 | 3.188 |
| x 1-3/8 | 3.506 |
| x 1-1/2 | 3.825 |
| x 1-3/4 | 4.463 |
| x 1-7/8 | 4.470 |
| x 2 | 5.100 |
| x 2-1/4 | 5.738 |
| x 2-1/2 | 6.375 |
| x 2-3/4 | 7.013 |
| x 3 | 7.650 |
| x 3-1/2 | 8.925 |
| x 4 | 10.200 |
| x 5 | 12.750 |
| x 6 | 15.300 |
| 7/8 x 1 | 2.975 |
| x 1-1/8 | 3.347 |
| x 1-1/4 | 3.719 |
| x 1-3/8 | 4.091 |
| x 1-1/2 | 4.463 |
| 7/8 x 1-3/4 | 5.206 |
| x 1-7/8 | 5.578 |
| x 2 | 5.950 |
| x 2-1/4 | 6.694 |
| x 2-1/2 | 7.438 |
| x 2-3/4 | 8.181 |
| x 3 | 8.925 |
| x 3-1/2 | 10.413 |
| x 4 | 11.900 |
| x 5 | 13.388 |
| x 6 | 17.850 |
| 1 x 1-1/8 | 3.825 |
| x 1-1/4 | 4.250 |
| x 1-3/8 | 4.675 |
| x 1-1/2 | 5.100 |
| x 1-3/4 | 5.950 |
| x 1-7/8 | 6.375 |
| x 2 | 6.800 |
| x 2-1/4 | 7.650 |
| x 2-1/2 | 8.500 |
| x 2-3/4 | 9.350 |
| x 3 | 10.200 |
| x 3-1/2 | 11.900 |
| x 4 | 13.600 |
| x 5 | 17.000 |
| x 6 | 20.400 |
| 1-1/4 x 1-3/8 | 5.844 |
| x 1-1/2 | 6.375 |
| x 1-3/4 | 7.438 |
| x 1-7/8 | 7.969 |
| x 2 | 8.500 |
| x 2-1/4 | 9.563 |
| x 2-1/2 | 10.625 |
| x 2-3/4 | 11.688 |
| x 3 | 12.750 |
| x 3-1/2 | 14.875 |
| x 4 | 17.000 |
| x 5 | 21.250 |
| x 6 | 25.500 |
| 1-1/2 x 1-3/4 | 8.925 |
| x 1-7/8 | 9.563 |
| x 2 | 10.200 |
| x 2-1/4 | 11.475 |
| x 2-1/2 | 12.750 |
| x 2-3/4 | 14.025 |
| x 3 | 15.300 |
| x 3-1/2 | 17.850 |
| x 4 | 20.400 |
| x 5 | 25.500 |
| x 6 | 30.600 |
| 2 x 2-1/4 | 15.300 |
| x 2-1/2 | 17.000 |
| x 2-3/4 | 18.700 |
| x 3 | 20.400 |
| x 3-1/2 | 23.800 |
| x 4 | 27.200 |
| x 5 | 34.000 |
| x 6 | 40.800 |
| 3 x 4 | 40.800 |
| x 5 | 51.000 |
| x 6 | 61.200 |

# SQUARE STEEL TUBING

| OD Size Inches (guage) | Weight Lbs/ foot |
|---|---|
| 1/2 x 0.049 (18) | 0.301 |
| x 0.065 (16) | 0.384 |
| 5/8 x 0.049 (18) | 0.384 |
| x 0.065 (16) | 0.495 |
| 3/4 x 0.035 (20) | 0.340 |
| x 0.049 (18) | 0.467 |
| x 0.065 (16) | 0.605 |
| x 0.120 (11) | 1.03 |
| 7/8 x 0.049 (18) | 0.550 |
| x 0.065 (16) | 0.716 |
| 1 x 0.035 (20) | 0.459 |
| x 0.049 (18) | 0.634 |
| x 0.065 (16) | 0.826 |
| x 0.073 | 0.920 |
| x 0.083 (14) | 1.04 |
| x 0.095 (13) | 1.09 |
| x 0.102 | 1.25 |
| x 0.109 (12) | 1.32 |
| x 0.120 (11) | 1.44 |
| 1–1/4 x 0.049 (18) | 0.800 |
| x 0.065 (16) | 1.05 |
| x 0.083 (14) | 1.32 |
| x 0.120 (11) | 1.84 |
| x 0.135 (10) | 1.96 |
| x 0.188 (3/16) | 2.62 |
| 1–1/2 x 0.049 (18) | 0.967 |
| x 0.065 (16) | 1.27 |
| x 0.083 (14) | 1.60 |
| x 0.120 (11) | 2.25 |
| x 0.140 | 2.50 |
| x 0.180 (7) | 3.23 |
| x 0.188 (3/16) | 3.23 |
| 1–3/4 x 0.065 (16) | 1.49 |
| x 0.083 (14) | 1.88 |
| x 0.120 (11) | 2.66 |
| 2 x 0.065 (16) | 1.71 |
| x 0.083 (14) | 2.16 |
| x 0.095 (13) | 2.46 |
| x 0.110 (7/64) | 2.69 |
| x 0.120 (11) | 3.07 |
| x 0.125 (1/8) | 3.24 |
| x 0.145 | 3.51 |
| x 0.188 (3/16) | 4.31 |
| x 0.250 (1/4) | 5.59 |
| 2–1/2 x 0.083 (14) | 2.73 |
| x 0.120 (11) | 3.88 |
| x 0.141 | 4.32 |
| 2–1/2 x 0.188 (3/16) | 5.59 |
| x 0.250 (1/4) | 7.10 |
| 3 x 0.083 (14) | 3.29 |
| x 0.120 (11) | 4.70 |
| x 0.155 (5/32) | 5.78 |
| x 0.188 (3/16) | 6.86 |
| x 0.250 (1/4) | 8.80 |
| 3–1/2 x 0.120 (11) | 5.52 |
| x 0.156 (5/32) | 6.88 |
| x 0.188 (3/16) | 8.14 |
| x 0.220 (5) | 9.81 |
| x 0.250 (1/4) | 10.50 |
| x 0.313 (5/16) | 12.69 |
| 4 x 0.120 (11) | 6.33 |
| x 0.188 (3/16) | 9.31 |
| x 0.250 (1/4) | 12.02 |
| x 0.313 (5/16) | 14.52 |
| x 0.375 (3/8) | 16.84 |
| x 0.500 (1/2) | 20.88 |
| 5 x 0.188 (3/16) | 11.86 |
| x 0.250 (1/4) | 15.42 |
| x 0.313 (5/16) | 18.77 |
| x 0.375 (3/8) | 21.94 |
| x 0.500 (1/2) | 27.68 |
| 6 x 0.188 (3/16) | 14.41 |
| x 0.250 (1/4) | 18.82 |
| x 0.313 (5/16) | 23.02 |
| x 0.375 (3/8) | 27.04 |
| x 0.500 (1/2) | 34.48 |
| 7x 0.188 (3/16) | 16.85 |
| x 0.250 (1/4) | 22.04 |
| x 0.313 (5/16) | 26.99 |
| x 0.375 (3/8) | 31.73 |
| x 0.500 (1/2) | 40.55 |
| 8 x 0.188 (1/4) | 25.44 |
| x 0.250 (5/16) | 31.24 |
| x 0.313 (3/8) | 36.83 |
| x 0.375 (1/2) | 47.35 |
| x 0.500 (5/8) | 56.89 |
| 10 x 0.188 (1/4) | 32.23 |
| x 0.250 (5/16) | 39.74 |
| x 0.313 (3/8) | 47.03 |
| x 0.375 (1/2) | 60.95 |
| x 0.500 (5/8) | 73.98 |

# RECTANGLE STEEL TUBING

| OD Size Inches (guage) | Weight Lbs/ foot |
|---|---|
| 1-1/2 x 1 x 0.083 (14) | 1.32 |
| x 0.120 (11) | 1.84 |
| 2 x 1 x 0.083 (14) | 1.60 |
| 2 x 1-1/4 x 0.083 (14) | 1.74 |
| 2 x 1-1/2 x 0.120 (11) | 2.66 |
| 2-1/2 x 1 x 0.083 (14) | 1.88 |
| 2-1/2 x 1-1/4 x 0.083(14) | 2.02 |
| 2-1/2 x 1-1/2 x 0.083(14) | 2.16 |
| x 0.145 | 3.51 |
| x 0.180 (7) | 4.46 |
| x 0.220 (5) | 5.33 |
| x 0.250 | 5.59 |
| 3 x 1 x 0.083 (14) | 2.16 |
| 3 x 1-1/2 x 0.083 (14) | 2.45 |
| x 0.120 (11) | 3.48 |
| x 0.180 (7) | 5.07 |
| 3 x 2 x 0.083 (14) | 2.73 |
| x 0.120 (11) | 3.88 |
| x 0.141 (9/64) | 4.32 |
| x 0.188 (3/16) | 5.59 |
| x 0.250 (1/4) | 7.10 |
| 4 x 2 x 0.083 (14) | 3.29 |
| x 0.120 (11) | 4.70 |
| x 0.155 (5/32) | 5.78 |
| x 0.188 (3/16) | 6.86 |
| x 0.250 (1/4) | 8.80 |
| 4 x 2-1/2 x 0.120 (11) | 5.11 |
| 4 x 3 x 0.120 (11) | 5.52 |
| x 0.156 (5/32) | 6.88 |
| x 0.188 (3/16) | 8.14 |
| x 0.250 (1/4) | 10.50 |
| x 0.313 (5/16) | 12.69 |
| 5 x 2 x 0.188 (3/16) | 8.43 |
| x 0.250 (1/4) | 10.50 |
| 5 x 2-1/2 x 0.120 (11) | 5.92 |
| x 0.180 (7) | 8.74 |
| 5 x 3 x 0.188 (3/16) | 9.31 |
| x 0.250 (1/4) | 12.02 |
| x 0.313 (5/16) | 14.52 |
| x 0.375 (3/8) | 16.84 |
| x 0.500 (1/2) | 20.88 |
| 6 x 2 x 0.188 (3/16) | 9.31 |
| x 0.250 (1/4) | 12.02 |
| 6 x 3 x 0.188 (3/16) | 10.97 |
| x 0.250 (1/4) | 14.45 |
| x 0.313 (5/16) | 16.65 |
| x 0.375 (3/8) | 19.39 |
| 6 x 3 x 0.500 (1/2) | 24.28 |
| 6 x 4 x 0.188 (3/16) | 11.86 |
| x 0.250 (1/4) | 15.42 |
| x 0.313 (5/16) | 18.77 |
| x 0.375 (3/8) | 21.94 |
| x 0.500 (1/2) | 27.68 |
| 7 x 5 x 0.188 (3/16) | 14.41 |
| x 0.250 (1/4) | 18.82 |
| x 0.313 (5/16) | 23.02 |
| x 0.375 (3/8) | 27.04 |
| x 0.500 (1/2) | 34.48 |
| 8 x 2 x 0.188 (3/16) | 11.86 |
| 8 x 3 x 0.188 (3/16) | 13.68 |
| x 0.250 (1/4) | 18.02 |
| 8 x 4 x 0.188 (3/16) | 14.41 |
| x 0.250 (1/4) | 18.82 |
| x 0.313 (5/16) | 23.02 |
| x 0.375 (3/8) | 27.04 |
| x 0.500 (1/2) | 34.48 |
| 8 x 6 x 0.188 (3/16) | 16.85 |
| x 0.250 (1/4) | 22.04 |
| x 0.313 (5/16) | 26.99 |
| x 0.375 (3/8) | 31.73 |
| x 0.500 (1/2) | 40.55 |
| 10 x 2 x 0.188 (3/16) | 14.42 |
| 10 x 4 x 0.188 (3/16) | 16.85 |
| x 0.250 (1/4) | 22.04 |
| 10 x 5 x 0.250 (1/4) | 24.70 |
| 10 x 6 x 0.250 (1/4) | 25.44 |
| x 0.313 (5/16) | 31.24 |
| x 0.375 (3/8) | 36.83 |
| x 0.500 (1/2) | 47.35 |
| 10 x 8 x 0.250 (1/4) | 28.83 |
| x 0.375 (3/8) | 41.93 |
| x 0.500 (1/2) | 54.15 |
| 12 x 2 x 0.188 (3/16) | 16.98 |
| 12 x 4 x 0.250 (1/4) | 26.03 |
| x 0.375 (3/8) | 38.55 |
| 12 x 6 x 0.250 (1/4) | 28.83 |
| x 0.375 (3/8) | 41.93 |
| x 0.500 (1/2) | 54.15 |

# ROUND STEEL TUBING

| OD Size Inches (guage) | Weight Lbs/ foot |
|---|---|
| 1/8 x 0.028 | 0.0290 |
| x 0.035 | 0.0336 |
| 5/32 x 0.028 | 0.0384 |
| x 0.035 | 0.0452 |
| 3/16 x 0.022 | 0.0390 |
| x 0.028 | 0.0478 |
| x 0.035 | 0.0572 |
| x 0.049 | 0.0727 |
| x 0.065 | 0.0854 |
| 1/4 x 0.194 | 0.0664 |
| x 0.035 | 0.0804 |
| x 0.049 | 0.1052 |
| x 0.065 | 0.1284 |
| 5/16 x 0.028 | 0.0852 |
| x 0.035 | 0.1039 |
| x 0.049 | 0.1382 |
| 3/8 x 0.028 | 0.1038 |
| x 0.035 | 0.1271 |
| x 0.049 | 0.1706 |
| x 0.065 | 0.2152 |
| 7/16 x 0.049 | 0.2036 |
| x 0.065 | 0.2589 |
| x 0.095 | 0.3480 |
| 1/2 x 0.035 | 0.1738 |
| x 0.042 | 0.2054 |
| x 0.049 | 0.2360 |
| x 0.065 | 0.3020 |
| x 0.083 | 0.3696 |
| x 0.095 | 0.4109 |
| 9/16 x 0.049 | 0.2690 |
| x 0.065 | 0.3457 |
| x 0.120 | 0.5677 |
| 5/8 x 0.035 | 0.2205 |
| x 0.049 | 0.3014 |
| x 0.065 | 0.3888 |
| x 0.095 | 0.5377 |
| x 0.120 | 0.6472 |
| 11/16 x 0.049 | 0.3344 |
| x 0.083 | 0.5363 |
| x 0.109 | 0.6740 |
| 3/4 x 0.035 | 0.2673 |
| x 0.049 | 0.3668 |
| x 0.065 | 0.4755 |
| x 0.083 | 0.5913 |
| x 0.109 | 0.7462 |
| x 0.120 | 0.8074 |
| x 0.250 | 1.3350 |
| 13/16 x 0.035 | 0.2908 |
| x 0.049 | 0.3998 |
| x 0.120 | 0.8881 |
| 7/8 x 0.035 | 0.3140 |
| x 0.049 | 0.4323 |
| x 0.065 | 0.5623 |

| OD Size Inches (guage) | Weight Lbs/ foot |
|---|---|
| 7/8 x 0.083 | 0.7021 |
| x 0.095 | 0.7914 |
| x 0.109 | 0.8917 |
| x 0.120 | 0.9676 |
| 15/16 x 0.049 | 0.4652 |
| x 0.095 | 0.8553 |
| 1 x 0.035 | 0.3670 |
| x 0.049 | 0.4977 |
| x 0.065 | 0.6491 |
| x 0.083 | 0.8129 |
| x 0.095 | 0.9182 |
| x 0.109 | 1.037 |
| x 0.120 | 1.128 |
| x 0.250 | 2.003 |
| 1–1/4 x 0.049 | 0.6285 |
| x 0.065 | 0.8226 |
| x 0.109 | 1.328 |
| x 0.120 | 1.448 |
| 1–1/2 x 0.049 | 0.7593 |
| x 0.065 | 0.9962 |
| x 0.095 | 1.426 |
| x 0.109 | 1.619 |
| x 0.250 | 3.338 |
| x 0.500 | 5.340 |
| 2 x 0.049 | 1.021 |
| x 0.065 | 1.343 |
| x 0.095 | 1.933 |
| x 0.120 | 2.409 |
| x 0.250 | 4.673 |
| x 0.500 | 8.010 |
| 2–1/2 x 0.065 | 1.690 |
| x 0.095 | 2.440 |
| x 0.120 | 3.050 |
| x 0.250 | 6.008 |
| x 0.500 | 10.68 |
| 3 x 0.065 | 2.037 |
| x 0.109 | 3.604 |
| x 0.120 | 3.691 |
| x 0.250 | 7.343 |
| x 0.500 | 13.35 |
| 4 x 0.065 | 2.732 |
| x 0.109 | 4.530 |
| x 0.120 | 4.973 |
| x 0.250 | 10.01 |
| x 0.500 | 18.69 |
| 5 x 0.109 | 5.694 |
| x 0.120 | 6.254 |
| x 0.250 | 12.68 |
| x 0.500 | 24.03 |
| 6 x 0.120 | 7.536 |
| x 0.250 | 15.35 |
| x 0.500 | 29.37 |
| x 1.000 | 53.40 |

# POCKET REF

# Survey & Mapping



(See also MATH, page 263, for Trig Functions)

# PERCENT GRADE TO DEGREES

| slope degrees | gradient 1:X | % grade |
|---|---|---|
| 0.1 | 573.1 | 0.2 |
| 0.2 | 286.5 | 0.3 |
| 0.3 | 191.0 | 0.5 |
| 0.4 | 143.3 | 0.7 |
| 0.5 | 114.6 | 0.9 |
| 0.6 | 95.5 | 1.0 |
| 0.7 | 81.9 | 1.2 |
| 0.8 | 71.6 | 1.4 |
| 0.9 | 63.7 | 1.6 |
| 1.0 | 57.3 | 1.7 |
| 2.0 | 28.6 | 3.5 |
| 3.0 | 19.1 | 5.2 |
| 4.0 | 14.3 | 7.0 |
| 5.0 | 11.4 | 8.7 |
| 6.0 | 9.5 | 10.5 |
| 7.0 | 8.1 | 12.3 |
| 8.0 | 7.1 | 14.1 |
| 9.0 | 6.3 | 15.8 |
| 10.0 | 5.7 | 17.6 |
| 11.0 | 5.1 | 19.4 |
| 12.0 | 4.7 | 21.3 |
| 13.0 | 4.3 | 23.1 |
| 14.0 | 4.0 | 24.9 |
| 15.0 | 3.7 | 26.8 |
| 16.0 | 3.5 | 28.7 |
| 17.0 | 3.3 | 30.6 |
| 18.0 | 3.1 | 32.5 |
| 19.0 | 2.9 | 34.4 |
| 20.0 | 2.7 | 36.4 |
| 21.0 | 2.6 | 38.4 |
| 22.0 | 2.5 | 40.4 |
| 23.0 | 2.4 | 42.4 |
| 24.0 | 2.2 | 44.5 |
| 25.0 | 2.1 | 46.6 |
| 26.0 | 2.1 | 48.8 |
| 27.0 | 2.0 | 50.9 |
| 28.0 | 1.9 | 53.2 |
| 29.0 | 1.8 | 55.4 |
| 30.0 | 1.7 | 57.7 |
| 31.0 | 1.7 | 60.1 |
| 32.0 | 1.6 | 62.5 |
| 33.0 | 1.5 | 64.9 |
| 34.0 | 1.5 | 67.4 |
| 35.0 | 1.4 | 70.0 |
| 36.0 | 1.4 | 72.6 |
| 37.0 | 1.3 | 75.3 |
| 38.0 | 1.3 | 78.1 |
| 39.0 | 1.2 | 81.0 |
| 40.0 | 1.2 | 83.9 |
| 41.0 | 1.2 | 86.9 |
| 42.0 | 1.1 | 90.0 |
| 43.0 | 1.1 | 93.2 |
| 44.0 | 1.0 | 96.5 |
| 45.0 | 1.0 | 100.0 |
| 46.0 | 1.0 | 103.5 |
| 47.0 | 0.9 | 107.2 |
| 48.0 | 0.9 | 111.0 |

# PERCENT GRADE TO DEGREES

| slope degrees | gradient 1:X | % grade |
|---|---|---|
| 49.0 | 0.9 | 115.0 |
| 50.0 | 0.8 | 119.1 |
| 51.0 | 0.8 | 123.4 |
| 52.0 | 0.8 | 127.9 |
| 53.0 | 0.8 | 132.7 |
| 54.0 | 0.7 | 137.6 |
| 55.0 | 0.7 | 142.8 |
| 56.0 | 0.7 | 148.2 |
| 57.0 | 0.6 | 153.9 |
| 58.0 | 0.6 | 160.0 |
| 59.0 | 0.6 | 166.4 |
| 60.0 | 0.6 | 173.1 |
| 61.0 | 0.6 | 180.3 |
| 62.0 | 0.5 | 188.0 |
| 63.0 | 0.5 | 196.2 |
| 64.0 | 0.5 | 204.9 |
| 65.0 | 0.5 | 214.3 |
| 66.0 | 0.4 | 224.5 |
| 67.0 | 0.4 | 235.4 |
| 68.0 | 0.4 | 247.3 |
| 69.0 | 0.4 | 260.3 |
| 70.0 | 0.4 | 274.6 |
| 71.0 | 0.3 | 290.2 |
| 72.0 | 0.3 | 307.5 |
| 73.0 | 0.3 | 326.8 |
| 74.0 | 0.3 | 348.4 |
| 75.0 | 0.3 | 372.8 |
| 76.0 | 0.2 | 400.7 |
| 77.0 | 0.2 | 432.6 |
| 78.0 | 0.2 | 469.9 |
| 79.0 | 0.2 | 513.7 |
| 80.0 | 0.2 | 566.3 |
| 81.0 | 0.2 | 630.3 |
| 82.0 | 0.1 | 710.1 |
| 83.0 | 0.1 | 812.6 |
| 84.0 | 0.1 | 948.9 |
| 85.0 | 0.1 | 1139.3 |
| 86.0 | 0.1 | 1424.3 |
| 87.0 | 0.1 | 1897.7 |
| 88.0 | 0.0 | 2840.0 |
| 89.0 | 0.0 | 5634.4 |
| 90.0 | 0.0 | 337465.3 |

$$\text{Tan [Slope Degrees]} = \frac{\text{Vertical Rise Distance}}{\text{Horizontal Distance}}$$

$$\text{Gradient} = 1 : \frac{\text{Horizontal Distance}}{\text{Vertical Rise Distance}}$$

$$\text{\% Grade is } 100 \text{ Tan [Slope]} = \frac{100 \times \text{Vertical Rise}}{\text{Horizontal Distance}}$$

## STADIA FORMULA

Most theodolites have an internal set of cross hairs that when used with a stadia rod, allows the calculation of slope distance. (magnification is normally 100 and slope distance = 100 x Stadia Rod Vertical Intercept). If angles are involved, the formula is:

$D = HI + [\text{Slope Distance} \times \sin(2V)/2] - M$

- D = Elevation difference between survey points
- HI = Instrument height above survey point
- V = Vertical angle (degrees) at the theodolite
- M = Mid point cross hair reading on stadia rod

## STADIA TABLE

Assume stadia slope distance of 100

| Slope Angle (°) | Vertical Distance | Horizontal Distance |
|---|---|---|
| 0.0 | 0.00 | 100.00 |
| 0.5 | 0.87 | 99.99 |
| 1.0 | 1.74 | 99.97 |
| 1.5 | 2.62 | 99.93 |
| 2.0 | 3.49 | 99.88 |
| 2.5 | 4.36 | 99.81 |
| 3.0 | 5.23 | 99.73 |
| 3.5 | 6.09 | 99.63 |
| 4.0 | 6.96 | 99.51 |
| 4.5 | 7.82 | 99.38 |
| 5.0 | 8.68 | 99.24 |
| 5.5 | 9.54 | 99.08 |
| 6.0 | 10.40 | 98.91 |
| 6.5 | 11.25 | 98.72 |
| 7.0 | 12.10 | 98.51 |
| 7.5 | 12.94 | 98.30 |
| 8.0 | 13.78 | 98.06 |
| 8.5 | 14.62 | 97.82 |
| 9.0 | 15.45 | 97.55 |
| 9.5 | 16.28 | 97.28 |
| 10.0 | 17.10 | 96.98 |
| 10.5 | 17.92 | 96.68 |
| 11.0 | 18.73 | 96.36 |
| 11.5 | 19.54 | 96.03 |
| 12.0 | 20.34 | 95.68 |
| 12.5 | 21.13 | 95.32 |
| 13.0 | 21.92 | 94.94 |
| 13.5 | 22.70 | 94.55 |
| 14.0 | 23.47 | 94.15 |

# STADIA TABLE

Assume stadia slope distance of 100

| Slope Angle (°) | Vertical Distance | Horizontal Distance |
|---|---|---|
| 14.5 | 24.24 | 93.73 |
| 15.0 | 25.00 | 93.30 |
| 15.5 | 25.75 | 92.86 |
| 16.0 | 26.50 | 92.40 |
| 16.5 | 27.23 | 91.93 |
| 17.0 | 27.96 | 91.45 |
| 17.5 | 28.68 | 90.96 |
| 18.0 | 29.39 | 90.45 |
| 18.5 | 30.09 | 89.93 |
| 19.0 | 30.78 | 89.40 |
| 19.5 | 31.47 | 88.86 |
| 20.0 | 32.14 | 88.30 |
| 20.5 | 32.80 | 87.74 |
| 21.0 | 33.46 | 87.16 |
| 21.5 | 34.10 | 86.57 |
| 22.0 | 34.73 | 85.97 |
| 22.5 | 35.36 | 85.36 |
| 23.0 | 35.97 | 84.73 |
| 23.5 | 36.57 | 84.10 |
| 24.0 | 37.16 | 83.46 |
| 24.5 | 37.74 | 82.80 |
| 25.0 | 38.30 | 82.14 |
| 25.5 | 38.86 | 81.47 |
| 26.0 | 39.40 | 80.78 |
| 26.5 | 39.93 | 80.09 |
| 27.0 | 40.45 | 79.39 |
| 27.5 | 40.96 | 78.68 |
| 28.0 | 41.45 | 77.96 |
| 28.5 | 41.93 | 77.23 |
| 29.0 | 42.40 | 76.50 |
| 29.5 | 42.86 | 75.75 |
| 30.0 | 43.30 | 75.00 |
| 30.5 | 43.73 | 74.24 |
| 31.0 | 44.15 | 73.47 |
| 31.5 | 44.55 | 72.70 |
| 32.0 | 44.94 | 71.92 |
| 32.5 | 45.32 | 71.13 |
| 33.0 | 45.68 | 70.34 |
| 33.5 | 46.03 | 69.54 |
| 34.0 | 46.36 | 68.73 |
| 34.5 | 46.68 | 67.92 |
| 35.0 | 46.98 | 67.10 |
| 35.5 | 47.28 | 66.28 |
| 36.0 | 47.55 | 65.45 |
| 36.5 | 47.82 | 64.62 |
| 37.0 | 48.06 | 63.78 |
| 37.5 | 48.30 | 62.94 |
| 38.0 | 48.51 | 62.10 |
| 38.5 | 48.72 | 61.25 |
| 39.0 | 48.91 | 60.40 |

# STADIA TABLE

Assume stadia slope distance of 100

| Slope Angle (°) | Vertical Distance | Horizontal Distance |
|---|---|---|
| 39.5 | 49.08 | 59.54 |
| 40.0 | 49.24 | 58.68 |
| 40.5 | 49.38 | 57.82 |
| 41.0 | 49.51 | 56.96 |
| 41.5 | 49.63 | 56.09 |
| 42.0 | 49.73 | 55.23 |
| 42.5 | 49.81 | 54.36 |
| 43.0 | 49.88 | 53.49 |
| 43.5 | 49.93 | 52.62 |
| 44.0 | 49.97 | 51.74 |
| 44.5 | 49.99 | 50.87 |
| 45.0 | 50.00 | 50.00 |
| 45.5 | 49.99 | 49.13 |
| 46.0 | 49.97 | 48.26 |
| 46.5 | 49.93 | 47.38 |
| 47.0 | 49.88 | 46.51 |
| 47.5 | 49.81 | 45.64 |
| 48.0 | 49.73 | 44.77 |
| 48.5 | 49.63 | 43.91 |
| 49.0 | 49.51 | 43.04 |
| 49.5 | 49.38 | 42.18 |
| 50.0 | 49.24 | 41.32 |
| 50.5 | 49.08 | 40.46 |
| 51.0 | 48.91 | 39.60 |
| 51.5 | 48.72 | 38.75 |
| 52.0 | 48.51 | 37.90 |
| 52.5 | 48.30 | 37.06 |
| 53.0 | 48.06 | 36.22 |
| 53.5 | 47.82 | 35.38 |
| 54.0 | 47.55 | 34.55 |
| 54.5 | 47.28 | 33.72 |
| 55.0 | 46.98 | 32.90 |
| 55.5 | 46.68 | 32.08 |
| 56.0 | 46.36 | 31.27 |
| 56.5 | 46.03 | 30.46 |
| 57.0 | 45.68 | 29.66 |
| 57.5 | 45.32 | 28.87 |
| 58.0 | 44.94 | 28.08 |
| 58.5 | 44.55 | 27.30 |
| 59.0 | 44.15 | 26.53 |
| 59.5 | 43.73 | 25.76 |
| 60.0 | 43.30 | 25.00 |

Horizontal Distance = Slope Distance x $\text{Cos}^2$(Slope Degrees)

Vertical Distance = (Slope Distance/2) x Sin(2 x Slope Degrees)

# MAPPING SCALES & AREAS

| scale 1:X | Feet/ Inch | Inch/ Mile | Acres/ Sq Inch | Sq Miles/ Sq Inch |
|---|---|---|---|---|
| 100 | 8.3 | 633.60 | 0.0016 | 0.000002 |
| 120 | 10.0 | 528.00 | 0.0023 | 0.000004 |
| 200 | 16.7 | 316.80 | 0.0064 | 0.000010 |
| 240 | 20.0 | 264.00 | 0.0092 | 0.000014 |
| 250 | 20.8 | 253.44 | 0.0100 | 0.000016 |
| 300 | 25.0 | 211.20 | 0.0143 | 0.000022 |
| 400 | 33.3 | 158.40 | 0.0255 | 0.000040 |
| 480 | 40.0 | 132.00 | 0.0367 | 0.000057 |
| 500 | 41.7 | 126.72 | 0.0399 | 0.000062 |
| 600 | 50.0 | 105.60 | 0.0574 | 0.000090 |
| 1000 | 83.3 | 63.36 | 0.1594 | 0.000249 |
| 1200 | 100.0 | 52.80 | 0.2296 | 0.000359 |
| 1500 | 125.0 | 42.24 | 0.3587 | 0.000560 |
| 2000 | 166.7 | 31.68 | 0.6377 | 0.000996 |
| 2400 | 200.0 | 26.40 | 0.9183 | 0.001435 |
| 2500 | 208.3 | 25.34 | 0.9964 | 0.001557 |
| 3000 | 250.0 | 21.12 | 1.4348 | 0.002242 |
| 3600 | 300.0 | 17.60 | 2.0661 | 0.003228 |
| 4000 | 333.3 | 15.84 | 2.5508 | 0.003986 |
| 4800 | 400.0 | 13.20 | 3.6731 | 0.005739 |
| 5000 | 416.7 | 12.67 | 3.9856 | 0.006227 |
| 6000 | 500.0 | 10.56 | 5.7392 | 0.008968 |
| 7000 | 583.3 | 9.05 | 7.8117 | 0.012206 |
| 7200 | 600.0 | 8.80 | 8.2645 | 0.012913 |
| 7920 | 660.0 | 8.00 | 10.0000 | 0.015625 |
| 8000 | 666.7 | 7.92 | 10.2030 | 0.015942 |
| 8400 | 700.0 | 7.54 | 11.2489 | 0.017576 |
| 9000 | 750.0 | 7.04 | 12.9132 | 0.020177 |
| 9600 | 800.0 | 6.60 | 14.6924 | 0.022957 |
| 10000 | 833.3 | 6.34 | 15.9423 | 0.024910 |
| 10800 | 900.0 | 5.87 | 18.5950 | 0.029055 |
| 12000 | 1000.0 | 5.28 | 22.9568 | 0.035870 |
| 13200 | 1100.0 | 4.80 | 27.7778 | 0.043403 |
| 14400 | 1200.0 | 4.40 | 33.0579 | 0.051653 |
| 15000 | 1250.0 | 4.22 | 35.8701 | 0.056047 |
| 15600 | 1300.0 | 4.06 | 38.7971 | 0.060620 |
| 15840 | 1320.0 | 4.00 | 40.0000 | 0.062500 |
| 16000 | 1333.3 | 3.96 | 40.8122 | 0.063769 |
| 16800 | 1400.0 | 3.77 | 44.9954 | 0.070305 |
| 18000 | 1500.0 | 3.52 | 51.6529 | 0.080708 |
| 19200 | 1600.0 | 3.30 | 58.7695 | 0.091827 |
| 20000 | 1666.7 | 3.17 | 63.7690 | 0.099639 |
| 20400 | 1700.0 | 3.11 | 66.3453 | 0.103664 |

# MAPPING SCALES & AREAS

| scale 1:X | Feet/ Inch | Inch/ Mile | Acres/ Sq Inch | Sq Miles/ Sq Inch |
|---|---|---|---|---|
| 21120 | 1760.0 | 3.00 | 71.1111 | 0.111111 |
| 21600 | 1800.0 | 2.93 | 74.3802 | 0.116219 |
| 22800 | 1900.0 | 2.78 | 82.8742 | 0.129491 |
| 24000 | 2000.0 | 2.64 | 91.8274 | 0.143480 |
| 25000 | 2083.3 | 2.53 | 99.6391 | 0.155686 |
| 30000 | 2500.0 | 2.11 | 143.4803 | 0.224188 |
| 31680 | 2640.0 | 2.00 | 160.0000 | 0.250000 |
| 40000 | 3333.3 | 1.58 | 255.0760 | 0.398556 |
| 45000 | 3750.0 | 1.41 | 322.8306 | 0.504423 |
| 48000 | 4000.0 | 1.32 | 367.3095 | 0.573921 |
| 50000 | 4166.7 | 1.27 | 398.5563 | 0.622744 |
| 60000 | 5000.0 | 1.06 | 573.9210 | 0.896752 |
| 62500 | 5208.3 | 1.01 | 622.7442 | 0.973038 |
| 63360 | 5280.0 | 1.00 | 640.0000 | 1.000000 |
| 80000 | 6666.7 | 0.79 | 1020.3041 | 1.594225 |
| 90000 | 7500.0 | 0.70 | 1291.3223 | 2.017691 |
| 96000 | 8000.0 | 0.66 | 1469.2378 | 2.295684 |
| 100000 | 8333.3 | 0.63 | 1594.2251 | 2.490977 |
| 125000 | 10416.7 | 0.51 | 2490.9767 | 3.892151 |
| 126720 | 10560.0 | 0.50 | 2560.0000 | 4.000000 |
| 200000 | 16666.7 | 0.32 | 6376.9003 | 9.963907 |
| 250000 | 20833.3 | 0.25 | 9963.9067 | 15.568604 |
| 253440 | 21120.0 | 0.25 | 10240.0000 | 16.000000 |
| 380160 | 31680.0 | 0.17 | 23040.0000 | 36.000000 |
| 500000 | 41666.7 | 0.13 | 39855.6270 | 62.274417 |
| 760320 | 63360.0 | 0.08 | 92160.0000 | 144.000000 |
| 1000000 | 83333.3 | 0.06 | 159422.5079 | 249.097669 |

Feet / Inch = Scale / 12

Meters / Inch = Scale / 39.37

Miles / Inch = Scale / 63,291.14

Chains / Inch = Scale / 792.08

Inch / Mile = Scale / 63360

Acres / Square Inch = $\text{Scale}^2$ / 6,272,640

Square Miles / Square Inch = $\text{Scale}^2$ / 4,014,489,600

# APPARENT DIP TABLE

| True Dip | Angle between Strike and direction of Cross Section 5° | 10° | 15° | 20° | 25° |
|---|---|---|---|---|---|
| 5° | 0.4 | 0.9 | 1.3 | 1.7 | 2.1 |
| 10° | 0.9 | 1.8 | 2.6 | 3.5 | 4.3 |
| 15° | 1.3 | 2.7 | 4.0 | 5.2 | 6.5 |
| 20° | 1.8 | 3.6 | 5.4 | 7.1 | 8.7 |
| 25° | 2.3 | 4.6 | 6.9 | 9.1 | 11.1 |
| 30° | 2.9 | 5.7 | 8.5 | 11.2 | 13.7 |
| 35° | 3.5 | 6.9 | 10.3 | 13.5 | 16.5 |
| 40° | 4.2 | 8.3 | 12.3 | 16.0 | 19.5 |
| 45° | 5.0 | 9.9 | 14.5 | 18.9 | 22.9 |
| 50° | 5.9 | 11.7 | 17.1 | 22.2 | 26.7 |
| 55° | 7.1 | 13.9 | 20.3 | 26.0 | 31.1 |
| 60° | 8.6 | 16.7 | 24.1 | 30.6 | 36.2 |
| 65° | 10.6 | 20.4 | 29.0 | 36.3 | 42.2 |
| 70° | 13.5 | 25.5 | 35.4 | 43.2 | 49.3 |
| 75° | 18.0 | 32.9 | 44.0 | 51.9 | 57.6 |
| 80° | 26.3 | 44.6 | 55.7 | 62.7 | 67.4 |
| 85° | 44.9 | 63.3 | 71.3 | 75.7 | 78.3 |

| True Dip | Angle between Strike and direction of Cross Section 30° | 35° | 40° | 45° | 50° |
|---|---|---|---|---|---|
| 5° | 2.5 | 2.9 | 3.2 | 3.5 | 3.8 |
| 10° | 5.0 | 5.8 | 6.5 | 7.1 | 7.7 |
| 15° | 7.6 | 8.7 | 9.8 | 10.7 | 11.6 |
| 20° | 10.3 | 11.8 | 13.2 | 14.4 | 15.6 |
| 25° | 13.1 | 15.0 | 16.7 | 18.2 | 19.7 |
| 30° | 16.1 | 18.3 | 20.4 | 22.2 | 23.9 |
| 35° | 19.3 | 21.9 | 24.2 | 26.3 | 28.2 |
| 40° | 22.8 | 25.7 | 28.3 | 30.7 | 32.7 |
| 45° | 26.6 | 29.8 | 32.7 | 35.3 | 37.5 |
| 50° | 30.8 | 34.4 | 37.5 | 40.1 | 42.4 |
| 55° | 35.5 | 39.3 | 42.6 | 45.3 | 47.6 |
| 60° | 40.9 | 44.8 | 48.1 | 50.8 | 53.0 |
| 65° | 47.0 | 50.9 | 54.0 | 56.6 | 58.7 |
| 70° | 53.9 | 57.6 | 60.5 | 62.8 | 64.6 |
| 75° | 61.8 | 65.0 | 67.4 | 69.2 | 70.7 |
| 80° | 70.6 | 72.9 | 74.7 | 76.0 | 77.0 |
| 85° | 80.1 | 81.3 | 82.2 | 82.9 | 83.5 |

# APPARENT DIP TABLE

| True Dip | Angle between Strike and direction of Cross Section 55° | 60° | 65° | 70° | 75° |
|---|---|---|---|---|---|
| 5° | 4.1 | 4.3 | 4.5 | 4.7 | 4.8 |
| 10° | 8.2 | 8.7 | 9.1 | 9.4 | 9.7 |
| 15° | 12.4 | 13.1 | 13.6 | 14.1 | 14.5 |
| 20° | 16.6 | 17.5 | 18.3 | 18.9 | 19.4 |
| 25° | 20.9 | 22.0 | 22.9 | 23.7 | 24.2 |
| 30° | 25.3 | 26.6 | 27.6 | 28.5 | 29.1 |
| 35° | 29.8 | 31.2 | 32.4 | 33.3 | 34.1 |
| 40° | 34.5 | 36.0 | 37.3 | 38.3 | 39.0 |
| 45° | 39.3 | 40.9 | 42.2 | 43.2 | 44.0 |
| 50° | 44.3 | 45.9 | 47.2 | 48.2 | 49.0 |
| 55° | 49.5 | 51.0 | 52.3 | 53.3 | 54.1 |
| 60° | 54.8 | 56.3 | 57.5 | 58.4 | 59.1 |
| 65° | 60.3 | 61.7 | 62.8 | 63.6 | 64.2 |
| 70° | 66.0 | 67.2 | 68.1 | 68.8 | 69.4 |
| 75° | 71.9 | 72.8 | 73.5 | 74.1 | 74.5 |
| 80° | 77.9 | 78.5 | 79.0 | 79.4 | 79.7 |
| 85° | 83.9 | 84.2 | 84.5 | 84.7 | 84.8 |

When a cross section cuts across the plane of a bedding surface, fault or topography (at any angle less than 90°), the observed dip (apparent dip) at the cross section intersection will always be less than the true dip. In order to use the above table, first, determine the "Angle Between Strike" value by subtracting the true strike (bearing) of the unit from the bearing of the cross section; second, measure the observed dip and locate that number in the body of the table; third, follow the row across to the left in order to determine the true dip. If you need a more precise value than listed in the table, use the following formula to calculate the answer:

Tan (Apparent Dip ) = Tan (True Dip ) x Sin (Angle Between )

Note that if the "Angle between Strike" value is 80° to 90°, the value of the true dip and apparent dip are nearly identical.

# THREE POINT PROBLEM

A common problem in both surveying and geology is the determination of the strike (bearing) of a bedding surface, fault, topographic surface, etc. when the location and elevation of three points on the surface are known.

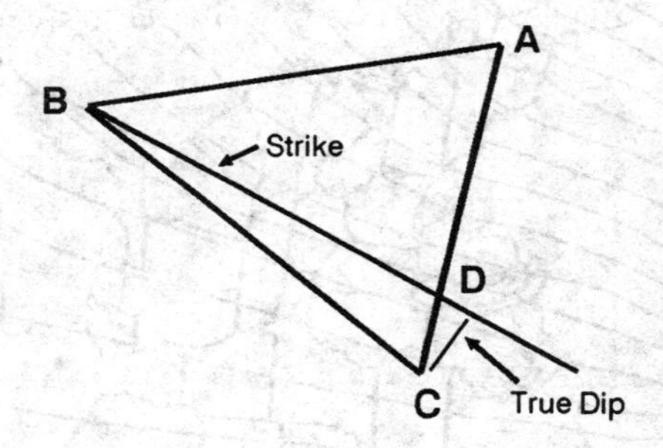


The elevation and location of points A, B, and C are known. If the elevation of point B is between the elevations of points A (the highest point) and C (the lowest point), then the strike (bearing) of the plane defined by ABC is determined by locating point D, on the line between A and C, and then drawing a strike line between B and D. Since strike is defined as the bearing of a horizontal line on the plane of ABC, the elevations of point B and D must be the same. The exact location of point D can be calculated from the following equation:

$$\text{Length AD} = \text{Length AC} \times \frac{\text{AB elevation difference}}{\text{AC elevation difference}}$$

Note that the true dip must always be at a right angle to the strike. The value of the true dip can be calculated with the following equation:

$$\text{Dip Angle } (^{\circ}) = \text{Arc Sin } ((\text{Elev D–C}) / \text{Length True Dip Line})$$

## MAGNETIC DECLINATION MAP

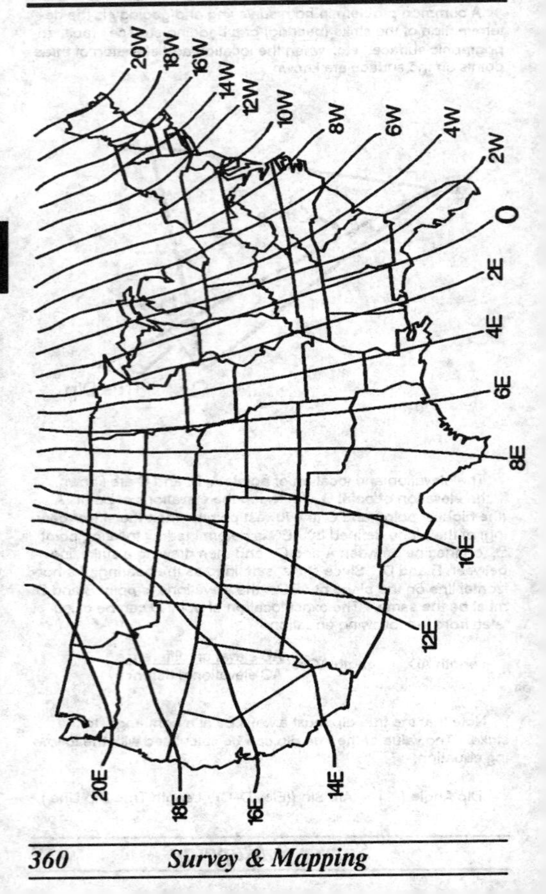

# POCKET REF

# Tools



(See WELDING, p. 397 for more tools & soldering)
(See also MATH, p.265 for drill number table)

# AMERICAN NATIONAL TAPS & DIES

| | Fine Threads | | | Coarse Threads | | |
|---|---|---|---|---|---|---|
| Thread | Threads / inch | Tap Drill | Tap Decimal inch | Threads / inch | Tap Drill | Tap Decimal inch |
| #0 | 80 | 3/64 | 0.0469 | ... | ... | ... |
| #1 | 72 | #53 | 0.0595 | 64 | #53 | 0.0595 |
| #2 | 64 | #50 | 0.0700 | 56 | #50 | 0.0700 |
| #3 | 56 | #45 | 0.0820 | 48 | #47 | 0.0785 |
| #4 | 48 | #42 | 0.0935 | 40 | #43 | 0.0890 |
| 1/8 | 40 | #38 | 0.1015 | 32 | 3/32 | 0.0937 |
| #5 | 44 | #37 | 0.1040 | 40 | #38 | 0.1015 |
| #6 | 40 | #33 | 0.1130 | 32 | #36 | 0.1065 |
| #8 | 36 | #29 | 0.1360 | 32 | #29 | 0.1360 |
| 3/16 | 32 | #22 | 0.1570 | 24 | #26 | 0.1470 |
| #10 | 32 | #21 | 0.1590 | 24 | #25 | 0.1495 |
| #12 | 28 | #14 | 0.1820 | 24 | #16 | 0.1770 |
| 1/4 | 28 | #3 | 0.2130 | 20 | #7 | 0.2010 |
| 5/16 | 24 | I | 0.2720 | 18 | F | 0.2570 |
| 3/8 | 24 | Q | 0.3320 | 16 | 5/16 | 0.3125 |
| 7/16 | 20 | 25/64 | 0.3906 | 14 | U | 0.3680 |
| 1/2 | 20 | 29/64 | 0.4531 | 13 | 27/64 | 0.4219 |
| 9/16 | 18 | 33/64 | 0.5156 | 12 | 31/64 | 0.4844 |
| 5/8 | 28 | 37/64 | 0.5781 | 11 | 17/32 | 0.5312 |
| 3/4 | 26 | 11/16 | 0.6875 | 10 | 21/32 | 0.6562 |
| 7/8 | 24 | 13/16 | 0.8125 | 9 | 49/64 | 0.7656 |
| 1 | 14 | 15/16 | 0.9375 | 8 | 7/8 | 0.8750 |
| 1-1/8 | 12 | 1-3/64 | 1.0469 | 7 | 63/64 | 0.9844 |
| 1-1/4 | 12 | 1-11/64 | 1.1719 | 7 | 1-7/64 | 1.1094 |
| 1-3/8 | 12 | 1-19/64 | 1.2969 | 6 | 1-7/32 | 1.2187 |
| 1-1/2 | 12 | 1-27/64 | 1.4219 | 6 | 1-11/32 | 1.3281 |
| 1-3/4 | ... | ... | ... | 5 | 1-9/16 | 1.5469 |
| 2 | ... | ... | ... | 4-1/2 | 1-25/32 | 1.7812 |
| 2-1/4 | ... | ... | ... | 4-1/2 | 2-1/32 | 2.0312 |
| 2-1/2 | ... | ... | ... | 4 | 2-1/4 | 2.2500 |
| 2-3/4 | ... | ... | ... | 4 | 2-1/2 | 2.5000 |
| 3 | ... | ... | ... | 4 | 2-3/4 | 2.7500 |
| 3-1/4 | ... | ... | ... | 4 | 3 | 3.0000 |
| 3-1/2 | ... | ... | ... | 4 | 3-1/4 | 3.2500 |
| 3-3/4 | ... | ... | ... | 4 | 3-1/2 | 3.5000 |
| 4 | ... | ... | ... | 4 | 3-3/4 | 3.7500 |

Note that there are literally hundreds of other sizes and thread per inch taps and dies available, e.g. 1/4 inch can have 10,12,14, 16,18,22,23,24,25,26,27,30,32,34,36,38,40,42,44,48,50,52,56,60, 64,72, and 80 threads/inch depending on the standard. Threads shown in the table above are simply the most common.

# METRIC TAPS AND DIES

| Thread Size | | Pitch in mm | | Tap Drill Size | |
|---|---|---|---|---|---|
| mm | Inches | French | International | mm | Inches |
| 1.5 | 0.0590 | 0.35 | ... | 1.10 | 0.0433 |
| 2 | 0.0787 | 0.45 | ... | 1.50 | 0.0590 |
| 2 | 0.0787 | ... | 0.40 | 1.60 | 0.0630 |
| 2.3 | 0.0895 | ... | 0.40 | 1.90 | 0.0748 |
| 2.5 | 0.0984 | 0.45 | ... | 2.00 | 0.0787 |
| 2.6 | 0.1124 | ... | 0.45 | 2.10 | 0.0827 |
| 3 | 0.1181 | ... | 0.5 | 2.50 | 0.0984 |
| 3 | 0.1181 | 0.60 | ... | 2.40 | 0.0945 |
| 3.5 | 0.1378 | 0.60 | 0.60 | 2.90 | 0.1142 |
| 4 | 0.1575 | 0.75 | ... | 3.25 | 0.1279 |
| 4 | 0.1575 | ... | 0.70 | 3.30 | 0.1299 |
| 4.5 | 0.1772 | 0.75 | 0.75 | 3.75 | 0.1476 |
| 5 | 0.1968 | 0.90 | ... | 4.10 | 0.1614 |
| 5 | 0.1968 | ... | 0.80 | 4.20 | 0.1653 |
| 5.5 | 0.2165 | 0.90 | 0.90 | 4.60 | 0.1811 |
| 6 | 0.2362 | 1.00 | 1.00 | 5.00 | 0.1968 |
| 7 | 0.2856 | 1.00 | 1.00 | 6.00 | 0.2362 |
| 8 | 0.3150 | 1.00 | ... | 7.00 | 0.2756 |
| 8 | 0.3150 | ... | 1.25 | 6.80 | 0.2677 |
| 9 | 0.3543 | 1.00 | ... | 8.00 | 0.3150 |
| 9 | 0.3543 | ... | 1.25 | 7.80 | 0.3071 |
| 10 | 0.3937 | 1.50 | 1.50 | 8.60 | 0.3386 |
| 11 | 0.3937 | ... | 1.50 | 9.60 | 0.3780 |
| 12 | 0.4624 | 1.50 | ... | 10.50 | 0.4134 |
| 12 | 0.4624 | ... | 1.75 | 10.50 | 0.4134 |
| 14 | 0.5512 | 2.00 | 2.00 | 12.00 | 0.4724 |
| 16 | 0.6299 | 2.00 | 2.00 | 14.00 | 0.5118 |
| 18 | 0.7087 | 2.50 | 2.50 | 15.50 | 0.6102 |
| 20 | 0.7974 | 2.50 | 2.50 | 17.50 | 0.6890 |
| 22 | 0.8771 | 2.50 | 2.50 | 19.50 | 0.7677 |
| 24 | 0.9449 | 3.00 | 3.00 | 21.00 | 0.8268 |
| 26 | 1.0236 | 3.00 | ... | 23.00 | 0.9055 |
| 27 | 1.0630 | ... | 3.00 | 24.00 | 0.9449 |
| 28 | 1.1024 | 3.00 | ... | 25.00 | 0.9842 |
| 30 | 1.1811 | 3.50 | 3.50 | 26.50 | 1.0433 |
| 32 | 1.2598 | 3.50 | ... | 28.50 | 1.1220 |
| 33 | 1.2992 | ... | 3.50 | 29.50 | 1.1614 |
| 34 | 1.3386 | 3.50 | ... | 30.50 | 1.2008 |
| 36 | 1.4173 | 4.00 | 4.00 | 32.00 | 1.2598 |
| 38 | 1.4961 | 4.00 | ... | 34.00 | 1.3386 |
| 39 | 1.5354 | ... | 4.00 | 35.00 | 1.3779 |
| 40 | 1.5748 | 4.00 | ... | 36.00 | 1.4173 |
| 42 | 1.6535 | 4.50 | 4.50 | 37.00 | 1.4567 |

## BRITISH TAPS & DIES

| Thread | British Std Whitworth Threads / inch | British Std Whitworth Tap Drill | British Standard Fine Threads / inch | British Standard Fine Tap Drill |
|---|---|---|---|---|
| 1/8 | 40 | 2.55mm | ... | ... |
| 3/16 | 24 | 3.70mm | 32 | 5/32 |
| 7/32 | ... | ... | 28 | 4.65mm |
| 1/4 | 20 | 5.10mm | 26 | 5.3mm |
| 9/32 | ... | ... | 26 | ... |
| 5/16 | 18 | 6.50mm | 22 | 6.75mm |
| 3/8 | 16 | 5/16 | 20 | 8.25mm |
| 7/16 | 14 | 9.25mm | 18 | 9.70mm |
| 1/2 | 12 | 10.5mm | 16 | 7/16 |
| 9/16 | 12 | 12.1mm | 16 | 1/2 |
| 5/8 | 11 | 13.5mm | 14 | 14mm |
| 11/16 | 11 | ... | 14 | ... |
| 3/4 | 10 | 41/64 | 12 | 16.75mm |
| 7/8 | 9 | 19.25mm | 11 | 25/32 |
| 1 | 8 | 22.00mm | 10 | 22.75 |
| 1-1/8 | 7 | 24.75mm | 9 | 25.50mm |
| 1-1/4 | 7 | 1-3/32 | 9 | 28.75mm |
| 1-3/8 | ... | ... | 8 | 31.50mm |
| 1-1/2 | 6 | 33.50mm | 8 | 1-23/64 |
| 1-3/4 | 5 | 39.00mm | ... | ... |
| 2 | 4.5 | 44.50mm | ... | ... |

## BRITISH ASSOC STD THREAD (B.A.)

| Thread | Threads / inch | Pitch mm | Major Diameter mm |
|---|---|---|---|
| 0 | 25.4 | 1.00 | 6.0 |
| 1 | 28.2 | 0.90 | 5.3 |
| 2 | 31.4 | 0.81 | 4.7 |
| 3 | 34.8 | 0.73 | 4.1 |
| 4 | 38.5 | 0.66 | 3.6 |
| 5 | 43.0 | 0.59 | 3.2 |
| 6 | 47.9 | 0.53 | 2.8 |
| 7 | 52.9 | 0.48 | 2.5 |
| 8 | 59.1 | 0.43 | 2.2 |
| 9 | 65.1 | 0.39 | 1.9 |
| 10 | 72.6 | 0.35 | 1.7 |
| 11 | 82.0 | 0.31 | 1.5 |
| 12 | 90.7 | 0.28 | 1.3 |
| 13 | 102.0 | 0.25 | 1.2 |
| 14 | 110.0 | 0.23 | 1.0 |
| 15 | 121.0 | 0.21 | 0.9 |
| 16 | 134.0 | 0.19 | 0.79 |

# AMERICAN STD TAPER PIPE

| Pipe Size inch | Threads per inch | Pipe Diameter inch | Tap Drill |
|---|---|---|---|
| 1/8 | 27 | 0.405 | R |
| 1/4 | 18 | 0.540 | 7/16 |
| 3/8 | 18 | 0.675 | 37/64 |
| 1/2 | 14 | 0.840 | 23/32 |
| 3/4 | 14 | 1.050 | 59/64 |
| 1 | 11.5 | 1.315 | 1-5/32 |
| 1-1/4 | 11.5 | 1.660 | 1-1/2 |
| 1-1/2 | 11.5 | 1.900 | 1-47/64 |
| 2 | 11.5 | 2.375 | 1-7/32 |
| 2-1/2 | 8 | 2.875 | 2-5/8 |
| 3 | 8 | 3.500 | 3-1/4 |
| 3-1/2 | 8 | 4.000 | 3-3/4 |
| 4 | 8 | 4.500 | 4-1/4 |
| 4-1/2 | 8 | 5.000 | 4-3/4 |
| 5 | 8 | 5.563 | 5-9/32 |
| 6 | 8 | 6.625 | 6-11/32 |
| 7 | 8 | 7.625 | ... |
| 8 | 8 | 8.625 | ... |
| 9 | 8 | 9.625 | ... |
| 10 | 8 | 10.750 | ... |
| 12 | 8 | 12.750 | ... |
| >14 OD | 8 | Same as Col 1 | ... |

# AMERICAN STD STRAIGHT PIPE

| Pipe Size inch | Threads per inch | Pipe Diameter inch | Tap Drill |
|---|---|---|---|
| 1/8 | 27 | 0.405 | S |
| 1/4 | 18 | 0.540 | 29/64 |
| 3/8 | 18 | 0.675 | 19/32 |
| 1/2 | 14 | 0.840 | 47/64 |
| 3/4 | 14 | 1.050 | 15/16 |
| 1 | 11.5 | 1.315 | 1-3/16 |
| 1-1/4 | 11.5 | 1.660 | 1-33/64 |
| 1-1/2 | 11.5 | 1.900 | 1-3/4 |
| 2 | 11.5 | 2.375 | 2-7/32 |
| 2-1/2 | 8 | 2.875 | 2-21/32 |
| 3 | 8 | 3.500 | 3-9/32 |
| 3-1/2 | 8 | 4.000 | 3-25/32 |
| 4 | 8 | 4.500 | 4-9/32 |
| 4-1/2 | 8 | 5.000 | 4-25/32 |
| 5 | 8 | 5.563 | 5-11/32 |
| 6 | 8 | 6.625 | 6-13/32 |

# DRILL & CUTTING LUBRICANTS

| Material to be Worked | Machine Process | | |
|---|---|---|---|
| | Drilling | Threading | Lathe |
| Aluminum | Soluble oil<br>Kerosene<br>Lard oil | Soluble oil,<br>Kerosene, &<br>Lard oil | Soluble oil |
| Brass | Dry<br>Soluble oil<br>Kerosene<br>Lard Oil | Soluble oil<br>Lard oil | Soluble oil |
| Bronze | Dry<br>Soluble oil<br>Mineral oil<br>Lard oil | Soluble oil<br>Lard oil | Soluble oil |
| Cast Iron | Dry<br>Air jet<br>Soluble oil | Dry<br>Sulphurized oil<br>Mineral lard oil | Dry<br>Soluble oil |
| Copper | Dry<br>Soluble oil<br>Mineral lard oil<br>Kerosene | Soluble oil<br>Lard oil | Soluble |
| Malleable Iron | Dry<br>Soda water | Lard oil<br>Soda water | Soluble oil<br>Soda water |
| Monel metal | Soluble oil<br>Lard oil | Lard oil | Soluble oil |
| Steel alloys | Soluble oil<br>Sulphurized oil<br>Mineral lard oil | Sulphurized oil<br>Lard oil | Soluble oil |
| Steel, machine | Soluble oil<br>Sulphurized oil<br>Lard oil<br>Mineral lard oil | Soluble oil<br>Mineral lard oil | Soluble oil |
| Steel, tool | Soluble oil<br>Sulphurized oil<br>Mineral lard oil | Sulphurized oil<br>Lard oil | Soluble oil |

*The above table of cutting fluids is courtesy of Cincinnati Milacron.*

# DRILLING SPEEDS vs MATERIAL

| Material | Speed rpm | Description |
|---|---|---|
| Cast Iron | 6000 to 6500 | 1/16 inch drill |
| | 3500 to 4500 | 1/8 inch drill |
| | 2500 to 3000 | 3/16 inch drill |
| | 2000 to 2500 | 1/4 inch drill |
| | 1500 to 2000 | 5/16 inch drill |
| | 1500 to 2000 | 3/8 inch drill |
| | 1000 to 1500 | > 7/16 inch drill |
| Glass | 700 | Special metal tube drilling |
| Plastics | 6000 to 6500 | 1/16 inch drill |
| | 5000 to 6000 | 1/8 inch drill |
| | 3500 to 4000 | 3/16 inch drill |
| | 3000 to 3500 | 1/4 inch drill |
| | 2000 to 2500 | 5/16 inch drill |
| | 1500 to 2000 | 3/8 inch drill |
| | 500 to 1000 | > 7/16 inch drill |
| Soft Metals (copper) | 6000 to 6500 | 1/16 inch drill |
| | 6000 to 6500 | 1/8 inch drill |
| | 5000 to 6000 | 3/16 inch drill |
| | 4500 to 5000 | 1/4 inch drill |
| | 3500 to 4000 | 5/16 inch drill |
| | 3000 to 3500 | 3/8 inch drill |
| | 1500 to 2500 | > 7/16 inch drill |
| Steel | 5000 to 6500 | 1/16 inch drill |
| | 3000 to 4000 | 1/8 inch drill |
| | 2000 to 2500 | 3/16 inch drill |
| | 1500 to 2000 | 1/4 inch drill |
| | 1000 to 1500 | 5/16 inch drill |
| | 1000 to 1500 | 3/8 inch drill |
| | 500 to 1000 | > 7/16 inch drill |
| Wood | 4000 to 6000 | Carving and routing |
| | 3800 to 4000 | All woods, 0 to 1/4 inch drills |
| | 3100 to 3800 | All woods, 1/4 to 1/2 inch drills |
| | 2300 to 3100 | All woods, 1/2 to 3/4 inch drills |
| | 2000 to 2300 | All woods, 3/4 to 1 inch drills |
| | 700 to 2000 | All woods, > 1 inch drills, fly cutters, |
| | < 700 | and multi–spur bits |

If in doubt about what speed to use, always select the slower speeds. Speeds for drill sizes not listed above can be estimated by looking at speeds for sizes one step over and one step under.

# FIRE EXTINGUISHERS

Fire extinguishers are an absolute must in any shop, garage, home, automobile, or business. Fire extinguishers are classified by the types of fires they will put out and the size of the fire they will put out. The basic types are as follows:

**TYPE A:** For wood, cloth, paper, trash and other common materials. These fires are put out by "heat absorbing" water or water based materials or smothered by dry chemicals.

**TYPE B:** For oil, gasoline, grease, paints & other flammable liquids. These fires are put out by smothering, preventing the release of combustible vapors, or stopping the combustion chain. Use Halon, dry chemicals, carbon dioxide, or foam.

**TYPE C:** For "live" electrical equipment. These fires are put out by the same process as TYPE B, but the extinguishing material must be electrically non-conductive. Use halon, dry chemicals, or carbon dioxide.

**TYPE D:** For combustible metals such as magnesium. These fires must be put out by heat absorption and smothering. Obtain specific information on these requirements from the fire department.

Combinations of the above letters indicate the extinguisher will put out more than one type of fire. For example, a Type ABC unit will put out all three types of fires. The "size" of the fire an extinguisher will put out is shown by a number in front of the Type, such as "10B". The base line numbers are as follows:

# FIRE EXTINGUISHERS

**Class "1A":** Will put out a stack of 50 burning sticks that are 20 inches long each.

**Class "1B":** Will put out an area of burning naptha that is 2.5 square feet in size.

Any number other than the "1" simply indicates the extinguisher will put out a fire that many times larger, for example "10A" will put out a fire 10 times larger than "1A".

Some general recommendations when purchasing a fire extinguisher are as follows:

1: Buy TYPE ABC so that you never have to think about what type of fire you are using it on. Also, buy a halon or carbon dioxide extinguisher so you don't damage electronic equipment and there is much less mess. The relative prices of extinguishers are Foam (very expensive – used on big fires such as aircraft), Carbon Dioxide (also expensive but leaves no mess), Halon (medium range prices), and Dry Chemical (very inexpensive, but leaves a mess).

2. Buy units with metal components and a gauge and are approved by Underwriters Labs or other testing group. Plastic units are generally poorly constructed and break easily; buy good extinguishers, your life and property may depend on it !

3. Buy more than one extinguisher and mount them on the wall near escape routes so that children can reach them.

4. Study the instructions when you get the unit, there may not be time after a fire has started.

# SANDPAPER & ABRASIVES

| Grit Current System | Grit Old System | Word Description | Use |
|---|---|---|---|
| 12 | 4-1/2 | Very Coarse | Very rough work, usually requires high speed, heavy machines. For unplanned woods, wood floors, rough cut. |
| 16 | 4 | Very Coarse | |
| 20 | 3-1/2 | Very Coarse | |
| 24 | 3 | Very Coarse | |
| 30 | 2-1/2 | Coarse | |
| 36 | 2 | Coarse | |
| 40 | 1-1/2 | Coarse | Rough wood work, #1 is coarsest for use with pad sander. |
| 50 | 1 | Coarse | |
| 60 | 1/2 | Medium | |
| 80 | 1/0 | Medium | General wood work, plaster patches, 1st smooth of old paint. |
| 100 | 2/0 | Medium | |
| 120 | 3/0 | Fine | |
| 150 | 4/0 | Fine | Hardwood prep, final for softwoods, old paint. |
| 180 | 5/0 | Fine | |
| 220 | 6/0 | Very Fine | Final sanding or between coats, won't show sand marks, dry sanding. |
| 240 | 7/0 | Very Fine | |
| 280 | 8/0 | Very Fine | |
| 320 | 9/0 | Extra Fine | Polish final coats, between coats, wet sand paints & varnishes, top coats. |
| 360 | | Extra Fine | |
| 400 | 10/0 | Extra Fine | |
| 500 | | Super Fine | Sand metal, plastic, & ceramics, wet sanding. |
| 600 | | Super Fine | |

## COATING TYPES:

Open coat: Grains cover 50% to 70% of the backing, which leaves a lot of space between each grain. The open space is necessary in applications where the material being sanded has a tendency to clog up or "load" the abrasive surface. The clogging drastically reduces the cutting ability of the abrasive and reduces the life of the abrasive. Less common than Closed coat.

Closed coat: Grains cover all of the backing. This type is much more efficient (removes material faster) than Open coat since there are more abrasion grains per square inch. Closed coat is the preferred coating type as long as grain clogging is not a problem.

## GLUE TYPES

Glues are generally restricted to a combination of "Hide" glues and resin based glues, depending on the application. Glues are usually applied as a two part process, the base coat and the top grain holding coat.

## BACKING TYPES:

Paper: Weights range from "A" through "F". "A" is the most flexible and is used mainly for finishing jobs and with small grain sizes abrasives. "A" is primarily used for hand sanding. "C" and "D" weights are stronger, less flexible, and used for both hand sanding and power sanders. "E" is much stronger, very tear resistant and much less flexible than "C" and "D" and is used mainly in belt and disc applications. "F" is the strongest paper and is used mainly for rolls and belts.

Cloth: Weights are "J" ( jeans) and "X" ( drills). "J" is used for finishing and polishing operations, particularly where contours are involved. "X" is less flexible than "J" but is stronger. "X" is used for heavy belt, disk and drum grinding and polishing.

Fiber: Composed of multiple layers of paper that has been chemically treated. These backings are very tough, heat resistant and used mostly for drum and disc applications, particularly high speed.

Combination: Composed of both paper and cloth or cloth and fiber, producing a very strong, flexible, non–stretching, and tear resistant backing. They are used mostly in high speed, drum sanding applications

## ABRASIVE TYPES:

Silicon Carbide: The hardest and sharpest of all abrasives but is more brittle than aluminum oxide. Color is blue–black and it is a manufactured abrasive. Typically used in applications such as finishing of soft metals, glass, ceramics, hard wood floors and plastics. It is very fast cutting and is therefore good for both material removal as well as polishing. This abrasive is very popular for sanding lacquered and enameled surfaces such as car paints. Durite by Norton is a common brand name. Grain sizes normally range from 600 to 12.

Aluminum Oxide: Extremely tough, grit is very sharp and much harder than flint, garnet and emery. Color is red to brown and it is a manufactured abrasive. More expensive than most other types but its toughness results in a longer lasting abrasive so its cost is actually equivalent. Recommended for metals and hard woods and is the preferred choice for power sanding. Norton brand names – Adalox or Metalite. Grain sizes normally range from 500 to 16.

Garnet: Much softer than the synthetic abrasives listed above but harder and sharper than flint. Garnet is a crushed natural mineral, red to brown in color. Used mainly in furniture finishing and woodworking. Yields an excellent wood finish. Grain sizes normally range from 280 to 20.

Flint: Generally poor cutting strength and durability so it is not used in production environments. Used in the leather industry and as a good non–clogging abrasive in paint removal and some woodwork. Flint is gray to white colored natural quartz mineral. Flint is non–conductive and therefore is also useful as an abrasive in the electronics industry. Inexpensive. Grain sizes normally range from 4/0 to 3.

Emery: Good polishing features but poor for material removal. Grains are round and black in color. Poor penetration but good for polishing metals. Poor for wood use. Grain sizes normally range from 3/0 to 3.

Crocus: Soft and short lived. Made of ferrous oxide (red color). Good for polishing, particularly soft metals like gold.

Pumice: Powdered volcanic glass that is commonly used to tone down a glossy finish to a satin, smooth surface. Grades range from 4–F ( the finest) through #7 ( the coarsest). Frequent inspections of the work surface should be made in order to prevent breaking through the surface.

Cork: Cork is sometimes used as a wet polishing media.

Rottenstone: Also referred to as diatomaceous earth. It is much softer and finer grained than pumice and is used in combination with water, solvents, or oils to produce a satin finish on woods.

Rubbing Compounds: Sometimes also referred to as "rouge" is normally used as a polish for enamel and lacquer paints. It is not for use on bare woods.

Steel Wool: Although not technically an abrasive, it is commonly used to remove rust, old finishes, and to smooth rough surfaces. Sizes range from 3/0 ( the finest) through #5 (the coarsest). With steel wool, it is important to rub the wood in the direction of the grain, not across it, so that the surface is not scratched.

Some of the above abrasive data is courtesy of *Norton, Coated Abrasives Division, Troy, New York.*

## Chain Saw Classification

Chain saws can be broadly grouped into the following four categories, based on ruggedness and size:

1. Mini–saw: Light weight (6 to 9 pounds), small engine size (1.8 to 2.5 cubic inches or electric ), and short bar lengths (8 to 12 inches, 1/4 inch pitch). Good for <3 cords per year.

2. Light–Duty: Light weight (9 to 13 pounds), small engine size (2.5 to 3.8 cubic inches or electric), and medium bar lengths (14 to 16 inches, 3/8 inch pitch). Good for 3 to 6 cords per year.

3. Medium–Duty: Medium weight (13 to 18 pounds), medium engine size (3.5 to 4.8 cubic inches), and medium–long bar lengths (16 to 24 inches, 3/8 to 0.404 inch pitch). Good for 6 to 10 cords per year. If money permits, this class of saw is probably the best choice for the average wood cutter, even if he does not cut the 6 to 10 cords per year. It is heavy duty enough to last a long time under light use.

4. Heavy–Duty: Heavy weight (over 18 pounds), large engine size (over 4.8 cubic inches), and long bar lengths (over 24 inches, 0.404 to 1/2 inch pitch). These heavy duty units are generally for the professional. They are heavy, expensive, and require more strength to use. They can be used continuously.

## Depth Capacity of Std Circular Power Handsaws

| Blade Diameter | Capacity @ 90° | Capacity @ 45° |
|---|---|---|
| 4-1/2 | 1-5/16 | 1-1/16 to 1-1/14 |
| 6-1/2 | 2-1/16 | 1-5/8 |
| 6-3/4 | 2-7/32 | 1-3/4 |
| 7-1/4 | 2-3/8 to 2-7/16 | 1-7/8 to 1-29/32 |
| 7-1/2 | 2-17/22 | 2-1/16 |
| 8-1/4 | 2-15/16 | 2-1/4 |
| 10-1/4 | 3-5/8 | 2-3/4 |
| 12 | 4-3/8 | 3-5/16 |

## Cut–off Wheel

Abrasive blade made of aluminum oxide (metal cutting) or silicon carbide (masonry cutting). Comes in standard sizes of 6, 7, and 8 inch diameter and with 1/2 or 5/8 diameter arbor. No teeth.

## Hollow–Ground

Quiet, accurate and leaves a very smooth finish. Designed especially for crosscuts and miters across wood grain. Acceptable for ripping, but it's not as fast as ripping blades. Hollow–Ground blades minimize chipping.

## Ripping Blade

These blades have large, set teeth with deep gullets. Designed especially for cutting fast in the direction of the wood grain. Minimum binding of blade. Very rough finish.

## Chisel–Tooth

General purpose, set–tooth blade. Good for both ripping and crosscuts and it cuts fast but leaves a rough cut. This is the most common blade used by contractors. Bevel–ground, carbide–tipped blades of the same basic design are among the most durable of the blades.

## Crosscut, Fine–Tooth, & Paneling Blades

All of these blades have a large number of small teeth that are very sharp. Crosscut has the least teeth, Fine–Tooth has more and Paneling has the most. Crosscut, as the name implies, was designed for cutting across the wood grain and leaving a smooth edge. It is also good for plywood. Fine–Tooth blades are also good for plywood, but are also used on fiber boards (Celotex), veneers, and thin plastics. Paneling blades have many, extra fine teeth. It is particularly useful in cutting paneling and laminates since the cut edge usually does not have to be touched up.

## Flattop–Ground Carbide Tipped

A fast cutting, long lasting blade used as a combo blade in construction. Good for ripping, crosscutting and mitering but does not leave a good smooth edge.

## Steel Cutting Blade

A unusually designed blade that is used to cut ferrous (iron and steel) sheetmetal that is up to 3/32 inch thick. This blade actually "burns" its way through the metal, leaving a clean edge.

# SABER SAW BLADES

| Teeth per inch | Blade Usage |
|---|---|
| 3 | Lumber up to 6 inches thick, fast cutting, very rough cut, good ripping blade. |
| 5 or 6 | Lumber up to 2 inches thick, fast cutting, rough cut, good ripping blade. |
| 7 or 8 | Best general purpose blade, relatively smooth cut. Good for lumber and fiber insulation board. |
| 10 | Good general purpose blade, smoother cut than the 7 or 8 blade. Use 10 through 14 for cutting hardwoods under 1/2 inch thick and for plastics, composition board, drywall, and plywood when a smooth edge is needed. If hard, abrasive materials are to be cut, such as laminates, use the metal cutting H.S.S. types. Good for some scrollwork. |
| 12 or 14 | Very smooth cutting blade but is also very slow cutting. Good for hardwoods, plywood, fiberglass, plastics, rubber, linoleum, laminates, and plexiglass. As with the 10 tpi blades, if the material is particularly hard or abrasive, use the metal cutting H.S.S. types instead. |
| Knife | These blades have either a knife edge or a sharp edge with an abrasive grit bonded to the blade. No grit blades are useful for cutting rubber, cork, leather, cardboard, styrofoam, & silicones. Grit blades come in fine, medium and coarse and can be used on fiberglass, epoxies, ceramic tile, stone, clay pipe, brick, steel, & veneer |
| | **H.S.S. METAL CUTTING BLADES** |
| 6 to 10 | Rough cutting for aluminum, brass, copper, laminates, hardwoods and other soft materials. Good up to 1/2 inch thickness. |
| 14 | Cuts same materials as 6 to 10 tpi plus mild steels and hardboards. Leaves a much smoother edge. Thickness should be 1/4 to 1/2 inch maximum. |
| 18 | Same as 14 tpi but maximum thickness 1/8 inch. |
| 24 | Smooth edge cutting for steel and sheet metal. Also good for other hard materials such as plastics, tile, and Bakelite. Maximum thickness should be 1/8 inch. |
| 32 | Very fine cuts for steel and thin wall tubing up to 1/16 inch thick. |

POCKET REF

# Water



(See also AIR chapter, p. 14 for more pollution data)

# FRICTION LOSS IN STEEL PIPE

| Flow GPM | Feet of Head Loss per 100 Feet @ Diameter (inches) 1/2 | 3/4 | 1 | 1–1/4 | 1–1/2 | 2 | 2–1/2 |
|---|---|---|---|---|---|---|---|
| 0.5 | 0.6 | ... | ... | ... | ... | ... | ... |
| 1 | 2.1 | ... | ... | ... | ... | ... | ... |
| 2 | 7.6 | 1.9 | 0.6 | ... | ... | ... | ... |
| 3 | 16.0 | 4.1 | 1.3 | ... | ... | ... | ... |
| 4 | 27.3 | 6.9 | 2.1 | 0.6 | ... | ... | ... |
| 5 | 41.2 | 10.5 | 3.4 | 0.9 | ... | ... | ... |
| 10 | 149 | 37.8 | 11.7 | 3.1 | 1.4 | 0.5 | ... |
| 15 | ... | 80.0 | 24.8 | 6.5 | 2.9 | 0.9 | ... |
| 20 | ... | 136 | 42.1 | 11.1 | 5.2 | 1.5 | 0.7 |
| 30 | ... | ... | 89.2 | 23.5 | 11.1 | 3.3 | 1.4 |
| 40 | ... | ... | 152 | 40.0 | 18.9 | 5.6 | 2.4 |
| 50 | ... | ... | ... | 60.4 | 28.5 | 8.5 | 3.6 |
| 60 | ... | ... | ... | 87.4 | 40.0 | 11.9 | 5.0 |
| 70 | ... | ... | ... | 144 | 53.2 | 15.8 | 6.6 |
| 80 | ... | ... | ... | ... | 68.1 | 20.2 | 8.5 |
| 90 | ... | ... | ... | ... | 84.7 | 25.1 | 10.6 |
| 100 | ... | ... | ... | ... | 103 | 30.5 | 12.8 |
| 150 | ... | ... | ... | ... | ... | 64.7 | 27.3 |
| 200 | ... | ... | ... | ... | ... | 110 | 46.3 |
| 250 | ... | ... | ... | ... | ... | ... | 81.7 |
| 300 | ... | ... | ... | ... | ... | ... | 98.1 |

| Flow GPM | Feet of Head Loss per 100 Feet @ Diameter (inches) 3 | 4 | 5 | 6 | 8 | 10 | 12 |
|---|---|---|---|---|---|---|---|
| 50 | 2.3 | 0.3 | ... | ... | ... | ... | ... |
| 75 | 3.3 | 0.7 | 1.2 | ... | ... | ... | ... |
| 100 | 4.5 | 2.6 | 1.6 | 0.2 | ... | ... | ... |
| 150 | 6.8 | 3.8 | 2.4 | 0.4 | ... | ... | ... |
| 200 | 9.1 | 5.1 | 3.3 | 2.3 | 0.1 | ... | ... |
| 250 | 11.3 | 6.4 | 4.1 | 2.8 | 0.2 | ... | ... |
| 300 | 13.6 | 7.7 | 4.9 | 3.4 | 0.3 | 0.1 | ... |
| 400 | 18.2 | 10.2 | 6.5 | 4.5 | 0.5 | 0.2 | 1.1 |
| 500 | 22.7 | 12.8 | 8.2 | 5.7 | 0.8 | 0.3 | 1.4 |
| 600 | ... | 15.3 | 9.8 | 6.8 | 1.2 | 0.4 | 1.7 |
| 700 | ... | 17.9 | 11.4 | 7.9 | 1.5 | 0.5 | 2.0 |
| 800 | ... | 20.4 | 13.1 | 9.1 | 2.0 | 0.7 | 2.3 |
| 900 | ... | 23.0 | 14.7 | 10.2 | 2.5 | 0.8 | 2.5 |
| 1000 | ... | ... | 16.3 | 11.4 | 3.0 | 1.0 | 2.8 |
| 1200 | ... | ... | 19.6 | 13.6 | 4.2 | 1.4 | 3.5 |
| 1500 | ... | ... | ... | 17.0 | 6.3 | 2.1 | 4.3 |
| 2000 | ... | ... | ... | ... | 10.8 | 3.6 | 5.2 |
| 3000 | ... | ... | ... | ... | ... | 7.7 | 8.5 |
| 4000 | ... | ... | ... | ... | ... | 13.1 | 11.4 |
| 5000 | ... | ... | ... | ... | ... | ... | 14.2 |

# FRICTION LOSS IN PLASTIC PIPE

| Flow | Feet of Head Loss per 100 Feet @ Diameter (inches) | | | | | | |
|---|---|---|---|---|---|---|---|
| GPM | 1/2 | 3/4 | 1 | 1-1/4 | 1-1/2 | 2 | 2-1/2 |
| 0.5 | 0.3 | ... | ... | ... | ... | ... | ... |
| 1 | 1.1 | ... | ... | ... | ... | ... | ... |
| 2 | 4.1 | 1.0 | 0.3 | ... | ... | ... | ... |
| 3 | 8.6 | 2.2 | 0.7 | ... | ... | ... | ... |
| 4 | 14.8 | 3.7 | 1.1 | 0.3 | ... | ... | ... |
| 5 | 22.2 | 5.7 | 1.7 | 0.5 | ... | ... | ... |
| 10 | 80.5 | 20.4 | 6.3 | 1.7 | 0.8 | 0.2 | ... |
| 15 | ... | 43.3 | 13.4 | 3.5 | 1.6 | 0.5 | ... |
| 20 | ... | 73.5 | 22.8 | 6.0 | 2.8 | 0.8 | 0.3 |
| 30 | ... | ... | 48.1 | 12.7 | 6.0 | 1.8 | 0.7 |
| 40 | ... | ... | 82.0 | 21.6 | 10.2 | 3.0 | 1.3 |
| 50 | ... | ... | ... | 32.6 | 15.4 | 4.6 | 1.9 |
| 60 | ... | ... | ... | 45.6 | 21.6 | 6.4 | 2.7 |
| 70 | ... | ... | ... | ... | 28.7 | 8.5 | 3.6 |
| 80 | ... | ... | ... | ... | 36.8 | 10.9 | 4.6 |
| 90 | ... | ... | ... | ... | 45.7 | 13.6 | 5.7 |
| 100 | ... | ... | ... | ... | 56.6 | 16.5 | 6.9 |
| 150 | ... | ... | ... | ... | ... | 35.0 | 14.7 |
| 200 | ... | ... | ... | ... | ... | 59.4 | 25.0 |
| 250 | ... | ... | ... | ... | ... | ... | 44.1 |

# FRICTION LOSS IN COPPER PIPE

| Flow | Feet of Head Loss per 100 Feet @ Diameter (inches) | | | | | | |
|---|---|---|---|---|---|---|---|
| GPM | 1/2 | 3/4 | 1 | 1-1/4 | 1-1/2 | 2 | 2-1/2 |
| 0.5 | 0.3 | ... | ... | ... | ... | ... | ... |
| 1 | 1.3 | ... | ... | ... | ... | ... | ... |
| 2 | 4.6 | 1.2 | 0.3 | ... | ... | ... | ... |
| 3 | 9.7 | 2.6 | 0.7 | ... | ... | ... | ... |
| 4 | 16.4 | 4.4 | 1.2 | 0.4 | ... | ... | ... |
| 5 | 24.8 | 6.6 | 1.9 | 0.6 | ... | ... | ... |
| 10 | 89.4 | 23.7 | 6.8 | 2.0 | 0.9 | 0.3 | ... |
| 15 | ... | 49.1 | 14.4 | 4.2 | 1.9 | 0.6 | ... |
| 20 | ... | 83.5 | 24.4 | 7.1 | 3.3 | 1.0 | 0.4 |
| 30 | ... | ... | 51.6 | 15.0 | 7.0 | 2.0 | 0.8 |
| 40 | ... | ... | 88.0 | 25.6 | 12.0 | 3.5 | 1.3 |
| 50 | ... | ... | ... | 38.7 | 14.9 | 5.2 | 2.0 |
| 60 | ... | ... | ... | 54.1 | 25.3 | 7.3 | 2.9 |
| 70 | ... | ... | ... | 73.2 | 33.8 | 9.8 | 3.8 |
| 80 | ... | ... | ... | 92.4 | 43.1 | 12.5 | 4.9 |
| 90 | ... | ... | ... | ... | 53.6 | 15.6 | 6.1 |
| 100 | ... | ... | ... | ... | 65.1 | 18.9 | 7.4 |
| 150 | ... | ... | ... | ... | ... | 40.1 | 15.6 |
| 200 | ... | ... | ... | ... | ... | 68.0 | 47.3 |
| 250 | ... | ... | ... | ... | ... | ... | 56.8 |

# FRICTION LOSS IN PIPE FITTINGS

| Steel/Copper Fitting | Feet of Head Loss per Joint @ Diameter (inches) 1/2 | 3/4 | 1 | 1–1/4 | 1–1/2 | 2 | 2–1/2 |
|---|---|---|---|---|---|---|---|
| 90° Std Elbow | 1.6 | 2.1 | 2.6 | 3.5 | 4.0 | 5.5 | 6.2 |
| 90° Long Elbow | 1 | 1.4 | 1.7 | 2.3 | 2.7 | 4.3 | 5.1 |
| 90° Street Elbow | 3 | 3.4 | 4.4 | 5.8 | 6.7 | 8.6 | 10.3 |
| 45° Std Elbow | 0.8 | 1.1 | 1.4 | 1.8 | 2.1 | 2.8 | 3.3 |
| 45° Street Elbow | 1 | 1.8 | 2.3 | 3.0 | 3.5 | 4.5 | 5.4 |
| Square Elbow | 3 | 3.9 | 5.0 | 6.5 | 7.6 | 9.8 | 11.7 |
| Std T Flow Run | 1 | 1.4 | 1.7 | 2.3 | 2.7 | 4.3 | 5.1 |
| Std T Flow Branch | 4 | 5.1 | 6.0 | 6.9 | 8.1 | 12.0 | 14.3 |
| Gate Valve–open | .7 | 0.9 | 1.1 | 1.5 | 1.7 | 2.2 | 2.7 |

# FRICTION LOSS IN PIPE FITTINGS

| Plastic Fitting | Feet of Head Loss per Joint @ Diameter (inches) 1/2 | 3/4 | 1 | 1–1/4 | 1–1/2 | 2 | 2–1/2 |
|---|---|---|---|---|---|---|---|
| 90° Std Elbow | 4 | 5 | 6 | 7 | 8 | 9 | 10 |
| Std T Flow Run | 4 | 4 | 4 | 5 | 6 | 7 | 8 |
| Std T Flow Branch | 7 | 8 | 9 | 12 | 13 | 17 | 20 |

# SUCTION, HEAD, VAPOR PRESSURE

| Water Temp °F | Vapor Pressure | Suction Lift or Head @ Altitude in Feet 0 | 2000 | 4000 | 8000 | 12000 |
|---|---|---|---|---|---|---|
| 60 | 0.6 ft water | 20 | 17.5 | 15.5 | 11.5 | 7.5 |
| 70 | 0.8 ft water | 19.5 | 17 | 15 | 11 | 7 |
| 80 | 1.2 ft water | 19.5 | 17 | 15 | 11 | 7 |
| 90 | 1.6 ft water | 19 | 16.5 | 14.5 | 10.5 | 6.5 |
| 100 | 2.2 ft water | 18.5 | 16 | 14 | 10 | 6 |
| 110 | 2.9 ft water | 17.5 | 15 | 13 | 9 | 5 |
| 120 | 3.9 ft water | 16.5 | 14 | 12 | 8 | 4 |
| 130 | 5.1 ft water | 15.5 | 13 | 11 | 7 | 3 |
| 140 | 6.7 ft water | 14 | 111.5 | 9.5 | 5.5 | 1.5 |
| 150 | 8.6 ft water | 12 | 9.5 | 7.5 | 3.5 | –0.5 |
| 160 | 10.9 ft water | 9.5 | 7 | 5 | 1 | ... |
| 170 | 13.8 ft water | 6.5 | 4 | 2 | –2 | ... |
| 180 | 17.3 ft water | 3 | 0.5 | –1.5 | ... | ... |
| 190 | 21.6 ft water | –1 | –3.5 | –5.5 | ... | ... |
| 200 | 26.6 ft water | –6 | –8.5 | ... | ... | ... |
| 210 | 32.6 ft water | –12 | ... | ... | ... | ... |
| 212 | 34.0 ft water | –13.5 | ... | ... | ... | ... |

+ values indicate suction lift, – values indicate suction head
See also Air and Water table on page 9.

# HORIZONTAL PIPE DISCHARGE

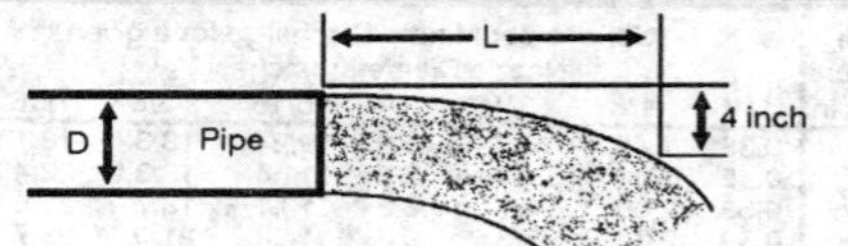


| L Distance inches | Gallons per Minute Discharge for a given Nominal Pipe Diameter D (Inches) | | | | | | |
|---|---|---|---|---|---|---|---|
| | 1 | 1–1/4 | 1–1/2 | 2 | 2–1/2 | 3 | 4 |
| 4 | 6 | 10 | 13 | 22 | 31 | 48 | 83 |
| 5 | 7 | 12 | 17 | 27 | 39 | 61 | 104 |
| 6 | 8 | 15 | 20 | 33 | 47 | 73 | 125 |
| 7 | 10 | 17 | 23 | 38 | 55 | 85 | 146 |
| 8 | 11 | 20 | 26 | 44 | 62 | 97 | 166 |
| 9 | 13 | 22 | 30 | 49 | 70 | 110 | 187 |
| 10 | 14 | 24 | 33 | 55 | 78 | 122 | 208 |
| 11 | 16 | 27 | 36 | 60 | 86 | 134 | 229 |
| 12 | 17 | 29 | 40 | 66 | 94 | 146 | 250 |
| 13 | 18 | 31 | 43 | 71 | 102 | 158 | 270 |
| 14 | 20 | 34 | 46 | 77 | 109 | 170 | 292 |
| 15 | 21 | 36 | 50 | 82 | 117 | 183 | 312 |
| 16 | 23 | 39 | 53 | 88 | 125 | 196 | 334 |
| 17 | ... | 41 | 56 | 93 | 133 | 207 | 355 |
| 18 | ... | ... | 60 | 99 | 144 | 220 | 375 |
| 19 | ... | ... | ... | 110 | 148 | 232 | 395 |
| 20 | ... | ... | ... | ... | 156 | 244 | 415 |
| 21 | ... | ... | ... | ... | ... | 256 | 435 |
| 22 | ... | ... | ... | ... | ... | ... | 460 |

| L Distance inches | Gallons per Minute Discharge for a given Nominal Pipe Diameter D (Inches) | | | | |
|---|---|---|---|---|---|
| | 5 | 6 | 8 | 10 | 12 |
| 5 | 163 | ... | ... | ... | ... |
| 6 | 195 | 285 | ... | ... | ... |
| 7 | 228 | 334 | 580 | ... | ... |
| 8 | 260 | 380 | 665 | 1060 | ... |
| 9 | 293 | 430 | 750 | 1190 | 1660 |
| 10 | 326 | 476 | 830 | 1330 | 1850 |
| 11 | 360 | 525 | 915 | 1460 | 2020 |
| 12 | 390 | 570 | 1000 | 1600 | 2220 |
| 13 | 425 | 620 | 1080 | 1730 | 2400 |
| 14 | 456 | 670 | 1160 | 1860 | 2590 |
| 15 | 490 | 710 | 1250 | 2000 | 2780 |
| 16 | 520 | 760 | 1330 | 2120 | 2960 |
| 17 | 550 | 810 | 1410 | 2260 | 3140 |
| 18 | 590 | 860 | 1500 | 2390 | 3330 |
| 19 | 620 | 910 | 1580 | 2520 | 3500 |
| 20 | 650 | 950 | 1660 | 2660 | 3700 |
| 21 | 685 | 1000 | 1750 | 2800 | 3890 |
| 22 | 720 | 1050 | 1830 | 2920 | 4060 |
| 23 | 750 | 1100 | 1910 | 3060 | 4250 |
| 24 | ... | 1140 | 2000 | 3200 | 4440 |

# NOZZLE DISCHARGE

| Nozzle Pressure lbs/sq in | Gallons per Minute Discharge for a given Nozzle Diameter (Inches) | | | | | | |
|---|---|---|---|---|---|---|---|
| | 1/16 | 1/8 | 3/16 | 1/4 | 5/16 | 3/8 | 7/16 |
| 10 | 0.38 | 1.48 | 3.3 | 5.9 | 9.24 | 13.3 | 18.1 |
| 15 | 0.45 | 1.81 | 4.1 | 7.2 | 11.4 | 16.3 | 22.4 |
| 20 | 0.53 | 2.09 | 4.7 | 8.3 | 13.1 | 18.7 | 25.6 |
| 25 | 0.59 | 2.34 | 5.3 | 9.3 | 14.6 | 21.0 | 28.7 |
| 30 | 0.64 | 2.56 | 5.8 | 10.2 | 16.0 | 23.1 | 31.4 |
| 35 | 0.69 | 2.78 | 6.2 | 11.1 | 17.1 | 25.0 | 33.8 |
| 40 | 0.74 | 2.96 | 6.7 | 11.7 | 18.4 | 26.6 | 36.2 |
| 45 | 0.79 | 3.14 | 7.1 | 12.6 | 19.5 | 28.2 | 38.3 |
| 50 | 0.83 | 3.30 | 7.4 | 13.2 | 20.6 | 29.9 | 40.5 |
| 60 | 0.90 | 3.62 | 8.2 | 14.5 | 22.6 | 32.6 | 44.3 |
| 70 | 0.98 | 3.91 | 8.8 | 15.7 | 24.4 | 35.3 | 47.9 |
| 80 | 1.05 | 4.19 | 9.4 | 16.8 | 26.1 | 37.6 | 51.2 |
| 90 | 1.11 | 4.43 | 10.0 | 17.7 | 27.8 | 40.1 | 54.5 |
| 100 | 1.17 | 4.67 | 10.4 | 18.7 | 29.2 | 42.2 | 57.3 |
| 120 | 1.23 | 5.17 | 11.5 | 20.4 | 31.8 | 46.0 | 62.4 |
| 140 | 1.28 | 5.70 | 12.4 | 22.1 | 34.4 | 49.8 | 67.6 |
| 160 | 1.32 | 6.30 | 13.3 | 23.6 | 36.9 | 53.3 | 72.3 |
| 180 | 1.36 | 6.92 | 14.1 | 25.0 | 39.0 | 56.4 | 76.5 |
| 200 | 1.38 | 7.52 | 14.9 | 26.4 | 41.1 | 59.5 | 81.6 |

| Nozzle Pressure lbs/sq in | Gallons per Minute Discharge for a given Nozzle Diameter (Inches) | | | | | | |
|---|---|---|---|---|---|---|---|
| | 1/2 | 9/16 | 5/8 | 3/4 | 7/8 | 1 | 1–1/8 |
| 10 | 23.6 | 30.2 | 36.9 | 53.3 | 72.5 | 94.8 | 120 |
| 15 | 28.9 | 36.7 | 45.2 | 65.1 | 88.7 | 116 | 147 |
| 20 | 33.4 | 42.4 | 52.2 | 75.4 | 102 | 134 | 169 |
| 25 | 37.3 | 47.3 | 58.2 | 84.0 | 115 | 149 | 189 |
| 30 | 40.9 | 51.9 | 63.9 | 92.2 | 126 | 164 | 208 |
| 35 | 44.2 | 56.1 | 69.0 | 99.8 | 136 | 177 | 224 |
| 40 | 47.3 | 59.9 | 73.8 | 106 | 145 | 189 | 239 |
| 45 | 50.1 | 63.4 | 78.2 | 113 | 153 | 200 | 254 |
| 50 | 52.8 | 67.0 | 82.5 | 119 | 162 | 211 | 268 |
| 60 | 57.9 | 73.3 | 90.4 | 130 | 177 | 232 | 293 |
| 70 | 62.6 | 79.3 | 97.8 | 141 | 192 | 251 | 317 |
| 80 | 66.8 | 84.8 | 105 | 151 | 205 | 268 | 339 |
| 90 | 70.8 | 90.3 | 111 | 160 | 218 | 285 | 360 |
| 100 | 74.9 | 95.0 | 117 | 169 | 229 | 300 | 379 |
| 120 | 81.8 | 103 | 128 | 184 | 250 | 327 | 413 |
| 140 | 88.3 | 112 | 138 | 199 | 271 | 354 | 447 |
| 160 | 94.6 | 120 | 148 | 213 | 289 | 378 | 478 |
| 180 | 100 | 127 | 156 | 225 | 306 | 400 | 506 |
| 200 | 106 | 134 | 165 | 238 | 323 | 423 | 535 |

NOTE: The above discharge rates are theoretical. Actual values will only be 95% of the above values, depending on such factors as shape of the nozzle, bore smoothness, etc.

# VERTICAL PIPE DISCHARGE

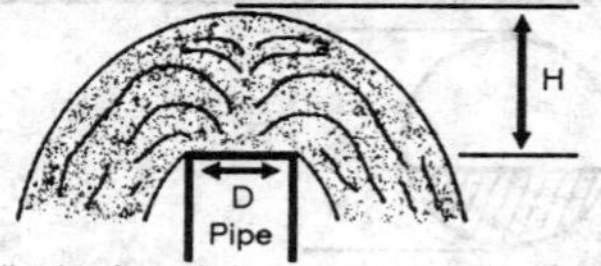


The following formula is an approximation of the output of a vertical pipe.

$$GPM = \sqrt{H} \times K \times D^2 \times 5.68$$

GPM = gallons per minute
H = height in inches
D = diameter of pipe in inches
K = constant from 0.87 to 0.97 for diameters of 2 to 6 inches and heights (H) up to 24 inches.

Example: 6 inch diameter with 10 inch height = ±626 gpm

# WEIR DISCHARGE VOLUMES

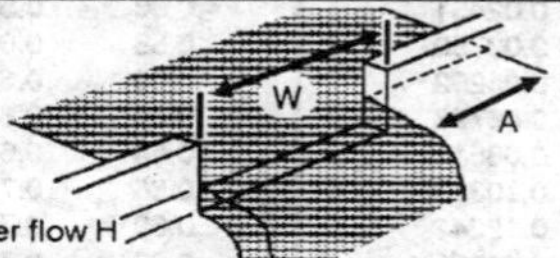


Height of water flow H

| Head Inches | GPM for Width of Weir in Feet 1 | 3 | 5 | gpm/foot over 5 feet wide |
|---|---|---|---|---|
| 1 | 35 | 107 | 180 | 36 |
| 1.5 | 65 | 197 | 330 | 66 |
| 2 | 98 | 302 | 506 | 102 |
| 2.5 | 136 | 422 | 706 | 143 |
| 3 | 178 | 552 | 926 | 187 |
| 4 | 269 | 846 | 1424 | 288 |
| 5 | 370 | 1175 | 1985 | 405 |
| 6 | 477 | 1535 | 2600 | 528 |
| 7 | ... | 1928 | 3260 | 668 |
| 8 | ... | 2338 | 3956 | 814 |
| 9 | ... | 2765 | 4699 | 970 |
| 10 | ... | 3216 | 5490 | 1136 |
| 12 | ... | 4185 | 7165 | 1495 |

Based on the Francis formula:

$$\text{Cu ft/sec water} = 3.33\,(W - 0.2\,H)\,H^{1.5}$$

(distance "A" should be at least 3 H)

# HORIZONTAL CYLINDER FILLAGE

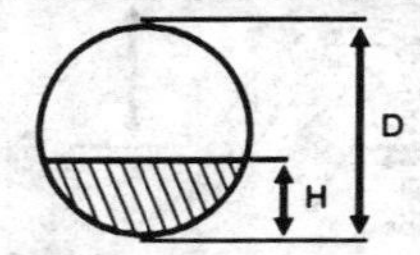


The following equation can be used to calculate the number of gallons remaining in a horizontal tank if the height of the liquid remaining in the tank and the diameter of the tank are known.

*Gallons Remaining = Depth Factor × Total Tank Gallons*

Use the formula $Ratio = \frac{H}{D}$ and then the following table in order to calculate the Depth Factor.

| Ratio | Depth Factor | Ratio | Depth Factor |
|---|---|---|---|
| 0.02 | 0.00480 | 0.52 | 0.52544 |
| 0.04 | 0.01348 | 0.54 | 0.55087 |
| 0.06 | 0.02451 | 0.56 | 0.57625 |
| 0.08 | 0.03750 | 0.58 | 0.60147 |
| 0.10 | 0.05202 | 0.60 | 0.62646 |
| 0.12 | 0.06798 | 0.62 | 0.65133 |
| 0.14 | 0.08511 | 0.64 | 0.67594 |
| 0.16 | 0.10323 | 0.66 | 0.70024 |
| 0.18 | 0.12242 | 0.68 | 0.72427 |
| 0.20 | 0.14235 | 0.70 | 0.74770 |
| 0.22 | 0.16308 | 0.72 | 0.77080 |
| 0.24 | 0.18447 | 0.74 | 0.79350 |
| 0.26 | 0.20650 | 0.76 | 0.81533 |
| 0.28 | 0.22919 | 0.78 | 0.83692 |
| 0.30 | 0.25230 | 0.80 | 0.85765 |
| 0.32 | 0.27573 | 0.82 | 0.87758 |
| 0.34 | 0.29976 | 0.84 | 0.89677 |
| 0.36 | 0.32406 | 0.86 | 0.91489 |
| 0.38 | 0.34867 | 0.88 | 0.93202 |
| 0.40 | 0.37354 | 0.90 | 0.94797 |
| 0.42 | 0.39852 | 0.92 | 0.96250 |
| 0.44 | 0.42375 | 0.94 | 0.97550 |
| 0.46 | 0.44913 | 0.96 | 0.98652 |
| 0.48 | 0.46456 | 0.98 | 0.99520 |
| 0.50 | 0.50000 | | |

## STEAM TABLE

| Gauge psi | Temp °F | Gauge psi | Temp °F | Gauge psi | Temp °F | Gauge psi | Temp °F |
|---|---|---|---|---|---|---|---|
| 5 | 228 | 55 | 302 | 110 | 344 | 210 | 391 |
| 6 | 230 | 56 | 303 | 112 | 345 | 212 | 392 |
| 7 | 233 | 57 | 304 | 114 | 346 | 214 | 393 |
| 8 | 235 | 58 | 305 | 116 | 348 | 216 | 394 |
| 9 | 237 | 59 | 306 | 118 | 349 | 218 | 394 |
| 10 | 240 | 60 | 307 | 120 | 350 | 220 | 395 |
| 11 | 242 | 61 | 308 | 122 | 351 | 222 | 396 |
| 12 | 244 | 62 | 309 | 124 | 352 | 224 | 397 |
| 13 | 246 | 63 | 310 | 126 | 353 | 226 | 397 |
| 14 | 248 | 64 | 311 | 128 | 354 | 228 | 398 |
| 15 | 250 | 65 | 312 | 130 | 355 | 230 | 399 |
| 16 | 252 | 66 | 312 | 132 | 356 | 232 | 400 |
| 17 | 254 | 67 | 313 | 134 | 357 | 234 | 400 |
| 18 | 255 | 68 | 314 | 136 | 359 | 235 | 401 |
| 19 | 257 | 69 | 315 | 138 | 360 | 237 | 401 |
| 20 | 259 | 70 | 316 | 140 | 361 | 239 | 402 |
| 21 | 261 | 71 | 317 | 142 | 362 | 241 | 403 |
| 22 | 262 | 72 | 317 | 144 | 363 | 243 | 403 |
| 23 | 264 | 73 | 318 | 146 | 364 | 245 | 404 |
| 24 | 265 | 74 | 319 | 148 | 365 | 247 | 405 |
| 25 | 267 | 75 | 320 | 150 | 366 | 249 | 405 |
| 26 | 268 | 76 | 321 | 152 | 367 | 251 | 406 |
| 27 | 270 | 77 | 321 | 154 | 368 | 253 | 407 |
| 28 | 271 | 78 | 322 | 156 | 368 | 255 | 407 |
| 29 | 273 | 79 | 323 | 158 | 369 | 257 | 408 |
| 30 | 274 | 80 | 324 | 160 | 370 | 259 | 409 |
| 31 | 275 | 81 | 324 | 162 | 371 | 261 | 409 |
| 32 | 277 | 82 | 325 | 164 | 372 | 263 | 410 |
| 33 | 278 | 83 | 326 | 166 | 373 | 265 | 411 |
| 34 | 279 | 84 | 327 | 168 | 374 | 267 | 411 |
| 35 | 281 | 85 | 327 | 170 | 375 | 269 | 412 |
| 36 | 282 | 86 | 328 | 172 | 376 | 271 | 413 |
| 37 | 283 | 87 | 329 | 174 | 377 | 273 | 413 |
| 38 | 284 | 88 | 330 | 176 | 378 | 275 | 414 |
| 39 | 285 | 89 | 330 | 178 | 378 | 277 | 415 |
| 40 | 287 | 90 | 331 | 180 | 379 | 279 | 415 |
| 41 | 288 | 91 | 332 | 182 | 380 | 281 | 416 |
| 42 | 289 | 92 | 332 | 184 | 381 | 283 | 416 |
| 43 | 290 | 93 | 333 | 186 | 382 | 285 | 417 |
| 44 | 291 | 94 | 334 | 188 | 383 | 295 | 420 |
| 45 | 292 | 95 | 334 | 190 | 384 | 305 | 423 |
| 46 | 293 | 96 | 335 | 192 | 384 | 355 | 437 |
| 47 | 294 | 97 | 336 | 194 | 385 | 375 | 442 |
| 48 | 295 | 98 | 336 | 196 | 386 | 385 | 444 |
| 49 | 297 | 99 | 337 | 198 | 387 | 405 | 449 |
| 50 | 298 | 100 | 338 | 200 | 388 | 455 | 460 |
| 51 | 299 | 102 | 339 | 202 | 388 | 510 | 472 |
| 52 | 300 | 104 | 340 | 204 | 389 | 560 | 481 |
| 53 | 301 | 106 | 341 | 206 | 390 | 585 | 486 |
| 54 | 302 | 108 | 343 | 208 | 391 | | |

# WATER POLLUTION

Drinking water standards as adopted by the EPA, Safe Drinking Water Act, and US Public Health Service in 1946 and adopted by the American Water Works Association are summarized below:

| Pollutant | Maximum Contaminant Level (MCL) mg/liter or ppm |
|---|---|
| Alkyl Benzene Sulfonate (ABS) | 0.5 |
| Arsenic | 0.05 |
| Barium | 1.00 |
| Cadmium | 0.01 |
| Chloride | 250.0 |
| Coliform bacteria | < 1/100 ml |
| Color (platinum–cobalt scale) | 20 units |
| Copper | 1.0 |
| Carbon Chloroform Extract (CCE) | 0.2 |
| Chromium (hexavalent) | 0.05 |
| Cyanide | 0.01 |
| Endrin (organic) | 0.0002 |
| Fluoride | 1.4 to 2.4 |
| Iron (> 0.3 makes red water) | 0.3 |
| Lead | 0.05 |
| Lindane (organic) | 0.004 |
| Manganese (> 0.1 forms brown-black stain) | 0.05 |
| Mercury | 0.002 |
| Methoxychlor (organic) | 0.1 |
| Nitrate | 45.0 |
| Phenols | 0.001 |
| Selenium | 0.01 |
| Silver | 0.05 |
| Sulfate ( SO4 ) (> 500 has a laxative effect) | 250.0 |
| Total Dissolved Solids | 500.0 |
| Toxaphene (organic) | 0.005 |
| Trihalomethanes (organic) | 0.1 |
| Turbidity (silica scale) | 10 ,1 to 5 TU |
| Zinc | 5.0 |
| 2, 4 – D (organic) | 0.1 |
| 2, 4, 5 – TP Silvex (organic) | 0.01 |
| RADIONUCLIDES: | |
| Gross Alpha particle activity | 15 pCi/l |
| Beta particle and photon radioactivity | 4 mrem |
| Radium–226 and Radium–228 | 5 pCi/l |

Exposures over safe limits can result in a variety of serious health problems ranging from liver and kidney damage, high cancer risk, nervous system disorders, skin discoloration, hypertension, and a variety of others. For specific information, refer to the the 1974, 1977, and 1986 versions of the Safe Drinking Water Act.

See also the AIR Chapter for more on pollution, page 14.

# WATER HARDNESS

Water hardness is a function of the amount of dissolved calcium salts, magnesium salts, iron and aluminum. The salts occur in a variety of forms but are typically calcium and magnesium bicarbonates (referred to as "temporary hardness") and sulphates and chlorides (referred to as "permanent hardness").

Although the most obvious effect of hard water is in preventing soap from lathering, most people cannot tolerate drinking water that exceeds 300 ppm carbonate, or 1500 ppm chloride, or 2000 ppm sulphate. As shown in the table to the left, >500 ppm sulphate can produce a laxative effect in the body. Livestock can usually tolerate much higher levels of hardness, but total dissolved solids >10,000 ppm will create problems.

The following formula is used to calculate total hardness:

Total Hardness in ppm Carbonate = (ppm Calcium x 2.497)
+ (ppm Magnesium x 4.115) + (ppm Iron x 1.792 +
+ (ppm Manganese x 1.822)

Hard water is treated by either a zeolite process (home water softeners) or a lime–soda ash process (large operations).

Hardness is also measured in "grains per gallon" and "degrees". Equivalents are as follows:

1 ppm = 0.058 grains/US gallon
1 ppm = 0.07 Clark degrees
1 ppm = 0.10 French degrees
1 ppm = 0.056 German degrees
1 French degree = 1 hydrotimetric degree
1 Clark degree = 1 grain / Imperial gallon as carbonate
1 French degree = 1 part / 100,000 calcium carbonate
1 German degree = 1 part / 100,000 calcium oxide
1 grain / US gallon = 17.1 ppm
1 grain / US gallon = 1.20 Clark degrees
1 grain / US gallon = 1.71 French degrees
1 grain / US gallon = 0.958 German degrees

# WATER DATA & FORMULAS

1 gallon water = 231 cubic inches = 8.333 pounds
1 po
1 cubic foot water = 7.5 gallons = 62.5 pounds (salt water weighs approximately 64.3 pounds per cubic foot )
pounds per square inch at bottom of a column of water = height of column in feet x 0.434
1 miner's inch = 9 to 12 gallons per minute

## Horsepower to Raise Water

$$Horsepower = \frac{gallons\ per\ minute \times Total\ Head\ in\ feet}{3960}$$

(if pumping a liquid other than water, multiply the gallons per minute above by the liquids specific gravity)

## Gallons Per Minute through a Pipe

$$GPM = 0.0408 \times Pipe\ Diameter\ inches^2 \times Feet/minute\ water\ velocity$$

## Weight of Water in a Pipe

$$Pounds\ Water = Pipe\ Length\ feet \times Pipe\ diameter\ inches^2 \times 0.34$$

## Gallons per Minute of a Slurry

$$GPM\ Slurry = GPM\ Water + \frac{4 \times Tons\ per\ hour\ of\ solids}{Specific\ Gravity\ of\ Solids}$$

## Cost to Pump Water – Electric

$$\$\ per\ hour = \frac{gpm \times Head\ in\ feet \times 0.746 \times Rate\ per\ KWH}{3960 \times Pump\ Efficiency \times Electric\ Motor\ Efficiency}$$

(70% Pump and 90% Motor Efficiency is a good average)

## Cost to Pump Water – Gasoline and Diesel

$$\$\ per\ hour = \frac{GPM \times Head\ in\ feet \times K \times \$\ per\ gallon\ fuel}{3960 \times Pump\ Efficiency}$$

K = 0.110 for gasoline or 0.065 for diesel
(K is actually gallons of fuel per horsepower)

(70% Pump Efficiency is a good average value)

# Weights of Materials

(See also GEOLOGY, page 197 for minerals)
(See also CARPENTRY, page 34 for woods)
(See also STEEL, page 333)

# WEIGHTS OF MATERIALS

| Material | Specific Gravity | Weights lbs per cu foot | lbs per cu yard | kgs per cu meter | Angle of Repose |
|---|---|---|---|---|---|
| Acetic acid, 90% | 1.06 | 66.3 | 1790 | 1062.0 | |
| Alcohol, ethyl | 0.789 | 49 | 1329 | 788.5 | |
| Alcohol, methyl | 0.791 | 49 | 1333 | 790.9 | |
| Alfalfa, ground | 0.26 | 16 | 432 | 256.3 | +45 |
| Alum, lumpy | 0.88 | 55 | 1485 | 881.0 | 30-45 |
| Alum, pulverized | 0.75 | 47 | 1269 | 752.9 | 30-45 |
| Alumina | 0.96 | 60 | 1620 | 961.1 | 30-45 |
| Aluminum, solid | 2.64 | 165 | 4455 | 2643.1 | |
| Aluminum oxide | 1.52 | 95 | 2565 | 1521.8 | 30 |
| Ammonia gas | 0.00 | 0.048 | 1.29 | 0.8 | |
| Ammonium sulfate | 0.83 | 52 | 1404 | 833.0 | |
| Andesite, solid | 2.77 | 173 | 4671 | 2771.2 | |
| Antimony, cast | 6.70 | 418 | 11286 | 6695.7 | |
| Apple wood, dry | 0.71 | 44 | 1188 | 704.8 | |
| Apples | 0.64 | 40 | 1080 | 640.7 | |
| Arsenic | 5.67 | 354 | 9558 | 5670.5 | |
| Asbestos, shredded | 0.35 | 22 | 594 | 352.4 | 30 |
| Asbestos, solid | 2.45 | 153 | 4131 | 2450.8 | |
| Ash wood, black, dry | 0.54 | 34 | 918 | 544.6 | |
| Ash wood, white, dry | 0.67 | 42 | 1134 | 672.8 | |
| Ashes | 0.66 | 41 | 1107 | 656.8 | |
| Aspen wood | 0.42 | 26 | 702 | 416.5 | |
| Asphalt, crushed | 0.72 | 45 | 1215 | 720.8 | 30-45 |
| Babbitt | 7.28 | 454 | 12258 | 7272.4 | |
| Bagasse | 0.12 | 7.5 | 202 | 120.1 | 45 |
| Bakelite, solid | 1.36 | 85 | 2295 | 1361.6 | |
| Baking powder | 0.72 | 45 | 1215 | 720.8 | 30-45 |
| Barium | 3.78 | 236 | 6372 | 3780.4 | |
| Bark, wood refuse | 0.24 | 15 | 405 | 240.3 | 45 |
| Barley | 0.61 | 38 | 1026 | 608.7 | |
| Barite, crushed | 2.88 | 180 | 4860 | 2883.3 | |
| Basalt, broken | 1.96 | 122 | 3294 | 1954.3 | |
| Basalt, solid | 3.01 | 188 | 5076 | 3011.5 | |
| Bauxite, crushed | 1.28 | 80 | 2160 | 1281.5 | 30-45 |
| Beans, castor | 0.58 | 36 | 972 | 576.7 | |
| Beans, cocoa | 0.59 | 37 | 999 | 592.7 | |
| Beans, navy | 0.80 | 50 | 1350 | 800.9 | |
| Beans,soy | 0.72 | 45 | 1215 | 720.8 | |
| Beeswax | 0.96 | 60 | 1620 | 961.1 | |
| Beets | 0.72 | 45 | 1215 | 720.8 | |
| Bentonite | 0.59 | 37 | 999 | 592.7 | 45 |
| Bicarbonate of Soda | 0.69 | 43 | 1161 | 688.8 | 42 |
| Birch wood, yellow | 0.71 | 44 | 1188 | 704.8 | |
| Bismuth | 9.79 | 611 | 16497 | 9787.3 | |
| Bones, pulverized | 0.88 | 55 | 1485 | 881.0 | |
| Borax, fine | 0.85 | 53 | 1431 | 849.0 | 30-45 |
| Bran | 0.26 | 16 | 432 | 256.3 | 30-45 |
| Brass, cast | 8.56 | 534 | 14418 | 8553.9 | |
| Brass, rolled | 8.56 | 534 | 14418 | 8553.9 | |
| Brewers grain | 0.43 | 27 | 729 | 432.5 | 45 |
| Brick, common red | 1.92 | 120 | 3240 | 1922.2 | |
| Brick, fire clay | 2.40 | 150 | 4050 | 2402.8 | |
| Brick, silica | 2.05 | 128 | 3456 | 2050.4 | |
| Brick, chrome | 2.80 | 175 | 4725 | 2803.2 | |
| Brick, magnesia | 2.56 | 160 | 4320 | 2563.0 | |

# WEIGHTS OF MATERIALS

| Material | Specific Gravity | Weights lbs per cu foot | lbs per cu yard | kgs per cu meter | Angle of Repose |
|---|---|---|---|---|---|
| Bronze | 8.16 | 509 | 13743 | 8153.4 | |
| Buckwheat | 0.66 | 41 | 1107 | 656.8 | |
| Butter | 0.87 | 54 | 1458 | 865.0 | |
| Cadmium | 8.65 | 540 | 14580 | 8650.0 | |
| Calcium carbide | 1.20 | 75 | 2025 | 1201.4 | 30-45 |
| Caliche | 1.44 | 90 | 2430 | 1441.7 | |
| Carbon, solid | 2.15 | 134 | 3618 | 2146.5 | |
| Carbon, powdered | 0.08 | 5 | 135 | 80.1 | |
| Carbon dioxide | 0.00 | 0.1234 | 3.3318 | 2.0 | |
| Carbon monoxide | 0.00 | 0.0781 | 2.1087 | 1.3 | |
| Cardboard | 0.69 | 43 | 1161 | 688.8 | |
| Cedar, red | 0.38 | 24 | 648 | 384.4 | |
| Cement, Portland | 1.60 | 100 | 2700 | 1601.8 | |
| Cement, Mortar | 2.16 | 135 | 3645 | 2162.5 | |
| Cement, slurry | 1.44 | 90 | 2430 | 1441.7 | |
| Chalk, solid | 2.50 | 156 | 4212 | 2498.9 | |
| Chalk, lumpy | 1.44 | 90 | 2430 | 1441.7 | 45 |
| Chalk, fine | 1.12 | 70 | 1890 | 1121.3 | 45 |
| Charcoal | 0.21 | 13 | 351 | 208.2 | |
| Cherry wood, dry | 0.56 | 35 | 945 | 560.6 | |
| Chestnut wood, dry | 0.48 | 30 | 810 | 480.6 | |
| Chloroform | 1.52 | 95 | 2565 | 1521.8 | |
| Chocolate, powder | 0.64 | 40 | 1080 | 640.7 | |
| Chromic acid, flake | 1.20 | 75 | 2025 | 1201.4 | 25 |
| Chromium | 6.86 | 428 | 11556 | 6855.9 | |
| Chromium ore | 2.16 | 135 | 3645 | 2162.5 | 30-45 |
| Cinders, Furnace | 0.91 | 57 | 1539 | 913.1 | |
| Cinders, Coal,ash | 0.64 | 40 | 1080 | 640.7 | 25-40 |
| Clay, Dry excavated | 1.09 | 68 | 1836 | 1089.3 | |
| Clay, Wet excavated | 1.83 | 114 | 3078 | 1826.1 | |
| Clay, Dry lump | 1.07 | 67 | 1809 | 1073.2 | 25-45 |
| Clay, fire | 1.36 | 85 | 2295 | 1361.6 | |
| Clay, Wet lump | 1.60 | 100 | 2700 | 1601.8 | |
| Clay, compacted | 1.75 | 109 | 2943 | 1746.0 | |
| Clover seed | 0.77 | 48 | 1296 | 768.9 | 28 |
| Coal, Anthracite,solid | 1.51 | 94 | 2538 | 1505.7 | |
| Coal, Anthracite,brkn | 1.11 | 69 | 1863 | 1105.3 | 30-45 |
| Coal, Bituminous,solid | 1.35 | 84 | 2268 | 1345.6 | |
| Coal, Bituminous,brkn | 0.83 | 52 | 1404 | 833.0 | 30-45 |
| Cobalt | 8.75 | 546 | 14742 | 8746.1 | |
| Coconut, meal | 0.51 | 32 | 864 | 512.6 | |
| Coconut, shredded | 0.35 | 22 | 594 | 352.4 | 45 |
| Coffee, fresh beans | 0.56 | 35 | 945 | 560.6 | 35-45 |
| Coffee, roast beans | 0.43 | 27 | 729 | 432.5 | |
| Coke | 0.42 | 26 | 702 | 416.5 | |
| Concrete, Asphaltic | 2.24 | 140 | 3780 | 2242.6 | |
| Concrete, Gravel | 2.40 | 150 | 4050 | 2402.8 | |
| Concrete, Limestone w/Portland | 2.37 | 148 | 3996 | 2370.7 | |
| Copper, cast | 8.69 | 542 | 14634 | 8682.0 | |
| Copper, rolled | 8.91 | 556 | 15012 | 8906.3 | |
| Copper sulphate, ground | 3.60 | 225 | 6073 | 3603.0 | |
| Copra, medium size | 0.53 | 33 | 891 | 528.6 | 20 |
| Copra, meal, ground | 0.64 | 40 | 1080 | 640.7 | 39 |

# WEIGHTS OF MATERIALS

| Material | Specific Gravity | Weights lbs per cu foot | lbs per cu yard | kgs per cu meter | Angle of Repose |
|---|---|---|---|---|---|
| Copra, expeller cake ground | 0.51 | 32 | 864 | 512.6 | 30 |
| Copra, expeller cake chopped | 0.46 | 29 | 783 | 464.5 | 20 |
| Cork, solid | 0.24 | 15 | 405 | 240.3 | |
| Cork, ground | 0.16 | 10 | 270 | 160.2 | 45 |
| Corn, on the cob | 0.72 | 45 | 1215 | 720.8 | |
| Corn, shelled | 0.72 | 45 | 1215 | 720.8 | |
| Corn, grits | 0.67 | 42 | 1134 | 672.8 | 30-45 |
| Cottonseed, dry, de-linted | 0.56 | 35 | 945 | 560.6 | 30-45 |
| Cottonseed, dry, not de-linted | 0.32 | 20 | 540 | 320.4 | 45 |
| Cottonseed, cake, lumpy | 0.67 | 42 | 1134 | 672.8 | 30-45 |
| Cottonseed, hulls | 0.19 | 12 | 324 | 192.2 | 45 |
| Cottonseed, meal | 0.59 | 37 | 999 | 592.7 | 30-45 |
| Cottonseed, meats | 0.64 | 40 | 1080 | 640.7 | 30-45 |
| Cottonwood | 0.42 | 26 | 702 | 416.5 | |
| Cryolite | 1.60 | 100 | 2700 | 1601.8 | 30-45 |
| Cullet | 1.60 | 100 | 2700 | 1601.8 | 30-45 |
| Culm | 0.75 | 47 | 1269 | 752.9 | |
| Cypress wood | 0.51 | 32 | 864 | 512.6 | |
| Dolomite, solid | 2.90 | 181 | 4887 | 2899.3 | |
| Dolomite, pulverized | 0.74 | 46 | 1242 | 736.9 | |
| Dolomite, lumpy | 1.52 | 95 | 2565 | 1521.8 | 30-45 |
| Earth, loam, dry, excavated | 1.25 | 78 | 2106 | 1249.4 | 30-45 |
| Earth, moist, excavated | 1.44 | 90 | 2430 | 1441.7 | 30-45 |
| Earth, wet, excavated | 1.60 | 100 | 2700 | 1601.8 | 30-45 |
| Earth, dense | 2.00 | 125 | 3375 | 2002.3 | 30-45 |
| Earth, soft loose mud | 1.73 | 108 | 2916 | 1730.0 | |
| Earth, packed | 1.52 | 95 | 2565 | 1521.8 | |
| Earth, Fullers, raw | 0.67 | 42 | 1134 | 672.8 | 35 |
| Ebony wood | 0.96 | 60 | 1620 | 961.1 | |
| Elm, dry | 0.56 | 35 | 945 | 560.6 | |
| Emery | 4.01 | 250 | 6750 | 4004.6 | |
| Ether | 0.74 | 46 | 1242 | 736.9 | |
| Feldspar, solid | 2.56 | 160 | 4320 | 2563.0 | |
| Feldspar, pulverized | 1.23 | 77 | 2079 | 1233.4 | 45 |
| Fertilizer, acid phosphate | 0.96 | 60 | 1620 | 961.1 | |
| Fir, Douglas | 0.53 | 33 | 891 | 528.6 | |
| Fish, scrap | 0.72 | 45 | 1215 | 720.8 | |
| Fish, meal | 0.59 | 37 | 999 | 592.7 | 45 |
| Flaxseed, whole | 0.72 | 45 | 1215 | 720.8 | |
| Flour, wheat | 0.59 | 37 | 999 | 592.7 | 45 |
| Fluorspar, solid | 3.21 | 200 | 5400 | 3203.7 | |
| Fluorspar, lumps | 1.60 | 100 | 2700 | 1601.8 | 45 |
| Fluorspar, pulverized | 1.44 | 90 | 2430 | 1441.7 | 45 |
| Garbage | 0.48 | 30 | 810 | 480.6 | |
| Glass, window | 2.58 | 161 | 4347 | 2579.0 | |
| Glue, animal, flaked | 0.56 | 35 | 945 | 560.6 | |
| Glue, vegetable, | | | | | |

# WEIGHTS OF MATERIALS

| Material | Specific Gravity | Weights lbs per cu foot | lbs per cu yard | kgs per cu meter | Angle of Repose |
|---|---|---|---|---|---|
| glue powdered | 0.64 | 40 | 1080 | 640.7 | |
| Gluten, meal | 0.63 | 39 | 1053 | 624.7 | 30-45 |
| Gneiss, bed in place | 2.87 | 179 | 4833 | 2867.3 | |
| Gneiss, broken | 1.86 | 116 | 3132 | 1858.1 | |
| Gold, pure 24 Kt | 19.29 | 1204 | 32508 | 19286.3 | |
| Granite, solid | 2.69 | 168 | 4536 | 2691.1 | |
| Granite, broken | 1.65 | 103 | 2781 | 1649.9 | |
| Graphite, flake | 0.64 | 40 | 1080 | 640.7 | 30-45 |
| Gravel, loose, dry | 1.52 | 95 | 2565 | 1521.8 | 30-45 |
| Gravel, w/ sand, natural | 1.92 | 120 | 3240 | 1922.2 | |
| Gravel, dry, 1/4 to 2 inch | 1.68 | 105 | 2835 | 1681.9 | |
| Gravel, wet, 1/4 to 2 inch | 2.00 | 125 | 3375 | 2002.3 | |
| Gypsum, solid | 2.79 | 174 | 4698 | 2787.2 | |
| Gypsum, broken | 1.81 | 113 | 3051 | 1810.1 | |
| Gypsum, crushed | 1.60 | 100 | 2700 | 1601.8 | |
| Gypsum, pulverized | 1.12 | 70 | 1890 | 1121.3 | 45 |
| Halite (salt), solid | 2.32 | 145 | 3915 | 2322.7 | |
| Halite (salt), broken | 1.51 | 94 | 2538 | 1505.7 | |
| Hay, pressed | 0.38 | 24 | 648 | 384.4 | |
| Hay, loose | 0.08 | 5 | 135 | 80.1 | |
| Hematite, solid | 4.90 | 306 | 8262 | 4901.7 | |
| Hematite, broken | 3.22 | 201 | 5427 | 3219.7 | |
| Hemlock, dry | 0.40 | 25 | 675 | 400.5 | |
| Hickory, dry | 0.85 | 53 | 1431 | 849.0 | |
| Hops, moist | 0.56 | 35 | 945 | 560.6 | 45 |
| Hydrochloric acid 40% | 1.20 | 75 | 2025 | 1201.4 | |
| Ice, solid | 0.92 | 57.4 | 1549.8 | 919.5 | |
| Ice, crushed | 0.59 | 37 | 999 | 592.7 | |
| Ilmenite | 2.31 | 144 | 3888 | 2306.7 | 30-45 |
| Iridium | 22.16 | 1383 | 37341 | 22153.6 | |
| Iron, cast | 7.21 | 450 | 12150 | 7208.3 | |
| Iron, wrought | 7.77 | 485 | 13095 | 7769.0 | |
| Iron oxide pigment | 0.40 | 25 | 675 | 400.5 | 40 |
| Ivory | 1.84 | 115 | 3105 | 1842.1 | |
| Kaolin, green crushed | 1.03 | 64 | 1728 | 1025.2 | 35 |
| Kaolin, pulverized | 0.35 | 22 | 594 | 352.4 | 45 |
| Lead, cast | 11.35 | 708 | 19116 | 11341.1 | |
| Lead, rolled | 11.39 | 711 | 19197 | 11389.1 | |
| Lead, red | 3.69 | 230 | 6210 | 3684.3 | |
| Lead, white pigment | 4.09 | 255 | 6885 | 4084.7 | |
| Leather | 0.95 | 59 | 1593 | 945.1 | |
| Lignite, dry | 0.80 | 50 | 1350 | 800.9 | 30-45 |
| Lignum Vitae, dry | 1.28 | 80 | 2160 | 1281.5 | |
| Lime, quick, lump | 0.85 | 53 | 1431 | 849.0 | |
| Lime, quick, fine | 1.20 | 75 | 2025 | 1201.4 | |
| Lime, stone, large | 2.69 | 168 | 4536 | 2691.1 | |
| Lime, stone, lump | 1.54 | 96 | 2592 | 1537.8 | |
| Lime, hydrated | 0.48 | 30 | 810 | 480.6 | 30-45 |
| Limonite, solid | 3.80 | 237 | 6399 | 3796.4 | |
| Limonite, broken | 2.47 | 154 | 4158 | 2466.8 | |
| Limestone, solid | 2.61 | 163 | 4401 | 2611.0 | |
| Limestone, broken | 1.55 | 97 | 2619 | 1553.8 | |

| Material | Specific Gravity | Weights lbs per cu foot | lbs per cu yard | kgs per cu meter | Angle of Repose |
|---|---|---|---|---|---|
| Limestone, pulverized | 1.39 | 87 | 2349 | 1393.6 | 45 |
| Linseed, whole | 0.75 | 47 | 1269 | 752.9 | |
| Linseed, meal | 0.51 | 32 | 864 | 512.6 | 30-45 |
| Locust, dry | 0.71 | 44 | 1188 | 704.8 | |
| Magnesite, solid | 3.01 | 188 | 5076 | 3011.5 | |
| Magnesium, solid | 1.75 | 109 | 2943 | 1746.0 | |
| Magnesium sulfate, crystal | 1.12 | 70 | 1890 | 1121.3 | |
| Magnetite, solid | 5.05 | 315 | 8505 | 5045.8 | |
| Magnetite, broken | 3.29 | 205 | 5535 | 3283.8 | |
| Mahogany, Spanish, dry | 0.85 | 53 | 1431 | 849.0 | |
| Mahogany, Honduras, dry | 0.54 | 34 | 918 | 544.6 | |
| Malt | 0.34 | 21 | 567 | 336.4 | 30-45 |
| Manganese, solid | 7.61 | 475 | 12825 | 7608.8 | |
| Manganese oxide | 1.92 | 120 | 3240 | 1922.2 | |
| Manure | 0.40 | 25 | 675 | 400.5 | |
| Maple, dry | 0.71 | 44 | 1188 | 704.8 | |
| Marble, solid | 2.56 | 160 | 4320 | 2563.0 | |
| Marble, broken | 1.57 | 98 | 2646 | 1569.8 | 30-45 |
| Marl, wet, excavated | 2.24 | 140 | 3780 | 2242.6 | |
| Mercury @ 32oF | 13.61 | 849 | 22923 | 13599.7 | |
| Mica, solid | 2.88 | 180 | 4860 | 2883.3 | |
| Mica, broken | 1.60 | 100 | 2700 | 1601.8 | 30-45 |
| Milk, powdered | 0.45 | 28 | 756 | 448.5 | 45 |
| Molybdenum | 10.19 | 636 | 17172 | 10187.8 | |
| Mortar, wet | 2.40 | 150 | 4050 | 2402.8 | |
| Mud, packed | 1.91 | 119 | 3213 | 1906.2 | |
| Mud, fluid | 1.73 | 108 | 2916 | 1730.0 | |
| Nickel, rolled | 8.67 | 541 | 14607 | 8666.0 | |
| Nickel silver | 8.45 | 527 | 14229 | 8441.7 | |
| Nitric acid, 91% | 1.51 | 94 | 2538 | 1505.7 | |
| Nitrogen | 0.00 | 0.0784 | 2.1168 | 1.3 | |
| Oak, live, dry | 0.95 | 59 | 1593 | 945.1 | |
| Oak, red | 0.71 | 44 | 1188 | 704.8 | |
| Oats | 0.43 | 27 | 729 | 432.5 | 32 |
| Oats, rolled | 0.30 | 19 | 513 | 304.4 | 30-45 |
| Oil Cake | 0.79 | 49 | 1323 | 784.9 | |
| Oil, linseed | 0.94 | 58.8 | 1587.6 | 941.9 | |
| Oil, petroleum | 0.88 | 55 | 1485 | 881.0 | |
| Oxygen | 0.00 | 0.0892 | 2.4084 | 1.4 | |
| Oyster shells, ground | 0.85 | 53 | 1431 | 849.0 | 30-45 |
| Paper, standard | 1.20 | 75 | 2025 | 1201.4 | |
| Paraffin | 0.72 | 45 | 1215 | 720.8 | |
| Peanuts, shelled | 0.64 | 40 | 1080 | 640.7 | 30-45 |
| Peanuts, not shelled | 0.27 | 17 | 459 | 272.3 | 30-45 |
| Peat, dry | 0.40 | 25 | 675 | 400.5 | |
| Peat, moist | 0.80 | 50 | 1350 | 800.9 | |
| Peat, wet | 1.12 | 70 | 1890 | 1121.3 | |
| Pecan wood | 0.75 | 47 | 1269 | 752.9 | |
| Phosphate Rock, broken | 1.76 | 110 | 2970 | 1762.0 | |
| Phosphorus | 2.34 | 146 | 3942 | 2338.7 | |
| Pine, White, dry | 0.42 | 26 | 702 | 416.5 | |

# WEIGHTS OF MATERIALS

| Material | Specific Gravity | Weights lbs per cu foot | lbs per cu yard | kgs per cu meter | Angle of Repose |
|---|---|---|---|---|---|
| Pine, Yellow Northern, dry | 0.54 | 34 | 918 | 544.6 | |
| Pine, Yellow Southern, dry | 0.72 | 45 | 1215 | 720.8 | |
| Pitch | 1.15 | 72 | 1935.9 | 1148.5 | |
| Plaster | 0.85 | 53 | 1431 | 849.0 | |
| Platinum | 21.51 | 1342 | 36234 | 21496.8 | |
| Porcelain | 2.40 | 150 | 4050 | 2402.8 | |
| Porphyry, solid | 2.55 | 159 | 4293 | 2546.9 | |
| Porphyry, broken | 1.65 | 103 | 2781 | 1649.9 | |
| Potash | 1.28 | 80 | 2160 | 1281.5 | |
| Potassium chloride | 2.00 | 125 | 3375 | 2002.3 | 30-45 |
| Potatoes, white | 0.77 | 48 | 1296 | 768.9 | |
| Pumice stone | 0.64 | 40 | 1080 | 640.7 | |
| Quartz, solid | 2.64 | 165 | 4455 | 2643.1 | |
| Quartz, lump | 1.55 | 97 | 2619 | 1553.8 | |
| Quartz sand | 1.20 | 75 | 2025 | 1201.4 | |
| Redwood, Calif, dry | 0.45 | 28 | 756 | 448.5 | |
| Resin,synthetic,crshd | 0.56 | 35 | 945 | 560.6 | |
| Rice, hulled | 0.75 | 47 | 1269 | 752.9 | |
| Rice, rough | 0.58 | 36 | 972 | 576.7 | 30-45 |
| Rice grits | 0.69 | 43 | 1161 | 688.8 | 30-45 |
| Rip-rap | 1.60 | 100 | 2700 | 1601.8 | |
| Rosin | 1.07 | 67 | 1809 | 1073.2 | |
| Rubber, caoutchouc | 0.95 | 59 | 1593 | 945.1 | |
| Rubber, mfged | 1.52 | 95 | 2565 | 1521.8 | |
| Rubber, ground scrap | 0.48 | 30 | 810 | 480.6 | 45 |
| Rye | 0.71 | 44 | 1188 | 704.8 | |
| Salt cake | 1.44 | 90 | 2430 | 1441.7 | 30-45 |
| Salt, coarse | 0.80 | 50 | 1350 | 800.9 | 30-45 |
| Salt, fine | 1.20 | 75 | 2025 | 1201.4 | 30-45 |
| Saltpeter | 1.20 | 75 | 2025 | 1201.4 | 30-45 |
| Sand, wet | 1.92 | 120 | 3240 | 1922.2 | 45 |
| Sand, dry | 1.60 | 100 | 2700 | 1601.8 | 34 |
| Sand, loose | 1.44 | 90 | 2430 | 1441.7 | 30-45 |
| Sand, rammed | 1.68 | 105 | 2835 | 1681.9 | |
| Sand, water filled | 1.92 | 120 | 3240 | 1922.2 | 15-30 |
| Sandstone, solid | 2.32 | 145 | 3915 | 2322.7 | |
| Sandstone, broken | 1.51 | 94 | 2538 | 1505.7 | |
| Sand, dry, loose | 1.60 | 100 | 2700 | 1601.8 | |
| Sand, damp | 1.92 | 120 | 3240 | 1922.2 | |
| Sand, wet | 2.08 | 130 | 3510 | 2082.4 | |
| Sand, wet packed | 2.08 | 130 | 3510 | 2082.4 | |
| Sand and Gravel, dry | 1.73 | 108 | 2916 | 1730.0 | |
| Sand and Gravel, wet | 2.00 | 125 | 3375 | 2002.3 | |
| Sawdust | 0.27 | 17 | 459 | 272.3 | |
| Sewage, sludge | 0.72 | 45 | 1215 | 720.8 | |
| Shale, solid | 2.68 | 167 | 4509 | 2675.1 | |
| Shale, broken | 1.59 | 99 | 2673 | 1585.8 | 30-45 |
| Silver | 10.46 | 653 | 17631 | 10460.1 | |
| Slag, solid | 2.12 | 132 | 3564 | 2114.4 | |
| Slag, broken | 1.76 | 110 | 2970 | 1762.0 | |
| Slag, crushed /4 inch | 1.19 | 74 | 1998 | 1185.4 | |
| Slag, furn. granulated | 0.96 | 60 | 1620 | 961.1 | |
| Slate, solid | 2.69 | 168 | 4536 | 2691.1 | |

# WEIGHTS OF MATERIALS

| Material | Specific Gravity | Weights lbs per cu foot | lbs per cu yard | kgs per cu meter | Angle of Repose |
|---|---|---|---|---|---|
| Slate, broken | 1.67 | 104 | 2808 | 1665.9 | |
| Slate, pulverized | 1.36 | 85 | 2295 | 1361.6 | 30-45 |
| Snow, freshly fallen | 0.16 | 10 | 270 | 160.2 | |
| Snow, compacted | 0.48 | 30 | 810 | 480.6 | |
| Soap, solid | 0.80 | 50 | 1350 | 800.9 | |
| Soap, chips | 0.16 | 10 | 270 | 160.2 | 30-45 |
| Soap, flakes | 0.16 | 10 | 270 | 160.2 | 30-45 |
| Soap, powder | 0.37 | 23 | 621 | 368.4 | 30-45 |
| Soda Ash, heavy | 0.96 | 60 | 1620 | 961.1 | 30-45 |
| Soda Ash, light | 0.43 | 27 | 729 | 432.5 | 30-45 |
| Sodium | 0.98 | 61 | 1647 | 977.1 | |
| Sodium Aluminate ground | 1.15 | 72 | 1944 | 1153.3 | |
| Sodium Nitrate,grnd | 1.20 | 75 | 2025 | 1201.4 | |
| Soybeans, whole | 0.75 | 47 | 1269 | 752.9 | |
| Spruce, Calif, dry | 0.45 | 28 | 756 | 448.5 | |
| Starch, powdered | 0.56 | 35 | 945 | 560.6 | |
| Steel, cast | 7.85 | 490 | 13230 | 7849.1 | |
| Steel, rolled | 7.93 | 495 | 13365 | 7929.2 | |
| Stone, crushed | 1.60 | 100 | 2700 | 1601.8 | |
| Sugar, brown | 0.72 | 45 | 1215 | 720.8 | |
| Sugar, powdered | 0.80 | 50 | 1350 | 800.9 | |
| Sugar, granulated | 0.85 | 53 | 1431 | 849.0 | 30-45 |
| Sugar, raw cane | 0.96 | 60 | 1620 | 961.1 | 45 |
| Sugarbeet pulp, dry | 0.21 | 13 | 351 | 208.2 | |
| Sugarbeet pulp, wet | 0.56 | 35 | 945 | 560.6 | |
| Sugarcane | 0.27 | 17 | 459 | 272.3 | 45 |
| Sulfur, solid | 2.00 | 125 | 3375 | 2002.3 | |
| Sulfur, lump | 1.31 | 82 | 2214 | 1313.5 | 30-45 |
| Sulfur, pulverized | 0.96 | 60 | 1620 | 961.1 | 30-45 |
| Sulfuric acid, 87% | 1.79 | 112 | 3024 | 1794.1 | |
| Sycamore, dry | 0.59 | 37 | 999 | 592.7 | |
| Taconite | 2.80 | 175 | 4725 | 2803.2 | |
| Talc, solid | 2.69 | 168 | 4536 | 2691.1 | |
| Talc, broken | 1.75 | 109 | 2943 | 1746.0 | |
| Tanbark, ground | 0.88 | 55 | 1485 | 881.0 | |
| Tankage | 0.96 | 60 | 1620 | 961.1 | |
| Tar | 1.15 | 72 | 1935.9 | 1148.5 | |
| Tin, cast | 7.36 | 459 | 12393 | 7352.5 | |
| Tobacco | 0.32 | 20 | 540 | 320.4 | 45 |
| Trap rock, solid | 2.88 | 180 | 4860 | 2883.3 | |
| Trap rock, broken | 1.75 | 109 | 2943 | 1746.0 | |
| Tungsten | 19.62 | 1224 | 33048 | 19606.6 | |
| Turf | 0.40 | 25 | 675 | 400.5 | |
| Turpentine | 0.87 | 54 | 1458 | 865.0 | |
| Vanadium | 5.50 | 343 | 9261 | 5494.3 | |
| Walnut, black, dry | 0.61 | 38 | 1026 | 608.7 | |
| Water, pure | 1.00 | 62.4 | 1684.8 | 999.6 | |
| Water, sea | 1.03 | 64.08 | 1730.16 | 1026.5 | |
| Wheat | 0.77 | 48 | 1296 | 768.9 | 28 |
| Wheat, cracked | 0.67 | 42 | 1134 | 672.8 | 30-45 |
| Willow wood | 0.42 | 26 | 702 | 416.5 | |
| Wool | 1.31 | 82 | 2214 | 1313.5 | |
| Zinc, cast | 7.05 | 440 | 11880 | 7048.1 | |
| Zinc oxide | 0.40 | 25 | 675 | 400.5 | 45 |

# POCKET REF

## Welding

## ARC ELECTRODES - MILD STEEL

| Electrode # | Description |
|---|---|
| ⇓ | This digit indicates the following: |
| Exx1z | All positions of welding |
| Exx2z | Flat and horizontal positions |
| Exx3z | Flat welding positions only |
| ⇓⇓ | These digits indicate the following: |
| Exx10 | DC, reverse polarity |
| Exx11 | AC or DC, reverse polarity |
| Exx12 | DC, straight polarity or AC |
| Exx13 | AC or DC, straight polarity |
| Exx14 | DC, either polarity or AC, iron powder |
| Exx15 | DC, reverse polarity, low hydrogen |
| Exx16 | AC or DC, reverse polarity, low hydrogen |
| Exx18 | AC or DC, reverse, iron powder, low hydrogen |
| Exx20 | DC, straight polarity, or AC for horizontal fillet welds; and DC either polarity, or AC, for flat position welding |
| Exx24 | DC, either polarity, or AC, iron powder |
| Exx27 | DC, straight polarity, or AC for horizontal fillet welding; and DC, either polarity, or AC, for flat position welding, iron powder |
| Exx28 | AC or DC, reverse polarity, iron powder, low hydrogen |

The "xx" shown above is a two digit number indicating the weld metal tensile strength in 1000psi increments. For example, E7018 is 70,000 psi weld metal.

# ELECTRODE AMPERAGES

| Type | Amperage Per Rod Diameter (inches) | | | | |
|---|---|---|---|---|---|
| | 1/16 | 5/64 | 3/32 | 1/8 | 5/32 |
| E6010 | | | 60-90 | 80-120 | 110-160 |
| E6011 | | | 50-90 | 80-130 | 120-180 |
| E6012 | | | 40-90 | 80-120 | 120-190 |
| E6013 | 20-40 | 25-60 | 30-80 | 80-120 | 120-190 |
| E7010-A1 | | | 30-80 | 70-120 | 100-160 |
| E7014 | | | 80-110 | 110-150 | 140-190 |
| E7016 | | | 75-105 | 100-150 | 140-190 |
| E7018 (& -A1) | | | 70-120 | 100-150 | 120-200 |
| E7020-A1 | | | | | |
| E7024 | | | 90-120 | 120-150 | 180-230 |
| E7028 | | | | | 175-250 |
| E8016-B2 | | | 60-100 | 80-120 | 140-190 |
| E8018-C3 | | | 70-120 | 100-150 | 120-200 |
| Stainless: | | | | | |
| 3xx AC-DC | 20-40 | 30-60 | 60-90 | 90-120 | 120-160 |
| 4xx AC-DC | 20-40 | 30-60 | 60-90 | 90-120 | 120-160 |
| 5xx AC-DC | 20-40 | 30-60 | 60-90 | 90-120 | 120-160 |

| Type | Amperage Per Rod Diameter (inches) | | | | |
|---|---|---|---|---|---|
| | 3/16 | 7/32 | 1/4 | 5/16 | 3/8 |
| E6010 | 150-200 | 175-250 | 225-300 | | |
| E6011 | 140-220 | 170-250 | 225-325 | | |
| E6012 | 140-240 | 180-315 | 225-350 | | |
| E6013 | 140-240 | 225-300 | 250-350 | | |
| E6020 | 175-250 | 225-325 | 250-350 | 325-450 | 450-600 |
| E6027 | 225-300 | 275-375 | 350-450 | | |
| E7010-A1 | 130-200 | | | | |
| E7014 | 180-260 | 250-325 | 300-400 | 400-500 | |
| E7016 | 190-250 | 250-300 | 300-375 | | |
| E7018 | 200-275 | 275-350 | | | |
| E7020-A1 | 250-350 | | | | |
| E7024 | 250-300 | 300-350 | 350-400 | 400-500 | |
| E7028 | 175-250 | 250-325 | 300-400 | | |
| E8016-B2 | 180-250 | 300-425 | | | |
| E8018-C3 | 200-275 | | | | |
| Stainless: | | | | | |
| 3xx AC-DC | 150-190 | 225-300 | | | |
| 4xx AC-DC | 150-190 | | | | |
| 5xx AC-DC | 150-190 | | 225-300 | | |

Note: All of the above ratings are estimates and you should always verify amperages with the manufacturer before you start a job.

## ELECTRODES - LOW ALLOY STEEL

Low Alloy Steel Specifications (American Welding Society Specification A 5.5-69) are coded the same as the Mild Steel Specification of two pages ago, except that the specification number is followed by a dash and then a letter-number code indicating the chemical composition of the weld metal.

For example, **E8016–C1**

The composition codes are as follows:

–A ....Carbon-molybdenum steel
–B ....Chromium-molybdenum steel
–C ....Nickel steel
–D ....Manganese-molybdenum steel
–G ....All other Low Alloy Steel Electrodes, with
....... minimums of 0.2% molybdenum, 0.3%
....... chromium, 1% manganese, 0.8% silicon,
....... 0.5% nickel, and 0.1% vanadium.
–M ....Military specification

The final digit of the composition code specifies the exact composition of the weld metal.

## ELECTRODES - STAINLESS STEEL

Stainless electrode specifications (AWS A5.4-62T) are coded with the American Iron and Steel Institute alloy type number followed by a dash and two digit number (either 15 or 16) indicating usability or a set of letters (AC, DC, AC–DC or ELC AC–DC) indicating the type of current to be used.

For example, **E308–15**

or **308 ELC AC–DC**

# ELECTRODE BRAND CONVERSION

| Make | E6010 | E6011 | E6012 |
|---|---|---|---|
| Airco | 6010 | 6011,C,LOC | 6012,C |
| Air Products | 6010IP | 6011,C | 6012GP,SF,IP |
| Arc Products | SW610,AP100 | SW14,IMP | SW612,PFA |
| Gen. Dynamics | 610,IP | 611,A | 612,A |
| Hobart Bros. | 10,IP | 335A | 12,212A,12A |
| Lincoln (Fleetweld) | 5,5P | 35,180,35LS | 7 |
| McKay Co | 6010,IP | 6011,IP | 6012 |
| Murex Weld Prod | Speedex 610 | Type A,611C | Type N13 |
| Reid Avery (Raco) | 6010 | 6011,IP | 6012,IP |
| Westinghouse | XL610,A | ACP611 | FP612,2-612 |

| Make | E6013 | E6020 | E6027 |
|---|---|---|---|
| Airco | 6013,C | 6020 | Easyarc 6027 |
| Air Products | 6013GP,SF | 6020 | 6027IP |
| Arc Products | SW16 | | DH27 |
| Gen. Dynamics | 613,A | 620 | IP627 |
| Hobart Bros. | 13A,447A,413 | | 27 |
| Lincoln (Fleetweld) | 37,57 | | Jetweld 2 |
| McKay Co | 6013 | 6020 | |
| Murex Weld Prod | Type U,U13 | Type D,FHP | Speedex 27 |
| Reid Avery Co | 6013 | | |
| Westinghouse | SW613,2M-613 | DH620 | ZIP 27 |

| Make | E7014 | E7016 | E7018 |
|---|---|---|---|
| Airco | Easyarc 7014 | 7016,M | 7018MR,C |
| Air Products | 7014IP | 7016,A | 7018,IP |
| Arc Products | SW15IP | 70LA-2 | 170LA,SW47 |
| Gen. Dynamics | IP714 | 716,A | IP718,A |
| Hobart Bros. | 14A | 16 | LH718 |
| Lincoln (Fleetweld) | 47 | | Jetweld-LH70 |
| McKay Co | 7014 | Puralloy 70AC | 7018 |
| Murex Weld Prod | Speedex U | Type HTS,18,180 | HTS,M,718 |
| Reid Avery Co | 7014 | 7016 | 7018 |
| Westinghouse | ZIP 14 | LOH-2-716 | WIZ-18 |

| Make | E7024 | E7028 |
|---|---|---|
| Airco | Easyarc 7024 | Easyarc 7028 |
| Air Products | 7024IP | 7028 |
| Arc Products | SW44 | DH170 |
| Gen. Dynamics | IP724 | IP728 |
| Hobart Bros. | 24 | |
| Lincoln | Jetweld 3,1 | Jetweld HL3800 |
| McKay Co | 7024 | |
| Murex Weld Prod | Speedex 24 | Speedex 28 |
| Reid Avery Co | | |
| Westinghouse | ZIP 24 | WIZ 28 |

## GAS WELDING RODS

| Rod Diameter | Rods per Pound (36 inch long) Steel | Brass | Aluminum | Cast Iron |
|---|---|---|---|---|
| 1/16 | 31 | 29 | 91 | NA |
| 3/32 | 14 | 13 | 41 | NA |
| 1/8 | 8 | 7 | 23 | NA |
| 5/32 | 5 | NA | NA | NA |
| 3/16 | 3-1/2 | 3 | 9 | 5-1/2 |
| 1/4 | 2 | 2 | 6 | 2-1/4 |
| 5/16 | 1-1/3 | NA | NA | 1/2 |
| 3/8 | 1 | 1 | NA | 1/4 |

## WELDING GASES

| Gas | Tank Sizes Cubic Ft | Comments |
|---|---|---|
| Acetylene | 300<br>100<br>75<br>40<br>10 | Formula – $C_2H_2$, explosive<br>Colorless, flammable gas, garlic-like odor, explosion danger if used in welding with gage pressures over 15 psig (30 psig absolute). |
| Oxygen | 244, 122,<br>80, 40, 20<br>4500 liquid | Formula – $O_2$, non-explosive<br>Colorless, odorless, tasteless.<br>Supports combustion in welding. |
| Nitrogen | 225, 113<br>80, 40, 20 | Formula – $N_2$, non-explosive<br>Colorless, odorless, tasteless, inert. |
| Argon | 330, 131<br>4754 liquid | Formula – Ar, non-explosive<br>Colorless, odorless, tasteless, inert. |
| Carbon Dioxide | 50 lbs<br>20 lbs | Formula – $CO_2$, non-explosive<br>Toxic in large quantities<br>Colorless, odorless, tasteless, inert. |
| Hydrogen | 191 | Formula – $H_2$, explosive<br>Colorless, odorless, tasteless<br>Lightest gas known. |
| Helium | 221 | Formula – He, non-explosive<br>Colorless, odorless, tasteless, inert. |

# HARD & SOFT SOLDER ALLOYS

| Metal to be Soldered | Alloy Component Percentage | | | | |
|---|---|---|---|---|---|
| | Tin | Lead | Zinc | Copper | Other |
| **SOFT SOLDER:** | | | | | |
| Aluminum | 70 | | 25 | | Al = 3,Pho = 2 |
| Bismuth | 33 | 33 | | | Bi = 34 |
| Block tin | 99 | 1 | | | |
| Brass | 66 | 34 | | | |
| Copper | 60 | 40 | | | |
| Gold | 67 | 33 | | | |
| Gun metal | 63 | 37 | | | |
| Iron & Steel | 50 | 50 | | | |
| Lead | 33 | 67 | | | |
| Pewter | 25 | 25 | | | Bi = 50 |
| Silver | 67 | 33 | | | |
| Steel, galvanized | 58 | 42 | | | |
| Steel, tinned | 64 | 36 | | | |
| Zinc | 55 | 45 | | | |
| **HARD SOLDER:** | | | | | |
| Brass, soft | | | 78 | 22 | |
| Brass, hard | | | 55 | 45 | |
| Copper | | | 50 | 50 | |
| Iron, cast | | | 45 | 55 | |
| Iron & Steel | | | 36 | 64 | |
| Gold | | | | 22 | Ag = 11, Au = 67 |
| Silver | | | 10 | 20 | Ag = 70 |

**FLUX:** . . . . . . . . . . . **Metal to be used on:**

Ammonia Chloride . Galvanized iron, iron, nickel, tin, zinc, brass, . . . . . . . . . . . . . . . . . copper, gun metal

Borax . . . . . . . . . . . For hard solders, brass, copper, gold, iron & . . . . . . . . . . . . . . . . . steel, silver

Cuprous oxide . . . . Cast iron

Hydrochloric Acid . . Galvanized iron and steel, tin, zinc

Organic . . . . . . . . . . Lead, pewter

Resin . . . . . . . . . . . Brass, bronze, cadmium, copper, lead, silver, . . . . . . . . . . . . . . . . . gun metal, tinned steel

Stainless Steel Flux Special for stainless steel only

Sterling . . . . . . . . . Silver

Tallow . . . . . . . . . . Lead, pewter

Zinc Chloride . . . . . Bismuth, tin, brass, copper, gold, silver, gun . . . . . . . . . . . . . . . . . metal, tinned steel

# TEMPERING COLOR FOR STEEL

| Heated Color of Carbon Steel | Temperature °F | Temper Item or Comment |
|---|---|---|
| Faint yellow | 420 | Knives, hammers |
| Very pale yellow | 430 | Reamers |
| Light yellow | 440 | Lathe tools, scrapers, milling cutters, reamers |
| Pale straw–yellow | 450 | Twist drills for hard use |
| Straw–yellow | 460 | Dies, punches, bits, reamers |
| Deep straw–yellow | 470 | |
| Dark yellow | 480 | Twist drills, large taps |
| | 485 | Knurls |
| Yellow–brown | 490 | |
| Brown–yellow | 500 | Axes, wood chisels, drifts, taps 1/2 inch or over, nut taps, thread dies |
| Spotted red–brown | 510 | |
| Brown–purple | 520 | Taps 1/4 inch and under |
| Light purple | 530 | |
| Full purple | 540 | Cold chisels, center punches |
| Dark purple | 550 | |
| Full blue | 560 | Screwdrivers, springs, gears |
| Dark blue | 570 | |
| Medium blue | 600 | Scrapers, spokeshaves |
| Light blue | 640 | |
| Red-visible at night | 750 | |
| Red-visible at twilight | 885 | |
| Red-visible in daylight | 975 | |
| Red-visible in sunlight | 1075 | |
| Dark red | 1290 | |
| Dull cherry red | 1475 | |
| Cherry red | 1650 | |
| Bright cherry red | 1830 | |
| Orange–red | 2010 | |
| Orange–yellow | 2190 | |
| Yellow–white | 2370 | |
| White | 2550 | |
| Brilliant white | 2730 | |
| Blue–white | 2900 | |
| Acetylene flame | 4080 | |
| Induction furnace | 5450 | |
| Electric arc light | 7200 | |

Tempering is commonly a two step process. Step 1: To harden the tool, heat the tool end to a bright red, quench tool end in cold water until it is cool to the touch, then sharpen or polish tool end. At this point the tool has been hardened but it is now brittle. Step 2: To temper the tool, heat the tool to the temperature indicated by its color in the above table, then quench the tool in water. The amount of temper is a function of what type of work the tool will be doing, so if your tool is not listed above, simply select one of the above tools that does similar work.

# Conversion Tables

NOTE: Conversions listed in these tables are not exact. Refer to sources such as *Handbook of Chemistry and Physics* and *C.R.C Standard Math Tables* by The Chemical Rubber Publishing Co, *Scientific Tables* by Ciba-Geigy Ltd, *Websters Desk Encyclopedia* by Grisewood & Dempsey, *Field Geologists Manual* by The Australian Institute of Mining & Metallurgy, *Conversion Factors* by Forney's Inc, *Conversions* by Cahn Instruments, and *Technical Reference Handbook* by E.P Rasis for more detailed conversions and specifications.

## TIPS ON CONVERSION FACTORS

The tables that follow contain some of the most commonly used conversion factors. If you can not locate the conversion you need (such as "Feet" to "Inches"), try looking up the reverse conversion ("Inches" to "Feet") and if it exists, divide that number into 1 to get your conversion. In order to save space, only one direction of conversion is listed in some cases.

## ABBREVIATIONS USED IN CONVERSION TABLES

abs .... absolute
apoth .. apothecary
atm .... atmosphere
avdp ... avoirdupois
Brit .... British
Btu .... British thermal unit
Cal .... calorie
cgs .... centimeter-gram-second
chem .. chemical
cm .... centimeter
cu ..... cubic
db ..... decibel
°C..... Degrees centigrade
°F ..... Degrees Fahrenheit
°K..... Degrees Kelvin
°R..... Degrees Reaumur
EM .... electromagnetic
Engl ... English
ES..... electrostatic
flu ..... fluid
ft ...... feet
Germ .. German
gm .... gram
hr ..... hour
Int ..... International
IST .... Intl. Steam Table
kg ..... kilogram
lbs .... pounds
liq ..... liquid
ln ..... logarithm (natural)
log .... logarithm (common)
mech .. mechanical
min .... minute
mm .... millimeter
Naut ... nautical
oz ..... ounces
petro ... petroleum
sec .... second
sq ..... square
US .... United States

| Convert From | Into | Multiply By |
|---|---|---|
| Abamperes | Amperes | 10 |
| | Faradays/sec (chem) | $1.03638 \times 10^{-4}$ |
| | Statamperes | $2.99793 \times 10^{10}$ |
| Abcoulombs | Ampere–hours | 0.00278 |
| | Coulombs | 10 |
| | Electronic charges | $6.24196 \times 10^{19}$ |
| | Faradays (chem) | $1.03638 \times 10^{-4}$ |
| | Statcoulombs | $2.99793 \times 10^{10}$ |
| Abfarads | Farads | $1 \times 10^{9}$ |
| | Microfarads | $1 \times 10^{15}$ |
| | Statfarads | $8.98758 \times 10^{20}$ |
| Abhenries | Henries | $1 \times 10^{-9}$ |
| Abmhos | Megamhos | 1000 |
| | Mhos | $1 \times 10^{9}$ |
| | Statmhos | $8.98758 \times 10^{20}$ |
| Abohms | Megohms | $1 \times 10^{-15}$ |
| | Microhms | 0.001 |
| | Ohms | $1 \times 10^{-9}$ |
| Abohm–cm | Ohm–cm | $1 \times 10^{-9}$ |
| Abvolts | Microvolts | 0.01 |
| | Millivolts | $1 \times 10^{-5}$ |
| | Volts | $1 \times 10^{-8}$ |
| Abvolts/cm | Volts/cm | $1 \times 10^{-8}$ |
| | Volts/inch | $2.54 \times 10^{-8}$ |
| Acres | Hectare or sq hectometer | 0.4047 |
| | Sq chains (Gunters) | 10 |
| | Sq cm | $4.04686 \times 10^{7}$ |
| | Sq feet | 43560 |
| | Sq feet (US Survey) | 43559.83 |
| | Sq inches | $6.27264 \times 10^{6}$ |
| | Sq kilometers | $4.04686 \times 10^{-3}$ |
| | Sq Gunter links | $1 \times 10^{5}$ |
| | Sq meters | 4046.86 |
| | Sq miles (statute) | $1.5625 \times 10^{-3}$ |
| | Sq perches | 160 |
| | Sq rods | 160 |
| | Sq yards | 4840 |
| Acre–feet | Cu feet | 43560 |
| | Cu meters | 1233.482 |
| | Cu yards | 1613.33 |
| | Gallons (US) | $3.259 \times 10^{5}$ |
| Acre–inches | Cu feet | 3630 |
| | Cu meters | 102.79033 |
| | Cu yards | 134.44975 |
| | Gallons (US) | 27154.286 |
| Almude, Portugal | Liters | 16.7 |
| Almude, Spain | Liters | 4.625 |
| Amma (Ancient Greece) | Orguias | 10 |
| | Stadion | 0.01 |
| Amma (Ancient Rome) | Digiti | 4 |
| Amperes | Abamperes | 0.1 |
| | Amperes (Int) | 1.00016 |
| | Coulombs/sec | 1 |
| | Faradays/sec (Chem) | $1.03638 \times 10^{-5}$ |
| | Statamperes | $2.99793 \times 10^{9}$ |

| Convert From | Into | Multiply By |
|---|---|---|
| Amperes (Int) | Amperes | 0.99984 |
| Ampere–hours | Abcoulombs | 360 |
| | Coulombs | 3600 |
| | Faradays (Chem) | 0.3731 |
| Amperes/sq cm | Amps/sq inch | 6.452 |
| | Amps/sq meter | $10^4$ |
| Amperes/sq inch | Amps/sq cm | 0.1550 |
| | Amps/sq meter | 1550.0 |
| Amperes/sq meter | Amps/sq cm | $10^{-4}$ |
| | Amps/sq inch | $6.452 \times 10^{-4}$ |
| Ampere–turns | Gilberts | 1.25664 |
| Angstrom units | Centimeters | $1 \times 10^{-8}$ |
| | Inches | $3.9370 \times 10^{-9}$ |
| | Meters | $1 \times 10^{-10}$ |
| | Microns | 0.0001 |
| | Millimicrons | 0.1 |
| Anker, Latvia | Liters | 38.256 |
| Anoman, Cylon | Bushels, US | 5.83 |
| Archin, Turkey | Meters | 1 |
| Ares | Acres | 0.024711 |
| | Sq dekameters | 1 |
| | Sq feet | 1076.39 |
| | Sq meters | 100 |
| | Sq miles | $3.86102 \times 10^{-5}$ |
| | Sq yards | 119.60 |
| Arpent (French) | Acre (see next line too) | ≈0.85 |
| Arpent (French) | Meters (see above line) | ≈58.6 |
| Arroba, Spain | Liters of wine | 16.14 |
| Artaba, Iran | Liters | 66 |
| Astronomical Unit | Kilometers | $1.459 \times 10^8$ |
| Atmospheres | Bars | 1.01325 |
| | Cm of Hg @ $0^0$C | 76 |
| | Cm of $H^2O$ @ $4^0$C | 1033.26 |
| | Dynes/sq cm | $1.01325 \times 10^6$ |
| | Ft of $H^2O$ @ $39.2^0$F | 33.8995 |
| | Grams/sq cm | 1033.23 |
| | In of Hg @ $32^0$F | 29.9213 |
| | Kg/sq cm | 1.0332 |
| | Kg/sq meter | 10332 |
| | Mm of Hg @ $0^0$C | 760 |
| | Pounds/sq inch | 14.6960 |
| | Tons (short)/sq inch | 0.00735 |
| | Tons (short)/sq foot | 1.05811 |
| | Torrs | 760 |
| Aune, France | Meters | 1.188 |
| Baht, Thailand | Grams | 15 |
| Barile, Rome | Liters | 58.34 |
| Barns | Sq cm | $1 \times 10^{-24}$ |
| Barrels (Brit) | Bags (Brit) | 1.5 |
| | Barrels (US dry) | 1.41540 |
| | Barrels (US liq) | 1.37251 |
| | Bushels (Brit) | 4.5 |
| | Bushels (US) | 4.64425 |
| | Cu feet | 5.77957 |
| | Cu meters | 0.16366 |

| Convert From | Into | Multiply By |
|---|---|---|
| Barrels (Brit) | Gallons (Brit) | 36 |
| | Liters | 163.6546 |
| Barrels (US oil) | Cu feet | 5.61458 |
| | Gallons (US) | 42 |
| | Liters | 158.9828 |
| Barrels (US dry) | Barrels (US liq) | 0.969696 |
| | Bushels (US) | 3.28122 |
| | Cu feet | 4.08333 |
| | Cu inches | 7056 |
| | Cu meters | 0.11563 |
| | Quarts (US dry) | 105 |
| Barrels (US liq) | Barrels (US dry) | 1.0312 |
| | Barrels (wine) | 1 |
| | Cu feet | 4.2104 |
| | Cu inches | 7276.5 |
| | Cu meters | 0.11924 |
| | Gallons (Brit) | 26.22925 |
| | Gallons (US liq) | 31.5 |
| | Liters | 119.2371 |
| Bars | Atmospheres | 0.98692 |
| | Baryes | $1 \times 10^6$ |
| | Cm of Hg @ $0^0$C | 75.0062 |
| | Dynes/sq cm | $1 \times 10^6$ |
| | Ft of H2O @ $60^0$F | 33.4883 |
| | Grams/sq cm | 1019.72 |
| | In of Hg @ $32^0$F | 29.530 |
| | Kg/sq cm | 1.01972 |
| | Millibars | 1000 |
| | Pounds/sq foot | 2089.0 |
| | Pounds/sq inch | 14.5038 |
| Baryl | Atmospheres | $9.8692 \times 10^{-7}$ |
| | Bars | $1 \times 10^{-6}$ |
| | Dynes/sq cm | 1 |
| | Grams/sq cm | 0.00102 |
| | Millibars | 0.001 |
| Baryl | Dyne/sq cm | 1.000 |
| Bath (Old Testament) | Liters | 22 |
| Bekah (Old Testament) | Grams | appx 5 |
| | Shekel | 0.5 |
| Bels | Decibels | 10 |
| Biot | Amperes | 0.10 |
| Bit | Byte (computers) | 1/8 |
| Board feet | Cu cm | 2359.74 |
| | Cu feet | 0.8333 |
| | Cu inches | 144 |
| Bolts of cloth | Ells | 32 |
| | Linear feet | 120 |
| | Meters | 36.576 |
| Bougie decimales | Candles (Int) | 1.0 |
| Btu | Btu (Int Steam Tab) | 0.99935 |
| | Btu (mean) | 0.99856 |
| | Btu @ $60^0$F | 0.99969 |
| | Cal, gm | 251.996 |
| | Cal, gm (Int Steam Tab) | 251.831 |
| | Cal, gm (mean) | 251.634 |

| Convert From | Into | Multiply By |
|---|---|---|
| Btu | Cal, gm @ 20$^0$C | 252.122 |
| | Ergs | $1.0543 \times 10^{10}$ |
| | Foot–poundals | 25020.1 |
| | Foot–pounds | 777.649 |
| | Gram–cm | $1.0751 \times 10^{7}$ |
| | Hp–hours | 0.00039 |
| | Joules | 1054.35 |
| | Joules (Int) | 1054.18 |
| | Kg–calories | 0.2520 |
| | Kg–meters | 107.514 |
| | Kw–hours | 0.00029287 |
| | Kw–hours (Int) | 0.00029283 |
| | Liter–atm | 10.4053 |
| | Therm | 0.00001 |
| | Watt–seconds | 1054.35 |
| | Watt–seconds (Int) | 1054.18 |
| Btu (IST) | Btu | 1.00065 |
| Btu (Mean) | Btu | 1.00144 |
| Btu @ 60$^0$F | Btu | 1.00031 |
| Btu/hr | Calorie–kg/hr | 0.252 |
| | Ergs/sec | $2.92875 \times 10^{5}$ |
| | Foot–pounds/hr | 777.649 |
| | Gram–cal/sec | 0.0700 |
| | Horsepower–hours | 0.00039 |
| | Kilowatts | 0.00029 |
| | Watts | 0.29287 |
| Btu/min | Foot–pounds/sec | 12.96 |
| | Horsepower | 0.02356 |
| | Kilowatts | 0.01757 |
| | Watts | 17.5725 |
| Btu/lb | Cal, gm/gram | 0.55555 |
| | Foot–pounds/lb | 777.649 |
| | Joules/gram | 2.3244 |
| Btu/sq ft/min | Watts/sq inch | 0.1221 |
| Buckets (Brit) | Cu cm | 18184.35 |
| | Gallons (Brit) | 4 |
| Bushels (Brit) | Bags (Brit) | 0.3333 |
| | Bushels (US) | 1.03206 |
| | Cu cm | 36368.7 |
| | Cu feet | 1.28435 |
| | Cu inches | 2219.35 |
| | Gallons (Brit) | 8 |
| | Liters | 36.3677 |
| Bushels (US) | Barrels (US dry) | 0.30476 |
| | Bushels (Brit) | 0.96894 |
| | Cu cm | 35239.07 |
| | Cu feet | 1.24446 |
| | Cu inches | 2150.42 |
| | Cu meters | 0.03524 |
| | Cu yards | 0.04609 |
| | Gallons (US dry) | 8 |
| | Gallons (US liq) | 9.30918 |
| | Liters | 35.23808 |
| | Ounces (US fluid) | 1191.57 |
| | Pecks (US) | 4 |

| Convert From | Into | Multiply By |
|---|---|---|
| Bushels (US) | Pints (US dry) | 64 |
| | Quarts (US dry) | 32 |
| | Quarts (US liq) | 37.23671 |
| Butts (Brit) | Bushels (US) | 13.53503 |
| | Cu feet | 16.84375 |
| | Cu meters | 0.47696 |
| | Gallons (US) | 126 |
| Byte | Bit (computers) | 8 |
| Cable (English) | Degrees latitude | 1/600th |
| | Meter | 185.37 |
| Cable lengths | Fathoms | 120 |
| | Feet | 720 |
| | Meters | 219.456 |
| Calories, gm | Btu | 0.003968 |
| | Btu (IST) | 0.003966 |
| | Btu (mean) | 0.00396 |
| | Btu @ 60$^0$F | 0.00397 |
| | Cal, gm (IST) | 0.99935 |
| | Cal, gm (mean) | 0.99856 |
| | Cal, gm @ 20$^0$C | 1.00050 |
| | Cu cm–atm | 41.2929 |
| | Cu ft–atm | 0.00146 |
| | Ergs | $4.184 \times 10^7$ |
| | Foot–poundals | 99.2878 |
| | Foot–pounds | 3.08596 |
| | Gram–cm | 42664.9 |
| | Hp–hours | $1.558 \times 10^{-6}$ |
| | Joules | 4.184 |
| | Joules (Int) | 4.1833 |
| | Kg–meters | 0.42665 |
| | Kw–hours | $1.162 \times 10^{-6}$ |
| | Liter–atm | 0.04129 |
| | Watt–hours | 0.00116 |
| | Watt–hours (Int) | 0.001162 |
| | Watt–seconds | 4.184 |
| Cal, gm (mean) | Btu | 0.00397 |
| | Cal, gm | 1.00144 |
| Cal, gm @ 20$^0$C | Btu | 0.00397 |
| | Cal, gm | 0.99949 |
| Calories, kg | Btu | 3.96832 |
| | Btu (IST) | 3.96573 |
| | Btu (mean) | 3.96262 |
| | Btu @ 60$^0$F | 3.96709 |
| | Cal, gm | 1000 |
| | Cal, kg (mean) | 0.99856 |
| | Cal, kg @ 20$^0$C | 1.0005 |
| | Cu cm–atm | 41292.9 |
| | Ergs | $4.184 \times 10^{10}$ |
| | Foot–poundals | 99287 |
| | Foot–pounds | 3085.96 |
| | Gram–cm | $4.266 \times 10^7$ |
| | Hp–hours | 0.00156 |
| | Joules | 4184 |
| | Kw–hours | 0.00116 |
| | Liter–atm | 41.292 |

| Convert From | Into | Multiply By |
|---|---|---|
| Calories, kg | Watt–hours | 1.1622 |
| Cal, kg (mean) | Btu | 3.974 |
| | Cal, gm | 1001.4 |
| Cal, gm/gm | Btu/lb | 1.8 |
| | Foot–pounds/lb | 1399.8 |
| | Joules/gram | 4.184 |
| Cal, gm/hr | Btu/hr | 0.00397 |
| | Ergs/sec | 11622 |
| | Watts | 0.00116 |
| Cal, kg/hr | Watts | 1.1622 |
| Cal, gm/min | Btu/min | 0.00397 |
| | Ergs/sec | 697333 |
| | Watts | 0.0697 |
| Cal, kg/min | Watts | 69.733 |
| Cal, gm/sec | Btu/sec | 0.00397 |
| | Foot–pounds/sec | 3.086 |
| | Horsepower | 0.0056 |
| | Watts | 4.184 |
| Cal, gm–sec | Planck's constant | $6.315 \times 10^{33}$ |
| Candles (Engl) | Candles (Int) | 1.04 |
| Candles (Germ) | Candles (Int) | 1.05 |
| Candles (Int) | Candles (Engl) | 0.96 |
| | Candles (Germ) | 0.95 |
| | Candles (pentane) | 1.00 |
| | Carcel units | 0.104 |
| | Hefner units | 1.11 |
| | Lumens/steradian | 1 |
| Candles/sq cm | Candles/sq in | 6.452 |
| | Candles/sq meter | 10000 |
| | Foot–lamberts | 2918 |
| | Lamberts | 3.1416 |
| Candles/sq in | Candles/sq cm | 0.155 |
| | Candles/sq ft | 144 |
| | Foot–lamberts | 452.39 |
| | Lamberts | 0.4869 |
| Candle power | Lumens | 12.566 |
| Cape foot (S. Africa) | Meter | 0.315 |
| Cape rood (S. Africa) | Meter | 3.788 |
| Carats (gold) | Milligrams/gram | 41.666 |
| Carats | Grains | 3.0865 |
| | Grams | 0.2 |
| | Milligrams | 200 |
| Carcel units | Candles (Int) | 9.61 |
| Centals | Kilograms | 45.359 |
| | Pounds | 100 |
| Centares | Ares | 0.01 |
| | Sq feet | 10.764 |
| | Sq inches | 1550 |
| | Sq meters | 1 |
| | Sq yards | 1.19599 |
| Centigrams | Grains | 0.15432 |
| | Grams | 0.01 |
| Centiliters | Cu cm | 10.00028 |
| | Cu inches | 0.61025 |
| | Drams | 2.705 |

| Convert From | Into | Multiply By |
|---|---|---|
| Centiliters | Liters | 0.01 |
| | Ounces (US fluid) | 0.33815 |
| Centimeters | Angstrom units | $1 \times 10^8$ |
| | Feet | 0.03281 |
| | Hands | 0.0984 |
| | Inches | 0.3937 |
| | Kilometers | $10^{-5}$ |
| | Links (Gunter's) | 0.0497 |
| | Links (Ramden's) | 0.0328 |
| | Meters | 0.01 |
| | Microns | 10000 |
| | Miles (Naut) | $5.3996 \times 10^{-6}$ |
| | Miles (statute) | $6.2137 \times 10^{-6}$ |
| | Millimeters | 10 |
| | Millimicrons | $1 \times 10^7$ |
| | Mils | 393.7 |
| | Picas (printers) | 2.371 |
| | Points (printers) | 28.4528 |
| | Rods | 0.00199 |
| | Yards | 0.01094 |
| Cm–dynes | Cm–grams | $1.02 \times 10^{-3}$ |
| | Meter–kgs | $1.02 \times 10^{-8}$ |
| | Pound–feet | $7.376 \times 10^{-8}$ |
| Cm–grams | Cm–dynes | 980.7 |
| | meter–kgs | $10^{-5}$ |
| | Pound–feet | $7.23 \times 10^{-5}$ |
| Cm of Hg $0^oC$ | Atmospheres | 0.01316 |
| | Bars | 0.01333 |
| | Dynes/sq cm | 13332 |
| | Ft of $H_2O$ @ $4^oC$ | 0.446 |
| | In Hg @ $0^oC$ | 0.3937 |
| | Kg/sq meter | 135.95 |
| | Lbs/sq ft | 27.845 |
| | Lbs/sq in | 0.1934 |
| | Torrs | 10 |
| Cm of $H_2O$ $4^oC$ | Atmospheres | 0.00097 |
| | Lbs/sq in | 0.1422 |
| Cm/sec | Feet/min | 1.9685 |
| | Feet/sec | 0.0328 |
| | Km/hr | 0.036 |
| | Km/min | 0.0006 |
| | Knots | 0.0194 |
| | Meters/min | 0.6 |
| | Miles/hr | 0.02237 |
| | Miles/min | 0.000373 |
| Cm/sec/sec | Ft/sec/sec | 0.0328 |
| | Km/hr/sec | 0.036 |
| | Meters/sec/sec | 0.01 |
| | Miles/hr/sec | 0.0224 |
| Cm/year | Inches/year | 0.3937 |
| Centipoises | Gms/cm/sec | 0.01 |
| | Poises | 0.01 |
| | Lbs/ft/hr | 2.4191 |
| | Lbs/ft/sec | 0.00067 |
| Chains (Gunter) | Centimeters | 2011.7 |

| Convert From | Into | Multiply By |
|---|---|---|
| Chains (Gunter) | Chains (Ramden) | 0.66 |
| | Feet | 66 |
| | Feet (US Survey) | 65.99 |
| | Furlongs | 0.1 |
| | Inches | 792 |
| | Links (Gunter) | 100 |
| | Links (Ramden) | 66 |
| | Meters | 20.117 |
| | Miles | 0.0125 |
| | Rods | 4 |
| | Yards | 22 |
| Chains (Ramden) | Chains (Gunter) | 1.5151 |
| | Feet | 100 |
| Chaldron, dry (English) | Bushels | 36 |
| Cheval vapeur | Horsepower | 1.0139 |
| Circles | Degrees | 360 |
| | Grades | 400 |
| | Minutes | 21600 |
| | Radians | 6.2832 |
| | Signs | 12 |
| Circular inches | Circular mm | 645.16 |
| | Sq cm | 5.067 |
| | Sq inches | 0.7854 |
| Circular mm | Sq cm | 0.00785 |
| | Sq inches | 0.00122 |
| | Sq mm | 0.7854 |
| Circular mils | Circular inches | $1 \times 10^{-6}$ |
| | Sq cm | 5.06707 |
| | Sq inches | $7.85398 \times 10^{-7}$ |
| | Sq mm | 0.000507 |
| | Sq mills | 0.7854 |
| Circumference | Degrees | 360 |
| | Radians | 6.28318 |
| Cords | Cord feet | 8 |
| | Cu feet | 128 |
| | Cu meters | 3.6246 |
| Cord–feet | Cords | 0.125 |
| | Cu feet | 16 |
| Coulombs | Abcoulombs | 0.1 |
| | Ampere–hours | 0.000278 |
| | Ampere–seconds | 1 |
| | Coulombs (Int) | 1.00016 |
| | Faradays (Chem) | $1.0364 \times 10^{-5}$ |
| | Faradays (Phys) | $1.0361 \times 10^{-5}$ |
| | Mks elec chg unit | 1 |
| | Statcoulombs | $2.9979 \times 10^{9}$ |
| Coulombs/sq cm | Coulombs/sq in | 64.52 |
| | Coulombs/sq meter | 10000 |
| Coulombs/sq in | Coulombs/sq cm | 0.1550 |
| | Coulombs/sq meter | 1550 |
| Coulombs/sq meter | Coulombs/sq cm | $10^{-4}$ |
| | Coulombs/sq inch | $6.452 \times 10^{-4}$ |
| Cu centimeters | Board feet | 0.00042 |
| | Bushels (Brit) | $2.7496 \times 10^{-5}$ |
| | Bushels (US) | $2.8378 \times 10^{-5}$ |

| Convert From | Into | Multiply By |
|---|---|---|
| Cu centimeters | Cu feet | $3.5315 \times 10^{-5}$ |
| | Cu inches | 0.06102 |
| | Cu meters | $1 \times 10^{-6}$ |
| | Cu yards | $1.308 \times 10^{-6}$ |
| | Drachms | 0.28156 |
| | Drams (US fluid) | 0.27051 |
| | Gallons (Brit) | 0.00022 |
| | Gallons (US dry) | 0.00023 |
| | Gallons (US liq) | 0.00026 |
| | Gills (Brit) | 0.00704 |
| | Gills (US) | 0.00845 |
| | Liters | 0.00099 |
| | Ounces (Brit liq) | 0.03519 |
| | Ounces (US liq) | 0.03381 |
| | Pints (US dry) | 0.00182 |
| | Pints (US liq) | 0.00211 |
| | Quarts (Brit) | 0.00088 |
| | Quarts (US dry) | 0.00091 |
| | Quarts (US liq) | 0.00106 |
| Cu cm/gram | Cu ft/lb | 0.01602 |
| Cu cm/sec | Cu ft/min | 0.00212 |
| | Gallons (US)/min | 0.01585 |
| | Gallons (US)/sec | 0.00026 |
| Cu cm-atm | Btu | $9.61 \times 10^{-5}$ |
| | Cal, gm | 0.02422 |
| | Joules | 0.10132 |
| | Watt-hours | $2.81 \times 10^{-5}$ |
| Cu decimeters | Cu cm | 1000 |
| | Cu feet | 0.03532 |
| | Cu inches | 61.0237 |
| | Cu meters | 0.001 |
| | Cu yards | 0.00131 |
| | Liters | 0.99997 |
| Cu dekameters | Cu decimeters | $1 \times 10^{6}$ |
| | Cu feet | 35314.7 |
| | Cu inches | $6.102 \times 10^{7}$ |
| | Cu meters | 1000 |
| | Liters | 999972 |
| Cu feet | Acre-feet | $2.296 \times 10^{-5}$ |
| | Board-feet | 12 |
| | Bushels (Brit) | 0.7786 |
| | Bushels (US) | 0.8036 |
| | Cords of wood | 0.00781 |
| | Cord-feet | 0.0625 |
| | Cu centimeters | 28316.8 |
| | Cu inches | 1728.0 |
| | Cu meters | 0.02832 |
| | Cu yards | 0.03704 |
| | Gallons (US dry) | 6.42851 |
| | Gallons (US liq) | 7.48052 |
| | Liters | 28.316 |
| | Ounces (Brit fluid) | 996.614 |
| | Ounces (US fluid) | 957.506 |
| | Pints (US dry) | 51.4281 |
| | Pints (US liq) | 59.8442 |

| Convert From | Into | Multiply By |
|---|---|---|
| Cu feet | Quarts (US dry) | 25.714 |
| | Quarts (US liq) | 29.922 |
| Cu ft $H_2O$ 60°F | Lbs of $H_2O$ | 63.367 |
| Cu ft/hr | Acre–feet/hr | $2.2957 \times 10^{-5}$ |
| | Cu cm/sec | 7.8658 |
| | Cu ft/hr | 60 |
| | Gallons (US)/hr | 7.4805 |
| | Liters/hr | 28.316 |
| Cu ft/min | Acre–feet/hr | 0.00138 |
| | Acre–feet/min | $2.2956 \times 10^{-5}$ |
| | Cu cm/sec | 471.95 |
| | Cu ft/hr | 60 |
| | Gallons (US)/min | 7.48052 |
| | Gallons (US)/sec | 0.1247 |
| | Liters/sec | 0.47193 |
| | Pounds of $H_2O$/min | 62.43 |
| Cu ft/pound | Cu cm/gram | 62.428 |
| | Mm/gram | 62.4262 |
| Cu ft/sec | Acre–inches/hr | 0.99173 |
| | Cu cm/sec | 28316.8 |
| | Cu yards/min | 2.2222 |
| | Gallons (US)/min | 448.83 |
| | Liters/min | 1698.96 |
| | Liters/sec | 28.316 |
| | Million gallons/day | 0.64632 |
| Cu ft $H_2O$/sec | Lbs $H_2O$/min | 3741.98 |
| Cu ft–atm | Btu | 2.7213 |
| | Cal, gm | 685.76 |
| | Foot–pounds | 2116.2 |
| | Hp–hours | 0.00107 |
| | Kg–meters | 292.58 |
| | Kw–hours | 0.000797 |
| Cubic inches | Barrels (Brit) | 0.0001 |
| | Barrels (US dry) | 0.0001417 |
| | Board feet | 0.00694 |
| | Bushels (Brit) | 0.00045 |
| | Bushels (US) | 0.00046 |
| | Cu cm | 16.3871 |
| | Cu feet | 0.000579 |
| | Cu meters | $1.639 \times 10^{-5}$ |
| | Cu yards | $2.143 \times 10^{-5}$ |
| | Drams (US fluid) | 4.43290 |
| | Gallons (Brit) | 0.00360 |
| | Gallons (US dry) | 0.00372 |
| | Gallons (US liq) | 0.00433 |
| | Liters | 0.01639 |
| | Milliliters | 16.3866 |
| | Ounces (Brit fluid) | 0.57674 |
| | Ounces (US fluid) | 0.55411 |
| | Pecks | 0.00186 |
| | Pints (US dry) | 0.02976 |
| | Pints (US liq) | 0.03463 |
| | Quarts (US dry) | 0.01488 |
| | Quarts (US liq) | 0.01732 |
| Cu in $H_2O$ 60°F | Lbs of $H_2O$ | 0.03609 |

| Convert From | Into | Multiply By |
|---|---|---|
| Cu meters | Acre–feet | 0.00081 |
| | Barrels (Brit) | 6.11026 |
| | Barrels (US dry) | 8.64849 |
| | Barrels (US liq) | 8.38641 |
| | Bushels (Brit) | 27.4962 |
| | Bushels (US) | 28.3776 |
| | Cu cm | $1 \times 10^6$ |
| | Cu feet | 35.3147 |
| | Cu inches | 61023.7 |
| | Cu yards | 1.30795 |
| | Gallons (Brit) | 219.969 |
| | Gallons (US liq) | 264.172 |
| | Hogshead | 4.1932 |
| | Liters | 999.97 |
| | Pints (US liq) | 2113.38 |
| | Quarts (US liq) | 1056.69 |
| | Steres | 1 |
| Cu meters/min | Gallons (Brit)/min | 219.969 |
| | Gallons (US)/min | 264.172 |
| | Liters/min | 999.97 |
| Cu mm | Cu cm | 0.001 |
| | Cu inches | $6.102 \times 10^{-5}$ |
| | Cu meters | $1 \times 10^{-9}$ |
| | Minims (Brit) | 0.01689 |
| | Minims (US) | 0.01623 |
| Cu yards | Bushels (Brit) | 21.0223 |
| | Bushels (US) | 21.6962 |
| | Cu cm | 764554.9 |
| | Cu feet | 27 |
| | Cu inches | 46.656 |
| | Cu meters | 0.76455 |
| | Gallons (Brit) | 168.179 |
| | Gallons (US dry) | 173.569 |
| | Gallons (US liq) | 201.974 |
| | Liters | 764.533 |
| | Prospecting dishes | 112 |
| | Quarts (Brit) | 672.715 |
| | Quarts (US dry) | 694.279 |
| | Quarts (US liq) | 807.896 |
| Cu yards/min | Cu ft/sec | 0.45 |
| | Gallons (US)/sec | 3.3662 |
| | Liters/sec | 12.742 |
| Cubits | Centimeters | 45.72 to 55.9 |
| | Feet | 1.5 |
| | Inches | 18 to 22 |

Cubit is the distance from the elbow to the finger tip:

| Convert From | Into | Multiply By |
|---|---|---|
| Cubit (Bible) | Inches | 21.8 |
| Cubit (Egypt 2650BC) | Inches | 20.6 |
| Cubit (Babylon 1500BC) | Inches | 20.9 |
| Cubit (Assyrian 700 BC) | Inches | 21.6 |
| Cubit (Jerusalem (1 AD) | Inches | 20.6 |
| Cubit (Druid Eng 1AD) | Inches | 20.4 |
| Cubit (Black, Arabia, 800 AD) | Inches | 21.3 |
| Cubit (Mexico–Aztec) | Inches | 20.7 |
| Cubit (Ancient China) | Inches | 20.9 |

| Convert From | Into | Multiply By |
|---|---|---|
| Cubit (Ancient Greece) | Inches | 18.2 |
| Cubit (England) | Inches | 18.0 |
| Cubit (Northern 3000BC to 1800AD) | Inches | 26.6 |
| Cup | Gallons | 0.0625 |
| | Gills | 2 |
| | Pint | 0.5 |
| | Milliliters | 284.13 |
| | Ounces, fluid | 8 |
| | Quarts | 0.25 |
| | Tablespoons | 16 |
| | Teaspoons | 48 |
| Cup, metric | Milliliters | 250.0 |
| Cup, tea | Pint | 0.25 |
| | Milliliters | 142.06 |
| Daltons (Chem) | Grams | $1.66 \times 10^{-24}$ |
| Daltons (Phys) | Grams | $1.659 \times 10^{-24}$ |
| Day– mean solar | Day (sidereal) | 1.0027379 |
| | Hours (mean solar) | 24 |
| | Hours (sidereal) | 24.06571 |
| | Years (calendar) | 0.0027397 |
| | Years (sidereal) | 0.002738 |
| | Years (tropical) | 0.002738 |
| Days (sidereal) | Days (mean solar) | 0.9972696 |
| | Hours (mean solar) | 23.93447 |
| | Hours (sidereal) | 24 |
| | Min (mean solar) | 1436.068 |
| | Min (sidereal) | 1440 |
| | Second (sidereal) | 86400 |
| | Years (calendar) | 0.002732 |
| | Years (sidereal) | 0.00273 |
| | Years (tropical) | 0.00273 |
| Decibels | Bels | 0.1 |
| Decigrams | Grams | 0.1 |
| Deciliters | Liters | 0.1 |
| Decimeters | Centimeters | 10 |
| | Feet | 0.32808 |
| | Inches | 3.937 |
| | Meters | 0.1 |
| Decisteres | Cu meters | 0.1 |
| Degrees | Circles | 0.00278 |
| | Minutes | 60 |
| | Quadrants | 0.01111 |
| | Radians | 0.01745 |
| | Seconds | 3600 |
| Degrees/cm | Radians/cm | 0.01745 |
| Degrees/foot | Radians/cm | 0.00056 |
| Degrees/inch | Radians/cm | 0.00687 |
| Degrees/min | Degrees/sec | 0.01667 |
| | Radians/sec | 0.00029 |
| | Revolutions/sec | $4.6296 \times 10^{-5}$ |
| Degrees/sec | Radians/sec | 0.01745 |
| | Revolutions/min | 0.16666 |
| | Revolutions/sec | 0.00278 |
| Dekagrams | Grams | 10.0 |
| Dekaliters | Liters | 10.0 |

| Convert From | Into | Multiply By |
|---|---|---|
| Dekaliters | Pecks | 1.1351 |
| | Pints (US dry) | 18.162 |
| Dekameters | Centimeters | 1000 |
| | Feet | 32.8084 |
| | Inches | 393.7008 |
| | Kilometers | 0.01 |
| | Meters | 10 |
| | Yards | 10.9361 |
| Demals | Gm–equiv/cu decim. | 1 |
| Digit (Ancient Greece) | Centimeters | 1.84 |
| | Inches | 0.72 |
| | Orguia | 0.01 |
| Digitus (Ancient Rome) | Centimeters | 1.84 |
| | Inches | 0.73 |
| | Palmus | 0.25 |
| Drachms | Cu centimeter | 3.55163 |
| | Cu inches | 0.21673 |
| | Drams | 0.96076 |
| | Grams | 4.2923 |
| | Milliliters | 3.55153 |
| | Scruples | 3 |
| Drachm, fluid | Cubic inches | 0.21673 |
| | Minims | 60 |
| | Milliliters | 3.55163 |
| Drams (troy) | Drams (avdp) | 2.19429 |
| | Grains | 60 |
| | Grams | 3887.93 |
| | Ounces (troy) | 0.125 |
| Drams (avdp) | Drams (troy) | 0.455729 |
| | Grains | 27.3437 |
| | Grams | 1.771845 |
| | Ounces (troy) | 0.056966 |
| | Ounces (avdp) | 0.0625 |
| | Pennyweights | 1.13932 |
| | Pounds (troy) | 0.004747 |
| | Pounds (avdp) | 0.00391 |
| | Scruples (apoth) | 1.36719 |
| Drams (US fluid) | Cu cm | 3.6967 |
| | Cu inches | 0.225586 |
| | Drachms | 1.04084 |
| | Gills (US) | 0.03125 |
| | Millimeters | 3.69659 |
| | Minims | 60 |
| | Ounces (US fluid) | 0.125 |
| | Pints (US liq) | 0.00781 |
| Dynes | Grains | 0.015737 |
| | Grams | 0.0010197 |
| | Joules/cm | $10^{-7}$ |
| | Joules/meter (newtons) | $10^{-5}$ |
| | Kilograms | $1.02 \times 10^{-6}$ |
| | Newtons | 0.00001 |
| | Poundals | $7.233 \times 10^{-5}$ |
| | Pounds | $2.248 \times 10^{-6}$ |
| Dynes/cm | Ergs/sq cm | 1 |
| | Ergs/sq mm | 0.01 |

| Convert From | Into | Multiply By |
|---|---|---|
| Dynes/cm | Grams/cm | 0.0010197 |
| | Poundals/inch | 0.0001837 |
| Dynes/cu cm | Grams/cu cm | 0.0010197 |
| | Poundals/cu inch | 0.001185 |
| Dynes/sq cm | Atmospheres | $9.869 \times 10^{-7}$ |
| | Bars | $1 \times 10^{-6}$ |
| | Baryes | 1 |
| | Cm of Hg @ 0°C | |
| | Cm of H20 @ 4°C | 0.00101975 |
| | Grams/sq cm | 0.00101972 |
| | In of Hg @ 32°F | $2.953 \times 10^{-5}$ |
| | In of H2O @ 4°C | 0.000401 |
| | Kg/sq meter | 0.010197 |
| | Poundals/sq in | 0.0004666 |
| | Pounds/sq in | $1.450 \times 10^{-5}$ |
| Dyne–Cm | Ergs | 1 |
| | Foot–poundals | $2.373 \times 10^{-6}$ |
| | Foot–pounds | $7.376 \times 10^{-5}$ |
| | Gram–cm | 0.00102 |
| | inch–pounds | $8.8507 \times 10^{-7}$ |
| | Kg–meters | $1.0197 \times 10^{-8}$ |
| | Newton–neters | $1 \times 10^{-7}$ |
| Electron Volts | Ergs | $1.6021 \times 10^{-12}$ |
| | kcal/mole | $1.602 \times 10^{-12}$ |
| Electronic charges | Abcoulombs | $1.6021 \times 10^{-20}$ |
| | Coulombs | $1.6021 \times 10^{-19}$ |
| | Statcoulombs | $4.803 \times 10^{-10}$ |
| Ells (cloth) | Cm | 114.3 |
| | Inches | 45 |
| Em, Pica | Inch | 0.167 |
| | Cm | 0.4233 |
| Ephah | Liters | 22 |
| | Omers | 10 |
| Ergs | Btu | $9.4845 \times 10^{-11}$ |
| | Gram calorie | $2.3901 \times 10^{-8}$ |
| | Kg calorie | $2.3901 \times 10^{-11}$ |
| | Cu cm–atmospheres | $9.8692 \times 10^{-7}$ |
| | Cu ft–atmospheres | $3.4853 \times 10^{-11}$ |
| | Cu ft–lbs/sq in | $5.122 \times 10^{-10}$ |
| | Dyne–cm | 1 |
| | Electron Volts | $6.242 \times 10^{11}$ |
| | Foot–poundals | $2.374 \times 10^{-6}$ |
| | Foot–pounds | $7.376 \times 10^{-8}$ |
| | Gram–calories | $0.239 \times 10^{-7}$ |
| | Gram–cm | 0.0010197 |
| | Horsepower–hours | $3.725 \times 10^{-14}$ |
| | Joules | $1 \times 10^{-7}$ |
| | Joules (Int) | $0.998 \times 10^{-7}$ |
| | Kw–hours | $2.778 \times 10^{-14}$ |
| | Kg–calories | $2.389 \times 10^{-11}$ |
| | Kg–meters | $1.0197 \times 10^{-8}$ |
| | Liter–atmospheres | $9.869 \times 10^{-10}$ |
| | Watt–hours | $0.278 \times 10^{-10}$ |
| | Watt–sec | $1 \times 10^{-7}$ |
| Ergs/sec | Btu/min | $5.691 \times 10^{-9}$ |

| Convert From | Into | Multiply By |
|---|---|---|
| Ergs/sec | Gram calorie/min | $1.434 \times 10^{-6}$ |
| | Dyne–cm/sec | 1 |
| | Foot–pounds/min | $4.425 \times 10^{-6}$ |
| | Foot–pounds/sec | $7.376 \times 10^{-8}$ |
| | Gram–cm/sec | 0.0010197 |
| | Horsepower | $1.341 \times 10^{-10}$ |
| | Joules/sec | $1 \times 10^{-7}$ |
| | Kg–calories/minute | $1.43 \times 10^{-9}$ |
| | Kilowatts | $1 \times 10^{-10}$ |
| | Watts | $1 \times 10^{-7}$ |
| Ergs/sq cm | Dynes/cm | 1 |
| | Ergs/sq mm | 0.01 |
| Ergs/sq mm | Dynes/cm | 100 |
| | Ergs/sq cm | 100 |
| Erg–sec | Planck's constant | $1.5093 \times 10^{26}$ |
| Faraday | Ampere–hours | 26.8 |
| | Coloumbs | $9.649 \times 10^{4}$ |
| Faraday/sec | Ampere (absolute) | $9.65 \times 10^{4}$ |
| Farads | Abfarads | $1 \times 10^{-9}$ |
| | Farads (Int) | 1.00049 |
| | Microfarads | $1 \times 10^{6}$ |
| | Statfarads | $8.98758 \times 10^{11}$ |
| Farads (Int) | Farads | 0.9995 |
| Fathoms | Centimeters | 182.88 |
| | Feet | 6 |
| | Furlongs | 0.1 |
| | Inches | 72 |
| | Meters | 1.8288 |
| | Miles (naut,Int) | 0.00098747 |
| | Miles (statute) | 0.0011363 |
| | Yards | 2 |
| Feet | Centimeters | 30.48 |
| | Chains (Gunter's) | 0.015151 |
| | Fathoms | 0.16666 |
| | Feet (US Survey) | 0.999998 |
| | Furlongs | 0.001515 |
| | Inches | 12 |
| | Kilometers | $3.048 \times 10^{-4}$ |
| | Meters | 0.3048 |
| | Microns | 304800 |
| | Miles (naut,Int) | 0.000165 |
| | Miles (statute) | 0.000189 |
| | Millimeters | 304.8 |
| | Mils | $1.2 \times 10^{4}$ |
| | Rods | 0.060606 |
| | Ropes (Brit) | 0.05 |
| | Yards | 0.333333 |
| Feet (US Survey) | Centimeters | 30.48006 |
| | Chains (Gunter's) | 0.015152 |
| | Chains (Ramden's) | 0.01000002 |
| | Feet | 1.000002 |
| | Inches | 12.000024 |
| | Links (Gunter's) | 1.515155 |
| | Links (Ramden's) | 1.000002 |
| | Meters | 0.304801 |

| Convert From | Into | Multiply By |
|---|---|---|
| Feet (US Survey) | Miles (statute) | 0.00018939 |
| | Rods | 0.06060618 |
| | Yards | 0.333334 |
| Feet (Athens History) | Inches | 12.44 |
| Feet (Aegina History) | Inches | 12.36 |
| Feet (Miletus History) | Inches | 12.52 |
| Feet (Olympia History) | Inches | 12.64 |
| Feet (Etruria History) | Inches | 12.44 |
| Feet (Rome History) | Inches | 11.66 |
| Feet (North History) | Inches | 13.19 |
| Feet (England History) | Inches | 13.19 |
| Feet (France History) | Inches | 12.79 |
| Feet (Moscos History) | Inches | 13.17 |
| Feet of Air @ 60°F | Atmospheres | $3.608 \times 10^{-5}$ |
| | Ft of Hg @ 32°F | 0.0009 |
| | Ft of $H_2O$ @ 60°F | 0.00122 |
| | In of Hg @ 32°F | 0.00108 |
| | Pounds/sq in | 0.00053 |
| Feet of Hg @ 32°F | Cm of Hg @ 0°C | 30.48 |
| | Ft of $H_2O$ @ 60°F | 13.608 |
| | In of $H_2O$ @ 60°F | 163.30 |
| | Ounces/sq in | 94.302 |
| | Pounds/sq in | 5.8938 |
| Feet of $H_2O$ @ 4°C | Atmospheres | 0.0295 |
| | Cm of Hg @ 0°C | 2.2419 |
| | Dynes/sq cm | 29889.8 |
| | Grams/sq cm | 30.479 |
| | In of Hg @ 32°F | 0.8826 |
| | Kg/sq meter | 304.79 |
| | Pounds/sq foot | 62.43 |
| | Pounds/sq inch | 0.43351 |
| Feet/hour | Cm/hr | 30.48 |
| | Cm/minute | 0.508 |
| | Cm/second | 0.00846 |
| | Feet/minute | 0.01666 |
| | Inches/hour | 12 |
| | Kilometers/hr | 0.0003 |
| | Kilometers/min | $5.08 \times 10^{-6}$ |
| | Knots (Int) | 0.00016458 |
| | Miles/hr | 0.0001894 |
| | Miles/min | $3.15656 \times 10^{-6}$ |
| | Miles/sec | $5.2609 \times 10^{-8}$ |
| Feet/minute | Cm/sec | 0.508 |
| | Feet/sec | 0.01666 |
| | Kilometers/hr | 0.018288 |
| | Meters/min | 0.3048 |
| | Meters/sec | 0.00508 |
| | Miles/hr | 0.011363 |
| Feet/second | Cm/sec | 30.48 |
| | Kilometers/hr | 1.09728 |
| | Kilometers/min | 0.01829 |
| | Knots | 0.5921 |
| | Meters/min | 18.288 |
| | Miles/hr | 0.681818 |
| | Miles/min | 0.011364 |

| Convert From | Into | Multiply By |
|---|---|---|
| Feet/(sec x sec) | Cm/(sec x sec) | 30.48 |
| | Km/(hr x sec) | 1.0973 |
| | Meters/(sec x sec) | 0.3048 |
| | Miles/(hr x sec) | 0.681818 |
| Feet/100 feet | Percent grade | 1 |
| Fifth | Jiggers | 38.4 |
| | Ounces, fluid | 25.6 |
| | Pints | 1.6 |
| | Pony | 76.9 |
| | Quart | 0.80 |
| | Shots | 25.6 |
| Firkins (Brit) | Bushels (Brit) | 1.125 |
| | Cu cm | 40914.8 |
| | Cu feet | 1.44489 |
| | Firkins (US) | 1.2009 |
| | Gallons (Brit) | 9 |
| | Liters | 40.91364 |
| | Pints (Brit) | 72 |
| Firkins (US) | Barrels (US dry) | 0.294643 |
| | Barrels (US liq) | 0.285715 |
| | Bushels (US) | 0.966788 |
| | Cu feet | 1.203125 |
| | Firkins (Brit) | 0.832675 |
| | Gallons (US liq) | 9 |
| | Liters | 34.0677 |
| | Pints (US liq) | 72 |
| Flask of mercury | Kilograms | 34.5 |
| Foot–candles | Lumens/sq ft | 1 |
| | Lumens/sq meter | 10.7639 |
| | Lux | 10.7639 |
| | Milliphots | 1.07639 |
| Foot–lamberts | Candles/sq cm | 0.00034 |
| | Candles/sq ft | 0.31831 |
| | Millilamberts | 1.07639 |
| | Lamberts | 0.0010764 |
| | Lumens/sq ft | 1 |
| Foot–poundals | Btu | $3.9968 \times 10^{-5}$ |
| | Btu (IST) | $3.9942 \times 10^{-5}$ |
| | Btu (mean) | $3.991 \times 10^{-5}$ |
| | Cal, gram | 0.010072 |
| | Cal, gram (IST) | 0.010065 |
| | Cal, gram (mean) | 0.010057 |
| | Cu cm–atmospheres | 0.41589 |
| | Cu ft–atmospheres | $1.4687 \times 10^{-5}$ |
| | Dyne–cm | $4.21401 \times 10^{5}$ |
| | Ergs | $4.21401 \times 10^{5}$ |
| | Foot–pounds | 0.03108 |
| | Hp–hours | $1.5697 \times 10^{-8}$ |
| | Joules | 0.0421401 |
| | Joules (Int) | 0.042133 |
| | Kg–meters | 0.004297 |
| | Kw–hours | $1.1706 \times 10^{-8}$ |
| | Liter–atmospheres | 0.0004159 |
| Foot–pounds | Btu | 0.001286 |
| | Btu (IST) | 0.001285 |

| Convert From | Into | Multiply By |
|---|---|---|
| Foot-pounds | Btu (mean) | 0.00128 |
| | Cal, gram | 0.32405 |
| | Cal, gram (IST) | 0.32384 |
| | Cal, gram (mean) | 0.32358 |
| | Cal, gram @ 20°C | 0.32421 |
| | Cal, kg | 0.00032 |
| | Cu ft-atmospheres | 0.00047 |
| | Dyne-cm | $1.3558 \times 10^{7}$ |
| | Ergs | $1.3558 \times 10^{7}$ |
| | Foot-poundals | 32.174 |
| | Gram-calories | 0.3238 |
| | Gram-cm | 13825.5 |
| | Hp-hours | $5.05 \times 10^{-7}$ |
| | Joules | 1.3558 |
| | Kg-calories | $3.24 \times 10^{-4}$ |
| | Kg-meters | 0.13825 |
| | Kw-hours | $3.766 \times 10^{-7}$ |
| | Kw-hours (Int) | $3.766 \times 10^{-7}$ |
| | Liter-atmospheres | 0.01338 |
| | Newton-meters | 1.35582 |
| | Watt-hours | 0.000377 |
| Foot-pounds/hr | Btu/min | $2.1432 \times 10^{-5}$ |
| | Btu (mean)/min | $2.1401 \times 10^{-5}$ |
| | Cal, gram/min | 0.0054 |
| | Cal, gram (mean)/min | 0.00539 |
| | Ergs/min | $2.2597 \times 10^{5}$ |
| | Foot-pounds/min | 0.01666 |
| | Horsepower | $5.0505 \times 10^{-7}$ |
| | Horsepower (metric) | $5.1205 \times 10^{-7}$ |
| | Kilowatts | $3.7662 \times 10^{-7}$ |
| | Watts | 0.0003766 |
| | Watts (Int) | 0.0003765 |
| Foot-pounds/min | Btu/minute | $1.286 \times 10^{-3}$ |
| | Btu/sec | $2.1432 \times 10^{-5}$ |
| | Btu (mean)/sec | $2.1401 \times 10^{-5}$ |
| | Cal, gm/sec | 0.00540 |
| | Cal, gm(mean)/sec | 0.00539 |
| | Ergs/sec | $2.2597 \times 10^{5}$ |
| | Foot-pounds/sec | 0.0166 |
| | Horsepower | $3.0303 \times 10^{-5}$ |
| | Horsepower (metric) | $3.0723 \times 10^{-5}$ |
| | Joules/sec | 0.0226 |
| | Joules (Int)/sec | 0.02259 |
| | Kilogram-calories/min | $3.24 \times 10^{-4}$ |
| | Kilowatts | $2.2597 \times 10^{-5}$ |
| | Watts | 0.022597 |
| Foot-pounds/lb | Btu/lb | 0.001286 |
| | Btu (IST)/lb | 0.001286 |
| | Btu (mean)/lb | 0.001284 |
| | Cal, gm/gm | 0.000714 |
| | Cal, gm (IST)/gm | 0.0007139 |
| | Cal, gm (mean)/gm | 0.0007134 |
| | Hp-hr/lb | $5.0505 \times 10^{-7}$ |
| | Joules/gram | 0.002989 |
| | Kg-meters/gram | 0.0003 |

| Convert From | Into | Multiply By |
|---|---|---|
| Foot-pounds/lb | Kw-hr/gram | $8.303 \times 10^{-10}$ |
| Foot-pounds/sec | Btu/hour | 4.6263 |
| | Btu/min | 0.077156 |
| | Btu (mean)/min | 0.07704 |
| | Btu/sec | 0.00129 |
| | Btu (mean)/sec | 0.00128 |
| | Cal, gm/sec | 0.32405 |
| | Cal, gm (mean)/sec | 0.32358 |
| | Ergs/sec | $1.3558 \times 10^{7}$ |
| | Gram-cm/sec | 13825.5 |
| | Horsepower | 0.001818 |
| | Joules/sec | 1.3558 |
| | Kg-calories/min | 0.01945 |
| | Kilowatts | 0.001356 |
| | Watts | 1.3558 |
| | Watts (Int) | 1.3556 |
| Furlongs | Centimeters | 20116 |
| | Chains (Gunter's) | 10 |
| | Chains (Ramden's) | 6.6 |
| | Feet | 660 |
| | Inches | 7920 |
| | Meters | 201.17 |
| | Miles (naut,Int) | 0.1086 |
| | Miles (statute) | 0.125 |
| | Rods | 40 |
| | Yards | 220 |
| Gallons (Brit) | Barrels (Brit) | 0.0277 |
| | Bushels (Brit) | 0.125 |
| | Cu centimeters | 4546.09 |
| | Cu feet | 0.1605 |
| | Cu inches | 277.419 |
| | Drachms (Brit flu) | 1280 |
| | Firkins (Brit) | 0.1111 |
| | Gallons (US liq) | 1.2009 |
| | Gills (Brit) | 32 |
| | Liters | 4.546 |
| | Minims (Brit) | 76800 |
| | Ounces (Brit flu) | 160 |
| | Ounces (US flu) | 153.721 |
| | Pecks (Brit) | 0.5 |
| | Lbs of $H_2O$ @ 62°F | 10 |
| Gallons (US dry) | Barrels (US dry) | 0.038096 |
| | Barrels (US liq) | 0.03694 |
| | Bushels (US) | 0.125 |
| | Cu centimeters | 4404.88 |
| | Cu feet | 0.15556 |
| | Cu inches | 268.8 |
| | Gallons (US liq) | 1.163647 |
| | Liters | 4.4048 |
| Gallons (US liq) | Acre-feet | $3.0688 \times 10^{-6}$ |
| | Barrels (US liq) | 0.031746 |
| | Barrels (US petro) | 0.023809 |
| | Bushels (US) | 0.10742 |
| | Cu centimeters | 3785.41 |
| | Cu feet | 0.13368 |

| Convert From | Into | Multiply By |
|---|---|---|
| Gallons (US liq) | Cu inches | 231 |
| | Cu meters | 0.00378 |
| | Cu yards | 0.00495 |
| | Gallons (Brit) | 0.83267 |
| | Gallons (US dry) | 0.85937 |
| | Gallons (wine) | 1 |
| | Gills (US) | 32 |
| | Liters | 3.7853 |
| | Minims (US) | 61440 |
| | Ounces (US flu) | 128 |
| | Pints (US liq) | 8 |
| | Quarts (US liq) | 4 |
| Gallons(US)$H_2O$ @4°C | Lb of $H_2O$ | 8.34517 |
| Gallons(US)$H_2O$@60°F | Lb of $H_2O$ | 8.33717 |
| Gallons (US)/day | Cu feet/hr | 0.00557 |
| Gallons (Brit)/hour | Cu meters/min | $7.5768 \times 10^{-5}$ |
| Gallons (US)/hour | Acre–feet/hr | $3.0689 \times 10^{-6}$ |
| | Cu feet/hr | 0.13368 |
| | Cu meters/min | $6.309 \times 10^{-5}$ |
| | Cu yards/min | $8.2519 \times 10^{-5}$ |
| | Liters/hr | 3.7853 |
| Gallons (US)/min | Liters/sec | 0.06308 |
| | Cu feet/sec | $2.228 \times 10^{-3}$ |
| | Cu feet/hour | 8.0208 |
| Gallons (Brit)/sec | Cu cm/sec | 4546.09 |
| Gallons (US)/sec | Cu cm/sec | 3785.4 |
| | Cu feet/min | 8.0208 |
| | Cu yards/min | 0.29707 |
| | Liters/min | 227.118 |
| Gammas | Grams | $1 \times 10^{-6}$ |
| | Micrograms | 1 |
| Gausses | Gausses (Int) | 0.9997 |
| | Maxwells/sq cm | 1 |
| | Lines/sq cm | 1 |
| | Lines/sq inch | 6.4516 |
| | Webers/sq cm | $10^{-8}$ |
| | Webers/sq inch | $6.452 \times 10^{-8}$ |
| | Webers/sq meter | $10^{-4}$ |
| Gausses (Int) | Gausses | 1.00033 |
| Geepounds | Slugs | 1 |
| | Kilograms | 14.594 |
| Gerah (Old Testament). | Bekahs | 0.1 |
| | Grams | appx 0.5 |
| | Shekels | 0.05 |
| Gigameters | Meters | $1 \times 10^{9}$ |
| Gilberts | Abampere–turns | 0.07958 |
| | Ampere–turns | 0.79577 |
| | Gilberts (Int) | 1.00016 |
| Gilberts (Int) | Gilberts | 0.99983 |
| Gilberts/cm | Ampere–turns/cm | 0.79577 |
| | Ampere–turns/in | 2.02127 |
| | Ampere–turns/meter | 79.58 |
| | Oersteds | 1 |
| Gilberts/maxwell | Ampere–turns/weber | $7.958 \times 10^{7}$ |
| Gills (Brit) | Cu centimeters | 142.06 |

| Convert From | Into | Multiply By |
|---|---|---|
| Gills (Brit) | Gallons (Brit) | 0.0312 |
| | Gills (US) | 1.2009 |
| | Liters | 0.142 |
| | Ounces (Brit flu) | 5 |
| | Ounces (US flu) | 4.8038 |
| | Pints (Brit) | 0.25 |
| Gills (US) | Cu centimeters | 118.29 |
| | Cu inches | 7.2187 |
| | Drams (US flu) | 32 |
| | Gallons (US liq) | 0.0312 |
| | Gills (Brit) | 0.8327 |
| | Liters | 0.1183 |
| | Minims (US) | 1920 |
| | Ounces (US flu) | 4 |
| | Pints (US liq) | 0.25 |
| | Quarts (US liq) | 0.125 |
| Grades | Circles | 0.0025 |
| | Circumferences | 0.0025 |
| | Degrees | 0.9 |
| | Minutes | 54 |
| | Radians | 0.01571 |
| | Revolutions | 0.0025 |
| | Seconds | 3240 |
| Grains | Carats (metric) | 0.32399 |
| | Drams (troy) | 0.01667 |
| | Drams (avdp) | 0.03657 |
| | Dynes | 63.546 |
| | Grams | 0.0648 |
| | Milligrams | 64.7989 |
| | Ounces (troy) | 0.00208 |
| | Ounces (avdp) | 0.00229 |
| | Pennyweights | 0.04167 |
| | Pounds (troy) | 0.00017 |
| | Pounds (avdp) | 0.00014 |
| | Scruples (apoth) | 0.05 |
| | Tons (metric) | $6.4799 \times 10^{-8}$ |
| Grains/cu ft | Grams/cu meter | 2.2883 |
| Grains/gal (US) | Parts/million (1gm/ml) | 17.118 |
| | Pounds/million gal | 142.86 |
| Grains/gal (Brit) | Parts/million | 14.286 |
| Grams | Carats (metric) | 5 |
| | Decigrams | 10 |
| | Dekagrams | 0.1 |
| | Drams (troy) | 0.2572 |
| | Drams (avdp) | 0.5644 |
| | Dynes | 980.66 |
| | Grains | 15.432 |
| | Joules/cm | $9.807 \times 10^{-5}$ |
| | Joules/meter (newtons) | $9.807 \times 10^{-3}$ |
| | Kilograms | 0.001 |
| | Micrograms | $1 \times 10^{6}$ |
| | Milligrams | 1000 |
| | Myriagrams | 0.0001 |
| | Ounces (troy) | 0.03215 |
| | Ounces (avdp) | 0.03527 |

| Convert From | Into | Multiply By |
|---|---|---|
| Grams | Pennyweights | 0.64301 |
| | Poundals | 0.07093 |
| | Pounds (troy) | 0.00268 |
| | Pounds (avdp) | 0.0022 |
| | Scruples (apoth) | 0.77162 |
| | Tons (metric) | $1 \times 10^{-6}$ |
| Grams/cm | Dynes/cm | 980.66 |
| | Grams/inch | 2.54 |
| | Kg/km | 100 |
| | Kg/meter | 0.1 |
| | Poundals/inch | 0.18017 |
| | Pounds/ft | 0.067197 |
| | Pounds/inch | 0.0056 |
| | Tons (metric)/km | 0.1 |
| Grams/(cm x sec) | Poises | 1 |
| | Lb/(ft x sec) | 0.0672 |
| Grams/cu cm | Dynes/cu cm | 980.66 |
| | Grains/milliliter | 15.4328 |
| | Grams/milliliter | 1.000028 |
| | Poundals/cu inch | 1.16236 |
| | Pounds/cu foot | 62.428 |
| | Pounds/cu inch | 0.0361 |
| | Pounds/gal (Brit) | 10.022 |
| | Pounds/gal (US dry) | 9.7111 |
| | Pounds/gal (US liq) | 8.3454 |
| Grams/cu meter | Grains/cu ft | 0.437 |
| Grams/liter | Grains/gallon | 58.417 |
| | Parts/million (1gm/ml) | 1000 |
| | Lbs/cu foot | 0.0624 |
| | Lbs/1000 gal (US) | 8.3452 |
| Grams/milliliter | Grams/cu cm | 0.99997 |
| | Pounds/cu foot | 62.426 |
| | Pounds/gallon (US) | 8.34517 |
| Grams/sq cm | Atmospheres | 0.00097 |
| | Bars | 0.00098 |
| | Cm of Hg @ 0°C | 0.07356 |
| | Dynes/sq cm | 980.66 |
| | In of Hg @ 32oF | 0.02896 |
| | Kg/sq meter | 10 |
| | Mm of Hg @ 0oC | 0.73556 |
| | Poundals/sq inch | 0.45762 |
| | Pounds/sq inch | 0.01422 |
| | Pounds/sq foot | 2.0481 |
| Grams/ton (long) | Milligrams/kg | 0.9842 |
| Grams/ton (short) | Milligrams/kg | 1.1023 |
| Gram–calories | Btu | $3.968 \times 10^{-3}$ |
| | Ergs | $4.1868 \times 10^{7}$ |
| | Foot/pounds | 3.088 |
| | Horsepower–hours | $1.5596 \times 10^{-6}$ |
| | Kilowatt–hours | $1.163 \times 10^{-6}$ |
| | Watt–hours | $1.163 \times 10^{-3}$ |
| Gram–calories/sec | Btu/hr | 14.286 |
| Grams–cm | Btu | $9.3 \times 10^{-8}$ |
| | Btu (IST) | $9.29 \times 10^{-8}$ |
| | Btu (mean) | $9.288 \times 10^{-8}$ |

| Convert From | Into | Multiply By |
|---|---|---|
| Grams–cm | Cal, gram | $2.344 \times 10^{-5}$ |
| | Cal, gram (IST) | $2.342 \times 10^{-5}$ |
| | Cal, gram (mean) | $2.34 \times 10^{-5}$ |
| | Cal, gram (15°C) | $2.343 \times 10^{-5}$ |
| | Cal, gram (20°C) | $2.345 \times 10^{-5}$ |
| | Cal, kg | $2.344 \times 10^{-8}$ |
| | Cal, kg (IST) | $2.342 \times 10^{-8}$ |
| | Cal, kg (mean) | $2.34 \times 10^{-8}$ |
| | Dyne–cm | 980.7 |
| | Ergs | 980.7 |
| | Foot–poundals | 0.00233 |
| | Foot–pounds | $7.233 \times 10^{-5}$ |
| | Hp–hours | $3.653 \times 10^{-11}$ |
| | Joules | $9.807 \times 10^{-5}$ |
| | Kw–hours | $2.724 \times 10^{-11}$ |
| | Kw–hours (Int) | $2.724 \times 10^{-11}$ |
| | Newton–meters | $9.807 \times 10^{-5}$ |
| | Watt–hours | $2.724 \times 10^{-8}$ |
| Gram–cm/sec | Btu/sec | $9.301 \times 10^{-8}$ |
| | Cal, gram/sec | $2.344 \times 10^{-5}$ |
| | Ergs–sec | 980.7 |
| | Foot–pounds/sec | $7.23 \times 10^{-5}$ |
| | Horsepower | $1.315 \times 10^{-7}$ |
| | Joules/sec | $9.807 \times 10^{-5}$ |
| | Kilowatts | $9.807 \times 10^{-8}$ |
| | Kilowatts (Int) | $9.805 \times 10^{-8}$ |
| | Watts | $9.807 \times 10^{-5}$ |
| Grams/sq cm | Pounds/sq inch | 0.00034 |
| Gram wt–sec/sq cm | Poises | 980.7 |
| Gravity constant | Cm/(sec x sec) | 980.6 |
| | Ft/(sec x sec) | 32.17 |
| Hands | Centimeters | 10.16 |
| | Feet | .3333 |
| | Inches | 4 |
| Hectares | Acres | 2.471 |
| | Ares | 100 |
| | Sq cm | $1 \times 10^{8}$ |
| | Sq feet | 107629 |
| | Sq meters | 10000 |
| | Sq miles | 0.00386 |
| | Sq rods | 395.369 |
| Hectograms | Grams | 100 |
| | Poundals | 7.0932 |
| | Pounds (apoth or troy) | 0.2679 |
| | Pounds (avdp) | 0.2205 |
| Hectoliters | Bushels (Brit) | 2.7497 |
| | Bushels (US) | 2.8378 |
| | Cu cm | $1.00028 \times 10^{5}$ |
| | Cu feet | 3.5316 |
| | Gallons (US liq) | 26.418 |
| | Liters | 100 |
| | Ounces (US flu) | 3381.5 |
| | Pecks (US) | 11.351 |
| Hectometers | Centimeters | 10000 |
| | Decimeters | 1000 |

| Convert From | Into | Multiply By |
|---|---|---|
| Hectometers | Dekameters | 10 |
| | Feet | 328.08 |
| | Meters | 100 |
| | Rods | 19.88 |
| | Yards | 109.36 |
| Hectowatts | Watts | 100 |
| Hefner units | Candles (English) | 0.86 |
| | Candles (German) | 0.85 |
| | Candles (Int) | 0.9 |
| | 10cp pentane candles | 0.09 |
| Henries | Abhenries | $1 \times 10^9$ |
| | Henries (Int) | 0.9995 |
| | Millihenries | 1000 |
| | Mks (r or nr) units | 1 |
| | Stathenries | $1.113 \times 10^{-12}$ |
| Henries (Int) | Henries | 1.0005 |
| Henries/meter | Gausses/oersted | 79577 |
| | Mks (nr) units | 0.0796 |
| | Mks (r) units | 1 |
| Hin (Old Testament) | Liters | 3.66 |
| Hogsheads | Butts (Brit) | 0.5 |
| | Cu feet | 8.4219 |
| | Cu inches | 14553 |
| | Cu meters | 0.238 |
| | Gallons (Brit) | 52.458 |
| | Gallons (US) | 63 |
| | Gallons (wine) | 63 |
| | Liters | 238.47 |
| Hogshead (Brit) | Cu feet | 10.11 |
| Homer (Old Testament) | Liters | 220 |
| Horsepower (mech) | Btu (mean)/hr | 2542.48 |
| | Btu/min | 42.436 |
| | Btu (mean)/sec | 0.706 |
| | Cal, gram/hr | $6.416 \times 10^5$ |
| | Cal, gram (IST)/hr | $6.412 \times 10^5$ |
| | Cal, gram (mean)/hr | $6.4069 \times 10^5$ |
| | Cal, gram/min | 10693 |
| | Cal, gram (IST)/min | 10686 |
| | Cal, gram (mean)/min | 10678 |
| | Ergs/sec | $7.457 \times 10^9$ |
| | Foot-pounds/hr | 1980000 |
| | Foot-pounds/min | 33000 |
| | Foot-pounds/sec | 550 |
| | Horsepower (boiler) | 0.076 |
| | Horsepower (electric) | 0.9996 |
| | Horsepower (metric) | 1.0139 |
| | Joules/sec | 745.7 |
| | Kg-calories/min | 10.68 |
| | Kilowatts | 0.7457 |
| | Kilowatts (Int) | 0.7456 |
| | Tons of refrigerant | 0.212 |
| | Watts | 745.7 |
| Horsepower (boiler) | Btu (mean)/hr | 33445 |
| | Cal, gram/min | 140671 |
| | Cal, gram (mean)/min | 140469 |

| Convert From | Into | Multiply By |
|---|---|---|
| Horsepower (boiler) | Cal, gram (20°C)/min | 140742 |
| | Ergs/sec | $9.809 \times 10^{10}$ |
| | Foot-pounds/min | 434107 |
| | Horsepower (mech) | 13.155 |
| | Horsepower (electric) | 13.149 |
| | Horsepower (metric) | 13.337 |
| | Horsepower (water) | 13.149 |
| | Joules/sec | 9809.5 |
| | Kilowatts | 9.809 |
| Horsepower (electric) | Btu/hr | 2547.16 |
| | Btu (IST)/hr | 2545.5 |
| | Btu (mean)/hr | 2543.5 |
| | Cal, gram/sec | 178.298 |
| | Cal, kg/hr | 641.87 |
| | Ergs/sec | $7.46 \times 10^{9}$ |
| | Foot-pounds/min | 33013 |
| | Foot-pounds/sec | 550.2 |
| | Horsepower (mech) | 1.0004 |
| | Horsepower (boiler) | 0.07605 |
| | Horsepower (metric) | 1.01428 |
| | Horsepower (water) | 0.99994 |
| | Joules/sec | 746 |
| | Kilowatts | 0.746 |
| | Watts | 746 |
| Horsepower (metric) | Btu/hr | 2511.3 |
| | Btu (IST)/hr | 2509.7 |
| | Btu (mean)/hr | 2507.7 |
| | Cal, gram/hr | $6.328 \times 10^{5}$ |
| | Cal, gram (IST)/hr | $6.324 \times 10^{5}$ |
| | Cal, gram (mean)/hr | $6.319 \times 10^{5}$ |
| | Ergs/sec | $7.355 \times 10^{9}$ |
| | Foot-pounds/min | 32548.6 |
| | Foot-pounds/sec | 542.476 |
| | Horsepower (mech) | 0.9863 |
| | Horsepower (boiler) | 0.07498 |
| | Horsepower (electric) | 0.9859 |
| | Horsepower (water) | 0.98587 |
| | Kg-meters/sec | 75 |
| | Kilowatts | 0.7355 |
| | Watts | 735.499 |
| Horsepower (water) | Foot-pounds/min | 33015 |
| | Horsepower (mech) | 1.00046 |
| | Horsepower (boiler) | 0.076 |
| | Horsepower (electric) | 1.00006 |
| | Horsepower (metric) | 1.0143 |
| | Kilowatts | 0.746 |
| Horsepower-hours | Btu | 2546.1 |
| | Btu (IST) | 2544.5 |
| | Btu (mean) | 2542.5 |
| | Cal, gram | 641616 |
| | Cal, gram (IST) | 641196 |
| | Cal, gram (mean) | 640693 |
| | Ergs | $2.684 \times 10^{13}$ |
| | Foot-pounds | $1.98 \times 10^{6}$ |
| | Gram-calories | 641190 |

| Convert From | Into | Multiply By |
|---|---|---|
| Horsepower–hours | Joules | $2.684 \times 10^6$ |
| | Kg–calories | 641.1 |
| | Kg–meters | 273745 |
| | Kw–hours | 0.7457 |
| | Watt–hours | 745700 |
| Horsepower–hr/lb | Btu/lb | 2546 |
| | Cal, gram/gram | 1414.5 |
| | Cu ft–(lb/sq in)/lb | 13750 |
| | Foot–pounds/lb | 1980000 |
| | Joules/gram | 5918.35 |
| Hours (mean solar) | Days (mean solar) | 0.0417 |
| | Days (sidereal) | 0.04178 |
| | Hours (sidereal) | 1.002738 |
| | Minutes (mean solar) | 60 |
| | Minutes (sidereal) | 60.164 |
| | Seconds (mean solar) | 3600 |
| | Seconds (sidereal) | 3609.86 |
| | Weeks (mean calendar) | 0.00595 |
| Hours (sidereal) | Days (mean solar) | 0.41553 |
| | Days (sidereal) | 0.04167 |
| | Hours (mean solar) | 0.99727 |
| | Minutes (mean solar) | 59.836 |
| | Minutes (sidereal) | 60 |
| Hundredweights (long) | Kilograms | 50.802 |
| | Pounds | 112 |
| | Quarters (Brit long) | 4 |
| | Quarters (US long) | 0.2 |
| | Tons (long) | 0.05 |
| Hundredweights (short) | Kilograms | 45.359 |
| | Ounces (advp) | 1600 |
| | Pounds (advp) | 100 |
| | Quarters (Brit short) | 4 |
| | Quarters (US short) | 0.2 |
| | Tons (long) | 0.04464 |
| | Tons (metric) | 0.04536 |
| | Tons (short) | 0.05 |
| Inches | Ångström units | $2.54 \times 10^8$ |
| | Centimeters | 2.54 |
| | Chains (Gunter's) | 0.001263 |
| | Cubits | 0.0555 |
| | Fathoms | 0.01388 |
| | Feet | 0.08333 |
| | Feet (US Survey) | 0.08333 |
| | Links (Gunter's) | 0.12626 |
| | Links (Ramden's) | 0.08333 |
| | Meters | 0.0254 |
| | Miles | $1.578 \times 10^{-5}$ |
| | Millimeters | 25.40 |
| | Mils | 1000 |
| | Picas (printer) | 6.0225 |
| | Points (printer) | 72.27 |
| | Yards | 0.0278 |
| Inches of Hg @ 32°F | Atmospheres | 0.03342 |
| | Bars | 0.03386 |
| | Dynes/sq cm | 33864 |

| Convert From | Into | Multiply By |
|---|---|---|
| Inches of Hg @ 32°F | Ft of air @ 1atm,60°F | 926.2 |
| | Ft of $H_2O$ @ 39.2°F | 1.1329 |
| | Grams/sq cm | 34.532 |
| | Kg/sq meter | 345.32 |
| | Mm of Hg @ 60°C | 25.4 |
| | Ounces/sq inch | 7.858 |
| Inches of Hg @ 32°F | Pounds/sq ft | 70.726 |
| Inches of Hg @ 60°F | Atmospheres | 0.033327 |
| | Dynes/sq cm | 39768 |
| | Grams/sq cm | 34.434 |
| | Mm of Hg @ 60°F | 25.4 |
| | Ounces/sq inch | 7.8363 |
| | Pounds/sq ft | 70.5269 |
| Inches of $H_2O$ @ 4°C | Atmospheres | 0.002458 |
| | Dynes/sq cm | 2490.8 |
| | Inch of Hg @ 32°F | 0.07355 |
| | Kg/sq meter | 25.399 |
| | Ounces/sq foot | 83.235 |
| | Ounces/sq inch | 0.57802 |
| | Pounds/sq foot | 5.20218 |
| | Pounds/sq inch | 0.03613 |
| Inch–Pounds | Foot–Pounds | 0.083 or 1/12 |
| Inches/hour | Cm/hour | 2.54 |
| | Feet/hour | 0.0833 |
| | Miles/hour | $1.5783 \times 10^{-5}$ |
| Inches/minute | Cm/hour | 152.4 |
| | Feet/hour | 5 |
| | Miles/hour | 0.000947 |
| Jiggers | Fifths | 0.026 |
| | Ounces, fluid | 1.5 |
| | Pints | 0.04166 |
| | Pony | 2 |
| | Shots | 1.5 |
| Joules (abs) | Btu | 0.0009484 |
| | Btu (IST) | 0.0009478 |
| | Btu (mean) | 0.0009471 |
| | Cal, gram | 0.23901 |
| | Cal, gram (IST) | 0.23885 |
| | Cal, gram (mean) | 0.23866 |
| | Cal, gram @ 20°C | 0.23913 |
| | Cal, kg (mean) | 0.000239 |
| | Cu ft–atmosphere | 0.000348 |
| | Ergs | $1 \times 10^{7}$ |
| | Foot–poundals | 23.73 |
| | Foot–pounds | 0.7376 |
| | Gram–cm | 10197.2 |
| | Horsepower–hours | $3.7251 \times 10^{-7}$ |
| | Joules (Int) | 0.9998 |
| | Kg–calories | $2.389 \times 10^{-4}$ |
| | Kg–meters | 0.10197 |
| | Kw–hours | $2.78 \times 10^{-7}$ |
| | Liter–atmospheres | 0.009869 |
| | Volt–coulombs (Int) | 0.999835 |
| | Watt–hours (abs) | 0.0002778 |
| | Watt–hours (Int) | 0.0002777 |

| Convert From | Into | Multiply By |
|---|---|---|
| Joules (abs) | Watt–seconds | 1 |
| | Watt–seconds (Int) | 0.9998 |
| Joules (Int) | Btu | 0.000949 |
| | Btu (IST) | 0.000948 |
| | Btu (mean) | 0.000947 |
| | Cal, gram | 0.23904 |
| | Cal, gram (IST) | 0.23888 |
| | Cal, gram (mean) | 0.2387 |
| | Cu cm–atmosphere | 9.87086 |
| | Cu ft–atmosphere | 0.000349 |
| | Dyne–cm | $1.00016 \times 10^7$ |
| | Ergs | $1.00016 \times 10^7$ |
| | Foot–poundals | 23.734 |
| | Foot–pounds | 0.7377 |
| | Gram–cm | 10199 |
| | Joules (abs) | 1.00016 |
| | Kw–hours | $2.778 \times 10^{-7}$ |
| | Liter–atmosphere | 0.00987 |
| | Volt–coulombs | 1.00016 |
| | Volt–coulombs (Int) | 1 |
| | Watt–second | 1.00016 |
| | Watt–second (Int) | 1 |
| Joules/ampere–hour | Joules/abcoulomb | 0.00278 |
| | Joules/statcoulomb | $9.266 \times 10^{-14}$ |
| Joules/coulomb | Joules/abcoulomb | 10 |
| | Volts | 1 |
| Joules/cm | Dynes | $10^7$ |
| | Grams | $1.020 \times 10^4$ |
| | Joules/meter (newtons) | 100 |
| | Poundals | 723.3 |
| | Pounds | 22.48 |
| Joules/second (abs) | Btu/min | 0.0569 |
| | Cal, gram/min | 14.34 |
| | Cal, kg/min | 0.01434 |
| | Cal, kg (mean)/min | 0.01432 |
| | Dyne–cm/sec | $1 \times 10^7$ |
| | Ergs/sec | $1 \times 10^7$ |
| | Foot–pounds/sec | 0.73756 |
| | Gram–cm/sec | 10197 |
| | Horsepower | 0.00134 |
| | Watts | 1 |
| | Watts (Int) | 0.9998 |
| Joules (Int)/sec | Btu/min | 0.05692 |
| | Btu (mean)/min | 0.05683 |
| | Cal, gram/min | 14.343 |
| | Cal, kg/min | 0.01434 |
| | Dyne–cm/sec | $1.00016 \times 10^7$ |
| | Ergs/sec | $1.00016 \times 10^7$ |
| | Foot–pounds/min | 44.26 |
| | Foot–pounds/sec | 0.73768 |
| | Gram–cm/sec | 10198 |
| | Horsepower | 0.00134 |
| | Watts | 1.00016 |
| | Watts (Int) | 1 |
| Kab (Old Testament) | Liters | 1.2 |

| Convert From | Into | Multiply By |
|---|---|---|
| Kantar (Egypt) | Pounds (avdp) | 99.094 |
| Kati (Malaysia) | Kilograms | 0.60 |
| | Pounds | 1.333 |
| Kilderkins (Brit) | Cu cm | 81829 |
| | Cu feet | 2.8898 |
| | Cu inches | 4993.5 |
| | Cu meters | 0.0818 |
| | Gallons (Brit) | 16 to 18 |
| | Liters | 72.7 to 81.8 |
| Kilograms | Drams (apoth or troy) | 257.21 |
| | Drams (avdp) | 564.38 |
| | Dynes | 980665 |
| | Grains | 15432.36 |
| | Grams | 1000 |
| | Hundredweights (long) | 0.019684 |
| | Hundredweights (short) | 0.022046 |
| | Joules/cm | 0.09807 |
| | Joules/meter (newtons) | 9.807 |
| | Ounces (apoth or troy) | 32.1507 |
| | Ounces (avdp) | 35.27396 |
| | Pennyweights | 643.0149 |
| | Poundals | 70.9316 |
| | Pounds (apoth or troy) | 2.67923 |
| | Pounds (avdp) | 2.20462 |
| | Quarters (Brit long) | 0.078736 |
| | Quarters (US long) | 0.003937 |
| | Scruples (apoth) | 771.6179 |
| | Slugs | 0.06852 |
| | Tons (long) | 0.00098 |
| | Tons (metric) | 0.001 |
| | Tons (short) | 0.001102 |
| Kilograms/cu meter | Grams/cu cm | 0.001 |
| | Lb/cu ft | 0.0624 |
| | Lb/cu inch | $3.6127 \times 10^{-5}$ |
| | Lb/mil-foot | $3.405 \times 10^{-10}$ |
| Kilograms/meter | Pounds/ft | 0.672 |
| Kilograms/sq cm | Atmospheres | 0.9678 |
| | Bars | 0.98066 |
| | Cm of Hg @ 0°C | 73.556 |
| | Dynes/sq cm | 98066 |
| | Ft of $H_2O$ @ 39.2°F | 32.809 |
| | In of Hg @ 32°F | 28.959 |
| | Pounds/sq inch | 14.22 |
| Kilograms/sq meter | Atmospheres | $9.678 \times 10^{-5}$ |
| | Bars | $9.8966 \times 10^{-5}$ |
| | Dynes/sq cm | 98.066 |
| | Ft of $H_2O$ @ 39.2°F | 0.00328 |
| | Grams/sq cm | 0.1 |
| | In of Hg @ 32°F | 0.0029 |
| | Mm of Hg @ 0°C | 0.07356 |
| | Pounds/sq foot | 0.20482 |
| | Pounds/sq inch | 0.00142 |
| Kilograms/sq mm | Pounds/sq ft | 204816 |
| | Pounds/sq in | 1422.3 |
| | Tons (short)/sq in | 0.71117 |

| Convert From | Into | Multiply By |
|---|---|---|
| Kilogram sq cm | Pounds sq ft | 0.00237 |
| | Pounds sq in | 0.34172 |
| Kilogram-meters | Btu (mean) | 0.00929 |
| | Cal, gram (mean) | 2.3405 |
| | Cal, kg (mean) | 0.00234 |
| | Cu ft-atmospheres | 0.003418 |
| | Dyne-cm | $9.807 \times 10^{7}$ |
| | Ergs | $9.807 \times 10^{7}$ |
| | Foot-poundals | 232.71 |
| | Foot-pounds | 7.233 |
| | Gram-cm | 100000 |
| | Horsepower-hours | $3.653 \times 10^{-6}$ |
| | Joules | 9.8066 |
| | Joules (Int) | 0.8050 |
| | Kw-hours | $2.724 \times 10^{-6}$ |
| | Liter-atmospheres | 0.0968 |
| | Newton-meters | 9.8066 |
| | Watt-hours | 0.002724 |
| | Watt-hours (Int) | 0.0027236 |
| Kilogram-meters/sec | Watts | 9.8066 |
| Kilolines | Maxwells | 1000 |
| | Webers | $1 \times 10^{-5}$ |
| Kiloliters | Cu centimeters | $1.000028 \times 10^{6}$ |
| | Cu feet | 35.3157 |
| | Cu inches | 61025.4 |
| | Cu meters | 1.000028 |
| | Cu yards | 1.30799 |
| | Gallons (Brit) | 219.975 |
| | Gallons (US dry) | 227.027 |
| | Gallons (US liq) | 264.179 |
| | Liters | 1000 |
| Kilometers | Astronomical units | $6.689 \times 10^{-9}$ |
| | Centimeters | 100000 |
| | Feet | 3280.84 |
| | Feet (US Survey) | 3280.83 |
| | Inches | $3.937 \times 10^{4}$ |
| | Light years | $1.057 \times 10^{-13}$ |
| | Meters | 1000 |
| | Miles (naut, Int) | 0.53996 |
| | Miles (statute) | 0.62137 |
| | Millimeters | $10^{6}$ |
| | Myriameters | 0.1 |
| | Rods | 198.839 |
| | Yards | 1093.61 |
| Kilometers/hr | Cm/sec | 27.778 |
| | Feet/hr | 3280.84 |
| | Feet/min | 54.6807 |
| | Knots (Int) | 0.53996 |
| | Meters/sec | 0.2778 |
| | Miles (statute)/hr | 0.62137 |
| Kilometers/hr/sec | Cm/sec/sec | 27.78 |
| | Ft/sec/sec | 0.9113 |
| | Meters/sec/sec | 0.2778 |
| | Miles/hr/sec | 0.6214 |
| Kilometers/min | Cm/sec | 1666.67 |

| Convert From | Into | Multiply By |
|---|---|---|
| Kilometers/min | Feet/min | 3280.8 |
| | Kilometers/hr | 60 |
| | Knots (Int) | 32.397 |
| | Miles/hr | 37.2823 |
| | Miles/min | 0.62137 |
| Kilovolts/cm | Abvolts/cm | $1 \times 10^{11}$ |
| | Microvolts/meter | $1 \times 10^{11}$ |
| | Millivolts/meter | $1 \times 10^{11}$ |
| | Statvolts/cm | 3.336 |
| | Volts/inch | 2540 |
| Kilowatts | Btu/hr | 3414.4 |
| | Btu (IST)/hr | 3412.2 |
| | Btu (mean)/hr | 3409.5 |
| | Btu (mean)/min | 56.82 |
| | Btu (mean)/sec | 0.9471 |
| | Cal, gram(mean)/hr | 859184 |
| | Cal, gram(mean)/min | 14319 |
| | Cal, gram(mean)/sec | 238.66 |
| | Cal, kg (mean)/hr | 859.18 |
| | Cal, kg (mean)/min | 14.32 |
| | Cal, kg (mean)/sec | 0.23866 |
| | Cu ft–atm/hr | 1254.7 |
| | Ergs/sec | $1 \times 10^{10}$ |
| | Foot–poundals/min | $1.424 \times 10^{6}$ |
| | Foot–pounds/hr | $2.655 \times 10^{6}$ |
| | Foot–pounds/min | 44253 |
| | Foot–pounds/sec | 737.56 |
| | Gram–cm/sec | $1.0197 \times 10^{7}$ |
| | Horsepower | 1.341 |
| | Horsepower (boiler) | 0.1019 |
| | Horsepower (electric) | 1.34 |
| | Horsepower (metric) | 1.3596 |
| | Joules/hr | $3.6 \times 10^{6}$ |
| | Joules (IST)/hr | $3.599 \times 10^{6}$ |
| | Joules/sec | 1000 |
| | Kg–meters/hr | $3.671 \times 10^{5}$ |
| | Kilowatts (Int) | 0.99983 |
| | Watts (Int) | 999.83 |
| Kilowatts (Int) | Btu/hr | 3414.99 |
| | Btu (IST)/hr | 3412.76 |
| | Btu (mean)/hr | 3410.08 |
| | Btu (mean)/min | 56.835 |
| | Btu (mean)/sec | 0.9472 |
| | Cal, gram (mean)/hr | 859326 |
| | Cal, gram (mean)/min | 14322 |
| | Cal, kg/hr | 860.56 |
| | Cal, kg (IST)/hr | 860 |
| | Cal, kg (mean)/hr | 859.3 |
| | Cu cm–atm/hr | $3.55 \times 10^{7}$ |
| | Cu ft–atm/hr | 1254.9 |
| | Ergs/sec | $1.00016 \times 10^{10}$ |
| | Foot–poundals/min | $1.424 \times 10^{6}$ |
| | Foot–pounds/min | 44261 |
| | Foot–pounds/sec | 737.68 |
| | Gram–cm/sec | $1.0199 \times 10^{7}$ |

| Convert From | Into | Multiply By |
|---|---|---|
| Kilowatts (Int) | Horsepower | 1.341 |
| | Horsepower (boiler) | 0.102 |
| | Horsepower (electric) | 1.341 |
| | Horsepower (metric) | 1.3598 |
| | Joules/hr | $3.6006 \times 10^6$ |
| | Joules (Int)/hr | $3.6 \times 10^6$ |
| | Kg–meters/hr | 367158 |
| | Kilowatts | 1.00016 |
| Kilowatt–hours | Btu (mean) | 3409.5 |
| | Cal, gram (mean) | 859184 |
| | Ergs | $3.6 \times 10^{13}$ |
| | Foot–pounds | $2.655 \times 10^6$ |
| | Gram–calories | 859850 |
| | Hp–hours | 1.341 |
| | Joules | $3.6 \times 10^6$ |
| | Kg–calories | 860.5 |
| | Kg–meters | 367098 |
| | Lb $H_2O$ evaporated from and at 212°F | 3.53 |
| | Lb $H_2O$ raised from 62° to 212°F | 22.75 |
| | Watt–hours | 1000 |
| | Watt–hours (Int) | 999.8 |
| Kilowatt–hours (Int) | Btu (mean) | 3410.1 |
| | Cal, gram (IST) | 860000 |
| | Cal, gram (mean) | 859326 |
| | Cu cm–atm | $3.553 \times 10^7$ |
| | Cu ft–atm | 1254.9 |
| | Foot–pounds | $2.656 \times 10^6$ |
| | Hp–hours | 1.3412 |
| | Joules | $3.6006 \times 10^6$ |
| | Joules (Int) | $3.6 \times 10^6$ |
| | Kg–meters | 367158 |
| Kw–hr/gram | Btu/lb | $1.549 \times 10^6$ |
| | Btu (IST)/lb | $1.548 \times 10^6$ |
| | Btu (mean)/lb | $1.546 \times 10^6$ |
| | Cal, gram/gram | 860421 |
| | Cal, gram (mean)/gram | 859184 |
| | Cu cm–atm/gram | $3.553 \times 10^7$ |
| | Cu ft–atm/gram | 569124 |
| | Hp–hr/lb | 608.28 |
| | Joules/gram | $3.6 \times 10^6$ |
| Kin (Japan) | Kilograms | 0.600 |
| Knots (Int) | Cm/sec | 51.44 |
| | Feet/hr | 6076.1 |
| | Feet/min | 101.269 |
| | Feet/sec | 1.688 |
| | Kilometers/hr | 1.852 |
| | Meters/min | 30.87 |
| | Meters/sec | 0.5144 |
| | Miles (naut,Int)/hr | 1 |
| | Miles (statute)/hr | 1.1508 |
| | Yards/hour | 2027 |
| Koku (Japan) | Liters | 180.39 |
| Kotyle (Greece) | Decileters | 1 |

| Convert From | Into | Multiply By |
|---|---|---|
| Kwan (Japan) | Kilograms | 3.75 |
| | Pounds (avdp) | 8.267 |
| Lamberts | Candles/sq cm | 0.31831 |
| | Candles/sq ft | 295.72 |
| | Candles/sq inch | 2.0536 |
| | Foot–lamberts | 929.03 |
| | Lumens/sq cm | 1 |
| Lasts (Brit) | Liters | 2909.4 |
| Leagues (naut, Brit) | Feet | 18239.76 |
| | Kilometers | 5.559 |
| | Leagues (naut, Int) | 1.0006 |
| | Leagues (statute) | 1.151 |
| | Miles (statute) | 3.4545 |
| Leagues (naut, Int) | Fathoms | 3038.06 |
| | Feet | 18228 |
| | Kilometers | 5.556 |
| | Leagues (statute) | 1.1508 |
| | Miles (statute) | 3.4523 |
| Leagues (statute) | Fathoms | 2640 |
| | Feet | 15840 |
| | Kilometers | 4.828 |
| | Leagues (naut, Int) | 0.86898 |
| | Miles (naut, Int) | 2.607 |
| | Miles (statute) | 3 |
| Li (China) | Miles | 0.333 |
| Libbra (Italy) | Kilograms | 1 |
| Light years | Astronomical units | 63279 |
| | Kilometers | $9.46 \times 10^{12}$ |
| | Miles (statute) | $5.878 \times 10^{12}$ |
| Lines | Maxwells | 1 |
| Lines (Brit) | Centimeters | 0.2117 |
| | Inches | 0.0833 |
| Lines/sq cm | Gausses | 1 |
| Lines/sq inch | Gausses | 0.155 |
| | Webers/sq cm | $1.55 \times 10^{-9}$ |
| | Webers/sq inch | $1 \times 10^{-8}$ |
| | Webers/sq meter | $1.55 \times 10^{-5}$ |
| Liniya (Russia) | Inches | 0.1 |
| Links (Gunters) | Chains (Gunters) | 0.01 |
| | Feet | 0.66 |
| | Feet (US Survey) | 0.659998 |
| | Inches | 7.92 |
| | Meters | 0.2012 |
| | Miles (statute) | 0.000125 |
| | Rods | 0.04 |
| Links (Ramdens) | Centimeters | 30.48 |
| | Chains (Ramdens) | 0.01 |
| | Feet | 1 |
| | Inches | 12 |
| Liters | Bath (Old Testament) | 0.0454 |
| | Bushels (Brit) | 0.0275 |
| | Bushels (US) | 0.02838 |
| | Cu cm | 1000.03 |
| | Cu feet | 0.03532 |
| | Cu inches | 61.002 |

| Convert From | Into | Multiply By |
|---|---|---|
| Liters | Cu meters | 0.00100003 |
| | Cu yards | 0.00131 |
| | Drams (US flu) | 270.52 |
| | Ephah (Old Testament) | 22 |
| | Gallons (Brit) | 0.21998 |
| | Gallons (US dry) | 0.22703 |
| | Gallons (US liq) | 0.26418 |
| | Gills (Brit) | 7.03922 |
| | Gills (US) | 8.4537 |
| | Hin (Old Testament) | 0.2732 |
| | Hogsheads | 0.00419 |
| | Homer (Old Testament) | 0.00454 |
| | Kab (Old Testament) | 0.833 |
| | Log (Old Testament) | 0.3 |
| | Minims (US) | 16231.2 |
| | Omer (Old Testament) | 2.2 |
| | Ounces (Brit flu) | 35.196 |
| | Ounces (US flu) | 33.81497 |
| | Pecks (Brit) | 0.10999 |
| | Pecks (US) | 0.1135 |
| | Pints (Brit) | 1.7598 |
| | Pints (US dry) | 1.8162 |
| | Pints (US liq) | 2.1134 |
| | Quarts (Brit) | 0.8799 |
| | Quarts (US dry) | 0.9081 |
| | Quarts (US liq) | 1.0567 |
| Liters/min | Cu ft/min | 0.0353 |
| | Cu ft/sec | 0.000588 |
| | Gal (US liq)/min | 0.26418 |
| | Gal (US liq)/sec | $4.403 \times 10^{-3}$ |
| Liters/sec | Cu ft/min | 2.1189 |
| | Cu ft/sec | 0.0353 |
| | Cu yards/min | 0.07848 |
| | Gal (US liq)/min | 15.8508 |
| | Gal (US liq)/sec | 0.26418 |
| Liter-atmospheres | Btu | 0.0961 |
| | Btu (IST) | 0.09604 |
| | Btu (mean) | 0.09597 |
| | Cal, gram | 24.2179 |
| | Cal, gram (IST) | 24.202 |
| | Cal, gram (mean) | 24.183 |
| | Cu ft-atm | 0.0353 |
| | Foot-poundals | 2404.5 |
| | Foot-pounds | 74.736 |
| | Hp-hours | $3.774 \times 10^{-5}$ |
| | Joules | 101.33 |
| | Joules (Int) | 101.31 |
| | Kg-meters | 10.33 |
| | Kw-hours | $2.815 \times 10^{-5}$ |
| Load (English) | Cu yards of alluvium | 1 |
| Log (Old Testament) | Liters | 0.3 |
| Lumens | Candle power | 0.07958 |
| | Watt | 0.0015 |
| Lumens/sq cm | Lamberts | 1 |
| | Phots | 1 |

| Convert From | Into | Multiply By |
|---|---|---|
| Lumens/sq ft | Foot–candles | 1 |
| | Foot–lamberts | 1 |
| | Lumens/sq meter | 10.7639 |
| Lumens/sq meter | Foot–candles | 0.0929 |
| | Lumens/sq ft | 0.0929 |
| | Phots | 0.0001 |
| Lux | Foot–candles | 0.0929 |
| | Lumens/sq meter | 1 |
| | Phots | 0.0001 |
| Maass (Germany) | Liters | 1.837 |
| Mace (China) | Grains | 58.33 |
| Magnum | Bottles of wine | 2.49797 |
| | Quarts | 2 |
| Marc (France) | Kilograms | 0.2448 |
| Maxwells | Gauss–sq cm | 1 |
| | Lines | 1 |
| | Maxwells (Int) | 0.99967 |
| | Volt–seconds | $1 \times 10^{-8}$ |
| | Webers | $1 \times 10^{-8}$ |
| Maxwells (Int) | Maxwells | 1.00033 |
| Maxwells/sq cm | Maxwells/sq in | 6.4516 |
| Maxwells/sq in | Maxwells/sq cm | 0.155 |
| Megabyte | Bytes (computers) | 1048576 |
| Megmhos/cm | Abmhos/cm | 0.001 |
| | Megmhos/inch | 2.54 |
| | $(\text{Microhm–cm})^{-1}$ | 1 |
| Megmhos/inch | megmhos/cm | 0.3937 |
| Megohms | Microhms | $1 \times 10^{12}$ |
| | Ohms | $1 \times 10^{6}$ |
| | Statohms | $1.1126 \times 10^{-6}$ |
| Meters | Angstrom units | $1 \times 10^{10}$ |
| | Centimeters | 100 |
| | Chains (Gunter's) | 0.04971 |
| | Chains (Ramden's) | 0.03281 |
| | Fathoms | 0.54681 |
| | Feet | 3.28084 |
| | Feet (US Survey) | 3.28083 |
| | Furlongs | 0.00497 |
| | Inches | 39.3701 |
| | Kilometers | 0.001 |
| | Links (Gunter's) | 4.97097 |
| | Links (Ramden's) | 3.2808 |
| | Megameters | $1 \times 10^{-6}$ |
| | Miles (Brit, naut) | 0.0005396 |
| | Miles (Int, naut) | 0.0005399 |
| | Miles (statute) | 0.000621 |
| | Millimeters | 1000 |
| | Millimicrons | $1 \times 10^{9}$ |
| | Mils | 39370.08 |
| | Rods | 0.1988 |
| | Yards | 1.0936 |
| Meters/hr | Feet/hr | 3.2808 |
| | Feet/min | 0.05468 |
| | Knots (Int) | 0.00054 |
| | Miles (statute)/hr | 0.000621 |

| Convert From | Into | Multiply By |
|---|---|---|
| Meters/min | Cm/sec | 1.66667 |
| | Feet/min | 3.2808 |
| | Feet/sec | 0.05468 |
| | Kilometers/hr | 0.06 |
| | Knots (Int) | 0.032397 |
| | Miles (statute)/hr | 0.03728 |
| Meters/sec | Feet/min | 196.85 |
| | Feet/sec | 3.2808 |
| | Kilometers/hr | 3.6 |
| | Kilometers/min | 0.06 |
| | Miles (statute)/hr | 2.2369 |
| | Miles (statute)/sec | 0.03728 |
| Meters/(sec x sec) | Cm/(sec x sec) | 100 |
| | Feet/(sec x sec) | 3.281 |
| | Kilometers/(hr x sec) | 3.6 |
| | Miles/(hr x sec) | 2.2369 |
| Meter–candles | Lumens/sq meter | 1 |
| Meter–kilograms | Cm–dynes | $9.807 \times 10^{7}$ |
| | Cm–grams | $10^{5}$ |
| | Lb–feet | 7.233 |
| Mhos | Abmhos | $1 \times 10^{-9}$ |
| | Mhos (Int) | 1.00049 |
| | Mks units | 1 |
| | Siemen's units | 1 |
| | Statmhos | $8.9876 \times 10^{11}$ |
| Mhos (Int) | Abmhos | $9.995 \times 10^{-10}$ |
| | Mhos | 0.9995 |
| Mhos/meter | Abmhos/cm | $1 \times 10^{-11}$ |
| | Mhos (Int)/meter | 1.00049 |
| Microfarads | Abfarads | $1 \times 10^{-15}$ |
| | Farads | $1 \times 10^{-6}$ |
| | Statfarads | $8.988 \times 10^{5}$ |
| | Picofarads | $1 \times 10^{6}$ |
| Micrograms | Grams | $1 \times 10^{-6}$ |
| | Milligrams | 0.001 |
| Microhenries | Henries | $1 \times 10^{-6}$ |
| | Stathenries | $1.113 \times 10^{-18}$ |
| Microhms | Abohms | 1000 |
| | Megohms | $1 \times 10^{-12}$ |
| | Ohms | $1 \times 10^{-6}$ |
| | Statohms | $1.113 \times 10^{-18}$ |
| Microhm–cm | Abohm–cm | 1000 |
| | Circ mil–ohms/ft | 6.015 |
| | Microhm–inches | 0.3937 |
| | Ohm–cm | $1 \times 10^{-6}$ |
| Microhm–inches | Circ mil–ohms/ft | 15.279 |
| | Microhm–cm | 2.54 |
| Micromicrofarads | Farads | $1 \times 10^{-12}$ |
| Micromicrons | Angstrom units | 0.01 |
| | Centimeters | $1 \times 10^{-10}$ |
| | Inches | $3.937 \times 10^{-11}$ |
| | Meters | $1 \times 10^{-12}$ |
| | Microns | $1 \times 10^{-6}$ |
| Microns | Angstrom units | 10000 |
| | Centimeters | 0.0001 |

| Convert From | Into | Multiply By |
|---|---|---|
| Microns | Feet | $3.2808 \times 10^{-6}$ |
| | Inches | $3.937 \times 10^{-5}$ |
| | Meters | $1 \times 10^{-6}$ |
| | Millimeters | 0.001 |
| | Millimicrons | 1000 |
| Miglio (Rome) | Miles | 0.925 |
| Miles (Naut,Brit) | Cable lengths (Brit) | 8.444 |
| | Fathoms | 1013.33 |
| | Feet | 6080 |
| | Meters | 1853.18 |
| | Miles (Naut,Int) | 1.00064 |
| | Miles (statute) | 1.1515 |
| Miles (Naut,Int) | Cable lengths | 8.439 |
| | Fathoms | 1012.69 |
| | Feet | 6067.11 |
| | Feet (US survey) | 6067.1 |
| | Kilometers | 1.852 |
| | Leagues (Naut,Int) | 0.3333 |
| | Meters | 1852 |
| | Miles (Naut,Brit) | 0.99936 |
| | Miles (Statute) | 1.15078 |
| Miles (Statute) | Centimeters | 160934 |
| | Chains (Gunter's) | 80 |
| | Chains (Ramden's) | 52.8 |
| | Feet | 5280 |
| | Feet (US Survey) | 5279.99 |
| | Furlongs | 8 |
| | Inches | 63360 |
| | Kilometers | 1.609344 |
| | Light years | $1.701 \times 10^{-12}$ |
| | Links (Gunter's) | 8000 |
| | Meters | 1609.34 |
| | Miles (Naut,Brit) | 0.8684 |
| | Miles (Naut,Int) | 0.86898 |
| | Myriameters | 0.16094 |
| | Rods | 320 |
| | Yards | 1760 |
| Miles/hr | Cm/second | 44.704 |
| | Feet/hour | 5280 |
| | Feet/minute | 88 |
| | Feet/second | 1.4667 |
| | Kilometers/hour | 1.6094 |
| | Kilometers/min | 0.0268 |
| | Knots (Int) | 0.86897 |
| | Meters/min | 26.822 |
| | Miles/min | 0.01667 |
| Miles/(hr x min) | Cm/(sec x sec) | 0.74507 |
| Miles/(hr x sec) | Cm/(sec x sec) | 44.704 |
| | Ft/(sec x sec) | 1.4667 |
| | Kilometers/(hr x sec) | 1.6093 |
| | Meters/(sec x sec) | 0.4470 |
| Miles/min | Cm/second | 2682.2 |
| | Feet/hr | 316800 |
| | Feet/sec | 88 |
| | Kilometers/min | 1.6093 |

| Convert From | Into | Multiply By |
|---|---|---|
| Miles/min | Knots (Int) | 52.1386 |
| | Meters/min | 1609.34 |
| | Miles/hr | 60 |
| Milion (New Testament) | Meters | 1478 |
| | Yards | 1618 |
| Milliar (Ancient Rome) | Miles | 0.92 |
| | Stadia | 8 |
| Millibars | Atmospheres | 0.000987 |
| | Bars | 0.001 |
| | Baryes | 1000 |
| | Dynes/sq cm | 1000 |
| | Grams/sq cm | 1.0197 |
| | In of Hg @ 32°F | 0.0295 |
| | Pounds/sq ft | 2.0885 |
| | Pounds/sq inch | 0.0145 |
| Milligrams | Carats (1877 defn) | 0.00487 |
| | Carats (metric) | 0.005 |
| | Drams (troy) | 0.000257 |
| | Drams (avdp) | 0.00056 |
| | Grains | 0.01543 |
| | Grams | 0.001 |
| | Ounces (troy) | $3.215 \times 10^{-5}$ |
| | Ounces (avdp) | $3.527 \times 10^{-5}$ |
| | Pennyweights | 0.000643 |
| | Pounds (troy) | $2.679 \times 10^{-6}$ |
| | Pounds (avdp) | $2.205 \times 10^{-6}$ |
| | Scruples (apoth) | 0.000772 |
| Milligrams/assay ton | Milligrams/kg | 34.2857 |
| | Ounces (troy)/ton (avdp) | 1 |
| Milligrams/gm | Carats | 0.024 |
| | Grams/ton (short) | 907.185 |
| | Milligrams/assay ton | 29.16667 |
| | Ounces (avdp)/ton (long) | 35.84 |
| | Ounces (avdp)/ton (short) | 32 |
| | Ounces (troy)/ton (long) | 32.6666 |
| | Ounces (troy)/ton (short) | 29.1666 |
| Milligrams/inch | Dynes/cm | 0.38609 |
| | Dynes/inch | 0.98066 |
| | Grams/cm | 0.000394 |
| | Grams/inch | 0.0001 |
| Milligrams/kg | Pounds (avdp)/ton (short) | 0.002 |
| Milligrams/liter | Grains/gallon | 0.05842 |
| | Grams/liter | 0.001 |
| | Parts/million | 1 |
| | Lb/cu ft | $6.2426 \times 10^{-5}$ |
| Milligrams/met. ton | Parts/billion | 1.1 |
| Milligrams/mm | Dynes/cm | 9.8066 |
| Millihenries | Abhenries | $1 \times 10^{6}$ |
| | Henries | 0.001 |
| | Stathenries | $1.1126 \times 10^{-15}$ |
| Millilamberts | Candles/sq cm | 0.000318 |
| | Candles/sq inch | 0.002054 |
| | Foot-lamberts | 0.929 |
| | Lamberts | 0.001 |
| | Lumens/sq cm | 0.001 |

| Convert From | Into | Multiply By |
|---|---|---|
| Millilamberts | Lumens/sq ft | 0.929 |
| Milliliters | Cu cm | 1.000028 |
| | Cu inches | 0.06102 |
| | Drams (US fluid) | 0.27052 |
| | Gills (US) | 0.00845 |
| | Liters | 0.001 |
| | Minims (US) | 16.231 |
| | Ounces (Brit,flu) | 0.035196 |
| | Ounces (US,flu) | 0.0338 |
| | Pints (Brit) | 0.00176 |
| | Pints (US liq) | 0.00211 |
| Millimeters | Angstrom units | $1 \times 10^{7}$ |
| | Centimeters | 0.1 |
| | Decimeters | 0.01 |
| | Dekameters | 0.0001 |
| | Feet | 0.00328 |
| | Inches | 0.03937 |
| | Kilometers | $10^{-6}$ |
| | Meters | 0.001 |
| | Microns | 1000 |
| | Mils | 39.37 |
| | Yards | $1.094 \times 10^{-3}$ |
| Millimeters Hg @ 0°C | Atmospheres | 0.001316 |
| | Bars | 0.00133 |
| | Dynes/sq cm | 1333.2 |
| | Grams/sq cm | 1.3595 |
| | Kg/sq meters | 13.595 |
| | Pounds/sq ft | 2.7845 |
| | Pounds/sq in | 0.0193 |
| | Torrs | 1 |
| Millimicrons | Angstrom units | 10 |
| | Centimeters | $1 \times 10^{-7}$ |
| | Inches | $3.937 \times 10^{-8}$ |
| | Meters | $1 \times 10^{-9}$ |
| | Microns | 0.001 |
| | Millimeters | $1 \times 10^{-6}$ |
| Million gal/day | cu ft/sec | 1.547 |
| Milliphots | Foot-candles | 0.929 |
| | Lumens/sq ft | 0.929 |
| | Lumens/sq meter | 10 |
| | Lux | 10 |
| | Phots | 0.001 |
| Millivolts | Statvolts | $3.336 \times 10^{-6}$ |
| | Volts | 0.001 |
| Mils | Centimeters | $2.540 \times 10^{-3}$ |
| | Feet | $8.333 \times 10^{-5}$ |
| | Inches | 0.001 |
| | Kilometers | $2.540 \times 10^{-8}$ |
| | Yards | $2.778 \times 10^{-5}$ |
| Mina | Pounds | 0.944 |
| Miners Inch | Cu feet/min | 1.2 to 1.56 |
| | Cu feet/sec | 0.02 to 0.026 |
| | Liters/sec | 0.57 to 0.74 |
| Minims (Brit) | Cu cm | 0.0592 |
| | Cu inches | 0.0036 |

| Convert From | Into | Multiply By |
|---|---|---|
| Minims (Brit) | Milliliters | 0.05919 |
| | Ounces (Brit,flu) | 0.00208 |
| | Scruples (Brit,flu) | 0.05 |
| Minims (US) | Cu cm | 0.06161 |
| | Cu inches | 0.00376 |
| | Drams (US,flu) | 0.01667 |
| | Gallons (US,liq) | $1.628 \times 10^{-5}$ |
| | Gills (US) | 0.00052 |
| | Liters | $6.16098 \times 10^{-5}$ |
| | Milliliters | 0.06161 |
| | Ounces (US,flu) | 0.00208 |
| | Pints (US,liq) | 0.00013 |
| Minutes (angle) | Degrees | 0.016667 |
| | Quadrants | 0.000185 |
| | Radians | 0.0002909 |
| | Seconds (angle) | 60 |
| Minutes (solar time) | Days (mean solar) | 0.000694 |
| | Days (sideral) | 0.000696 |
| | Hours (mean solar) | 0.016667 |
| | Hours (sideral) | 0.016712 |
| | Minutes (sideral) | 1.002738 |
| Minutes (sideral) | Days (mean solar) | 0.000693 |
| | Minutes (mean solar) | 0.99727 |
| | Months (mean calendar) | $2.2769 \times 10^{-5}$ |
| | Seconds (sideral) | 60 |
| Minutes(angle)/cm | Radians/cm | 0.0002909 |
| Mna (Greece) | Kilograms | 1.5 |
| Momme (Japan) | Grams | 3.75 |
| | Kwan | 0.001 |
| Months (lunar) | Days (mean solar) | 29.5306 |
| | Hours (mean solar) | 708.734 |
| | Minutes (mean solar) | 42524.05 |
| | Second (mean solar) | $2.551 \times 10^{6}$ |
| | Weeks (mean calendar) | 4.2186 |
| Months (mean calend.) | Days (mean solar) | 30.4167 |
| | Hours (mean solar) | 730 |
| | Months (lunar) | 1.030005 |
| | Weeks (mean calendar) | 4.34524 |
| | Years (calendar) | 0.08333 |
| | Years (sideral) | 0.08327 |
| | Years (tropical) | 0.08328 |
| Morgan (S. Africa) | Acres | 2.1165 |
| Mou (China) | Square yards | 806.65 |
| Myriagrams | Grams | 10000 |
| | Kilograms | 10 |
| | Pounds (avdp) | 22.046 |
| Myriawatts | Kilowatts | 10 |
| Nail (Old English) | Inches | 2.25 |
| Nepers | Decibels | 8.686 |
| Newtons | Dynes | 100000 |
| | Pounds | 0.2248089 |
| Newton–meters | Dyne–cm | $1 \times 10^{7}$ |
| | Gram–cm | 10197 |
| | Pound–inch | 8.8507 |
| | Kg–meters | 0.10197 |

| Convert From | Into | Multiply By |
|---|---|---|
| Newton–meters | Pound–feet | 0.73756 |
| Noggins (Brit) | Cm cm | 142.06 |
| | Gallons (Brit) | 0.0312 |
| | Gills (Brit) | 1 |
| Obolos (Greece) | Grams | 0.1 |
| Oersteds | Ampere–turns/inch | 2.0213 |
| | Ampere–turns/meter | 79.577 |
| | Gilberts/cm | 1 |
| | Oersteds (Int) | 1.00016 |
| Oersteds (Int) | Oersteds | 0.9998 |
| Ohms | Abohms | $1 \times 10^{9}$ |
| | Megohms | $1 \times 10^{-6}$ |
| | Microhms | $1 \times 10^{6}$ |
| | Ohms (Int) | 0.9995 |
| | Statohms | $1.1126 \times 10^{-12}$ |
| Ohms (Int) | Ohms | 1.00049 |
| Ohm–cm | Circ mil–ohms/ft | $6.015 \times 10^{6}$ |
| | Microhm–cm | $1 \times 10^{6}$ |
| | Ohm–inches | 0.3937 |
| Ohm–inches | Ohm–cm | 2.54 |
| Ohm–meters | Abohm–cm | $1 \times 10^{11}$ |
| | Statohm–cm | $1.113 \times 10^{-10}$ |
| Omer | Ephah | 0.1 |
| | Liters (Old Testament) | 2.2 |
| | Liters | 3.964 |
| Orguia (Ancient Greece) | Amma | 0.01 |
| | Digits | 100 |
| | Feet | 6 |
| Ounces (apoth or troy) | Dekagrams | 1.7554 |
| | Drams (apoth or troy) | 8 |
| | Drams (advp) | 17.554 |
| | Grains | 480 |
| | Grams | 31.103 |
| | Milligrams | 31103 |
| | Ounces (advp) | 1.0971 |
| | Pennyweights | 20 |
| | Pounds (apoth or troy) | 0.0833 |
| | Pounds (advp) | 0.06857 |
| | Scruples (apoth) | 24 |
| | Tons (short) | $3.429 \times 10^{-5}$ |
| Ounces (advp) | Drams (apoth or troy) | 7.2917 |
| | Drams (advp) | 16 |
| | Grains | 437.5 |
| | Grams | 28.349 |
| | Hundredweights (long) | 0.000558 |
| | Hundredweights (shrt) | 0.00062 |
| | Ounces (apoth or troy) | 0.91146 |
| | Pennyweights | 18.229 |
| | Pounds (apoth or troy) | 0.0759 |
| | Pounds (advp) | 0.0625 |
| | Scruples (apoth) | 21.875 |
| | Tons (long) | $2.79 \times 10^{-5}$ |
| | Tons (metric) | $2.835 \times 10^{-5}$ |
| | Tons (short) | $3.125 \times 10^{-5}$ |
| Ounces (Brit,flu) | Cu cm | 28.413 |

| Convert From | Into | Multiply By |
|---|---|---|
| Ounces (Brit,flu) | Cu inches | 1.7339 |
| | Drachms (Brit,flu) | 8 |
| | Drams (US,flu) | 7.6861 |
| | Gallons (Brit) | 0.00625 |
| | Milliliters | 28.412 |
| | Minims (Brit) | 480 |
| | Ounces (US,flu) | 0.96076 |
| Ounces (US,flu) | Cu cm | 29.5737 |
| | Cu inches | 1.80469 |
| | Cups | 0.1698 |
| | Cu meters | $2.9574 \times 10^{-5}$ |
| | Drops | 360.14 |
| | Drams (US,flu) | 8 |
| | Fifths | 0.039 |
| | Gallons (US,dry) | 0.006714 |
| | Gallons (US,liq) | 0.00781 |
| | Gills (US) | 0.25 |
| | Jiggers | 0.6666 |
| | Liters | 0.02957 |
| | Minims (US) | 480 |
| | Ounces (Brit,flu) | 1.0408 |
| | Pints (US,liq) | 0.0625 |
| | Ponys | 1.333 |
| | Quarts (US,liq) | 0.0312 |
| | Teaspoons | 2 |
| | Tablespoons | 6 |
| Ounces/sq inch | Dynes/sq cm | 4309.2 |
| | Grams/sq cm | 4.39419 |
| | In of $H_2O$ @ 39.2°F | 1.73004 |
| | In of $H_2O$ @ 60°F | 1.73166 |
| | Pounds/sq foot | 9 |
| | Pounds/sq inch | 0.0625 |
| Ounces (advp)/ton (L) | Milligrams/kg | 27.9018 |
| Ounces (advp)/ton (S) | Milligrams/kg | 31.25 |
| Ounces (troy)/ton (L) | Ounces (troy)/ton (Met) | 0.984 |
| | Ounces (troy)/ton (S) | 0.8929 |
| | Parts per million | 30.612 |
| Ounces (troy)/ton (S) | Ounces (troy)/ton (L) | 1.120 |
| | Ounces (troy)/ton (Met) | 1.1023 |
| | Parts per million | 34.286 |
| Ounces (troy)/ton (met) | Ounces (troy)/ton (L) | 1.106 |
| | Ounces (troy)/ton (S) | 0.9072 |
| | Parts per million | 31.104 |
| Paces | Centimeters | 76.2 |
| | Chains (Gunter's) | 0.03788 |
| | Chains (Ramden's) | 0.025 |
| | Feet | 2.5 |
| | Hands | 7.5 |
| | Inches | 30 |
| | Ropes (Brit) | 0.125 |
| Palmi (Ancient Rome) | Inches | 2.9 |
| | Digiti | 4 |
| | Pes | 0.25 |
| Palms | Centimeters | 7.62 |
| | Chains (Ramden's) | 0.0025 |

| Convert From | Into | Multiply By |
|---|---|---|
| Palms | Cubits | 0.16667 |
| | Feet | 0.25 |
| | Hands | 0.75 |
| | Inches | 3 |
| Parsecs | Kilometers | $3.084 \times 10^{13}$ |
| | Miles (statute) | $1.916 \times 10^{13}$ |
| Parts/billion | Milligrams/metric ton | 0.90909 |
| Parts/million | Grains/gal (Brit) | 0.070155 |
| | Grains/gal (US) | 0.05842 |
| | Grams/liter | 0.001 |
| | Grams/ton (Met) | 1 |
| | Milligrams/liter | 1 |
| | Ounces (troy)/ton (S) | 0.0292 |
| | Percent | 0.0001 |
| | Pounds/million gal | 8.345 |
| Passus (Ancient Rome) | Feet | 4.86 |
| | Pes | 5 |
| | Stadium | 0.008 |
| Pecks (Brit) | Bushels (Brit) | 0.25 |
| | Coombs (Brit) | 0.0625 |
| | Cu cm | 9092.17 |
| | Cu inches | 554.84 |
| | Gallons (Brit) | 2 |
| | Gills (Brit) | 64 |
| | Hogsheads | 0.03812 |
| | Kilderkins (Brit) | 0.1111 |
| | Liters | 0.90192 |
| | Pints (Brit) | 16 |
| | Quarterns (Brit,dry) | 4 |
| | Quarters (Brit,dry) | 0.03125 |
| | Quarts (Brit) | 8 |
| | Quarts (US,dry) | 8.2564 |
| Pecks (US) | Barrels (US,dry) | 0.07619 |
| | Bushels (US) | 0.25 |
| | Cu cm | 8809.77 |
| | Cu feet | 0.31111 |
| | Cu inches | 537.6 |
| | Gallons,(US,dry) | 2 |
| | Gallons (US,liq) | 2.3273 |
| | Liters | 8.8095 |
| | Pints (US,dry) | 16 |
| | Quarts (US,dry) | 8 |
| Pennyweights | Drams (apoth or troy) | 0.4 |
| | Drams (avdp) | 0.87771 |
| | Grains | 24 |
| | Grams | 1.55517 |
| | Ounces (apoth or troy) | 0.05 |
| | Ounces (avdp) | 0.05486 |
| | Pounds (apoth or troy) | 0.00417 |
| | Pounds (advp) | 0.003429 |
| Perch (Masonry) | Stone 12 in x 12 in x 16.5 feet long | |
| Perches | Cu feet | 24.75 |
| Pes (Ancient Rome) | Inches | 11.7 |
| | Palmi | 4 |
| | Passus | 0.2 |

| Convert From | Into | Multiply By |
|---|---|---|
| Petrograd standard | Cu feet | 165 |
| Phots | Foot–candles | 929.03 |
| | Lumens/sq cm | 1 |
| | Lumens/sq meter | 10000 |
| | Lux | 10000 |
| Picas (printing) | Centimeters | 0.42175 |
| | Inches | 0.16604 or 1/6 |
| | Points | 12 |
| Picofarads | Farads | $1 \times 10^{-12}$ |
| | Microfarads | $1 \times 10^{-6}$ |
| Picul (Malaysia) | Katis | 100 |
| | Kilogram | 60.48 |
| | Pound | 133.33 |
| Pie (Rome) | Inches | 11.73 |
| Pinch | Teaspoon | 1/3 to 1/4 |
| Pints (Brit) | Cu cm | 568.261 |
| | Gallons (Brit) | 0.125 |
| | Gills (Brit) | 4 |
| | Gills (US) | 4.8038 |
| | Liters | 0.5682 |
| | Minims (Brit) | 9600 |
| | Ounces (Brit,flu) | 20 |
| | Pints (US,dry) | 1.03206 |
| | Pints (US,liq) | 1.2009 |
| | Quarts (Brit) | 0.5 |
| | Scruples (Brit,flu) | 480 |
| Pints (US,dry) | Bushels (US) | 0.0156 |
| | Cu cm | 550.61 |
| | Cu inches | 33.6003 |
| | Gallons, (US,dry) | 0.125 |
| | Gallons (US,liq) | 0.14546 |
| | Liters | 0.5506 |
| | Pecks (US) | 0.625 |
| | Pints (US,liq) | 1.1636 |
| | Quarts (US,dry) | 0.5 |
| Pints (US,liq) | Cu cm | 473.176 |
| | Cu feet | 0.01671 |
| | Cu inches | 28.875 |
| | Cu yards | 0.000619 |
| | Cups | 2 |
| | Drams (US,flu) | 128 |
| | Fifths | 0.625 |
| | Gallons (US,liq) | 0.125 |
| | Gills (US) | 4 |
| | Jiggers | 24 |
| | Liters | 0.47316 |
| | Milliliters | 473.163 |
| | Minims (US) | 7680 |
| | Ounces (US,flu) | 16 |
| | Pints (Brit) | 0.8327 |
| | Pints (US,dry) | 0.85937 |
| | Ponys | 48 |
| | Quarts (US,liq) | 0.5 |
| | Shots | 16 |
| | Teaspoons | 96 |

| Convert From | Into | Multiply By |
|---|---|---|
| Pints (US,liq) | Tablespoons | 32 |
| Pipe (English, wine) | Gallons | 105 |
| | Liters | 477.34 |
| Planck's constant | Erg–seconds | $6.625 \times 10^{-27}$ |
| | Joule–seconds | $6.625 \times 10^{-34}$ |
| | Joule–sec/Avagad. # | $3.99 \times 10^{-10}$ |
| Points (printing) | Centimeters | 0.0351 |
| | Inches | 0.01384 or 1/72 |
| | Picas | 0.0833 |
| Poises | Grams/(cm x sec) | 1 |
| Poise–cu cm/gram | Sq cm/sec | 1 |
| Poise–cu ft/lb | Sq cm/sec | 62.428 |
| Poise–cu in/gram | Sq cm/sec | 16.3871 |
| Ponys | Fifths | 0.013 |
| | Jiggers | 0.5 |
| | Ounces, (US,fluid) | 0.75 |
| | Pints | 0.0208 |
| | Shots | 0.75 |
| Pottles (Brit) | Gallons (Brit) | 0.5 |
| | Liters | 2.273 |
| Poud (Russia) | Pounds | 36.113 |
| Pounce (France) | Millimeters | 27.07 |
| Poundals | Dynes | 13825 |
| | Grams | 14.098 |
| | Joules/cm | $1.383 \times 10^{-3}$ |
| | Joules/meter (newtons) | 0.1383 |
| | Kilograms | 0.0141 |
| | Pounds (avdp) | 0.03108 |
| Pounds (apoth or troy) | Drams (apoth or troy) | 96 |
| | Drams (avdp) | 210.65 |
| | Grains | 5760 |
| | Grams | 373.24 |
| | Kilograms | 0.37324 |
| | Ounces (apoth or troy) | 12 |
| | Ounces (avdp) | 13.166 |
| | Pennyweights | 240 |
| | Pounds (avdp) | 0.82286 |
| | Scruples (apoth) | 288 |
| | Tons (long) | 0.000367 |
| | Tons (metric) | 0.000373 |
| | Tons (short) | 0.0004114 |
| Pounds (avdp) | Drams (apoth or troy) | 116.667 |
| | Drams (avdp) | 256 |
| | Dynes | $44.48 \times 10^{4}$ |
| | Grains | 7000 |
| | Grams | 453.59 |
| | Hundredweights (long) | 0.008929 |
| | Hundredweights (shrt) | 0.01 |
| | Joules/cm | 0.04448 |
| | Joules/meter | 4.448 |
| | Kilograms | 0.4536 |
| | Ounces (apoth or troy) | 14.583 |
| | Ounces (avdp) | 16 |
| | Pennyweights | 291.667 |
| | Poundals | 32.174 |

| Convert From | Into | Multiply By |
|---|---|---|
| Pounds (avdp) | Pounds (apoth or troy) | 1.21528 |
| | Scruples (apoth) | 350 |
| | Slugs | 0.03108 |
| | Tons (long) | 0.0004464 |
| | Tons (metric) | 0.0004536 |
| | Tons (short) | 0.0005 |
| Pounds/cu foot | Grams/cu cm | 0.016018 |
| | Kg/cu meter | 16.018 |
| | Lbs/cu inch | $5.787 \times 10^{-4}$ |
| | Lbs/cu yard | 27 |
| | Lbs/mil–foot | $5.456 \times 10^{-9}$ |
| Pounds/cu inch | Grams/cu cm | 27.6799 |
| | Grams/liter | 27.6807 |
| | Kg/cu meter | 27679.9 |
| | Lbs/cu foot | 1728 |
| | Lbs/mil–foot | $9.425 \times 10^{-6}$ |
| Pounds/cu yard | Lbs/cu foot | 0.037 |
| Pounds/foot | Kilograms/meter | 1.488 |
| Pounds/gal (Brit) | Pounds/cu ft | 6.2288 |
| Pounds/gal (US,liq) | Grams/cu cm | 0.11983 |
| | Pounds/cu ft | 7.48052 |
| Pounds/inch | Grams/cm | 178.5797 |
| | Grams/ft | 5443.11 |
| | Grams/inch | 453.592 |
| | Ounces/cm | 6.2992 |
| | Ounces/inch | 16 |
| | Pounds/meter | 39.37008 |
| Pounds/mil–foot | Gms/cu cm | $2.306 \times 10^{6}$ |
| Pounds/minute | Kilograms/hr | 27.2155 |
| | Kilograms/min | 0.45359 |
| Pounds/sq ft | Atmospheres | 0.00047 |
| | Bars | 0.000479 |
| | Cm of Hg @ 0°C | 0.03591 |
| | Dynes/sq cm | 478.803 |
| | Feet of water | 0.01602 |
| | Grams/sq cm | 0.48824 |
| | In of Hg @ 32°F | 0.014139 |
| | In of $H_2O$ @ 39.2°F | 0.19223 |
| | Kg/sq meter | 4.88243 |
| | Lbs/sq inch | 0.00694 |
| | Mm of Hg @ 0°C | 0.35913 |
| Pounds/sq in | Atmospheres | 0.06805 |
| | Bars | 0.06895 |
| | Cm of Hg @ 0°C | 5.17149 |
| | Cm of $H_2O$ @ 4°C | 70.3089 |
| | Dynes/sq cm | 68947 |
| | Feet of water | 2.307 |
| | Grams/sq cm | 70.307 |
| | In of Hg @ 32°F | 2.036 |
| | In of $H_2O$ @ 39.2°F | 27.681 |
| | Kg/sq cm | 0.07031 |
| | Lbs/sq foot | 144 |
| | Mm of Hg @ 0°C | 51.715 |
| Pounds of water | Cu feet | 0.01602 |
| | Cu inches | 27.68 |

| Convert From | Into | Multiply By |
|---|---|---|
| Pounds of water | Gallons | 0.1198 |
| Pounds of water/min | cu ft/sec | $2.67 \times 10^{-4}$ |
| Pound–feet | Cm–dynes | $1.356 \times 10^{7}$ |
| | Cm–grams | 13825 |
| | Meter–kilograms | 0.1383 |
| PPM | See parts/million | |
| Prospecting dish | Gallons | 2 |
| | Cu Yards | 008929 |
| Pu (China) | Inches | 70.5 |
| Puncheons (Brit) | Cu meters | 0.31797 |
| | Gallons (Brit) | 69.9447 |
| | Gallons (US) | 84 |
| Quadrants | Degrees | 90 |
| | Minutes | 5400 |
| | Radians | 1.5708 |
| | Seconds | $3.24 \times 10^{5}$ |
| Quarterns (Brit,dry) | Buckets (Brit) | 0.125 |
| | Bushels (Brit) | 0.0625 |
| | Cu cm | 2273.04 |
| | Gallons (Brit) | 0.5 |
| | Liters | 2.27298 |
| | Pecks (Brit) | 0.25 |
| Quarterns (Brit,liq) | Cu cm | 142.065 |
| | Gallons (Brit) | 0.03125 |
| | Liters | 0.14206 |
| Quarters (US,long) | Kilograms | 254.0117 |
| | Pounds (avdp) | 560 |
| Quarters (US,shrt) | Kilograms | 226.796 |
| | Pounds (avdp) | 500 |
| Quarts (Brit) | Cu cm | 1136.52 |
| | Cu inches | 69.355 |
| | Gallons (Brit) | 0.25 |
| | Gallons (US,liq) | 0.30024 |
| | Liters | 1.1365 |
| | Quarts (US,dry) | 1.0321 |
| | Quarts (US,liq) | 1.2009 |
| Quarts (US,dry) | Bushels (US) | 0.03125 |
| | Cu cm | 1101.2 |
| | Cu feet | 0.03889 |
| | Cu inches | 67.2006 |
| | Gallons (US,dry) | 0.25 |
| | Gallons (US,liq) | 0.29091 |
| | Liters | 1.10119 |
| | Pecks (US) | 0.125 |
| | Pints (US,dry) | 2 |
| Quarts (US,liq) | Cu cm | 946.353 |
| | Cu feet | 0.0334 |
| | Cu inches | 57.75 |
| | Cu meters | $9.464 \times 10^{-4}$ |
| | Cu yards | $1.238 \times 10^{-3}$ |
| | Drams (US,flu) | 256 |
| | Fifth | 1.25 |
| | Gallons (US,dry) | 0.2148 |
| | Gallons (US,liq) | 0.25 |
| | Gills (US) | 8 |

| Convert From | Into | Multiply By |
|---|---|---|
| Quarts (US,liq) | Liters | 0.9463 |
| | Magnums | 0.5 |
| | Ounces (US,flu) | 32 |
| | Pints (US,liq) | 2 |
| | Quarts (Brit) | 0.83267 |
| | Quarts (US,dry) | 0.859367 |
| | Shots | 32 |
| Quintals (metric) | Grams | 100000 |
| | Hundredweights (long) | 1.9684 |
| | Kilograms | 100 |
| | Pounds (avdp) | 220.462 |
| Quintal (USA, old) | Kilograms | 45.36 |
| | Pounds | 100 |
| Quires | Ream | 0.05 |
| | Sheets | 24 |
| Radians | Circumferences | 0.15915 |
| | Degrees | 57.29578 |
| | Minutes | 3437.747 |
| | Quadrants | 0.63662 |
| | Revolutions | 0.15915 |
| | Seconds | 206264 |
| Radians/cm | Degrees/cm | 57.2958 |
| | Degrees/ft | 1746.37 |
| | Degrees/in | 145.531 |
| | Minutes/cm | 3437.75 |
| Radians/sec | Degrees/sec | 57.2958 |
| | Revolutions/min | 9.5493 |
| | Revolutions/sec | 0.15915 |
| Radians/(sec x sec) | Revolutions/(min x min) | 572.96 |
| | Revolutions/(min x sec) | 9.549297 |
| | Revolutions/(sec x sec) | 0.15915 |
| Rattel (Arabia) | Pounds (avdp) | 1.02 |
| Ream | Quires | 20 |
| | Sheets | 480 |
| Register tons | Cu feet | 100 |
| | Cu meters | 2.8317 |
| Revolutions | Degrees | 360 |
| | Grades | 400 |
| | Quadrants | 4 |
| | Radians | 6.2832 |
| Revolutions/min | Degrees/sec | 6 |
| | Radians/sec | 0.1047 |
| | Revolutions/sec | 0.01667 |
| Revolutions/(min x min) | Radians/(sec x sec) | $1.745 \times 10^{-3}$ |
| | Revolutions/(min x sec) | 0.01667 |
| | Revolutions/(sec x sec) | $2.778 \times 10^{-4}$ |
| Revolutions/sec | Degrees/sec | 360 |
| | Radians/sec | 6.283 |
| | Revolutions/min | 60 |
| Revolutions/(sec x sec) | Radians/(sec x sec) | 6.283 |
| | Revolutions/(min x min) | 3600 |
| | Revolutions/(min x sec) | 60 |
| Reyns | Centipoises | $6.8948 \times 10^{6}$ |
| Rhes | $\text{Poises}^{-1}$ | 1 |
| Ri (Japan) | Miles | 2.440 |

| Convert From | Into | Multiply By |
|---|---|---|
| Rods | Centimeters | 502.92 |
| | Chains (Gunter's) | 0.25 |
| | Chains (Ramden's) | 0.165 |
| | Feet | 16.5 |
| | Feet (US Survey) | 16.49997 |
| | Furlongs | 0.025 |
| | Inches | 198 |
| | Links (Gunter's) | 25 |
| | Links (Ramden's) | 16.5 |
| | Meters | 5.0292 |
| | Miles (statute) | 0.00312 |
| | Perches | 1 |
| | Yards | 5.5 |
| Ropes (Brit) | Feet | 20 |
| | Meters | 6.096 |
| | Yards | 6.66667 |
| Sabbath's Day Journey | Cubits | 2000 |
| | Yards | 1000 |
| Schoppen (Germany) | Liters | 0.5 |
| Score | | 20 |
| Scruples (apoth) | Drams (apoth or troy) | 0.3333 |
| | Drams (avdp) | 0.73143 |
| | Grains | 20 |
| | Grams | 1.29598 |
| | Ounces (apoth or troy) | 0.04167 |
| | Ounces (avdp) | 0.45714 |
| | Pennyweights | 0.8333 |
| | Pounds (apoth or troy) | 0.00347 |
| | Pounds (avdp) | 0.002857 |
| Scruples (Brit,flu) | Grains | 20 |
| | Minims (Brit) | 20 |
| Se (Japan) | Square Yards | 118.615 |
| Sea Mile | Degree of latitude | 1/60th |
| Seah (Old Testament) | Liters | 7.3 |
| Seams (Brit) | Bushels (Brit) | 8 |
| | Cu feet | 10.2 |
| | Liters | 290.94 |
| Seconds (angle) | Degrees | 0.000278 |
| | Minutes | 0.016667 |
| | Quadrants | $3.087 \times 10^{-6}$ |
| | Radians | $4.8481 \times 10^{-6}$ |
| Seconds (mean solar) | Days (mean solar) | $1.1574 \times 10^{-5}$ |
| | Days (sideral) | $1.1606 \times 10^{-5}$ |
| | Hours (mean solar) | 0.000278 |
| | Hours (sideral) | 0.000278 |
| | Minutes (mean solar) | 0.0166667 |
| | Minutes (sideral) | 0.016712 |
| | Seconds (sideral) | 1.002738 |
| Seconds (sideral) | Days (mean solar) | $1.1542 \times 10^{-5}$ |
| | Days (sideral) | $1.1574 \times 10^{-5}$ |
| | Hours (mean solar) | 0.000277 |
| | Hours (sideral) | 0.0002778 |
| | Minutes (mean solar) | 0.016621 |
| | Minutes (sideral) | 0.0166667 |
| | Seconds (mean solar) | 0.9972696 |

| Convert From | Into | Multiply By |
|---|---|---|
| Seer (India) | Pounds (avdp) | 2.057 |
| Shaku (Japan) | Meters | 10/33 |
| Shekel (Old Palestine) | Grains | 252 |
| | Grams | 16.33 |
| | Pounds (avdp) | 0.035999 |
| | Talents | 0.00033 |
| Shekel (Old Testament) | Bekah | 2 |
| | Grams | 11 |
| | Mina | 0.02 |
| | Talent | 0.00033 |
| Sheng (China) | Liters | 1.035 |
| Shih (China) | Pounds | 157.89 |
| Sho (Japan) | Liters | 1.804 |
| Shots | Fifths | 0.039 |
| | Jiggers | 0.666 |
| | Ounces, (US, fluid) | 1 |
| | Pints | 0.0625 |
| | Ponys | 1.333 |
| | Quarts | 0.03125 |
| Skeins | Feet | 360 |
| | Meters | 109.728 |
| Slugs | Geepounds | 1 |
| | Kilograms | 154.594 |
| | Pounds (avdp) | 32.174 |
| Slugs/cu ft | Grams/cu cm | 0.51538 |
| Spans | Centimeters | 22.86 |
| | Fathoms | 0.125 |
| | Feet | 0.75 |
| | Inches | 9 |
| | Quarters (Brit) | 1 |
| Span (Old English) | Inches | 6 |
| Sphere | Steradians | 12.57 |
| Sq centimeters | Ares | $1 \times 10^{-6}$ |
| | Circ mm | 127.324 |
| | Circ mils | 197352 |
| | Sq chains (Gunter's) | $2.471 \times 10^{-7}$ |
| | Sq chains (Ramden's) | $1.0764 \times 10^{-7}$ |
| | Sq decimeters | 0.01 |
| | Sq feet | 0.001076 |
| | Sq feet (US Survey) | 0.001076 |
| | Sq inches | 0.155 |
| | Sq meters | 0.0001 |
| | Sq mm | 100 |
| | Sq miles | $3.861 \times 10^{-11}$ |
| | Sq mils | 155000 |
| | Sq rods | $3.9537 \times 10^{-6}$ |
| | Sq yards | 0.0001196 |
| Sq chains (Gunter's) | Acres | 0.1 |
| | Sq feet | 4356 |
| | Sq feet (US Survey) | 4355.98 |
| | Sq inches | 627264 |
| | Sq links (Gunter's) | 10000 |
| | Sq meters | 404.69 |
| | Sq miles | 0.000156 |
| | Sq rods | 16 |

| Convert From | Into | Multiply By |
|---|---|---|
| Sq chains (Gunter's) | Sq yards | 484 |
| Sq chains (Ramden's) | Acres | 0.22957 |
| | Sq feet | 10000 |
| | Sq feet (US Survey) | 9999.96 |
| | Sq inches | $1.44 \times 10^6$ |
| | Sq links (Ramden's) | 10000 |
| | Sq meters | 929.03 |
| | Sq miles | 0.000305 |
| | Sq rods | 36.7309 |
| | Sq yards | 1111.11 |
| Sq decimeters | Sq cm | 100 |
| | Sq inches | 15.50003 |
| Sq degrees | Steradians | 0.0003 |
| Sq dekameters | Acres | 0.02471 |
| | Ares | 1 |
| | Sq meters | 100 |
| | Sq yards | 119.599 |
| Sq feet | Acres | $2.2957 \times 10^{-5}$ |
| | Ares | 0.000929 |
| | Circular mils | $1.833 \times 10^8$ |
| | Sq cm | 929.03 |
| | Sq chains (Gunter's) | 0.0002296 |
| | Sq feet (US Survey) | 0.999996 |
| | Sq inches | 144 |
| | Sq links (Gunter's) | 2.2957 |
| | Sq meters | 0.0929 |
| | Sq miles | $3.58701 \times 10^{-8}$ |
| | Sq millimeters | $9.290 \times 10^4$ |
| | Sq rods | 0.003673 |
| | Sq yards | 0.1111 |
| Sq feet (US Survey) | Acres | $2.2957 \times 10^{-5}$ |
| | Sq centimeters | 929.034 |
| | Sq chains (Ramden's) | 0.0001 |
| | Sq feet | 1.0000040 |
| Sq hectometers | Sq meters | 10000 |
| Sq inches | Circular mils | 1273239 |
| | Sq cm | 6.4516 |
| | Sq chains (Gunter's) | $1.5942 \times 10^{-6}$ |
| | Sq decimeters | 0.064516 |
| | Sq feet | 0.00694 |
| | Sq ft (US Survey) | 0.00694 |
| | Sq links (Gunter's) | 0.01594 |
| | Sq meters | 0.000645 |
| | Sq miles | $2.491 \times 10^{-10}$ |
| | Sq mm | 645.16 |
| | Sq mils | $1 \times 10^6$ |
| Sq inches/sec | Sq cm/hour | 23226 |
| | Sq cm/sec | 6.4516 |
| | Sq ft/min | 0.41667 |
| Sq kilometers | Acres | 247.105 |
| | Sq centimeters | $10^{10}$ |
| | Sq feet | $1.07639 \times 10^7$ |
| | Sq feet (US Survey) | $1.07639 \times 10^7$ |
| | Sq inches | $1.550003 \times 10^9$ |
| | Sq meters | $1 \times 10^6$ |

| Convert From | Into | Multiply By |
|---|---|---|
| Sq kilometers | Sq miles | 0.3861 |
| | Sq yards | $1.196 \times 10^6$ |
| Sq links (Gunter's) | Acres | $1 \times 10^{-5}$ |
| | Sq cm | 404.686 |
| | Sq chains (Gunter's) | 0.0001 |
| | Sq feet | 0.4356 |
| | Sq feet (US Survey) | 0.4356 |
| | Sq inches | 62.726 |
| Sq links (Ramden's) | Acres | $2.2957 \times 10^{-5}$ |
| | Sq feet | 1 |
| Sq meters | Acres | 0.000247 |
| | Ares | 0.01 |
| | Hectares | 0.0001 |
| | Sq cm | 10000 |
| | Sq feet | 10.7639 |
| | Sq inches | 1550.003 |
| | Sq kilometers | $1 \times 10^{-6}$ |
| | Sq links (Gunter's) | 24.71054 |
| | Sq links (Ramden's) | 10.764 |
| | Sq miles | $3.861 \times 10^{-7}$ |
| | Sq mm | $1 \times 10^6$ |
| | Sq rods | 0.03954 |
| | Sq yards | 1.10599 |
| Sq miles | Acres | 640 |
| | Hectares | 258.999 |
| | Sq chains (Gunter's) | 6400 |
| | Sq feet | $2.78783 \times 10^7$ |
| | Sq feet (US Survey) | $2.7829 \times 10^7$ |
| | Sq kilometers | 2.58999 |
| | Sq meters | 258999 |
| | Sq rods | 102400 |
| | Sq yards | $3.098 \times 10^6$ |
| Sq millimeters | Circular mm | 1.2732 |
| | Circular mils | 1973.5 |
| | Sq cm | 0.01 |
| | Sq feet | $1.076 \times 10^{-5}$ |
| | Sq inches | 0.00155 |
| | Sq meters | $1 \times 10^{-6}$ |
| Sq mils | Circular mils | 1.273 |
| | Sq cm | $6.452 \times 10^{-6}$ |
| | Sq inches | $1 \times 10^{-6}$ |
| | Sq mm | 0.000645 |
| Sq rods | Acres | 0.00625 |
| | Ares | 0.25293 |
| | Hectares | 0.00253 |
| | Sq cm | 252928 |
| | Sq feet | 272.25 |
| | Sq feet (US Survey) | 272.249 |
| | Sq inches | 39204 |
| | Sq links (Gunter's) | 625 |
| | Sq links (Ramden's) | 272.25 |
| | Sq meters | 25.293 |
| | Sq miles | $9.7656 \times 10^{-6}$ |
| | Sq yards | 30.25 |
| Sq yards | Acres | 0.000207 |

| Convert From | Into | Multiply By |
|---|---|---|
| Sq yards | Ares | 0.00836 |
| | Hectares | $8.3613 \times 10^{-5}$ |
| | Sq cm | 8361.27 |
| | Sq chains (Gunter's) | 0.002066 |
| | Sq chains (Ramden's) | 0.0009 |
| | Sq feet | 9 |
| | Sq feet (US Survey) | 8.99996 |
| | Sq inches | 1296 |
| | Sq links (Gunter's) | 20.661 |
| | Sq links (Ramden's) | 9 |
| | Sq meters | 0.8361 |
| | Sq miles | $3.228 \times 10^{-7}$ |
| | Sq millimeters | $8.361 \times 10^{5}$ |
| | Sq rods | 0.03306 |
| Stadia (Ancient Rome) | Miles | 0.92 |
| | Milliar | 0.125 |
| Stadion (Ancient Greece) | Ammas | 10 |
| | Yards | 200 |
| Stadium (Ancient Rome) | Passus | 125 |
| | Yards | 202.3 |
| Statamperes | Abamperes | $3.3356 \times 10^{-11}$ |
| | Amperes | $3.3356 \times 10^{-10}$ |
| Statcoulombs | Ampere–hours | $9.2656 \times 10^{-14}$ |
| | Coulombs | $3.3356 \times 10^{-10}$ |
| | Elecltronic charges | $2.082 \times 10^{9}$ |
| Statfarads | Farads | $1.1126 \times 10^{-12}$ |
| | Microfarads | $1.1126 \times 10^{-6}$ |
| Stathenries | Abhenries | $8.9876 \times 10^{29}$ |
| | Henries | $8.9876 \times 10^{11}$ |
| | Millihenries | $8.9876 \times 10^{14}$ |
| Statohms | Abohms | $8.9876 \times 10^{20}$ |
| | Ohms | $8.9876 \times 10^{11}$ |
| Statvolts | Abvolts | $2.9979 \times 10^{10}$ |
| | Volts | 299.79 |
| Statvolts/cm | Volts/cm | 299.79 |
| | Volts/inch | 761.47 |
| Statvolts/inch | Volts/cm | 118.029 |
| Steradians | Hemispheres | 0.15915 |
| | Solid angles | 0.079577 |
| | Spheres | 0.079577 |
| | Spher. right angles | 0.63662 |
| | Square degrees | 3282.81 |
| Steres | Cubic meters | 1 |
| | Decisters | 10 |
| | Dekasteres | 0.1 |
| | Liters | 999.97 |
| Stilbs | Candles/sq cm | 1 |
| | Candles/sq inch | 6.4516 |
| | Lamberts | 3.14159 |
| Stokes | Sq cm/sec | 1 |
| | Sq inches/sec | 0.1550003 |
| | Poise cu cm/gram | 1 |
| Stones (Brit,legal) | Centals (Brit) | 0.14 |
| | Pounds | 14 |
| Talent (Old Palestine) | Kilograms | 20 to 40 |

| Convert From | Into | Multiply By |
|---|---|---|
| Talent (Old Palestine) | Pounds | 107.9999 |
| | Mina | 60 |
| | Shekel | 3000 |
| Tablespoons | Cups | 0.0625 |
| | Drops | 180 |
| | Gills | 0.125 |
| | Ounces (US, fluid) | 0.5 |
| | Quarts | 0.01562 |
| | Teaspoons | 3 |
| Teaspoons | Cups | 0.02083 |
| | Drops | 60 |
| | Gills | 0.0416 |
| | Ounces (US, fluid) | 0.1666 |
| | Pinch | 3 to 4 |
| | Pints | 0.01042 |
| | Quarts | 0.00521 |
| | Tablespoons | 0.3333 |
| Therm | Btu | 100000 |
| Tierce | Gallons | 42 |
| To (Japan) | Liters | 18.039 |
| Toise (France) | Meters | 1.949 |
| Tonos (Greece) | Pounds (avdp) | 3307 |
| Tons (long) | Dynes | $9.964 \times 10^8$ |
| | Hundredweights (long) | 20 |
| | Hundredweights (shrt) | 22.4 |
| | Kilograms | 1016.05 |
| | Ounces (avdp) | 35840 |
| | Pounds (apoth or troy) | 2722.2 |
| | Pounds (avdp) | 2240 |
| | Tons (metric) | 1.10605 |
| | Tons (short) | 1.12 |
| Tons (metric) | Dynes | $9.807 \times 10^8$ |
| | Grams | $1. \times 10^6$ |
| | Hundredweights (shrt) | 22.0462 |
| | Kilograms | 1000 |
| | Ounces (avdp) | 35273.96 |
| | Pounds (apoth or troy) | 2679.23 |
| | Pounds (avdp) | 2204.62 |
| | Tons (long) | 0.98421 |
| | Tons (short) | 1.1023 |
| Tons (short) | Dynes | $8.8964 \times 10^8$ |
| | Hundredweights (shrt) | 20 |
| | Kilograms | 907.18 |
| | Ounces (avdp) | 32000 |
| | Pounds (apoth or troy) | 2430.55 |
| | Pounds (avdp) | 2000 |
| | Tons (long) | 0.89286 |
| | Tons (metric) | 0.90718 |
| Tons (long)/sq ft | Atmospheres | 1.0585 |
| | Dynes/sq cm | $1.0725 \times 10^6$ |
| | Grams/sq cm | 1093.66 |
| | Pounds/sq ft | 2240 |
| Tons (short)/sq ft | Atmospheres | 0.9451 |
| | Dynes/sq cm | 957.6 |
| | Grams/sq cm | 976.49 |

| Convert From | Into | Multiply By |
|---|---|---|
| Tons (short)/sq ft | Pounds/sq inch | 13.889 |
| Tons (long)/sq in | Atmospheres | 152.423 |
| | Dynes/sq cm | $1.544 \times 10^8$ |
| | Grams/sq cm | 15749 |
| Tons (short)/sq in | Dynes/sq cm | $1.3789 \times 10^8$ |
| | Kg/sq mm | 1406.14 |
| | Pounds/sq inch | 2000 |
| Tons of water/24 hrs | Pounds of water/hr | 83.333 |
| | Gallons/min | 0.16643 |
| | Cu ft/hr | 1.3349 |
| Torrs | Mm of Hg @ $0^oC$ | 1 |
| Townships | Acres | 23040 |
| | Sections | 36 |
| | Sq miles | 36 |
| Tu (China) | Miles | 100.142 |
| Tuns | Gallons (US) | 252 |
| | Hogsheads | 4 |
| | Pipes | 2 |
| | Puncheons | 3 |
| Vara (Old Spanish) | Feet | 2.6816 |
| | Meters | 0.8359 |
| Vara (S. America) | Meters | 0.8 to 1.1 |
| Vedro (Russia) | Liters | 12.3 |
| Verst (Russia) | Feet | 3500 |
| | Meters | 1067.07 |
| Volts | Abvolts | $1 \times 10^8$ |
| | Statvolts | 0.00333 |
| | Volts (Int) | 0.99967 |
| Volts (Int) | Volts | 1.0003 |
| Volt–coulombs | Joules (Int) | 0.9998 |
| Volt–seconds | Maxwells | $1 \times 10^8$ |
| Watts | Btu/hr | 3.4144 |
| | Btu (mean)/hr | 3.4095 |
| | Btu (mean)/min | 0.056825 |
| | Btu/sec | 0.000948 |
| | Btu (mean)/sec | 0.000947 |
| | Cal,gm/hr | 860.42 |
| | Cal,gm (mean)/hr | 859.18 |
| | Cal,gm (@$20^oC$)/hr | 860.85 |
| | Cal,gm/min | 14.34 |
| | Cal,gm (IST)/min | 14.331 |
| | Cal,gm (mean)/min | 14.3197 |
| | Cal,kg/min | 0.01434 |
| | Cal,kg (IST)/min | 0.01433 |
| | Cal,kg (mean)/min | 0.01432 |
| | Ergs/sec | $1 \times 10^7$ |
| | Foot–pounds/min | 44.2537 |
| | Foot–pounds/sec | 0.7378 |
| | Horsepower | 0.00134 |
| | Horsepower (boiler) | 0.0001 |
| | Horsepower (electric) | 0.00134 |
| | Horsepower (metric) | 0.0013596 |
| | Joules/sec | 1 |
| | Kilogram–calories/min | 0.01433 |
| | Kilowatts | 0.001 |

| Convert From | Into | Multiply By |
|---|---|---|
| Watts | Liter–atmosphere/hr | 35.528 |
| | Watts (Int) | 0.9998 |
| Watts (Int) | Btu/hr | 3.41499 |
| | Btu (mean)/hr | 3.41008 |
| | Btu/min | 0.56916 |
| | Btu (mean)/min | 0.0568 |
| | Cal,gm/hr | 860.56 |
| | Cal,gm (mean)/hr | 859.326 |
| | Cal,kg/min | 0.0143 |
| | Cal,kg (IST)/min | 0.01433 |
| | Cal,kg (mean)/min | 0.01432 |
| | Ergs/sec | $1.00016 \times 10^7$ |
| | Joules (Int)/sec | 1 |
| | Watts | 1.00016 |
| Watts/sq cm | Btu/(hr x sq ft) | 3172.1 |
| | Cal,gm/(hr x sq cm) | 860.421 |
| | Ft–lb/(min x sq ft) | 41113 |
| Watts/sq in | Btu/(hr x sq ft) | 491.68 |
| | Cal,gm/(hr x sq cm) | 133.36 |
| | Ft–lb/(min x sq ft) | 6372.5 |
| Watt–hours | Btu | 3.4144 |
| | Btu (mean) | 3.4095 |
| | Cal,gm | 860.42 |
| | Cal,kg (mean) | 0.85918 |
| | Cal, gm (mean) | 859.18 |
| | Ergs | $3.60 \times 10^{10}$ |
| | Foot–pounds | 2655.22 |
| | Hp–hours | 0.00134 |
| | Joules | 3600 |
| | Joules (Int) | 3599.41 |
| | Kg–calories | 0.8605 |
| | Kg–meters | 367.098 |
| | Kw–hours | 0.001 |
| | Watt–hours (Int) | 0.9998 |
| Watt–sec | Foot–pounds | 0.73756 |
| | Gram–cm | 10197.2 |
| | Joules | 1 |
| | Liter–atmospheres | 0.00987 |
| | Volt–coulombs | 1 |
| Webers | Kilolines | $1 \times 10^5$ |
| | Lines | $1 \times 10^8$ |
| | Maxwells | $1 \times 10^8$ |
| | Volt–seconds | 1 |
| Webers/sq cm | Gausses | $1 \times 10^8$ |
| | Lines | $1 \times 10^8$ |
| | Lines/sq in | $6.4515 \times 10^8$ |
| Webers/sq in | Gausses | $1.550003 \times 10^7$ |
| | Lines/sq inch | $10^8$ |
| | Webers/sq cm | 0.1550 |
| | Webers/sq meter | 1550 |
| Webers/sq meter | Gausses | $10^4$ |
| | Lines/sq inch | $6.452 \times 10^4$ |
| | Webers/sq cm | $10^{-4}$ |
| | Webers/sq inch | $6.452 \times 10^{-4}$ |
| Weeks (mean calendar) | Days (mean solar) | 7 |

| Convert From | Into | Multiply By |
|---|---|---|
| Weeks (mean calendar) | Days (sideral) | 7.01916 |
| | Hours (mean solar) | 168 |
| | Hours (sideral) | 168.46 |
| | Minutes (mean solar) | 10080 |
| | Minutes (sideral) | 10107.6 |
| | Months (lunar) | 0.237042 |
| | Months (mean calendar) | 0.230137 |
| | Years (calendar) | 0.019178 |
| | Years (sideral) | 0.0191646 |
| | Years (tropical) | 0.019165 |
| Weys (Brit) | Pounds (avdp) | 252 |
| Yards | Centimeters | 91.44 |
| | Chains (Gunter's) | 0.454545 |
| | Chains (Ramden's) | 0.03 |
| | Cubits | 2 |
| | Fathoms | 0.5 |
| | Feet | 3 |
| | Feet (US Survey) | 2.999994 |
| | Furlongs | 0.004545 |
| | Inches | 36 |
| | Kilometers | $9.144 \times 10^{-4}$ |
| | Meters | 0.9144 |
| | Miles (naut) | $4.934 \times 10^{-4}$ |
| | Miles (statute) | $5.682 \times 10^{-4}$ |
| | Millimeters | 914.4 |
| | Poles (Brit) | 0.181818 |
| | Quarters (Brit) | 4 |
| | Rods | 0.181818 |
| | Spans | 4 |
| Years (calendar) | Days (mean solar) | 365 |
| | Hours (mean solar) | 8760 |
| | Minutes (mean solar) | 525600 |
| | Months (lunar) | 12.36006 |
| | Months (mean calendar) | 12 |
| | Seconds (mean solar) | $3.1536 \times 10^{7}$ |
| | Weeks (mean calendar) | 52.14286 |
| | Years (sideral) | 0.999298 |
| | Years (tropical) | 0.999337 |
| Years (leap) | days (mean solar) | 366 |
| Years (sideral) | Days (mean solar) | 365.2564 |
| | Days (sideral) | 366.2564 |
| | Years (calendar) | 1.000702 |
| | Years (tropical) | 1.000039 |
| Years (tropical) | Days (mean solar) | 365.242 |
| | Days (sideral) | 366.242 |
| | Hours (mean solar) | 8765.81 |
| | Hours (sideral) | 8789.81 |
| | Months (mean calendar) | 12.00796 |
| | Seconds (mean solar) | $3.15569 \times 10^{7}$ |
| | Seconds (sideral) | $3.16433 \times 10^{7}$ |
| | Weeks (mean calendar) | 52.17746 |
| | Years (calendar) | 1.000663 |
| | Years (sideral) | 0.99996 |
| Zoll (Switzerland) | Centimeters | 3 |
| Zolotnik (Russia) | Grains | 65.8306 |

# Index

# NOTES

# NOTES

# NOTES

# NOTES